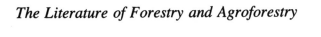

The Literature of Forestry and Agroforestry

A volume in the series

The Literature of the Agricultural Sciences

WALLACE C. OLSEN, series editor

Agricultural Economics and Rural Sociology: The Contemporary Core Literature
By Wallace C. Olsen

The Literature of Agricultural Engineering
Edited by Carl W. Hall and Wallace C. Olsen

The Literature of Animal Science and Health
Edited by Wallace C. Olsen

The Literature of Soil Science
Edited by Peter McDonald

The Literature of Crop Science
Edited by Wallace C. Olsen

The Contemporary and Historical Literature of Food Science and Human Nutrition
Edited by Jennie Brogdon and Wallace C. Olsen

The Literature of Forestry and Agroforestry
Edited by Peter McDonald and James Lassoie

THE LITERATURE OF FORESTRY AND AGROFORESTRY

EDITED BY

Peter McDonald and
James Lassoie

Cornell University Press

ITHACA AND LONDON

This book was typeset from disks supplied by the staff of the Core Agricultural Literature Project, Albert R. Mann Library, Cornell University. Nicole Kasmer Kresock prepared the machine-readable text, and Eveline Ferretti made final corrections. The research was financially supported by Cornell Agricultural Experiment Station; National Agricultural Library, United States Department of Agriculture; and the Rockefeller Foundation.

First published 1996 by Cornell University Press.

Library of Congress Cataloging-in-Publication Data

The Literature of forestry and agroforestry / edited by Peter McDonald and James Lassoie.
    p.   cm. — (The Literature of the agricultural sciences)
    Includes bibliographical references and index.
    ISBN 0-8014-3181-6 (alk. paper)
    1. Forestry literature. 2. Agroforestry literature.
  I. McDonald, Peter, 1952–   . II. Lassoie, J. P. III. Series.
  SD387.D6L58 1996
634.9—dc20   95-39328

Contents

Preface

Human use of forests and their various products predates written history. This important natural resource base has long provided the raw materials necessary to help develop societies around the world. It has been only within the last century that humankind has developed other primary sources for its fuel and construction materials, and such sources are still largely unavailable to many of the world's agriculturally based nations. Our forest resources played a very important role in the development of the United States, just as they are now doing in many parts of the developing world today. However, there is currently a rapidly growing understanding that we must better consider the long-term maintenance of environmental quality if we are to enjoy the economic, social, and ecological benefits of forests on into the future. Such an understanding is the essence of modern society's new quest for "sustainable" approaches to its land use practices.

It is certain that humans have manipulated the forest environment for its products for millennia. However, the science, art, and practice of managing and using the forest for human benefits, that is, professional forestry, is relatively new. Central Europe, especially Germany, gave rise to the birth of modern forestry during the 1700s, a result of its growing population and shrinking land base. As with any applied discipline, the need for practitioners stimulated the development of education and training opportunities and a literature needed to support study and practice. Much of the early literature was adapted directly from the botanical and physical sciences, but this new discipline soon started producing its own unique literature. The United States experienced its forest conservation movement and saw its first professional forestry schools near the end of the nineteenth century. Without question, the evolution of professional forestry can be traced to the interdependence between practice and education and the literature that supports and reinforces them.

Although the evolution of agriculture as a scientifically based land use practice predates that of forestry, patterns of development in the two fields were quite similar. It is interesting that in the developed world agriculture and forestry have their own infrastructures typically involving separate educational institutions and state and federal agencies. The two land uses are carried out on completely different pieces of land and usually by very different types of practitioners advised and supported by very different professionals. This separation is an artifact of the developed world which can afford the time, money, and space to deal with food and fiber production separately. In stark contrast, most of the world's rural people do not separate these practices, being forced by limited resources to obtain food, forage, and fiber from the same pieces of land. This ancient land use has recently been discovered by the scientific community and termed agroforestry. Although early studies of this practice focused on its use for improving the quality of life in the developing world, there is currently interest in agroforestry as one approach to sustainable agriculture in developed countries like the United States. Over the past ten years there has been a rapid expansion of educational programs and a growing literature associated with the study and practice of agroforestry.

This volume discusses the evolution of forestry and agroforestry and presents the core literature supporting these complex fields of study. It puts in perspective the major scholarly activities which record this evolution from nature appreciation, to exploitation, to sustained yield, to multiple use, to sustainable forestry. The British scholar N. D. G. James provides a chapter detailing the growth of forestry as a science from the publication of the first literature in Germany in 1713 through the utilization of wood and forestry issues in the early twentieth century. This is information of value to today's scholars in placing current controversies and patterns in a realistic and long-term frame of reference.

This volume builds on the traditional subject areas of forestry and expands into those coming into prominence today. The place of basic plant science as the beginning of forest science is emphasized by N. D. G. James as well as John C. Gordon. Gordon chronologically outlines the primary activities, influences, and changes in forest science in the twentieth century in a systematic and knowledgeable manner. P. K. R. Nair covers the development of agroforestry literature supporting this emerging discipline. Of particular importance for today's scholars is the role that professional societies and the U.S. Forest Service played in dispersing information and data, and their continuing influences. Both of these are explained in complementary chapters by L. W. Tombaugh and J. S. Spencer.

Current changes in subject matter and methodologies of great interest to forest scientists and professional foresters are also documented: James Coufal and Donald Webster examine the emergence of sustainable forestry as a viable and necessary field of study. The place of geographic information systems in forest management is presented by J. K. Berry and W. J. Ripple, and R. A. Skok contributes an historical discussion of forestry education in the United States. T. T. Kozlowski, the "father" of forest tree physiology, reflects on his long career as a researcher, writer, and editor in another chapter.

The Core Agricultural Literature Project at the Albert R. Mann Library, Cornell University, in 1988 began to identify the most useful current literature for university-level instruction and research. Portions of this volume are conclusions from that work. The Project was funded primarily by the Rockefeller Foundation, which wished to identify this literature for the developing world and then transfer the full texts onto compact disks for easy distribution and use in remote areas. The Project staff worked with Steering Committees for each of the seven subject areas that were studied intensively and reported in this series of books. The preceding six volumes published are identified opposite the title page. The Core Agricultural Literature Project took four years to survey, evaluate, and write the results of these deliberations. The procedures included evaluation and counsel by scientists in all seven disciplines to the level of ranking of titles in lengthy monograph lists. These 500 participants are identified in the volumes; without their deep interest and commitment to scholarship, the core literature could not have been identified with such certainty. A historical literature component was added beginning with the second volume; its purpose was to identify the most valuable American literature for immediate preservation. In this volume on forestry and agroforestry, this information is presented in five chapters written by the senior editor, Peter McDonald.

This thorough assessment of the literature of forestry and agroforestry is valuable at this time because of the increasing globalization of agriculture and forestry. Books and journals originating in developed countries are vital sources of information for education and research in the developing world. Increasingly, professionals in developing countries are adding to this literature, and its importance to the rest of the world is understandably growing. Scholars and literature collectors throughout the world will find extensive assistance in this book for evaluating the strengths of their literature collections and determining the merits of their journal and report holdings. Forest scientists and practicing professional foresters will also find this volume a useful reference in their search for relevant literature.

The Steering Committee for this volume on core literature, in addition to myself, were James Coufal and Donald F. Webster, State University of New York, College of Environmental Science and Forestry, Syracuse, N.Y.; and P. K. R. Nair, University of Florida, Gainesville.

JAMES LASSOIE

Ithaca, New York
October 1994

1. Trends and Developments in Modern Forestry

JOHN C. GORDON

Yale School of Forestry and Environmental Studies

Forests and their use by people, the subjects of forestry, have been written about since classical times. In every era, in every literate culture, forests have been practically useful, worshipped by some, and often feared, giving rise to a rich and varied literature and store of knowledge and myth. The practice of tending trees and forests is similarly ancient; Babylonians, Egyptians, Greeks, and Romans were all concerned about wood supply, watershed protection, and the beauty of forests.[1] At the same time, the history of "civilization" has been a history of deforestation and the conversion of forests and woodlands to agriculture and urban space. This conversion continues to this day, primarily in the tropics and subtropics. Forest extent has stabilized in much of the northern hemisphere, and in North America and Western Europe wood growth exceeds the rate at which wood is cut.[2] But concern about forest loss and degradation, and the obvious utility of wood and forests, have been and continue to be the long-term drivers of scientific forestry and its literature.

A. The Rise of Scientific Forestry

Although "scientific" forestry has its roots in antiquity, the beginnings of its modern form is largely a nineteenth-century phenomenon. The great expansion of European practical science of the last century coincided with the development of European colonies in forested regions. Thus, the first modern forestry school with instruction in English was located at Dehra Dun in what was then British India. It was no accident that the Rector at Dehra

---

1. J. Perlin, *A Forest Journey: The Role of Wood in the Development of Civilization* (Cambridge: Harvard University Press, 1989).
2. J. Laarman and R. Sedjo, *Global Forests: Issues for Six Billion People* (New York: McGraw-Hill, 1992).

1

Dun was a German named Brandis. Germans, described by Julius Caesar as a "forest-dwelling people,"[3] were the first to develop scientific forestry and institutions that treated forestry as an advanced discipline. Up until the time of World War II, much of the serious forestry literature of the world was written in German, although British, French, Scandinavian, Russian, and, after the American Civil War, increasingly American authors made significant contributions. The rich traditional forest literature of China and Japan, and other Asian cultures, is only now being "discovered" by Western foresters. However, it is fair to say that all reasonably economically developed countries had a science-based forestry literature by the early years of the twentieth century.[4]

Much of this early scientific forestry literature dealt with determining the wood content of forests or its value, or with methods to ensure a continuous supply of wood for use on farms and by manufacturing industries. Articles and books treating the values of forest conservation, from water to aesthetics, also began to proliferate in the late nineteenth and early twentieth centuries.[5] As forest loss and wood scarcity became increasingly matters of concern, a literature of forest growth and silvicultural enhancement began to increase, and has now grown numerically to dominate the scientific literature about forests.

The creation of vast forest reserves beginning in 1891 in the United States as advocated by the American Forestry Association (the oldest continuing citizens' forestry organization), the establishment of the Society of American Foresters in 1900, the founding of the Yale Forest School the same year (the oldest surviving such school in North America), and the creation of the modern U.S. Forest Service three years later, signaled the emergence of a new and eventually dominant actor on the world forestry stage.[6] Norbert Weiner suggested after World War II that the language of science had changed from mathematics to broken English. After World War II, in the culmination of economic and intellectual processes begun much earlier, the language of forestry literature changed from German to American English. This did not, however, keep the center of forestry literature management from developing and remaining in Oxford, England.[7] Nor did

3. J. G. Frazer, *The Golden Bough*, Vol.1 (New York: Macmillan, 1950), p. 757.

4. E. Johnson, "Forestry and Forest Products," in *Guide to Sources for Agricultural and Biological Research*, ed. J. Richard Blanchard and L. Farrell (Berkeley: University of California Press, 1981).

5. J. Matthews, "The Development of Forest Science," in *Applied Biology*, ed. T. Croaker (New York: Academic Press, 1976).

6. H. Clepper, and A. Meyer, eds., *American Forestry; Six Decades of Growth* (Washington, D.C.: Society of American Foresters, 1960).

7. Johnson, "Forestry and Forest Products."

it stop forestry graduate students in America from being admonished, "If you think you have found out something new, it probably means that you don't read German."

B. Forestry Challenges: The Last Forty Years

Over the last forty years, forestry has changed rapidly. These changes are reflected in the scientific and professional literature of forestry, which has become broader, more disciplinary (specialized), and more rigorous. The emergence of *Forest Science* and *Forest Science Monographs,* published by the Society of American Foresters, exemplifies this trend. *Forest Science* also signaled the intent to create an integrated science of forestry by accepting articles from nearly the full range of forest-related disciplines.

The scope of forestry, in terms of both disciplines and concepts, has greatly increased in response to changes in economics, demography, and politically expressed societal needs and desires. Whereas the mapping, measurement (mensuration), and utilization of wood dominated forestry literature until the 1950s, that decade showed a rapid shift toward the biology and tending (silviculture) of forests. Forest genetics, physiology, and ecology emerged as disciplines with their own journals and appeared in university forestry curricula. Later in this period, but just as inexorably, economics, sociology, and anthropology have emerged. One of the most interesting contrasts in this regard is with agriculture, which early in this century embraced basic plant science and economics research, but kept them relatively separate in curriculum and in practice. Forestry, a later entrant in both fields, has evolved in a more integrated way. Modern silviculture, and to a considerable degree, forest ecology, are regarded as having foundations in both natural and social science.[8] Forest management is no longer regarded as primarily the tending of forests, but as the achievement of the objectives of human forest owners and stakeholders.[9]

Forests as Strategic Resources: The Cold War

At the beginning of the 1950s, forestry texts and journals were remarkably similar to those of the previous three decades. Worldwide concern for economic and social recovery from World War II dominated forestry thought and concentrated attention on wood supply to recreate or expand

8. D. Smith, *The Practice of Silviculture* (New York: Wiley, 1986).
9. L. Davis and K. Johnson, *Forest Management* (New York: McGraw-Hill, 1987).

housing stocks; paper use increased dramatically as postwar (particularly European) economies recovered and then boomed. Thus, forests were seen as strategic resources primarily for commodity production, and the modern practice and literature of forest economics grew rapidly, first and most vigorously in North America.[10]

At the same time, concern about wood supply, and about the maintenance of healthy forests for both wood production and environmental purposes gave a great boost to research in forest biology (genetics, physiology, and ecology) and in forest protection (fire, entomology, and pathology). In the latter areas, modern concepts of pest management, now widely applied in both agriculture and forestry evolved, in part, from the extensive and ecologically based methods of early forest pathologists and entomologists.[11]

Forest Regeneration

As mature forests were harvested at an increasing rate, forest regeneration became a central issue, both to ensure that new forests followed old in the supply continuum, and to reassure a skittish public that forestry "worked." Complete harvesting of existing stands, often by clearcutting, followed by planting with trees raised in nurseries became standard practice on many forests during the 1960s and 1970s.[12] This system of management posed many new problems, from maintenance of soil productivity[13] to the creation of genetically and physiologically improved nursery stock,[14] and at the same time created the cash flow to support more intensive research. Industry/university/government research cooperatives were formed, first in the southeastern United States and then in the Pacific Northwest. Forest engineering research focused on more efficient, less disruptive harvesting systems that minimized soil disturbance and compaction, and on creating road systems that were cost-effective and that reduced erosion and consequent water pollution.[15] Clearcutting, a time-honored, but previously not-

10. M. Clawson, *The Federal Lands Revisited* (Washington, D.C.: Research Future, 1983).

11. (a) R. Hartig, *Important Diseases of Forest Trees: Contributions to Mycology and Pathology for Botantists and Foresters,* trans. W. Merrill, D. Lambert, and W. Liese (St. Paul, Minn.: American Phytopathological Society, 1975). (b) R. Coulson and J. Witter, *Forest Entomology: Ecology and Management* (New York: Wiley, 1984).

12. P. Wakeley, *Planting the Southern Pines* (Washington, D.C.: Forest Service, 1954 [*USDA Agriculture Monograph* 18])

13. A. Leaf, *Impact of Intensive Harvesting on Forest Nutrient Cycling* (Syracuse, N.Y.: SUNY College of Environmental Science and Forestry, 1979).

14. (a) K. Dorman, *The Genetics and Breeding of Southern Pines* (Washington, D.C.: Forest Service, 1976 [*USDA Handbook* no. 471]). (b) M. Duryea and T. Landis, *Forest Nursery Manual: Production of Bareroot Seedlings* (The Hague: M. Nijhoff, 1984).

15. *Symposium on Engineering Systems for Forest Regeneration; Forest Regeneration, the Proceedings* (St. Joseph, Mich.: American Society of Agricultural Engineers, 1981 [*ASAE Pub.* 10–81]).

widely-practiced silvicultural regeneration system, became controversial because of its unsightliness and presumed adverse environmental effects.[16]

Forests and the Environment

The passage of the National Environmental Protection Act in the United States in 1970, and other forest-related legislation of roughly the same period, called for a huge increase in foresters' ability to predict the environmental consequences of their actions (such as clearcutting) and enhanced the debate over whether specific forestry actions were sustainable. The focus of forest biology and management began to shift from the stand, a usually small, relatively homogenous group of trees, to watersheds and landscapes.[17] Watershed protection, one of the oldest overt human uses of forests, became a more central concern as western populations increased, and water yield and quality became increasingly contentious issues. Anadromous fish emerged as a major forestry issue in the Pacific Northwest, as it became clear that decreasing stocks were linked to degraded forest habitat. During the 1970s and 1980s, great strides were made in understanding the relationships between forest and stream conditions.[18] Many old notions were discarded. For example, it had been standard practice to remove downed logs from small streams, but subsequent research showed that the logs were highly beneficial in creating fish habitat and did not impede fish on their way upstream.

International Forestry

Other major world trends and events also began early in the period to move forestry beyond its traditional intellectual base and commodity focus and its predominantly north temperate locus. Colonial powers began rapidly to lose their colonies; as new nations asserted themselves, the Third World and "foreign aid" were created. These were to have profound effects on forests, forestry practice, and forestry thought and literature throughout the world. Tropical deforestation as a world issue, social forestry, and science-based agroforestry are examples of the results of the reordering of the priorities and economies of former colonies and emergent regional and world

16. W. Dietrich, *The Final Forest: The Battle for the Last Great Trees of the Pacific Northwest* (New York: Simon & Schuster, 1989).

17. F. Bormann and G. Likens, *Pattern and Process in a Forested Ecosystem: Disturbance, Development, and the Steady State Based on the Hubbard Brook Ecosystem Study* (New York: Springer-Verlag, 1979).

18. W. Meehan, ed., *Influences of Forest and Rangeland Management on Salmonid Fishes and Their Habitats* (Bethesda, Md.: American Fisherman's Society, 1991 [*Special Pub.* no. 19]).

powers (e.g., India, China, Indonesia, and Brazil) and the attempts of developed countries to help or manipulate them.

Several key concepts have been particularly prominent in the literature of forestry applied to economic development and forest protection problems in developing countries, and have begun to affect forestry in the developed world. Perhaps the most interesting is the commingling of forests, tree plantations, and agriculture to meet specific goals. Although agroforestry is an ancient practice in many places, and indigenous and colonial foresters wrote early treatises on it, scientific evaluation of its potential was not undertaken on even a modest scale until the 1970s. This has begun a reevaluation of the assumptions and concepts underlying both agricultural and forest science that will surely be beneficial for both.[19] This continues an inquiry formally begun by soil scientists' observations that agricultural and forest soils were different, and that forestry required a different approach to both the study and management of soils from that used in agriculture.[20]

Interest in "nitrogen-fixing trees" and "multi-purpose trees" has intensified during the 1970s.[21] This was again primarily a result of tropical rural development interests, although the energy crises of the period created a bubble of interest and literature on trees capable of symbiotic nitrogen fixation as sources of both wood and nitrogen in the temperate zone. Current interest in sustainable forestry (see Chapter 8) has rekindled interest in the role of microorganisms in forest nutrition and health,[22] and the concept of "multi-purpose" trees and forests,[23] as previously implied in the phrase "multiple use."

Attempts to preserve tropical forest ecosystems and the biological diversity they contain has alerted foresters everywhere to the apparent truth that people who live near and in the forests to be preserved need attention if preservation is to be successful. The idea of managing across ownership boundaries, in ways that are inclusive and cooperative rather than authoritarian, has thus gained a research base and credibility in temperate as well as tropical forestry.[24]

19. J. Gordon and W. Bentley, *A Handbook on the Management of Agroforestry Research* (New Delhi: WI, Oxford & IBH Pub. Co., 1990).

20. B. Warkentin, "Overview and Trends in Soil Science," in *The Literature of Soil Science*, ed. Peter McDonald (Ithaca, N.Y.: Cornell University Press, 1994).

21. J. Gordon and C. Wheeler, eds., *Biological Nitrogen Fixation in Forest Ecosystems: Foundations and Applications* (The Hague: Martinus Nijhoff/W. Junk, 1983).

22. Y. R. Dommergues and H. G. Diem, eds., *Microbiology of Tropical Soils and Plant Productivity* (The Hague: Martinus Nijhoff, 1982).

23. J. Burley and L. Stewart, eds., *Increasing Productivity of Multipurpose Species* (Vienna: IUFRO, 1985).

24. L. Fortman and J. Bruce, eds., *Whose Trees?: Proprietary Dimensions of Forestry* (Boulder, Colo.: Westview Press, 1988).

Forests and Recreation

In a similar vein, it was widely assumed in the late 1950s and early 1960s that technology and increased industrial production efficiency would create a leisure society, and indeed leisure time increased as weekends and the forty-hour week became widespread. This gave rise to forest recreation literature and practice; concepts like "wilderness" and "dispersed recreation" were examined in detail for their economic and social costs and benefits.[25] Recreation effects on forests and the effects of forest recreationists on each other were measured and debated, and the literature of forest recreation, outdoor recreation, and leisure studies took shape and grew.

Forests, Wildlife, and Politics

The wave of environmental legislation affecting forests that surrounded Earth Day, 1970, exemplified most prominently by the United States Endangered Species Act, linked public and private forestry and the industries dependent on forests for raw material directly to the maintenance of populations of forest-dependent wildlife. This fundamental change has been most clearly illustrated by the controversies over the Northern spotted owl in the Pacific Northwest, and the red-cockaded woodpecker in the Southeast. This linkage also eventually caused a surge in research and publication on forest wildlife. Before that time, much of the forestry literature on wildlife focused on animal damage. The shift to understanding and predicting optimum habitat for and behavior of potentially threatened and endangered species was swift and dramatic, but it took place against a backgound of ignorance and neglect.[26]

The new legislation also caused a renaissance in forest policy research and literature. The "taking" issue in relation to private property rights quickly emerged as central to the environmental regulation of forests and their contents, particularly wildlife, and has remained on center stage. Political science became a discipline in forestry schools, and the forest policy literature grew rapidly as the laws and their consequences were evaluated. During the 1960s and 1970s, partially at least in response to the Viet Nam War and Watergate, distrust of traditional leaders and leadership paradigms increased. Public and pressure groups formed and became adept at being heard by public agencies and forest products businesses that had been dis-

25. M. Clawson and J. Knetsch, *Economics of Outdoor Recreation* (Baltimore, Md.: Johns Hopkins Press, 1966).
26. J. Thomas, *Wildlife Habitats in Managed Forests: The Blue Mountains of Oregon and Washington* (Washington, D.C.: USDA, 1979 [*USDA Agriculture Handbook* no. 553]).

tanced from publicity and public debate. The era of public involvement in forestry decisions arrived, and social science research and publication on forestry matters increased accordingly.[27] Practicing foresters now often indicate that access to policy and social science information is one of their highest priorities.

Forest decline, presumably due to air pollution ("acid rain"), publicized in Germany and Central Europe as *Waldsterben* (forest death) became an issue in North America during the 1970s.[28] The largest policy-driven environmental research program ever undertaken in the United States, the "National Acid Precipitation Assessment Program" (1980–1990), had a relatively large forest component. This ten-year program found that acid rain was not an extensive problem in U.S. forests, at least as far as could be determined from relatively short-term studies, but that air-borne oxidants ("ozone") could be causing extensive damage to forests, particularly in the Northeast.[29] The program also exposed a major weakness in the monitoring of U.S forests and their immediate environment; few pollution-monitoring stations are located in forests, and forest growth information is not usually sufficiently accurate, nor are enough measurements taken, to expose long-term decline and growth loss.

Human-caused global environmental change became an issue of general concern in this period; global warming and tropical deforestation were conceptually linked, and an era of estimation of the effects of forests on carbon sequestration and the production of greenhouse gases began. The energy shocks of the 1970s had engendered a flood of research and publication on biomass fuels, biological nitrogen fixation, and energy efficiency in forestry operations and manufacturing. The forest carbon cycle information thus generated was used in the 1980s, particularly in the search for the "missing" carbon (that which should be in the atmosphere based on emissions calculations but wasn't).[30] Recent articles have suggested that much of this carbon may be sequestered in the excess of growth over cut in the temperate forests of North America and Europe.[31]

27. P. Ellefson, *Forest Resource Policy: Process, Participants, and Programs* (New York: McGraw-Hill, 1992).

28. E. D. Shulze et al., eds., *Forest Decline and Air Pollution: A Study of Spruce on Acid Soils* (Berlin and New York: Springer, 1989).

29. D. Shriner et al., "Response of Vegetation to Atmospheric Deposition and Air Pollution." in *Acidic Deposition: State of Science and Technology*, vol. 3 (Washington, D.C.: National Acid Precipitation Assessment Program, 1990 [NAPAP SOS/T Report 18]).

30. W. Shands and J. Hoffman, eds., *The Greenhouse Effect, Climate Change, and United States Forests* (Washington, D.C.: Conservation Foundation, 1987).

31. P. Kauppi et al., "Biomass and Carbon Budget of European Forests 1971–1990," *Science* 256 (1992): 70–74.

Forests and the Computer Revolution

Digital computers, satellites, and improved remote sensing techniques, and later the ubiquity of personal computers, put technological tools in the hands of foresters that allowed the possiblity of greatly enhanced and more accurate estimates and models of forests and forest processes. This vastly improved mapping, modeling, and analysis technology had an enormous impact on the practice of forestry and the conduct of forestry research. Just as forest science and practice were broadened, these tools promised the ability to describe, analyze, and prescribe for complex systems and often conflicting objectives. This capability was exemplified first in harvest scheduling and management planning models such as FORPLAN, adopted and widely used by the U.S. Forest Service, and later by the widespread adoption of geographic information systems (GIS) by most public and private forest management organizations.[32] Computer science and GIS became integral parts of forestry curricula, and forest management gained enormous analytical power.

Forests and the Science of Ecology

Large-scale ecosystem studies were begun in the 1970s in a few locations, such as Hubbard Brook in New Hampshire, Coweeta in North Carolina, and the Andrews Experimental Forest in Oregon. These three are all on National Forests, and were made possible by the cooperation of the research branch of the U.S. Forest Service, National Forests, the National Science Foundation, and universities. These studies aimed at quantifying the flows of energy and nutrients through forests, usually after massive manipulations such as clearcutting. Although they were all based on calibrated watersheds, they quickly moved beyond classical forestry watershed studies, and permanently changed the way forests, particularly old ones, are viewed.[33] For example, the role of minor species in nutrient cycling had not been appreciated earlier, nor had the close and intricate relationships between forest condition and succession and the water chemistry of forest streams.

These studies were made feasible by better technology for gathering and handling data, but the central conceptual breakthrough was the view of the

32. Thomas W. Hoekstra, A. A. Dyer, and Dennis C. LeMaster, eds., *FORPLAN: An Evaluation of a Forest Planning Tool; Proceedings of a Symposium, November 1986* (Fort Collins, Colo.: UDSA Forest Service, 1987).

33. J. Franklin et al., *Ecological Characteristics of Old-Growth Douglas-Fir Forests* (Portland, Ore.: USDA Forest Service, 1981 [*USDA Forest Service General Technical Report* PNW-118]).

forest ecosystem as an integral whole, rather than a collection of organisms and processes. This conceptual spadework, coupled with the hard data from the large-scale studies, laid the groundwork for the emergence of "ecosystem management" as an approach to managing forests on a sustained basis.[34] The previously dominant concept in forest management was sustained yield, the notion that an even, non-declining flow of wood indicated optimum treatment of forests. Forests were now seen to be both too complex and too different, biologically and in terms of human expectations, to be managed by that simple criterion.

Plantation Forestry

At the same time plantation forestry, the so-called "industrial forestry model" incorporated sophisticated technology and biological information, and became relatively widely practiced and accepted in places as diverse as New Zealand, the southeastern United States, and Brazil.[35] Stunning increases in the yield of wood per unit land area were achieved, and the growing and processing of wood were linked more rationally than in the past. It became clear that trees could be treated as a crop on a large scale, although nagging questions about sustainability and biological diversity continued to arise. The economics of intensive forestry developed rapidly, as did earnest debate over the profitability of these techniques. However, the wide adoption of plantation techniques indicates a high degree of economically informed faith in their fiscal soundness.

Forests and Public Perception

During this period, the public view of forests as reservoirs of commodities and game began to shift to a more complex view in which the forest itself became the object of interest. People became concerned about forests through the widely publicized issues of the time: acid rain, tropical deforestation, loss of species, global warming, and even, in some places, loss of economic activity and jobs. These issues exposed a gap between foresters and environmental groups. Foresters found themselves under adverse scrutiny over such practices as clearcutting, deforestation, and biological diversity loss. It became clear that not only had we, as foresters, neglected

---

34. J. Gordon, "Ecosystem Management: An Idiosyncratic Overview," in *Sustainable Forestry*, ed. A. Sample and J. Olsen (Washington, D.C.: Island Press, 1993).

35. E. Ford et al., eds., *The Ecology of Even-Aged Plantations; Proceedings of IUFRO Div. 1 Meeting* (Cambridge, U.K.: Institute of Terrestrial Ecology, Natural Environment Research Council, 1979).

to communicate with the public, but that we were operating from a knowledge base inadequate to solve modern problems.[36]

C. Forestry Responds

Forestry intitutions began slowly to respond with new or revived interest in conservation, sustainability, and environmental ethics. These shifts were on an apparent collision course with industrial forestry models, which by then were widely adopted, at least in principle, on public and some non-industrial private lands, as well as on industry land, in the United States, Canada, parts of Europe, Australia and New Zealand, and many tropical countries. Foresters, maintaining that what they were already doing was good stewardship, tended to respond defensively to public criticism. However, some changes were profound and effective. Public forest management agencies, particularly the U.S. Forest Service, greatly augmented their staffs to include more wildlife biologists, watershed specialists, landscape architects, and other professionals to serve a broader vision of what public forests should produce and be.

How Should Forests Be Managed?

Two emerging ideas about how forests "ought" to be managed are amply reflected in the literature of the late 1980s and early 1990s: I call them the "biotechnology" view and the "ecosystem management" view. Neither maps well onto the classical American forestry concepts of sustained yield and multiple use, nor are they necessarily mutually exclusive or even competitive with each other, although they are often so portrayed.

The biotechnology view, supporting the industrial forestry model, holds that wood yield per unit land area can be greatly increased, and invokes reductionist science to supply more wood from a constant land base to serve expanding world populations and markets. Because of the relative abundance and general good stewardship of forests in Europe and North America (where considerably more wood is grown than harvested), the notion of timber famine is largely discredited. Thus the increased productivity per unit land area promised by the biotechnology approach has not been regarded as particularly important, despite the fact that the world outlook is for demand considerably to exceed supply within the next fifty years. Nei-

---

36. National Academy of Sciences, Committee on Forestry Research, *Forestry Research: A Mandate for Change* (Washington, D.C.: National Research Council, 1990).

ther has the conservation argument for the biotechnology view been taken very seriously, although it is scientifically correct: if more wood can be grown in some forests, others can be primarily or wholly dedicated to purposes other than wood production, notably conservation of biological diversity. Sustainability, a concept given wide notice by the report of the United Nations Commission on Environment and Development, the Brundtland Report, is often invoked in arguments against wide adoption of the biotechnology view, on the grounds that intensive techniques will inevitably degrade the productive potential of forest sites.[37]

The ecosystem management view, in its forestry incarnation in the United States and Europe, derives from ecosystem ecology's notion of the wholeness of the functioning of a forest, and sets forest health, rather than the yield of any forest component as the primary indicator of good forest management. Most of the action in ecosystem management has been on public forest lands in the western United States where controversies over endangered species and the harvest of old-growth forests have been intense for two decades. Ecosystem management is seen as a way to secure the future health of the forest while still using it, with the price being the non-optimization of any forest product or attribute. Advocates of this approach are particularly sensitive to the maintenance of biological diversity at the species level in forests.

Two allied but separate and potentially incompatible concepts about forests have risen with the ecosystem management view. One, "landscape ecology" linked to management at "landscape" and regional scales, insists that in the past forestry has focused on areas too small for the effective management of all forest attributes, for example, wildlife migration corridors and patterns. The second, codified by the Leopold Commission on the U. S. National Parks as "Vignettes of Primitive America," argues that a much greater portion of the forested or potentially forested landscape should be restored to an estimate of what they were like in pre-Columbian times. Their potential incompatibility with the ecosystem management view is that they pose objectives that could be at odds with the notion of overall forest health.

Temperate and Tropical Forestry

As temperate regions examine forest management, the debate broadens in both scope and participation. Temperate and boreal forests are left less to

---

37. United Nations Commission on Environment and Development, *Our Common Future* (The Brundtland Report") (New York: United Nations, 1986).

management decisions of foresters and those who live in and near the forests as controversies over clearcutting, preservation of old growth, native peoples' rights, changing agricultural practices, and rapidly changing political systems and boundaries are voiced. Tropical deforestation remains an accelerating reality as well. Although many innovative ideas and practices have been introduced into tropical forestry (beginning with the development of the "taungya" system in the nineteenth century and continuing through social, community, and scientific agroforestry in this century) with some local successes, most tropical deforestation is caused by factors usually regarded as beyond the ken and control of foresters. These include urban and rural overpopulation and poverty and attendant human migration, government and multinational economic development policies and schemes, and absent or imperfect institutions to define and attack the root causes as well as the repair of the forests.

Forestry Milestones

Forestry's response to world trends has been uneven, although major changes are occurring. Many of these are reflected in the titles of forestry documents (e.g., *Forestry Research: A Mandate for Change; New Forests; Journal of Sustainable Forestry*). Several illustrative major changes occurring since the 1950s include:

1. The increasing adaptation of modern biological science concepts and methods into forestry science and practice; tree physiology, forest genetics, and forest ecology became part of forestry curricula and spawned journals; forest practices have become increasingly science-based, particularly forest regeneration and genetic tree improvement.
2. An even more radical departure from classical curricular and research directions has been the meteoric rise in the quantity of time and effort given to the social sciences; at first, economics, and later sociology, political science, social psychology, and anthropology.
3. The increasing awareness of the "wholeness" of forests and the less obvious environmental benefits they provide has produced a great increase in the literature of such subjects as forest hydrology, and forest influences on air quality, noise, urban spaces, and wildlife habitat.

D. Forestry in the Next Century

As forestry responds to the trends of the last half of the twentieth century, it may be preparing itself for a resolution of the diverging views of biotechnology and ecosystem management. Forests are still the least known of human terrestial habitats and surroundings, and as the science base grows, new solutions to seemingly unresolvable conflicts will emerge. Foresters will continue to broaden their educational experience, and like other professionals, progress to graduate degrees and lifelong learning. Similarly, foresters will begin to use their ability to cope with long time horizons and whole systems for the management of entities other than forests. They have been the closest of any modern profession to the human environmental conundrum of "use it but keep it," and may well extract the sustainability rabbit from the economic development hat during the next century. But two barriers and an enormous unknown currently block, or at least obscure, that path.

Though increasing, the science base of world forestry is terribly weak in comparison to other science-based human activities, including agriculture. Until a serious effort is made to remedy this, forestry solutions to sustainablility, biodiversity, or any other problem will be severely limited by available knowledge.

The second barrier stems from the still sequestered nature of foresters and forestry. Forestry's ability and presence at the national and international policy level have retreated even as awareness of and concern over forests has advanced. Foresters must become much more sophisticated in communicating with policy makers, and in particular, in relating forests and forest issues to broad societal themes and concerns.

The enormous unknown is the now rapidly expanding human population and its effects and demands on forests. The prognosis is not comforting for current notions of forestry that are dependent on relatively light use and extensive management of large tracts of forest.

Forestry and its literature are in a period of massive growth and change. It is perhaps the most exciting time for foresters since the turn of the nineteenth century. A tremendous increase in technical capability, with roots in physical science (computers and GIS), social science (new economics and sociology), and biological science (biotechnology, ecosystem ecology), exists uneasily with a huge challenge to manage forests for more goals for more people. Whether forestry institutions, including its literature, are evolving rapidly enough to effectively focus the technology on the challenge, should be apparent fairly soon.

2. A History of Forestry and Monographic Forestry Literature in Germany, France, and the United Kingdom

N. D. G. JAMES*

Forestry Historian, United Kingdom

*deceased 1993

This chapter covers a period of almost four hundred and forty years from 1500 to 1939. The first date has been chosen because forestry, as it is understood today, did not exist before 1500 while 1939 marked the end of the historical period of forest science. During the years of the World War II and its aftermath, little progress or development took place in either forestry or forest literature, so the 1940s have been excluded from this study. By 1950, forestry had entered the modern era, a subject covered in Chapters 1, 3, and 8.

Although Germany, France, and the United Kingdom are considered individually in this chapter, certain features in their respective forest histories are common to all. Early forests were areas of land over which kings or their equivalents exercised the right to hunt: these lands were not necessarily entirely covered with trees but often included cultivated fields, wastes, villages, and scattered woodland. However, in some parts of France and Germany there were large areas of natural forest which at first met the needs of these countries. But in course of time, as the countries of Europe developed and achieved economic progress, the demand for timber increased, putting great pressures on these remaining forests.

As populations increased during the Middle Ages, more houses and buildings were constructed. Overseas trade was also in development and the

---

Editors' note: N. D. G. James died in March 1993. According to his son, T. D. R. James, this was his final writing. There was no opportunity to have Dr. James read the editorial changes. Any incorrect statements or references must be ascribed to the editors.

need for ships, both navy and merchant fleets, steadily increased. Again, as iron use increased, the fuel requirements of the iron-founders for smelting rapidly expanded. At the same time the demand for domestic wood fuel gained momentum. Food production became increasingly important so that existing forests were cleared for agriculture. For a time the Black Death slowed this trend but even with one third to a half of Europe's population dead of the plague, by the early fifteenth century forests were once again under population pressure.[1]

A. Early German Forestry and Literature

As demand for timber reasserted itself in the fifteenth century, it became apparent in Germany that the continuing destruction of forests was seriously affecting the country's timber supplies. A further depletion of the forests was brought about by the Thirty Years War (1618–48); although the conflict was fought principally on German soil, other countries were ultimately involved in the widespread devastation.

At this time Germany was composed of several independent states, principalities, and palatinates which had their own legislative bodies and traditions. Consequently, when the need for legislation arose, it originated in these independent states and not in a national government as was the case in France and England. By the seventeenth and eighteenth centuries the need for a policy of organized and regulated forestry had become so pressing that a number of forest ordinances or *Forstordnungen*, were issued. These laid down procedures and practices and provided for the supervision of forests by foresters appointed by the state or palatinate. One of the earliest was the Hesse-Kassel Ordinance of 1711 which was followed by the Hesse-Nassau Ordinance in 1736 and a further Ordinance of Hesse-Kassel in 1761. At the same time, steps were taken with a view to dealing with the shortage of timber which Germany was facing. As a result, areas of land which had not previously carried a crop of trees were planted, the principal species being spruce, pine, and oak.

There were large areas of woodland which consisted of either simple coppice or coppice-with-standards often in the vicinity of towns, the chief purpose of which was to provide domestic fuel. Standards were set for trees which were allowed to grow to timber size in the coppice but it was found that these standards adversely affected the coppice and were gradually eliminated. As the use of coal became more general, the market for coppice

---

1. Rowland Ernle, *English Farming, Past and Present*, 4th ed. (London and New York: Longmans, Green, 1927), p. 20.

began to decline at the same time that the demand for timber increased so that the question of growing more timber and less coppice was hotly debated.

Broadly speaking, timber can be grown either by planting young trees or by making use of self-sown seedlings, collectively referred to as "natural regeneration." In many German forests, several species regenerate naturally under suitable conditions, including oak, beech, silver fir, spruce, and pine. In the course of time, different methods of raising forest trees evolved, which are the basis of all modern silvicultural systems. Examples of these are the selection, group, strip, uniform, and clearcutting systems, the first four being based on the use of natural regeneration to provide young trees for the succeeding crop. A great deal of thought and research was given to the initiation and development of these silvicultural systems and many German foresters were pioneers in this work. As silviculture developed, so did the need for forest management and this in turn made it necessary to measure stands of timber and to calculate their volume, growth, increment, and yield. Accurate forest surveys were essential for efficient management as well as for the construction of roads to access and extract timber.

By 1770, the Germans were gaining a reputation as foresters. The reasons are not difficult to recognize. Germany was in the forefront of a more technical approach to forestry. An increasing number of books on forestry matters were being published in Germany which provided valuable sources of information as well as a means of disseminating new ideas. The importance of education was realized and a number of forestry schools were formed, several of which subsequently merged with universities.

First signs of the remarkable forest literature appeared in the eighteenth and nineteenth centuries. As early as 1576, Noe Meurer, considered to be the first writer on German forestry, published a book on game law of which a part dealt with forestry.[2] However, the first book concerned solely with forestry, as opposed to one including agriculture, hunting, and other rural matters, did not appear until 1713. This was *Sylvicultura Oeconomica* [The economics of silviculture] by Hans Carl von Carlowitz which was chiefly concerned with silviculture, but also dealt with forest management.[3] In 1759, J. G. Beckmann (1700–77) wrote his *Anweisung zu einer pfleglichen Forstwirtschaft* [Instructions on cultivation in forest management].[4]

Other leading authorities during the eighteenth century were Johann

---

2. *Forestry in the Federal Republic of Germany*, 4th English ed. (Bonn: Ministry of Food, Agriculture, and Forestry, 1962), p. 54.

3. Hans Carl von Carlowitz, *Sylvicultura Oeconomica* (Leipzig: Johann Friedrich Braun, 1713).

4. Johann Gottlieb Beckmann, *Anweisung zu einer pfleglichen Forstwirtschaft* (Chemnitz: Stössel, 1759).

Georg von Langen and his pupil Hans Dietrich von Zanthier whom Bernhard E. Fernow described as the two most eminent practitioners of the period.[5] Both were members of prominent German families and were well
educated. They managed forests which belonged to the duke of Brunswick
and the count of Stolberg-Wernigerode in the Harz Mountains. Although
von Langen did not write books, the outstanding reports which he produced
and the working plans which he drew up were far in advance of his contemporaries. Von Zanthier wrote a number of books of which *Kurzer System-
atischer Grundiss der Praktischen Forstwissenschaft* [A short concise guide
to practical forest science] published in 1764 is one of his best. He also
founded the first forestry school at Wernigerode in 1763.[6]

Another writer who made a valuable contribution to contemporary forestry knowledge was J. A. Cramer, whose book *Anleitung zum Forstwesen*
[Instructions on forestry] published in 1766 contained much information on
silviculture.[7] A steady flow of books covering many aspects of forestry
continued until the end of the nineteenth century, but space only allows a
limited number of these to be mentioned. Toward the end of the century
two authors appeared who are still considered to be among the most outstanding writers on German forestry.[8] The first, Georg Ludwig Hartig
(1764–1837), was trained at the University of Giessen, and in 1789 established a forestry school at Hungen which was later transferred to Stuttgart.
He was the author of a large number of publications, one of the most important being *Answeisung zur Holzzucht für Foerster* [Instructions in silviculture for foresters] which he wrote in 1791.[9] In 1811, he was appointed
head of the Prussian Forest Administration and carried out many improvements in the organization of the forest services especially on financial matters. At the same time he became a lecturer at the University of Berlin, a
position which he held until his death in 1837. He is said to have been one
of the outstanding foresters of all time.

The other authority was Heinrich von Cotta (1763–1844) who, after
studying at the University of Jena, chose forestry as a profession and
worked in Thuringia. He began instruction on a private basis, and in 1795
founded a forestry school at Zillbach, but in 1811 he was made Director of

5. Bernhard E. Fernow, *A Brief History of Forestry in Europe, the United States, and Other
Countries*, 3d ed. (Toronto: University Press; Washington, D. C.: American Forestry Association,
1913), p. 66.
6. Ibid., p. 88.
7. Johann A. Cramer, *Anleitung zum Forstwesen* (Braunschweig: Im Verlage der Fürstl. Wais-
enhaus-Buchh., 1766).
8. *Forestry in the Federal Republic*, p. 54.
9. (a) Fernow, *A Brief History of Forestry*, p. 98. (b) Georg Ludwig Hartig, *Anweisung zur
Holzzucht für Foerster* (Marburg: Neunen Akademischen Buchhandlung, 1804).

Forest Surveys in Saxony. On taking up this appointment he moved his school to Tharandt and began a new *forstlehranstalt* or school of forestry there. In 1816 he was chosen to be Director of the Bureau of Forest Management and the school became the State Forest Academy of Saxony with Cotta as the director, a position which he held until his death in 1844.[10] A man of great originality who was also a plant physiologist, he made important contributions to silviculture and forest management.[11] His book *Answeisung zum Waldbau* [Instructions in silviculture] published in 1817 is an excellent account of the subject.[12] During 1929, the Forest Academy of Saxony was combined with the Dresden Polytechnical University (or *Technische Hochschule*) and Heske considered that through Cotta's influence it had become "the birthplace of scientific forestry."[13]

Two others in the early years of the nineteenth century who greatly enhanced the standing of German forestry were Johann Christian Hundeshagen (1783–1834) and Gottlob Konig (1776–1849). After studying at Heidelberg, Hundeshagen spent some years gaining practical silviculture experience, and later became professor of forestry at Tübingen University in 1817. In 1825, he moved to the University of Giessen, again as professor of forestry, where he specialized in the scientific aspects of forestry. Although Konig did not have the benefit of a university education, his views were largely influenced by Cotta. After being appointed as head of the forest administration for a private estate at Eisenbach near Kassel, he established his own forest school there. He was an expert in mathematics as applied to forestry, especially forest mensuration. His book *Anleitung zur Holztaxation* [Instructions on the taxation of timber] was published in 1813.[14] Mention must also be made of Carl J. Heyer (1797–1856) who was professor of forestry at the University of Giessen in 1835. Among his books, which were noted for their clarity, were *Die Waldertrags-Regelung* [The profitable management of forests] in 1841 and *Waldbau* [Silviculture] in 1854.[15] Both of these were described by Fernow as classics.[16]

In 1855, H. Burckhardt who three years previously had published a set of forest yield tables, wrote *Säen und Pflanzen* [Sowing and planting][17] which

10. N. V. Brasnett, *Planned Management of Forests* (London: Allen & Unwin, 1953), p. 184.
11. Ibid.
12. Heinrich von Cotta, *Anweisung zum Waldbau* (Dresden: Arnold, 1817).
13. Franz Heske, *German Forestry* (New Haven: Yale University Press, and London: H. Milford, Oxford University Press, 1938), p. 213.
14. Fernow, *A Brief History of Forestry*, p. 100.
15. (a) Carl J. Heyer, *Die Waldertrags-Regelung* (Giessen: 1841). (b) Carl J. Heyer, *Der Waldbau oder, die Forstproductenzucht* (Leipzig: B. G. Teubner, 1854), in his *Encyclopädie der Forstwissenschaft*, 4.
16. Fernow, *A Brief History of Forestry*, p. 101.
17. Heinrich Burckhardt, *Säen und Pflanzen* (Hannover: C. Rumpler, 1855).

met with unqualified approval.[18] Thirty years later, Richard Hess who became professor of forestry at the University of Giessen, published his *Encyclopädie und Methodologie der Forstwissenschaft* [An encyclopedia on the science of forestry]. In 1895 an English translation of *Der Forstschutz* [Forest protection] which Hess had written in 1878, was made by W. R. Fisher.[19] This became Volume 4 of *Schlich's Manual of Forestry* under the subtitle of *Forest Protection* (1907).[20] In the following year, Fisher completed a translation of *Die Forstbenutzung* [Forest utilization] which Johann Karl Gayer (1822–1907) had written in 1863, the eighth and last edition of which appeared in 1894.[21] This translation was adopted as Volume 5 : *Forest Utilization* in *Schlich's Manual*.[22] Gayer was appointed professor of forestry at Aschaffenburg in 1855 and when part of the school was transferred to the University of Munich in 1878, he became a professor there. He also wrote *Der Waldbau* [Silviculture] in 1880.[23]

One of the first writers to deal with the management and organization of forest nurseries was H. Furst, director of the Forest School at Aschaffenburg, who wrote *Pflanzenzucht im Walde* [The cultivation of trees in forests] in 1882.[24] Franz Baur, a professor at University of Munich, published *Handbuch der Waldwertberechnung* [A handbook of forest valuations] in 1886 on a subject not often touched during this period.[25] A fourth edition of his *Die Holzmesskunde* [The measurement of trees] appeared in 1891.[26] Another book translated from the German was *Forstwissenschaft* [The science of forestry] by A. Schwappach, professor of forestry at the University of Giessen.[27] The English translation by Fraser Story and E. A. Nobbs was published in 1904 with the title *Forestry*. One of the last of the many distinguished foresters of the nineteenth century was J. F. Judeich (1824–1894) who was professor and director of the Forest Academy at Tharandt. One of his books, *Die Forsteinrichtung* [Forest organization] published in 1893, enjoyed wide circulation over a number of years; the sixth edition was published in 1904.[28]

18. Fernow, *A Brief History of Forestry*, p. 131.

19. Richard A. Hess, *Der Forstschutz* (Leipzig: B. G. Teubner, 1878).

20. W. Schlich, *Schlich's Manual of Forestry*, Vol. 4: *Forest Protection*, by Richard Hess, 2d ed. (London: Bradbury and Agnew, 1907), pp. ix–x.

21. Johann K. Gayer, *Die Forstbenutzung* (Aschaffenburg, 1863; 13th ed., Berlin: P. Parey, 1935).

22. W. Schlich, *Schlich's Manual of Forestry*, Vol. 5: *Forest Utilization*, by Johann K. Gayer. 2d ed. (London: Bradbury and Agnew, 1908), pp. v–vi.

23. Johann K. Gayer, *Der Waldbau* (Berlin: P. Parey, 1880).

24. Hermann H. von Furst, *Pflanzenzucht im Walde: Ein Hanbuch fur Forstwirthe, Waldbesitzer und Studierende* (Berlin: 1882).

25. *Forestry in the Federal Republic*, p. 54.

26. Franz Baur, *Die Holzmesskunde*, 4th ed. (Berlin: P. Parey, 1891).

27. Adam F. Schwappach, *Forestry* (London: J. M. Dent, 1904).

28. Friedrich Judeich, *Die Forsteinrichtung* (Dresden: G. Schonfeld, 1893); 6th ed. edited by Max Neumeister (Berlin: P. Parey, 1904).

One of the most prolific writers in the early part of the twentieth century was Karl Dannecker whose books included *Der Waldwirt* [The forest manager] (1926),[29] *Der Plenterwald, Einst und Jetzt* [The selection forest, past and present] (1929) and *Das Wesen der Kontrollwirtschaffe* [The nature of management control] (1932). The persons mentioned here are only a small selection of those who by their work, books, and research have contributed to the knowledge of forestry.

A notable feature of German forestry was the early realization of the need to provide instruction and training for those who wished to become foresters. During the eighteenth and nineteenth centuries, three channels of education evolved: private schools, forestry academies or institutions, and universities. Private schools known as "master schools" or *Meisterschule* were formed by some of the leading figures in forestry. They were essentially personal to the founder and were often discontinued when he died. If the "master" took up a new appointment in a different locality, his school usually moved with him. In some cases a school might be taken over by the state and advanced to the status of a forestry academy. The first master school was established by H. D. von Zanthier in 1763 at Wernigerode on the northern boundary of the Harz Forest. It was later transferred to a few miles from Ilsenberg, but it closed when von Zanthier died in 1778.[30]

In 1789, G. L. Hartig formed a school at Hungen, southeast of Giessen, but in 1807 moved with his school to Stuttgart. When he was appointed head of the Prussian Forest Service in 1811 he proceeded with the school to Berlin and was granted the rank of professor in the university.[31] In 1821, the first state forestry academy was established in Berlin in association with the university. However, in 1830, the academy moved to Eberswalde, and was renamed the High Institution of Forest Science.[32] At the same time Hartig continued to give lectures in Berlin until his death in 1837.[33]

During the nineteenth century other centers for forestry education were established including a school at Ruhla founded by G. Konig in 1808, which in 1830 became a state forestry institution at Eisenach.[34] In 1868, a forest academy was established at Münden; the first director was Gustav Heyer, the son of C. J. Heyer who had been a professor at the University of Giessen.[35]

29. Karl Dannecker, *The Forest Manager; A Handbook for Farm Woodland Owners Who Manage Their Own Woodlands* (Washington, D. C.: American Forestry Association, 1939). (Translation of his *Der Waldwirt*, 1926.)

30. Fernow, *History of Forestry*, p. 83.

31. Ibid., p. 98.

32. F. B. Hough, *Report upon Forestry* (Washington, D. C.: GPO, 1878), p. 614.

33. Fernow, *History of Forestry*, p. 146.

34. Hough, *Report upon Forestry*, p. 622.

35. Fernow, *History of Forestry*, p. 146.

Several universities were concerned with forestry to some extent. A school of forestry was established in 1825 at Giessen, under the direction of J. C. Hundeshagen.[36] However, in 1831 its status was changed and it became the Forestry Institute of the University of Giessen. Over the course of years the staff included such eminent foresters as Carl Heyer, Gustav Heyer, J. L. Klaupretch, R. Hess, A. Schwappach, and T. Lorey.[37]

By the middle of the nineteenth century, the importance of forestry was beginning to be recognized by many countries. Not surprisingly, the knowledge and expertise of German foresters were sought. The Indian government recognized the need for a forest policy and established a small Forest Department in 1847, which became very much larger during the following fifteen years.[38] The first inspector-general of forests appointed in 1864 was Dietrich Brandis (1824–1907) who had been born and trained in Germany and had been Superintendent of Forests in Burma.[39] He laid the foundations of the Forest Service in India and was instrumental in establishing the Imperial Forest College at Dehra Dun in 1878. In 1866, he recruited two more German foresters to the Indian Forest Service: W. Schlich who succeeded him as Inspector-General in 1883 and Berthold Ribbentrop who followed Schlich as Inspector-General in 1888.[40]

W. Schlich (1840–1925) had studied at the University of Giessen under Gustav Heyer and at the age of twenty-seven went to the Forest Service in India. In 1885 he returned to the United Kingdom upon government invitation to organize the Forestry Branch of the Royal Indian Engineering College at Cooper's Hill. When this Forestry Branch was transferred to Oxford University in 1905, Schlich was appointed professor of forestry. His best known publication, *Schlich's Manual of Forestry,* consisted of five volumes which appeared over a number of years, the first in 1889.[41] Schlich was appointed a Knight Commander of the Indian Empire in 1909.[42] When Schlich completed his term as Inspector-General in 1888, Ribbentrop, also born and trained in Germany, was appointed as his successor. His book, *Forestry in British India* (1900), provides a valuable account of forestry in India until that date. It should be noted however, that forest officers for the Indian Forest Service were normally recruited in the United Kingdom. Dur-

---

36. Hough, *Report upon Forestry*, p. 620.

37. Heske, *German Forestry*, p. 215.

38. B. Ribbentrop, *Forestry in British India* (Calcutta: Office of the Superintendent of Government Printing, 1900), pp. 75–76.

39. Ibid., p. 72.

40. "Obituary Notices; Sir Dietrich Brandis," *Quarterly Journal of Forestry* 1 (1907): 317–318.

41. William Schlich, *A Manual of Forestry* (London: Bradbury and Agnew, 1889–1896).

42. R. S. Troup, "Sir William Schlich's Work in India," *Quarterly Journal of Forestry* 19 (1925): 9–11.

ing the eighteen years from 1869 to 1886, ninety-five officers were accepted from this source.[43]

During this same period two German foresters made a noticeable impact on forestry in the United States. The first was Bernhard E. Fernow, born in 1851 at Inowraclaw in what was then Prussia but is now Poland. He attended the Forest Academy at Münden in 1870 where he was a pupil of Gustav Heyer. After emigrating, he became chief of the Forestry Division of the United States Department of Agriculture in 1886, but left that position in 1899 to organize the New York State College of Forestry at Cornell University and became its first director, an appointment he held until 1903.[44] In 1907 he became dean of the Faculty of Forestry in the University of Toronto.[45] His important work, *The Economics of Forestry*, was followed by *A Brief History of Forestry in Europe, the United States, and Other Countries*.[46] He died in 1923.

Carl A. Schenck was the second of these two German foresters to emigrate to the United States after studying at the universities in Tübingen and Giessen. He was recommended by Dietrich Brandis to manage the forests of G. W. Vanderbilt's Biltmore estate in North Carolina.[47] He engaged two apprentices who were later followed by others, in effect creating the equivalent of the old master school. This was replaced in 1898 by a formal school, the Biltmore Forest School, but in 1909 Schenck retired from Biltmore.[48] He continued to conduct his school from various localities until it closed in 1913 when he returned to Germany, served in the army during World War I, and in 1922, retired to Lindenfels in the Odenwald, near Darmstadt.[49]

B. Early French Forestry and Literature

During the Middle Ages, certain features were common to the forests of Germany, France, and England. In these countries, areas of land designated

43. Ribbentrop, *Forestry in British India*, p. 227.

44. A. D. Rogers, *Bernhard Eduard Fernow, A Story of North American Forestry* (Princeton: Princeton University Press, 1951), p. 108.

45. Ibid., p. 403.

46. (a) Bernhard E. Fernow, *Economics of Forestry* (New York: T. Y. Crowell, 1902). (b) Bernhard E. Fernow, *A Brief History of Forestry in Europe, the United States and Other Countries*, 1st ed. (New Haven: Price, Lee & Atkins, 1907).

47. Rogers, *Bernhard Eduard Fernow*, p. 247.

48. C. Clepper, *Professional Forestry in the United States* (Baltimore: Johns Hopkins Press, 1971), p. 35.

49. Ibid., pp. 35–36.

as forests were not necessarily covered by trees. The value of many of the royal forests lay in providing sport for the nobility rather than the production of timber. Forestry did not exist as practiced in the twentieth century. By the middle of the fourteenth century, the principal forest owners in France were the crown, ecclesiastical bodies, communes, and private individuals.

By the beginning of the thirteenth century in France, woodland crafts were appearing where conditions were favorable and where supplies of suitable material were available. These were largely dependant on coppice and included the production of clogs and sabots, charcoal, bark for use in tanning leather, and fuel for domestic use. Increasingly large areas were felled and in order to control and regulate this harvesting, legislation was introduced. This became forest ordinances and some of the earliest were those of 1280, 1318, and 1346 but the most important was the Ordinance of Mélun which became law in 1376.

Amongst its various provisions was one which directed that when a virgin forest was felled, the area cleared must be restricted to between 25 and 37 acres (10 to 15 hectares); further, before felling commenced the area must be surveyed, clearly marked and on completion of the work, satisfactorily enclosed. Trees were marked with a stamping hammer, on the head of which the insignia of authority was impressed, the procedure being known as *martilage* or "marking trees for felling." This work was carried out by a *garde-marteau* (literally a "hammer guard") who was the responsible forester. It was also specified that 6 to 8 trees must be reserved as seed bearers on each *arpent* which was approximately equal to one acre. This gave rise to a system known as *tire et aire* under which annual fellings or coupes were fixed by area (*aire*) and were adjacent to each other (*à tire*). It was applied to both high forest and coppice but in the case of the latter, produced a result which was not unlike coppice-with-standards.[50]

By the beginning of the seventeenth century, French forests had deteriorated because of the ever-increasing demand for timber, the effect of the Thirty Years War, failure to maintain the productivity of the forests, government corruption, and a state of general instability. Anxiety concerning future timber supplies especially as regards ship-building for the navy caused Louis XIV in 1661 to instruct his Controller of Finance, Jean-Baptiste Colbert (1619–83), to set up a commission to examine the problem. Investigations were carried out mainly by envoys and special commissioners, one of whom was Louis de Froidour who wrote *Instructions pour la Vente des Bois du Roy* [Directions for the sale of the king's woods] in

---

50. F. C. Osmaston, *The Management of Forests* (New York: Hafner, 1968), p. 315.

1668. He followed this in 1685 with *Règlement Concernant les Forêts du Pays de Bigorre* [Regulations relating to the forests in the District of Bigorre].[51] In 1680 he introduced to the Pyrenees the method of management termed *jardinae* but now known as the selection system, or *Plenterbetrieb* in Germany. Under this system, trees are removed singly or in small groups scattered throughout the area consequently producing an uneven-aged type of forest.[52]

As a result of Colbert's commission, the Ordinance of 1669 was enacted which established a new forest code which applied to state and private forests. It reduced the number of *grand-maîtres* or grand masters in the forest administration, many of whom had little, if any, knowledge of forestry matters. A classic example was the celebrated French writer, Jean de la Fontaine (1621–95), who had been appointed on his literary merits.[53] Those found guilty of corruption or negligence were dismissed and severely punished. The ordinance laid down the minimum rotations for coppice and high forest, stipulated the number of seed-bearing trees to be retained during felling and gave detailed directions on various silvicultural matters.[54] In the case of degenerated coppice, rejuvenation was to be effected by means of *sartage* which entailed burning brushwood on top of deteriorating coppice stools in order to encourage the growth of fresh shoots from below and the production of new roots.[55] In spite of these reforms, results were not always as beneficial as might have been expected because the fear of being accused of malpractices had a restricting effect. As a result, some silvicultural operations, such as thinning and improvement fellings, were sometimes reduced to a minimum or omitted altogether leading to a decline of the forest.

Shortly after the beginning of the eighteenth century, a number of books on different aspects of forestry began to appear. One published in 1721 was *Réflexions sur l'État des Bois Royaume* [Thoughts on the condition of the royal forests] by R. A. F. de Reaumur (1683–1757).[56] He was described in the *Biographie Universelle Ancienne et Moderne*, as: "l'un des Plus Ingenieux Naturalists et Physiciens que la France Ait Produits," one of the cleverest naturalists and physicists that France has produced.[57] The Reaumur thermometer was named after him. In 1739 a memoire on preserving and repairing forests was published by Georges Louis Leclerc, comte de Buf-

51. (a) Louis de Froidour, *Instructions pour la Vente des Bois Roy* (Toulouse: 1668). (b) Louis de Froidour, *Règlement Concernant les Forêts du Pays de Bigorre* (1685).

52. J. L. Reed, *Forests of France* (London: Faber & Faber, 1954), p. 47.

53. Ibid., p. 41.

54. Osmaston, *The Management of Forests*, p. 317.

55. Brasnett, *Planned Management of Forests*, p. 175.

56. R. A. F. de Reaumur, *Réflexions sur l'État des Bois Royaume* (Paris: 1721).

57. *Biographie Universelle, Ancienne et Moderne* (Paris: Michaud Frères, 1811–1862).

fon, an eminent naturalist, born in 1707 at Montbard, northwest of Dijon, and who died in 1788.[58] This work appeared in English and was bound with *A Treatise on . . . Raising Forest Trees* by the earl of Haddington (1761) which also included a bound-in tract by Buffon entitled *Memorial on the Culture of Forests* dated 1742.[59] This was a translation of his French article which had appeared in *Memoires de l'Academie des Sciences* (pages 233–46) in 1742.[60] It is very probable that the 1739 memorial had a similar source.

The first two volumes of a series of outstanding books by Henri Louis Duhamel du Monceau (1700–82) were published in 1755 in the comprehensive *Traité Complet des Bois et des Forêts* [A comprehensive treatise on woods and forests]. It comprised eight volumes although they were not numbered consecutively as such. The first two were followed in 1758 by two more volumes, *La Physique des Arbres* [The structure of trees], followed by a single volume, and in 1764 *De l'Exploitation des Bois* [The utilization of woodlands] in two more volumes. The subtitle of the latter focuses on advantageous ways of coppicing young and high forests, and of making accurate valuations. His last volume (1767) was titled *Du Transport*.[61] In addition to being clear and scholarly texts, these books contained a large number of excellent illustrations. Duhamel, generally considered the originator of the scientific approach to forestry in France, was a landowner, and doubtless the woodlands on his estate encouraged a practical approach to silviculture and woodland management. He was also a keen naturalist

58. H. Nadault de Buffon, *Buffon, sa Famille ses Callaborateurs et ses Familiers* (Paris: J. Renouard, 1863), p. 3; 139.

59. *A Treatise on the Manner of Raising Forest Trees, & c. in a Letter from the Right Honourable the Earl of Haddington to His Grandson. To Which are Added Two Memoirs; the One on Preserving and Repairing Forests; the Other on the Culture of Forests. Both translated from the French of M. de Buffon of the Royal Academy of Paris* (Edinburgh: G. Hamilton and J. Balfour, 1761).

60. Arnold Arboretum Library, *Catalogue of the Library of the Arnold Arboretum of Harvard University*, Vol. 2: *Forestry and Forest Description* (Cambridge: Cosmos Press, 1927), p. 433.

61. (a) Henri Louis Duhamel du Monceau, *Traité des Arbres et Arbustes qui se Cultivent en France en Pleine Terre* (Paris: H. L. Guerin & L. F. Delatour, 1755). (b) Henri Louis Duhamel du Monceau, *La Physique des Arbres* (Paris: H. L. Guerin & L. F. Delatour, 1758). (c) Henri Louis Duhamel du Monceau, *Des Semis et Plantations, ou, Methodes pour Multiplier et Élever les Arbres, les Planter en Massifs & en Avenues* [Sowing and planting, or, methods for multiplying and raising trees, planting along walls and avenues] (Paris: H. L. Guerin & L. F. Delatour, 1760). (d) Henri Louis Duhamel du Monceau, *De l'Exploitation des Bois, ou, Moyens de Tirer un Parti Avantageux des Taillis, Demi-Futaies, et Hautes-Futaies, et d'En Faire une Juste Estimation* [The utilization of woodlands, or ways to turn to advantage coppice, young high forest, and high forest and make an accurate valuation] (Paris: H. L Guerin & L. F. Delatour, 1764). (e) Henri Louis Duhamel du Monceau, *Du Transport, de la Conservation de la Force des Bois: ou, l'On Trouvera des Moyens d'Attendrir les Bois, de Leur Donner Diverses Courures, sur-tout pour Suppléer au Défaut des Pièces Simples* [On transporting, conserving wood strength] (Paris: L. F. Delatour, 1767).

and wrote several books on agriculture and fruit growing. In 1732 he was appointed Inspecteur Général de la Marine.

Another book containing a great deal of useful information was the *Manuel Forestier et Portatif* [Forester's manual and pocket book]. This was written by Guoit of whom little appears to have been recorded except that the title page termed him a *garde-marteau* in the Forêt de Rambouillet.[62] The title page also states that the book is taken to a large extent from the general account of forests by M. Duhamel du Monceau. The first part describes sixty-seven species of trees and shrubs growing in the royal forests; the second deals with sowing and planting, forest management, amenity planting in parks, marking standards, valuations, sales, utilization, and forest inventories.

C. P. J. Balland published *Observations sur l'Administration des Forêts* in 1791. A year later, P. C. M. Varennes de Fenille published a report on management of forests.[63] The century closed with J. B. Bridel's book, *Manuel Practique du Forestier*.[64]

Although the majority of French forests were under some form of regular management by 1760, the naval losses in the Seven Years War (1756–63) necessitated heavy fellings, especially of oak. In addition, France entered a period of social upheaval and war. This began with the French Revolution (1789–94) and continued until the Battle of Waterloo, and the Second Treaty of Paris in 1815. The Revolution had two important forestry results: the dissolution of the forest service, and secondly the change in land ownership. Although some remained unaffected, many private owners were dispossessed of their estates and ecclesiastical property was confiscated; in both cases, forests and woodlands were involved. Communal forests, on the other hand, received very different treatment; feudal dues and payments were abolished, and in some cases their area was increased by the transfer of woodlands from monasteries and church authorities. The overall effect on forestry was to produce a period of stagnation during which German forestry exerted an increasing influence.

An encouraging sign of a return to more normal conditions was that some thought was being given to forestry education. Several schools were estab-

62. Guoit, *Manuel Forestier et Portatif* (Paris, 1770).

63. P. C. M. Varennes de Fenille, *Mémoires sur l'Aménagement des Forêts Nationales, sur l'Administration Forestière, sur les Qualités Individual des Bois Indigènes* . . . [Reports on the management of the national forests, on forest administration, on the individual characteristics of the timber of native trees], 2d ed. (Paris: A. J. Marchant, 1808).

64. J. B. Bridel, *Manuel Practique du Forestier; Ouvrage dans Lequel On Traite de l'Estimation, Exploitation, Conservation, Repeuplement, des Semise Plantations des Forêts* [The forester's practical handbook; a work dealing with forest valuation, utilization, conservation, restocking, sowing and plantations] (1790).

lished in Germany during the latter part of the eighteenth century, but not until 1824 did France take any positive steps. The establishment of a school was largely due to the persistent efforts of J. J. Baudrillart, a senior officer in l'Administration Générale des Forêts, and also a professor of political economy.[65] In 1803, he wrote *Traité de l'Aménagement des Forêts* [Textbook on forest management]. From 1823 to 1825 his *Dictionnaire* was published in two volumes.[66] L'École Nationale des Eaux et Forêts was founded in 1824 at Nancy. The first director was Bernard Lorentz (1775–1865) who held this position until 1834.[67]

Early in the French Revolution, Lorentz escaped to Germany where he lived for eight years during which he became a close friend of G. L. Hartig and was greatly impressed by his views and ideas. Hartig was an enthusiastic advocate of the uniform system which involved the raising of a young crop through the medium of three regeneration fellings. These comprised a preliminary or seeding felling to encourage seed result. Lorentz returned to France and endeavored to install the uniform system. His reasons were twofold: first, he considered that it was technically wrong to grow coppice and timber on the same site; second, he felt that as the use of coal became more widespread the demand for fuel wood would decrease. Lorentz promptly found himself in conflict with two powerful bodies. First, those who had commercial interests which utilized coppice, such as the iron smelters, feared a shortage in the supply of fuel wood. Second, the Treasury foresaw a substantial fall in revenue from the forests. So powerful was the opposition that in 1839, Lorentz was forced to resign as head of the forest administration. A full account of this will be found in *Une Grande Querelle Forestière —la Conversion* [A great forestry dispute—conversion] by Roger Blais.[68]

More settled conditions were exemplified by a new *Code Forestier* passed in 1821; the forest service was reinstated, and new books began to appear. In 1800 a new publication, based on that of Duhamel who had died in 1782, was issued as *Traité des Arbres et Arbustes que l'On Cultive en France* [Treatise on trees and shrubs that can be grown in France]. This comprised seven volumes which became known as the *Nouveau Duhamel*

65. Fernow, *History of Forestry*, p. 245.

66. (a) Jacques J. Baudrillart, *Traité de l'Aménagement des Forêts* (1803). (b) Jacques J. Baudrillart, *Dictionnaire Général Raisonné et Historique des Eaux et Forêts* [General descriptive and historical dictionary of waters and forests] (Paris: A. Bertrand, 1823–1825).

67. Fernow, *History of Forestry*, p. 242.

68. Roger Blais, *Une Grande Querelle Forestière: la Conversion* (Paris: Presses Universitaires de France, 1936).

[The new Duhamel] edited by E. Michel with assistance from several contributors. [69]

After serving in the army, L. de Perthuis de Laillevant (1757–1818) returned to civilian life in 1791, became an engineer and land manager and began writing on rural matters. In 1803 he published *Traité de l'Aménagement et de la Restauration des Bois et Forêts de la France* [Treatise on the management and restoration of the woods and forests of France].[70] Three years later, E. Chevalier also wrote a restoration and management work.[71] Two other books concerned with the reestablishment of the forests were *Les Forêts de la France* written by the Baron de Rougier de la Bergerie in 1817, and *La Régénération de la Nature Végétale* [The regeneration of wild plants] by R. Rauch. [72]

After Lorentz had become director of L'École Nationale at Nancy in 1824, he appointed A. L. F. Parade (1802–65) the following year as lecturer in silviculture. Parade had studied at Tharandt with H. von Cotta who held similar views to Hartig. Parade became friendly with Lorentz and was director of the school from 1835 to 1864; some sources give his name as Parade-Soubeirol.[73] Lorentz with Parade wrote *Cours Élémentaire de Culture des Bois* [Elementary course on the cultivation of woods] (1837). This ran to six editions, the fifth being published in 1867 under the names of Lorentz and H. Nanquette, although Lorentz had died in 1865 and Parade a year before him. Lorentz and Louis Tassy were named authors for the sixth and last edition (1882) which was in six volumes.[74] Kitchingman considered it the most important book on French silviculture published in the nineteenth century.[75]

In addition to work with Lorentz, L. Tassy also wrote *Études sur l'Aménagement des Forêts* [Studies in forest management] (1858), a treatise on the reorganisation of the French forest service, and *États des Forêts en*

69. M. Duhamel du Monceau, *Traité des Arbes et Arbustes l'On Cultive en France en Pleine Terre*, New ed. (Paris: Michel, 1800–1819).

70. L. de Perthuis de Laillevant, *Traité de l'Aménagement et de la Restauration des Bois et Forêts de la France* (Paris: Imprimerie, Librairie de Madame Huzard, 1803).

71. E. Chevalier, *Restauration et Aménagement des Forêts et des Bois Particuliers* [The restoration and management of private woods and forests] (Paris, 1806).

72. (a) Baron de Rougier de la Bergerie, *Les Forêts de la France* (Paris: A. Bertrand, 1817). (b) R. Rauch, *La Régénération de la Nature Végétale* (Paris: 1815).

73. R. S. Troup, *Silvicultural Systems* (Oxford: Clarendon Press, 1928), p. 61.

74. Bernard Lorentz and A. L. F. Parade, *Cours Élémentaire de Culture des Bois* (Paris: A. Parade, 1837); 5th ed. by Bernhard Lorentz, A. L. F. Parade and H. Nanquette; edited by Adolphe Lorentz (Paris: 1867); 6th ed. by Bernard Lorentz, A. L. F. Parade, and L. Tassy; edited by Adolphe Lorentz (Paris: Octave Doin, 1882).

75. G. D. Kitchingman, "An Outline of French Silvicultural Literature," *Forestry; Journal of the Society of Foresters of Great Britain*, 25 (1952): 136.

France [The condition of the forests in France] in 1887.[76] Henri Nanquette professor of forestry economy at Nancy and a subsequent director, provided two important books: *Exploitation, Debit et Estimation des Bois* [Utilisation, conversion and valuation of woods] in 1859 and *Cours d'Aménagement des Forêts* [Course of forest management] the following year.[77]

One of the most successful French books was *Guide du Forestier* [The forester's guide] by A. Bouquet de la Grye in 1859. It reached its fourteenth edition by 1947, although the title had been changed to *Technique Forestière* [Practical forestry].[78] Bouquet wrote several other books, the most significant being: *La Surveillance des Forêts* [The supervision of forests], the sixth edition of which was published in 1872, and *Le Régime Forestier, Appliqué aux Bois des Communes et des Établissements Publics* [State regulations applied to communal woods and publicly owned forests] (1883).[79] Adolphe Gurnaud (1825–98) was a prolific writer who became a *ĝarde général* or ranger in the conifer forests of the Jura after studying at Nancy. He later resigned from the forest service. Knuchel cites eighteen of his publications which appeared between 1865 and 1898, and considered that *Cahier d'Aménagement sur la Forêt des Esperons* [Manual of management for the application of the area method as used in Des Esperons Forest] was one of his best.[80]

In 1873, G. Bagneris, inspector of forests and professor of silviculture at Nancy, wrote *Manuel de Sylviculture* with a second edition in 1878.[81] It was translated into English by E. E. Fernandez and A. Smythies, who were in the Indian Forest Service, and published in India in 1876 and in England in 1882 as *Elements of Sylviculture*.[82] C. Broillard (1831–1910), a writer of this period, was on the staff of L'École Nationale from 1865 to 1878. His

76. Louis Tassy, *Études sur l'Aménagement des Forêts* (Paris: J. Rothschild, 1858). (b) Louis Tassy, *Réorganisation du Service Forestier* [The reorganisation of the forest service] (Paris, 1879). (c) Louis Tassy, *États des Forêts en France* (Paris, 1887).

77. (a) Henri Nanquette, *Exploitation, Débit et Estimation des Bois* (Nancy: Ve Raybois, 1859). (b) Henri Nanquette, *Cours d'Aménagement des Forêts* (Paris: Me ve Bouchard-Huzard, 1860).

78. Amédeé Bouquet de la Grye, *Guide du Gard Forestier* (Paris: 1859); 14th ed. with title *Technique Forestière* (1947).

79. (a) Amédée Bouquet de la Grye, *La Surveillance des Forêts*, 6th ed. (Paris: J. Rothschild, 1872) (His *Guide du Forestier*, pt. 2, 1882). (b) Amédée Bouquet de la Grye, *Le Régime Forestier Appliqué aux Bois des Communes et des Établissements Publics* (Paris: Librairie Agricole de la Maison Rustique, 1883).

80. H. Knuchel, *Planning and Control in the Managed Forest*, English edition translated by M. L. Anderson (Edinburgh: Oliver and Boyd, 1953), pp. 172, 355.

81. Gustave Bagneris, *Manuel de Sylviculture* (Paris: Berger-Levrault, 1873); 2d ed. (Paris: Berger-Levrault, 1878).

82. Gustave Bangeris, *Elements of Sylviculture* (translated from French 2d ed., by E. E. Fernandez and A. Smythies) (London: W. Rider & Son, 1882).

Cours d'Aménagement des Forêts [Course of forest management] (1878) was followed in 1881 by *Traitement des Bois en France*.[83]

Lucien Boppe (1834–1907) followed Bagneris as professor of sylviculture at Nancy and was director of L'École Nationale (1893–1898). He was an authority on broadleaved trees and, like Lorentz, was greatly in favor of converting coppice-with-standards to high forest. He was also a prominent advocate of crown thinning or *éclaircie par le haut*. In 1889 he wrote *Traité de Sylviculture* [Treatise on silviculture], and in collaboration with A. Jolyet, *Les Forêts; Traité Pratique de Sylviculture* [A practical treatise on silviculture].[84] Antoine Jolyet (1867–1942) had succeeded Boppe as professor of silviculture at Nancy in 1898, and in 1916 he wrote *Traité Pratique de Sylviculture*. This was the subtitle of *Les Forêts* and although Jolyet maintained that it was a new edition of *Les Forêts*, it was considered an entirely different book.[85] One of the last books on French forestry written in the nineteenth century was *L'Enseignement Forestier en France* [Forestry education in France] by Charles Guyot.[86] He was professor of law at Nancy and later became director. He wrote a two volume *Cours de Droit Forestier* [Course of forest law] (1908), and earlier in 1904 completely revised A. Puton's *Code de la Législation Forestière* originally published in 1883.[87]

In 1904, the first of three volumes comprising *Économie Forestière* [Forest economics] by Gustave Huffel (1859–1935) was published. Second editions of individual volumes were available between 1910 and 1927.[88] Huffel was exceptionally gifted as a forester and writer and at the age of twenty-nine was appointed professor of forest economy and management at L'École Nationale, where he remained for the next thirty-seven years. The first volume dealt with the economic and practical value of forests, forest ownership and laws, forest policy, protection forests, and taxation. The second volume covered forest mensuration, production, and valuations, while the third volume was concerned with forest management, working plans, yields, and management methods, past and present.

83. (a) Charles Broillard, *Cours d'Aménagement des Forêts Enseigné à l'Ecole Forestière* (Paris: 1878). (b) Charles Broillard, *Les Traitement des Bois en France à l'Usage des Particuliers* (Paris: Berger-Levrault, 1881).

84. (a) Lucien Boppe, *Traité de Sylviculture* (Paris: Berger-Levrault, 1889). (b) Lucien Boppe and A. Jolyet, *Les Forêts; Traité Pratique de Sylviculture* (Paris: J. B. Bailliere et fils, 1901).

85. Antoine Jolyet, *Traité Pratique de Sylviculture* (Paris: J. B. Bailliere et fils, 1916).

86. Charles Guyot, *L'Enseignement Forestier en France: l'École de Nancy* (Nancy: Crepin-Leblond, 1898).

87. (a) Charles Guyot, *Cours de Droit Forestier* (Paris: L. Laveur, 1908–1912). (b) Charles Guyot, *Code de la Législation Forestière*, Revised and updated (Paris: L. Laveur 1904). (Originally by Alfred Puton, Paris: Rothschild, 1883)

88. Gustave Huffel, *Économie Forestière* (Paris: 1904–1907); 2d ed. (Paris: L. Laveur, 1910–1926).

Another early-twentieth-century book which made a valuable contribution to forestry literature was *Incendies en Forêt* [Forest fires], published in 1903 by André Jacquot (1856–1944). It covered a larger range of subjects than suggested by its title, which Fernow described as misleading. An English translation was published in Calcutta in 1910.[89] Jacquot followed this with *La Forêt, son Rôle dans la Nature et les Sociétés* [Forests: their role in nature and in society], and in 1913 with *Sylviculture*.[90] Once again, the title was considered inadequate by some since it covered all aspects of forestry, not just silviculture. Although a number of prominent French authors have been mentioned in the preceding pages, these are but a few of the many who could have been included.

C. Early British Forestry and Literature

During the Middle Ages, the area referred to as the British Isles comprised England, Wales, Scotland, and Ireland. The economic position of the forests in England was not unlike that which existed in some parts of France and Germany. In Wales, the English cleared substantial areas of woodland in their endeavors to subdue the Welsh. In his journeys through Wales between 1534 and 1543, John Leland remarked that "many hilles thereabout hath bene well woodid . . . but now in them is almost no woode." He did not, however, attribute this state of affairs to the English: "The caussses be these: First the wood cut down was never copisid and this hath bene a great cause of destruction of wood through Wales. Secondly after cutting down of wooddys the gottys (goats) hath so bytten the young spring that it never grew but like shrubbes. Thirddley men . . . destroied the great woddis that thei should not harborow thieves."[91]

In early times, extensive forests covered a substantial part of Scotland but, as was the case in several other countries, large areas had been cleared by the end of the fifteenth century. In Ireland, until the beginning of the seventeenth century, there were many woodlands which, for the most part, lay north of a line from Donegal to Belfast and south of a line from Galway to Wicklow. However, these woodlands were considerably reduced after 1641 as noted by an anonymous author of Dublin in 1705: "Within the

89. Andre Jacquot, *Incendies en Forêt*, 2d ed., (Paris: Berger-Levrault, 1904); Translation by C. E. C. Fischer, with title: *Incendies en Forêt; Forest Fires* (Calcutta: Superintendent Government Printing, India, 1910).

90. (a) André Jacquot, *La Forêt, son Rôle dans la Nature et les Sociétés* (Paris: Berger-Levrault, 1911). (b) André Jacquot, *Sylviculture* (Paris: Berger-Levrault, 1913).

91. J. Leland, *The Itinerary of John Leland the Antiquary*, 3d ed., Vol. 5 (Oxford: J. Fletcher and J. Pote, 1769), pp. 81–82, 75–76.

memory of some now living and for some time after the Rebellion of 1641, great tracts and scopes of ground have been clear'd, and made fit for the plough and scythe, that were taken up before and covered with vast thickets and forests of oak trees."[92]

From a commercial aspect, the needs of the British Isles were basically the same as those of France and Germany, with one great difference which had an important bearing on foreign trade: France and Germany had direct access to neighboring countries by land. In addition the coast lines varied in miles to a very great extent as can be seen from the following figures:

British Isles	12,090 miles (19,470km)
France	3,420 miles (5,500km)
Germany	1,070 miles (1,720km)

The successful defense of so long a seaboard required a large navy which also was needed to protect an expanding overseas trade. Consequently, the constant demand for oak became increasingly difficult to meet.

By the fifteenth century, signs of timber shortages were becoming apparent. In 1482 an act was passed to allow the enclosure of any private woods which lay within the bounds of a royal forest.[93] Its purpose was to protect the young growth of stools and any natural regeneration from being eaten by "all manner of beasts and cattle" after the area had been felled. This was probably the earliest legislation concerned with safeguarding timber and coppice. In 1543, further steps were taken with the object of increasing the supply of timber, by the passing of an Act for the Preservation of Woods. The situation which led to this is recorded in the opening words of the act:

The King our Sovereign Lord perceiving and right well knowing the great decay of timber and woods universally within this his Realm of England to be such, that unless speedy remedy in that behalf be provided, there is a great and manifest likelihood of scarcity and lack, as well of timber for building, making, repairing and maintaining of houses and ships and also for fuel and firewood . . .[94]

Whether this statute had any beneficial effect on the country's timber supplies is a matter of doubt. Three subsequent acts passed in 1558, 1581, and 1585 prohibited or restricted the felling of trees for use in iron smelting in certain localities. As time passed, books began to appear which drew

92. *A Short Treatise of Firr Trees* (Dublin, 1705).
93. G. B. *An Act for Inclosing of Woods in Forests, Chases and Purlieus*, 1482, 22 Edward 4, c.7.
94. G. B. *An Act for the Preservation of Woods*, 1543, 35 Henry 8, c.17.

attention to the diminishing timber supplies and the need for action. The authors of such books differed noticeably from their German and French contemporaries since their interests covered a much wider background. In Germany, authors were usually professional foresters, or professors, or on the staff of academic institutions. Writers in France tended to be private landowners or simply those who were interested in forests until after the founding in 1824 of L'École Nationale when they were largely members of the school. In contrast, in the British Isles, authors might be connected with the royal forests, be landowners, writers on agriculture and rural matters in general, land agents (managers of landed estates), or foresters.

One of the first writers on matters concerned with forests and raising trees was Roger Taverner who wrote his *Book of Survey* (1565) which was followed in 1584 by another of the same title written by his son, John.[95] They both held the position of Surveyor General of Woods. Under the Act of 1541, it was laid down that the Court of the General Surveyor of the King's Lands should include the Surveyor General and Master of the Woods.[96] Toward the end of the sixteenth century a book appeared entitled *A Brefe Collection of the Lawes of the Forest* which dealt with the many legal aspects of the royal forests in medieval times. John Manwood, the author and a member of the legal profession, wrote and printed this for private circulation in 1592 while the first copies for sale to the public appeared in 1598. An enlarged edition with an amended title became a classic authority for the Middle Ages ending with a fifth edition in 1741.[97]

Arthur Standish reflected the general concern at the increasing shortage of timber in *The Commons Complaint* which contained two special grievances, as noted in the subtitles: "The first, general destruction and waste of woods in this Kingdom with a remedy for the same: Also how to plant wood according to the nature of every soile."[98] The second concerned "the extreme dearth of victuals" and was to be remedied by planting fruit trees, breeding more poultry, and destroying vermin. *New Directions of Experience to the Commons Complaint for the Planting of Timber and Firewood* was aimed at improving timber supplies. This was followed in 1615 by *New*

95. (a) Roger Taverner, *Book of Survey* (1565). (b) John Taverner, *Book of Survey* (1584).

96. G. B. *An Act Concerning the Erection of the Court of Surveyors of the King's Lands and the Names of the Officers There and Their Authority*, 1541, 32 Henry 8, c.39.

97. John Manwood, *A Treatise and Discourse of the Lawes of the Forrest* (London: Thomas Wright and Bonham Norton, 1598); original unpublished version had title *A Brefe Collection of the Lawes of the Forest* (London, 1592).

98. Arthur Standish, *The Commons Complaint. Wherein is Contained Two Speciall Grievances. The First, the General Destruction and Waste of Woods in This Kingdome, with a Remedy for the Same. The Second Grievance is, the Extreme Dearth of Victuals. Four Remedies for the Same* (London: W. Stansby, 1611), p. 1.

Directions of Experience . . . for the Increasing of Timber and Fire-Wood
which advocated the planting of 250,000 acres (11,000 hectares) of woodland as a contribution to the country's timber supply.[99] In 1612 a book
appeared under the lengthy title of *An Olde Thrift Newly Revived; Wherein
is Declared the Manner of Planting, Preserving and Husbanding Young
Trees of Divers Kindes for Timber and Fuell.* It was written as a conversa-
tion "betweene a surveyour, woodward, gentleman and a farmer" by Rocke
Church although only his initials, "R. C." appear on the title page.[100]

Probably the best known book on trees and woodlands in the English
language was published in 1664: *Sylva; Or a Discourse of Forest-Trees* by
John Evelyn (1620–1706) who was a country landowner, a fellow of the
Royal Society, and the author of some twenty books.[101] Many of these were
on rural subjects including cidermaking, fruit trees, gardening, and vine-
yards. There were five editions of *Sylva* issued between 1664 and 1729. Its
purpose was to encourage landowners to plant trees and so contribute to the
nation's timber supply. Whether it achieved the desired effect is difficult
to decide, but it became a landmark in English forest literature to which
many subsequently referred. Seventy years after Evelyn's death, Alexander
Hunter produced a revised *Silva* to which he added extensive notes culled
from various sources. This version also ran to five editions, the last in
1825.[102]

Moses Cook wrote *The Manner of Raising, Ordering, and Improving
Forest-Trees* (1676) of which there were three editions.[103] The term "ordering" which often appears in books of this period may be defined as management. Cook was gardener for the earl of Essex at Cassiobury Park near
Watford, Hertfordshire, and his book contained much useful information.

The demand for timber continued to increase while supplies diminished

99. (a) Arthur Standish, *New Directions of Experience to the Commons Complaint by the
Incouragement of the Kings Most Excellent Majesty As May Appear, for the Planting of Timber and
Fire-Wood* (London, 1613). (b) Arthur Standish, *New Directions of Experience to the Commons
Complaint Authorized by the Kings Most Excellent Majesty, As May Appeare, For the Increasing of
Timber and Fire-Wood, With the Least Waste and Losse of Ground* (London, 1615).

100. R. C., *An Olde Thrift Newly Revived: Wherein Is Declared the Manner of Planting, Preserving, and Husbanding Yong Trees of Divers Kindes for Timber and Fuell* (London: Richard
Moore, 1612).

101. John Evelyn, *Sylva, Or A Discourse of Forest-Trees* (London: J. Martyn and J. Allestry,
1664).

102. (a) John Evelyn, with notes by Alexander Hunter, *Silva; Or, a Discourse of Forest-Trees,
and the Propagation of Timber in His Majesty's Dominions* (York: Printed by A. Ward for J.
Dodsley, 1776). (b) John Evelyn, with notes by Alexander Hunter, *Silva, Or A Discourse of
Forest-Trees, and the Propagation of Timber in His Majesty's Dominions*, 5th ed. with the editor's
last corrections (London: H. Colburn, 1825).

103. Moses Cook, *The Manner of Raising, Ordering, and Improving Forest Trees* (London:
Peter Parker, 1676); 3d, corrected ed. (London: Eliz. Bell, 1724).

with several factors contributing to this state of affairs. Shipyards provided a market for oak that was seldom, if ever, satisfied, with much of the timber coming from the royal forests, more particularly the New Forest in Hampshire and the Forest of Dean in Gloucestershire. Little was done to replace trees, and even when replanted, young trees were often neglected and damaged by cattle or deer. No action was taken to provide instruction in forestry matters or to train those who worked in the woods. Due largely to a lack of supervision, low wages, and various undesirable forms of commissions on sales, corruption and dishonesty were common. With a view to overcoming some of these problems, two Acts were passed to further the increase and preservation of timber in the Dean and New Forests in 1668 and 1698 respectively.[104] Timber for ship building had to conform to rigorous specifications and could only be obtained from large, older trees. The government would buy oak from private woodland owners but few owners were prepared to wait until their oak had reached the required sizes. The situation was further confounded by the low prices offered by the government. It was also difficult to transport timber to the dockyards. The New Forest was so valuable because it adjoined several shipyards. Elsewhere, timber might be carried by navigable and suitably located rivers, but in the absence of hard roads, haulage overland could be a formidable task.

The eighteenth century had a considerable increase in the number of books relating to trees and forestry. *A Short Treatise of Firr-Trees* (1705) was published anonymously in Dublin, and is one of the earliest Irish books on trees and planting.[105] There followed other books concerning Irish woodlands: *Some Hints on Planting; By a Planter* (1773) is thought to have been written by James Fortescue of Ravensdale Park near Newry.[106] In 1783 another Irish book with an anonymous writer was published, *An Account of the Method of Raising and Planting the Pinus Sylvestris, That Is, Scotch Fir or Pine*. It is possible that this was written by the second earl of Clanbrassil (1730–98) who was a cousin of James Fortescue.[107] An unnumbered edition of *A Treatise on Forest-Trees* by William Boutcher first was published in Edinburgh in 1775, but the second and third editions in Dublin.[108] The last of these eighteenth-century Irish books was *A Practical Treatise on Plant-*

---

104. (a) G. B. *An Act for the Increase and Preservation of Timber Within the Forest of Dean,* 1668, 20 Charles 2, c.3. (b) G. B. *An Act for the Increase and Preservation of Timber in the New Forest in the County of Southamptom,* 1698, 9&10 William 3, c.36.

105. *Short Treatise of Firr-Trees* (Dublin, 1705).

106. James Fortescue, *Some Hints on Planting; by a Planter* (Newry: G. Stevenson, 1773).

107. *An Account of the Method of Raising and Planting the Pinus Sylvestris, That Is, Scotch Fir or Pine* (1783).

108. William Boutcher, *A Treatise on Forest-Trees* (Edinburgh: R. Fleming, 1775); 3d ed. (Dublin: Printed for William Wilson and John Exshaw, 1784).

ing by Samuel Hayes (1794) who was a member of the Royal Irish Academy and also of the Dublin Society.[109] The author is identified on the title page only by his initials, S. H.

The Manner of Making Plantations Either for Pleasure or Profit was written by Richard Bradley in 1733.[110] He was a Fellow of the Royal Society and in 1724 he was appointed as the first professor of botany at Cambridge University. He wrote a number of books on botany, horticulture, and farming. Two other writers, most of whose books, like those of Bradley, were concerned with matters other than raising trees, were Batty Langley and William Ellis. Langley (1696–1751) was an architect, surveyor, gardener, and landscape designer. He wrote a book on planting and trees entitled *A Sure Method of Improving Estates by Plantations of Oak, Elm, Ash, Beech, and Other Timber Trees* (1728). A second edition appeared in 1740 with the slightly amended title *A Sure and Easy Method* followed by the names of eighteen species of trees.[111]

William Ellis, a farmer living near Hemel Hempstead in Hertfordshire, wrote a book on farming and another on cider, but the one with which he is usually associated is *The Timber-Tree Improved* in 1738 which was followed by three further editions, the last in 1745. In 1742 he wrote a second part to this title.[112] Between 1744 and 1747, he produced eight volumes as *Modern Husbandman* in which he incorporated some of his earlier books.[113]

After 1750, the number of forestry books increased dramatically, coming from a variety of authors including clergymen. One of these was the Rev. William Watkins, curate of Hay-on-Wye in Breconshire, whose *Treatise on Forest-Trees* was published in 1753.[114] Rev. William Hanbury, rector of Church Langton, Leicestershire, wrote *A Complete Body of Planting and Gardening* in 1770 in two folio volumes. The title page stated that it contained natural history, culture and management of deciduous and evergreen forest-trees, with practical directions for raising and improving woods, nurseries, seminaries, and plantations. The volumes also contained exten-

109. Samuel Hayes, *A Practical Treatise on Planting; And the Management of Woods and Coppices* (Dublin: Printed by W. Sleater and sold by Allen & West, 1794).

110. Richard Bradley, *The Manner of Making Plantations Either for Pleasure or Profit* (London, 1733).

111. Batty Langley, *A Sure Method of Improving Estates by Plantations of Oak, Elm, Ash, Beech, and other Timber Trees, etc.* (London: Francis Clay, 1728); 2d ed. has title: *A Sure and Easy Method of Improving Estates by Plantation of Oak, Elm, Ash, Beech* (London, Noble, 1740).

112. William Ellis. *A Timber-Tree Improved, Or, The Best Practical Methods of Improving Different Land with Proper Timber, and Those Fruit-Trees Whose Woods Make the Most Profitable Returns to Their Owners* (London: J. & J. Fox, E. Withers, and T. Cooper, 1738); Pt. 2 (London: T. Osborne and M. Cooper, 1742); 4th ed. (London: T. Osborne and M. Cooper, 1742–1745).

113. William Ellis, *The Modern Husbandman* (London: D. Browne, 1744–1750).

114. William Watkins, *A Treatise on Forest Trees* (London: J. and J. Rivington, 1753).

sive sections on gardening.[115] In 1794, Rev. Phillip Le Brocq wrote *Outlines of a Plan for Making the Tract of Land Called the New Forest a Real Forest.*[116]

During the latter half of the eighteenth century a spirit of enthusiasm for tree planting emerged and soon became popular amongst landowners. Their interest may have been awakened by the large number of books which were then becoming available on planting and other forestry matters, coupled with a sense of duty to grow oak for the navy. Considerable interest was created by the Society of Arts which is now the Royal Society for the Encouragement of Arts, Manufactures, and Commerce, which conceived the idea in 1758 of presenting medals for the establishment of young plantations of outstanding excellence. An example was the award of a gold medal to Mr. Fenwick of Edlingham in Northumberland, who planted 102,00 Scotch pine in 1764, and a further 100,000 in 1766. In cases which did not reach the highest standards, a silver medal would be presented.

Several books concerned with Scottish forestry were published after 1750, one of the best being *A Treatise on the Manner of Raising Forest Trees* written by the earl of Haddington in 1761.[117] James Anderson, under the pseudonym Agricola, wrote *Miscellaneous Observations on Planting and Training Timber-Trees.*[118] Another Scotsman, John Kennedy, wrote *Treatise Upon Planting, Gardening and the Management of the Hot-House* of which half was concerned with trees and their maintenance.[119] The last eighteenth century Scottish book on such matters was *The Practical Planter, Or A Treatise on Forest Planting* by Walter Nicol. Both the first edition (1799) and the second (1803) were based on forestry practice in Scotland.[120] Nicol began another book in 1812, *The Planter's Kalendar*, but died before it was finished; it was subsequently completed by Edward Sang, one of his relations, and issued in a second edition (1820).[121]

115. William Hanbury, *A Complete Body of Planting and Gardening; Containing the Natural History, Culture, and Management of Deciduous and Evergreen Forest-Trees* (London: Printed for the author, 1770–1771).

116. Phillip Le Brocq, *Outlines of a Plan for Making the Tract of Land Called the New Forest a Real Forest, and for Various Other Purposes of the First National Importance* (1794).

117. *A Treatise on the Manner of Raising Forest Trees, & c. in a Letter from the Right Honourable the Earl of Haddington to his Grandson. To Which Are Added Two Memoirs. Both Trans. from the French of M. de Buffon* (Edinburgh: G. Hamilton and J. Balfour, 1761).

118. Agricola [i.e., James Anderson], *Miscellaneous Observations on Planting and Training Timber-Trees Particularly Calculated for the Climate of Scotland* (Edinburgh: Elliot, 1777).

119. John Kennedy, *A Treatise Upon Planting, Gardening and the Management of the Hot-House*, 1st ed. (York, Engl.: Printed by A. Ward, 1776). (Later edition, Dublin: Printed for W. Wilson, 1784).

120. Walter Nicol, *The Practical Planter, Or A Treatise on Forest Planting* (Edinburgh: Nicol, 1799); 2d ed. (London: J. Scatcherd, 1803).

121. Walter Nicol, *The Planter's Kalender, or The Nurseryman's & Forester's Guide*, edited and completed by Edward Sang (Edinburgh: A. Constable, 1812); 2d improved and enlarged ed.

In 1785, William Marshall's *Planting and Ornamental Gardening* was published; the second (1796) and third (1803) editions appeared under the amended title *Planting and Rural Ornament*.[122] Each edition was in two volumes, the first dealing with planting and rural ornamentation in approximately equal proportions, while the second provided a descriptive list of trees and shrubs. Two small but very sound books by Thomas Nichols, who was the purveyor of the Navy for Portsmouth Dock Yard, were *Observations on the Propagation and Management of Oak Trees in General* and *Methods Proposed for Decreasing the Consumption of Timber in the Navy*, in 1791 and 1793 respectively.[123] Both were in the form of a letter addressed to the earl of Chatham who was the First Lord Commissioner of the Admiralty.

During the nineteenth century, more than sixty books or booklets dealing with various aspects of forestry were published but space only allows for reference to be made to a limited number of these. In 1800, William Pontey, who described himself as "nurseryman and ornamental gardener," wrote *The Profitable Planter* which ran to five editions, although the last, dated 1828, is described as an augmented fourth edition.[124] He followed this in 1805 with *The Forest Pruner, or Timber Owner's Assistant* of which the fourth and last edition was issued in 1826.[125] Between 1808 and 1817 a series of reports on the counties of England were published by the Board of Agriculture. For this purpose the country was divided into six departments, namely the northern, western, eastern, midland, southern, and peninsular. These reports were written by William Marshall, the author of *Planting and Rural Ornament* and included useful information on the woodlands and forestry of the English counties.[126]

In 1820, a Scotsman, Robert Monteath of Stirling, wrote *The Forester's Guide* on the title page of which he described himself as "wood surveyor

(Edinburgh: Printed for Archibald Constable, and London: Longman, Hurst, Rees, Orme, and Brown, 1820).

122. William Marshall, *Planting and Ornamental Gardening: A Practical Treatise* (London: J. Dodsley, 1785); 2d ed. has title: *Planting and Rural Ornament* (London: Printed for G. Nicol by G. G. and J. Robinson and J. Debrett, 1796); 3d ed. has title: *On Planting and Rural Ornament; A Practical Treatise* (London: Printed by W. Bulmer for G. and W. Nicol, 1803).

123. (a) Thomas Nichols, *Observations on the Propagation and Management of Oak Trees in General* (Southampton: Printed by T. Baker, 1791). (b) Thomas Nichols, *Methods Proposed for Decreasing the Consumption of Timber in the Navy by Means of Prolonging the Duration of our Ships of War* (Southampton: Printed and sold by T. Baker, 1793).

124. William Pontey, *The Profitable Planter: A Treatise on the Theory and Practice of Planting Forest Trees*, 4th enlarged ed. (London: Ridgway, 1828). Earlier enlarged ed. also published as 4th in 1814.

125. William Pontey, *The Forest Pruner, or Timber Owner's Assistant* (Huddersfield, Engl. T. Smart, 1805; 4th ed., 1826).

126. An exemplary title is: W. Marshall, *A Review of the Reports to the Board of Agriculture from the Northern Department of England* (London: Longman, Hurst, Rees, and Orme, 1808).

and valuator." In the second (1824) and third editions (1836) the title was altered to *The Forester's Guide and Profitable Planter*.[127] In the third edition Monteath was stated to be "Forester to His Majesty," and in 1827 his *Miscellaneous Reports on Woods and Plantation* was published.[128] A book which had considerable influence on different methods of raising young oak was written by William Billington in 1825.[129] This was largely the result of his experiences in the Forest of Dean where he superintended the planting of some 11,000 acres (4,450 hectares) of oak. During 1847, *The Forester* by James Brown was published, and four years later a second and much enlarged edition appeared; altogether, six editions were issued, each one larger than its predecessor.[130] The fifth (1882) was largely revised by his son G. E. Brown, while the sixth, published in 1894, was edited by John Nisbet. When the first and second editions were printed, Brown was forester at Arniston near Edinburgh. His last appointment was that of Inspector of Woods and Forests for Ontario, Canada.

Two developments had taken place by 1863 that were ultimately to replace traditional forestry in Britain. Since medieval times, ships had been built largely of oak but timber suitable for the navy had long been in short supply, and by 1850 a major crises could not be averted. The problem was solved by chance in 1862 by a naval encounter which took place in Hampton Roads, near Norfolk, Virginia, during the American Civil War. The *Merrimac*, an ironclad, engaged two timber-built ships, the *Congress* and the *Cumberland* and sank them both, their guns being ineffective against the armor of the *Merrimac*. This event signed the death warrant for timber-built ships in the Royal Navy and eliminated the demand for oak by the dockyards. This battle had far-reaching effects on forestry throughout the United Kingdom.

The second development was the introduction to the British Isles of fast-growing conifers which were of considerable value as timber trees. This

127. Robert Monteath, *The Forester's Guide; or, A Practical Treatise on the Training and Pruning of Forest Trees* (Edinburgh: Stirling, J. Forman, 1820); 2d ed. has title: *Forester's Guide and Profitable Planter* (Edinburgh: Stirling and Kenney, 1824); 3d ed. (London: T. Tegg & Son, 1836).

128. Robert Monteath, *Miscellaneous Reports on Woods and Plantations* (Dundee: J. Chambers, 1827).

129. William Billington, *A Series of Facts, Hints, Observations, and Experiments on the Different Modes of Raising Young Plantations of Oaks* (London: Sold by Baldwin, Cradock & Joy, 1825).

130. James Brown, *The Forester: A Practical Treatise on the Planting, Rearing, and General Management of Forest Trees* (Edinburgh: 1847); 2d enl. ed. (Edinburgh: W. Blackwood, 1851); 5th ed. enlarged by G. E. Brown (Edinburgh: W. Blackwood & Sons, 1882); 6th ed. has title: *The Forester: A Practical Treatise on the Planting and Tending of Forest Trees and the General Management of Woodland Estates*, edited by John Nisbet (Edinburgh: W. Blackwood, 1894).

was the result of expeditions by plant collectors, such as David Douglas and John Jeffreys, who operated mainly in western North America. The Douglas fir *(Pseudotsuga menziesii)* and Sitka spruce *(Picea sitchensis)*, which were introduced to Britain by Douglas in 1827 and 1831 respectively, became of great commercial importance.

During the second half of the nineteenth century, some writers began to turn their thoughts to a more precise and accurate approach to forestry. G. Cree produced his *Essays on the Scientific Management of Forest Trees* (1851), while in 1883, J. L. L. MacGregor's book, *The Organization and Valuation of Forests* was published.[131] The first of the five volumes of *Schlich's Manual of Forestry* was available in 1889 and although largely based on German forestry practice, it marked a turning point in British forestry literature. Another supporter of the scientific approach was John Nisbet who, after serving in the Indian Forest Service, was appointed professor of forestry at the West of Scotland Agricultural College. In 1893, his *British Forest Trees and Their Sylvicultural Characteristics and Treatment* was published; this was followed by several others including *Studies in Forestry* (1894) and *The Elements of British Forestry* (1911).[132] In 1905 he wrote *The Forester* in two volumes which was based on the sixth edition of James Brown's book, although it is generally regarded as a separate publication.[133]

Unfortunately the United Kingdom trailed far behind Germany and France as regards forestry education. Although a forestry faculty had been formed in the Royal Indian Engineering College in 1885, training was provided only for prospective officers in the Indian Forest Service. Since the College did not teach nor was much concerned with British forestry, there was no forestry education concerned with the United Kingdom. The first step was a lectureship in forestry at Edinburgh University in 1888, although it was not until 1912 that the university granted a degree in forestry. In 1900 a school of forestry was established at Aberdeen University but owing to World War I, degrees in forestry were not conferred until 1920. A chair of forestry was established in 1925.

A forestry department was formed in 1904 at the University College of

131. (a) Gavin Cree, *Essays on the Scientific Management of Forest Trees* (Lanark, Engl.: Printed by R. Wood, 1851). (b) J. L. L. MacGregor, *The Organization and Valuation of Forests on the Continental System in Theory and Practice* (London: Wyman & Sons, 1883).

132. (a) John Nisbet, *British Forest Trees and Their Sylvicultural Characteristics and Treatment* (London and New York: Macmillan, 1893). (b) John Nisbet, *Studies in Forestry* (Oxford: Clarendon Press, 1894). (c) John Nisbet, *The Elements of British Forestry: A Handbook for Forest Apprentices and Students of Forestry* (Edinburgh and London: W. Blackwood & Sons, 1911).

133. John Nisbet, *The Forester, a Practical Treatise on British Forestry and Aboriculture for Landowners, Land Agents and Foresters* (Edinburgh and London: W. Blackwood, 1905).

North Wales in Bangor, bit was closed from 1916 to 1919 and reopened in 1920. A year before the Royal Indian Engineering College closed in 1906, the forestry faculty was transferred to Oxford University where a school of forestry was established under the direction of Schlich. In 1919, a university chair of forestry was founded and in 1939 the school was combined with the Imperial Forestry Institute to form the university's Department of Forestry. A forestry school was also established at Cambridge University (1907) and although a degree in forestry was approved in 1919, the school closed in 1932.

Apart from the universities, less intensive forestry training was available at some agricultural colleges, more particularly at the Royal Agricultural College at Cirecncester in Gloucestershire and the East of Scotland and West of Scotland Agricultural Colleges. Training for government foresters was provided at the School of Forestry for Woodmen at Parkend in the Forest of Dean which opened in 1904. In Scotland training schools were established at Birnam and Beaufort shortly after the end of the First World War, but were later replaced by the Benmore Forester Training School which operated between 1929 and 1964.

One of the first twentieth-century books was *English Estate Forestry* by A. C. Forbes which appeared in 1904.[134] After working as a head forester he became lecturer in forestry at Durham College of Science and later appointed Chief Forestry Inspector to the Department of Agriculture in Ireland. He also wrote a major history of British forestry.[135] In 1908 the first book written solely on the insects found in British forests was published. This was *Forest Entomology*, by A. T. Gillanders, which remained the standard textbook on the subject until 1937 when R. N. Chrystal wrote *Insects of the British Woodlands*.[136]

In 1809, Charles Waistell produced a small booklet entitled *The Method of Ascertaining the Value of Growing Timber Trees*.[137] This was one of the first attempts to calculate the value and yield of stands of growing timber. Although the third edition of volume 3 of *Schlich's Manual of Forestry* contained yield tables for eight different species, these had been based on various districts in Germany. A book which had a bearing on this matter was written in 1909 by P. T. Maw who was a land agent (estate manager) and had been professor of forestry at the Royal Agricultural College. *The*

134. A. C. Forbes, *English Estate Forestry* (London: E. Arnold, 1904).
135. A. C. Forbes, *The Development of British Forestry* (London: E. Arnold, 1910).
136. (a) Alexander T. Gillanders, *Forest Entomology* (Edinburgh: W. Blackwood, 1908). (b) Robert N. Chrystal, *Insects of the British Woodlands* (London and New York: F. Warne, 1937).
137. Charles Waistell, *The Method of Ascertaining the Value of Growing Timber Trees at Different and Distant Periods of Time With Observations on the Growth of Timber* (1809).

Practice of Forestry, Concerning also the Financial Aspect of Afforestation included measurement and interest tables.[138] *Complete Yield Tables for British Woodlands*, also by Maw, was published in 1912.[139] These were not replaced until 1920 when the Forestry Commission began to issue yield tables.

A small and invaluable book, *Forestry for Woodmen* (1911), helped many who wished to take up forestry as a career. This was authored by C. O. Hanson who had been a deputy conservator in India, and later appointed an instructor at the School of Forestry for Woodmen in the Forest of Dean. A testimony of its value is that it ran to a second edition published in 1921.[140]

Largely as a result of extensive fellings during the First World War, the United Kingdom decided to take action to restore the country's woodlands and to create a national forest service. In 1919 the Forestry Commission was established by Act of Parliament, one of the most important events in the history of British forestry.

One of the distinguished figures in forestry after the First World War was W. E. Hiley who began his career as a botanist in the School of Forestry at Oxford, and whose first book, *The Fungal Diseases of the Common Larch*, was published in 1919.[141] He subsequently transferred to economics and became a lecturer in forest economics while at Oxford. In 1930 his *Economics of Forestry* became available. He left Oxford in 1931 to take over management of the woodlands on a private estate in Devon where he remained until his death in 1961. He wrote a number of books including *Improvement of Woodlands*, *Woodland Management*, and *Economics of Plantations*.[142]

In 1928 an outstanding book entitled *Silvicultural Systems* was written by R. S. Troup who was a professor at Oxford and who, like many others, had been a member of the Indian Forest Service.[143] It was the first comprehensive book in English devoted entirely to silvicultural systems. A typical book of the period was *Practical British Forestry* by C. P. Ackers.[144] He

138. Percival T. Maw, *The Practice of Forestry, Concerning also the Financial Aspects of Afforestation* (Brockenhurst: Hants, Walter and Walter, 1909).
139. Percival T. Maw, *Complete Yield Tables for British Woodlands and the Finance of British Forestry* (London: C. Lockwood and Son, 1912).
140. C. O. Hanson, *Forestry for Woodmen* (Oxford: Clarendon Press, 1911); 2d ed. (Oxford, Clarendon Press, 1921).
141. Wilfrid E. Hiley, *The Fungal Diseases of the Common Larch* (Oxford: Clarendon Press, 1919).
142. (a) Wilfred E. Hiley, *The Economics of Forestry* (Oxford: Clarendon Press, 1930). (b) Wilfred E. Hiley, *Improvement of Woodlands* (London: Country Life, 1931). (c) Wilfred E. Hiley, *Woodland Management* (London: Faber and Faber, 1954). (d) Wilfred E. Hiley, *Economics of Plantations* (London: Faber, 1956).
143. Robert S. Troup, *Silvicultural Systems* (Oxford: Clarendon Press, 1928).
144. C. P. Ackers, *Practical British Forestry* (London: Oxford University Press, 1938).

was the owner of a large woodland estate in Gloucestershire and devoted much of his time to its management. His book reflects his considerable knowledge of forestry and the great interest that he took in his woods.

D. Concluding Observations

The simple beginnings of forestry in Germany, France, and the United Kingdom follow closely the patterns of intellectual growth in other scientific and social sectors. These early patterns of work demonstrate the slow growth of knowledge during this period with the methodical accumulation of information and data from a variety of sources. Breakthroughs in some fields often came via one or two people such as with Pasteur in biology. These companion areas have become interrelated as findings in biology, geology, ecology, and agriculture have provided new knowledge and advancements which could be utilized in forestry. During the last forty years, interdependence has grown even as each field has become more fractionated and specialized.

Early forestry in Germany engendered the scientific and intellectual basis for advancements and knowledge in many other countries. Probably the greatest parallel activity built on the German base was that of the Scandinavian countries with their immense forest resources and intellectual interest. Challenges to forestry practices and philosophy abounded during the last quarter of the twentieth century. The place of forestry in society will continue to evolve under growing pressures of population and environmental concerns.

3. Characteristics of Forestry and Agroforestry Literature

PETER MCDONALD

New York State Agricultural Experiment Station, Geneva

The Core Agricultural Literature Project at the Albert R. Mann Library, Cornell University, was coordinator for this volume, and its primary effort was identifying the core literature of forestry and agroforestry of primary importance in university instruction and research today. This identification covered forestry in both temperate and tropical zones for developed and developing countries.

Scientific tropical forestry, almost exclusively the province of the Third World, has a heritage dating back more than a hundred years. The challenge facing the Core Agricultural Literature Project was to identify the most current and useful trends and literature in tropical forestry and bring that knowledge to a wide audience. To this end, a decision to include agroforestry with tropical forestry for developing countries was made by the International Advisory Board for the Core Agricultural Literature Project when meeting in Nairobi in 1989. Agroforestry, an emerging system of sustainable forest practice, expresses an integrated farming system capable of immediate and beneficial application in the Third World. As P. K. R. Nair points out in Chapter 4, agroforestry technologies are expected to be of great value in alleviating immediate human needs among farmers in the Third World. The analyses of the Project included agroforestry and sustainable forestry practices in the tropics, and related ecology and preservation of rainforest ecosystems.

In this volume, works which dealt primarily with the international policy aspects of tropical forestry and agroforestry were considered as developed countries' literature. Except for these and a few texts on rainforest ecosystems, the forestry literature of developed countries has been analyzed separately from that for the Third World. Because the information needs of low latitude countries are growing, there is a correlation between the forest science literature applicable to developed countries with that of importance for the Third World.

In this study, wildlife management was encompassed in the broad topic of general forest management. Literature was included if it clearly deals with the relationship of wildlife in forested ecosystems, particularly their management. General literature on wildlife biology, biodiversity, animal ecology, species extinction, and most systematics were excluded, as were national and international laws dealing with forestry, patents on forest products, and product procedures. In a previous volume of the *Literature of the Agricultural Sciences* series, forest engineering was extensively covered so it is dealt with only peripherally in this volume.[1] The study of forest soils is seen as an important part of forestry practice, and is fully included in this volume. Finally, works about nut and fruit trees, palms, rubber, cork, and woody plant products other than lumber are excluded here, but are included in the series volume on crop science. These subject parameters apply in this chapter as well as throughout the book.

Forest science is in transition and its subject emphases are shifting. Schools of forestry linked closely to timber production have evolved in the past twenty years to a broader concept of forestry which includes holistic ecosystem management, biodiversity, ecology, environmental science, and sustainable development. These changes are discussed by J. Gordon in Chapter 1, and by J. Coufal and D. Webster in Chapter 8, and are predicated on an understanding of the historical record of forestry from which they evolved. The literary heritage of forestry is vast, reaching back to antiquity. Historical developments in Europe are extensively treated in Chapter 2 by N. D. G. James, and United States historical literature in Chapter 14. This chapter touches only briefly on the rich literary antecedents, but provides an analysis of the trends in forestry and agroforestry literature of the past twenty years.

A. Early Modern Trends

Early nineteenth-century forestry included silviculture, dendrology, woody plant biology, forest fauna and flora, engineering, timber economics, and forest mensuration, and emerged as a distinct science by the last quarter of the century. By the second decade of the twentieth century forest science was fully incorporated into university curricula with instruction and re-

---

1. J. L. Fridley, J. A. Miles, and F. E. Greulich, "Development of Forest Engineering and Its Literature," in *The Literature of Agricultural Engineering*, ed. Carl H. Hall and Wallace C. Olsen (Ithaca, N.Y.: Cornell University Press, 1993), pp. 126–143.

search. Rev. J. C. Brown, in the "Advertisement" or preface to *The Forests of England*, provided this insight of the literature of forestry in late nineteenth century Germany:

> In Herman Schmidt's *Fach Katalogue* . . . there are given the titles, &c., of German works in *Forst und Jagd-Literatur,* published from 1870 to 1875 inclusive, amounting in all to 650 (references) . . . classified thus:—General Forest Economy, 93; Forest Botany, 60; Forest History and Statistics, 50; Forest Legislation and Game Laws, 56; Forest (Mensuration), 25; Forest Tables and Measurements, 148; Forest Technology, 6; Forest Zoology, 19; Peat and Bog Treatment, 14; Forest Union and Yearbooks, 13; Game, 135. Upwards of a hundred new works had been published annually.[2]

Literature analysis is a means of measuring how the published record of forestry has evolved over time. In 1903, when Bernhard Fernow, Dean of the Faculty of Forestry at the University of Toronto, published *A Brief History of Forestry in Europe, the United States, and Other Countries*, forest science was an established discipline with a vocal international professional society.[3] Fernow examined in detail the status of forestry literature in twenty-three countries. What emerges from his examination is a realization that as early as the turn of the century, "forestry science (had) grown in all directions; schools, associations, journals and prolific literature attesting the complete establishment of the profession and practice."[4]

The number of forestry serials published worldwide was reported at near 120 in 1900, including report series and newsletters.[5] A comprehensive list of published forestry works from France for the year 1900 gives seventy-six citations, although it is not clear how many of these citations are journal articles or monographs.[6] By 1925, one compendium of current and important monographic United States imprints listed 185 titles under these headings: General Forestry, Forest Esthetics, Forest Botany, Forest Research Methodologies, Forest Influences, Forest Mensuration, Silviculture and Management, Forest Planting, Forest Protection, Forest Policy and Administration, Forest Utilization, Wood Technology, and Wood Preservation and

2. J. C. Brown, *Forests of England and Management of Them in By-Gone Times* (Edinburgh: Oliver & Boyd, 1883), p. vii.

3. This was the International Union of Forestry Research Organizations founded in Vienna in 1892.

4. Bernhard Fernow, *A Brief History of Forestry in Europe, the United States, and Other Countries* (New Haven: Lee & Adkins, 1907), p. 145.

5. Helen Moore, *Union Checklist of Forestry Serials* (Washington, D.C.: Dept. of Agriculture, Forest Service Library, 1936).

6. Theodore Woolsey, *Studies in French Forestry* (New York: J. Wiley & Sons, 1920), Appendix H.

Seasoning.[7] Notably absent are works on forest ecology, tropical timbers, and wildlife.

As early as 1920, the systematic referencing of current forest literature was begun by the Oxford University School of Forestry which continued jointly with the Imperial Forestry Institute after its establishment in 1924. Oxford University published periodic bibliographies, and in 1936 its *Forest Bibliography to 31st December, 1933*, a chronological listing of forestry citations dating back to the 1820s.[8] The bibliography is in three broad subjects: General Forestry (by country), Silviculture (with seven subheadings), and Forest Protection. Over 11,000 citations are given from about 750 serials, although there are few entries for either forest economics or forest products. This bibliography reveals a sixfold increase in serial articles in just thirty years.

In 1916, *The Bradley Bibliography* was published at Harvard University, and like the bibliographies compiled at Oxford, the intention was to provide a comprehensive guide to forestry literature containing "the titles of all publications relating to forestry and silviculture" available worldwide to that time.[9] The *Bradley Bibliography* contains about 25,000 citations, covering 750 serial titles, references to codes, laws, treatises, pamphlets, and monographs. This large number of citations reveals the importance of timber production and forestry by 1915.

Biological Abstracts (BA) was established in 1926 and the first volume had about 370 citations to forestry including the biology and pathology of specific tree species.[10] *BA* listed almost a threefold increase in the number of forestry-related citations by 1939, when *Forestry Abstracts* (Imperial Forestry Bureau, Oxford) began publication.

It was under the auspices of the Imperial Agricultural Bureaux (the precursor of today's CAB International) that the Imperial Forestry Bureau at Oxford began the first abstracting journal in forestry on the eve of the Second World War. Volume 1 of *Forestry Abstracts* (1939) listed over 1,200 citations divided into these categories: General Forestry; Fundamental Natural Laws of the Forest; Silviculture; Forest Utilization and Technology; Forest Injuries and Protection; Mensuration/Increment/Yield; Forest Management and Forest Policy. Except for Fundamental Natural Laws of the

---

7. Miss Stockbridge, "Books on Forestry in English," in *Forestry Almanac, Semicentennial Edition* (Washington, D.C.: American Tree Association, 1926), pp. 326–333.

8. Oxford University, Department of Forestry, *Forest Bibliography to 31st December, 1933*, 3 vols., (Oxford: Oxford University, 1936).

9. Alfred Rehder, *The Bradley Bibliography, A Guide to the Literature*, Vol. 4: *Forestry* (Cambridge, Mass.: Riverside Press, 1916), Introduction.

10. Union of American Biological Societies, *Biological Abstracts*, 1926+ (Philadelphia: UABS, University of Pennsylvania).

Forest, which has today become forest ecology, the subdivision of *Forestry Abstracts* has changed little in this indexing journal.

Forestry did not receive such an auspicious beginning in the U.S. Department of Agriculture's *Bibliography of Agriculture* in its inaugural issue of August 1942.[11] The first four sections of the bibliography had gone to press before a fifth section on forestry was added. Differences in subject coverage were apparent from the first between the British abstracting journal and its American counterpart. *Forestry Abstracts* has always stressed forest biology in the broadest sense over forestry as a business enterprise where economics, utilization, and products dominate, which is the American forestry emphasis. In 1923 there were over forty timber trade associations, large and small, in the United States publishing as many trade journals (see "Popular Journal Literature" in Chapter 14.).[12] In the U.S. Forest Service, utilization of wood was stressed along with timber extraction, wood products, and engineering technologies. The subject divisions of Section E of the USDA *Bibliography of Agriculture* in 1942 included: General Forestry; Forest Administration and Policy; Planting; Forest Fire Protection; Harvest Cutting and Stand Improvement; Forest Measurements; Forest Recreation; Forest Resources; Lumbering; Log Transportation; Lumber Drying; Products Utilization; Wood as Fuel; Wood Preservation; Wood Technology; Range Management; and Wildlife. A quick comparison between these broad groupings and those from *Forestry Abstracts* reveals the penchant in the *Bibliography of Agriculture* to stress utilization of forests as opposed to the basic sciences of tree breeding, forest biology, and ecology. These differing emphases in indexing coverage have changed some in the last fifty years and show in comparisons of current bibliographic databases. CABI transferred forest products items out of *Forestry Abstracts* and into *Forest Products Abstracts* which began as a separate publication in 1978.[13]

During the late nineteenth and early twentieth centuries, the forestry literature overwhelmingly reflected a temperate bias, in part because vast tracts of virgin forest remained in northern latitudes. But as the ancient forests were successively logged and replaced by pasture or monocultures, a corresponding change occurred in the literature mirroring changes in popular perceptions of the natural world. One such trend involved a growing interest in international forest trade and economics, driven in part, by the rubber booms in tropical Brazil and India at the beginning of the twentieth century. There was also a growing interest in forest conservation and the protection

11. *Bibliography of Agriculture*, 1942+ (Washington, D.C.: U.S. Department of Agriculture Library).

12. Nelson Brown, *The American Lumber Industry* (New York: J. Wiley & Sons, 1923).

13. *Forest Products Abstracts*, 1978+ (Wallingford, U.K.: CAB International).

of wildlands, subjects which were eloquently championed by poets of the Romantic Movement in Europe and naturalists such as John Muir and William Burroughs in the United States. Later works by James Toumey[14] and Russell Smith[15] spoke to the need to put silviculture on an ecological basis. Influential works such as Aldo Leopold's *Sand County Almanac*,[16] Edward Graham's *The Land and Wildlife*,[17] and William Vogt's *Road to Survival*[18] presaged a change in perceptions and the importance of healthy forest ecosystems. By the 1960s, the concept of natural resources and ecosystem management as regenerative national assets became a paradigm of college instruction, and began to supersede concepts of timber management championed by earlier men such as Pinchot.[19] Much of this change resulted from the growing influence of wilderness and conservation societies. (See Chapter 5.)

Such men as the irrepressible Rev. Frederick Starr thundered against the destruction of American forests as early as 1865 in several USDA reports, particularly against the wastage of timber in fueling the ever-expanding railroads.[20] Today, no less strident voices decry the destruction of tropical rainforests.[21] In 1988, no fewer than sixty-seven monographs of fifty pages or more were published worldwide on the subject of rainforests and their value. This one-year count was equal to the sum of all previous monographs on the subject published between 1984 and 1987.[22]

This interest in tropical forestry is not new, for it dates at least to the late nineteenth century. In 1912, Broun published his seminal *Sylviculture in the Tropics*,[23] and the Yale Forestry School inaugurated a journal, *Tropical Woods*, in 1925. By the 1930s almost every tropical country could boast some type of serial report on the state of its forests. Some countries, such as India and Brazil, were leaders in the tropical forestry vanguard publishing

14. James Toumey, *Foundations of Silviculture upon an Ecological Basis* (Ann Arbor, Mich.: Edwards Bros., 1925).

15. Russell Smith, *Tree Crops: A Permanent Agriculture* (New York: Harcourt Brace, 1929).

16. Aldo Leopold, *Sand County Almanac* (New York: Oxford University Press, 1949).

17. Edward Graham, *The Land and Wildlife* (Oxford, U.K.: Oxford University Press, 1947).

18. William Vogt, *Road to Survival* (New York: Sloane Associates, 1948).

19. Gifford Pinchot, *The Training of a Forester*, 3d ed. (Philadelphia and London: J.B. Lippincott, 1917).

20. (a) Frederick Starr, "American Forests: Their Destruction and Preservation," in *The Report of the Commissioner of Agriculture for the Year 1865* (Washington, D.C.: GPO, 1866), pp. 210–234. (b) R. F. Hammatt, *Forestry and Permanent Prosperity* (Washington, D.C.: USDA Forest Service, 1936 [*Miscellaneous Publication* No. 247]).

21. (a) O. E. Wilson, *The Diversity of Life* (Cambridge, Mass.: Belknap Press of Harvard University Press, 1992). (b) Dionella Meadows et al., *Beyond the Limits: Confronting Global Collapse, Envisioning a Sustainable Future* (Post Mills, Vt.: Chelsea Free, 1992).

22. From a composite search of *Books-in-Print* (New York: Bowker, 1988), and RLIN, OCLC, and the Cornell Online Catalog.

23. Alfred Broun, *Sylviculture in the Tropics* (London: Macmillan, 1912).

extensively on a variety of forestry subjects. These historical precedents, early interest, and the rich literary heritage which exists in tropical forestry should be kept in mind as a counterpoint to the analysis of agroforestry literature presented in this chapter and by P. K. R. Nair in Chapter 4.

B. Forest Science Bibliographic Databases

An aim of the Core Agricultural Literature Project is to identify the core literature of forestry and agroforestry of primary importance in university instruction and research today for use in developed countries and in the Third World. A second objective was to transfer the image text of this primary, core literature onto compact disk (CD-ROM) technology for distribution to academic institutions and research establishments of developing countries. This full-text computer library for all of agriculture, including forestry, is planned to include 8,400 monographs and 360 journals covering five years. The systematic and comprehensive analysis of this scientific discipline is described in detail in Chapters 11 and 12.

This study of forestry and agroforestry was divided into five major subject areas.

1) Agroforestry and Tropical Silviculture, Rainforest Ecology, International Timber Trade
2) Forest Biology, Breeding, Genetics, and Ecology
3) Forest Pathology and Protection
4) Forest Management, Silviculture, Wood Products, Recreation, Wildlife Management, Mensuration
5) Forest Policy and Economics, General Forestry, International Development, Forestry Education, Social Issues.

This subject taxonomy was derived from a combination of sources, but primarily from the Society of American Foresters' *Forestry Handbook*,[24] and the journal *Forest Science*.[25] The literature of this discipline was further divided into that pertaining to developed regions and that pertaining to developing countries most commonly associated with the tropics.

The study of forest science literature divided by region and by subject helped to facilitate comparative statistical analysis. Today, most agricultural databases, such as *AGRIS*, the bibliographic databases of the Food and Ag-

24. Karl Wenger and the Society of American Foresters, *Forestry Handbook* (New York: J. Wiley, 1984).
25. *Forest Science* (Washington, D.C.: Society of American Foresters, 1955 +).

riculture Organization of the United Nations, and *CAB Abstracts*, the bibliographic database compiled by CAB International, divide forest science by subject or category codes enabling searchers to gather data specific to clearly defined subject subsets. The needs of researchers in developed countries differ both in scope and application from those in the Third World, though these differences are diminishing. When this literature is analyzed separately, a clearer comparison between them can be made, enabling distinct correlations with the evolutionary trends in the literature in both temperate and tropical zones.

The universe of published material in forest science is immense and defies a precise count, in part because forest science today lacks clear boundaries and blends into a variety of scientific, economic, and sociological disciplines, particularly the environmental sciences and natural resources. The use of agricultural and forestry bibliographic databases, such as *CAB Abstracts* and *AGRICOLA*, the indexing database of the U.S. National Agricultural Library, is indispensable for literature analysis, but the reader should keep in mind that data collected from such sources do not constitute the full extent of forest science literature. All subject searching in this chapter was done in 1993, except on pages 58 and 59.

Minor problems are overcome by the sheer quantity of data available in these databases. *AGRICOLA* has over 2,750,000 citations as of 1993, *CAB Abstracts* a half million more. Considered in aggregate, these numbers stand up to rigorous examination and are useful in providing scale rankings by percentages. Nevertheless, data problems in these files persist. For one, citation counts in these bibliographic databases overwhelmingly reflect research published in journals; monographs are less commonly indexed at the chapter level and usually only when they are edited editions with multiple authors. Secondly, both *AGRICOLA* and *CAB Abstracts* tend to index English-language material predominantly with other languages sparsely handled. In a 1983 study, forestry articles from German-language serials indexed by *CAB Abstracts* were 30% of the total German output.[26] Locally published manuals and how-to pamphlets are indexed selectively in *CAB Abstracts* as in most agricultural bibliographic databases. These types of "help" publications which are widely used in Third World agriculture may never be listed or indexed. These databases cover the literature of devel-

26. E. Stage, "ELFIS—An Agricultural Bibliographic Database for Forestry-Related Information," in *Information Systems for Forestry-Related Subjects: Access, Search Techniques and User Needs*, ed. Alois Kempf and Rosmarie Louis (Birmensdorf, G.D.R.: International Union of Forest Research Organizations, 1988), pp. 63–69.

oped nations and publications in English, which is largely a reflection of general publishing.[27]

Other bibliometric tools such as the *Science Citation Index*[28] did not regularly index or analyze monographs until 1992, although they have indexed the proceedings of selected conferences in the past.[29] Established subjects covered in publications such as textbooks are under-represented in the citation counts reflected in bibliographic databases because chapters are rarely indexed. These and similar caveats present skews in bibliometric analysis and must be addressed.

C. Forestry Database Analyses

Tables 3.1 and 3.2 present a numerical analysis by database of forestry citations over a ten-year period divided by subject categories. In the *AGRICOLA* bibliographic database, thirty-eight subject categories relate to the total scope of forestry in this study. Forestry was searched online for numbers of citations both in *CAB Abstracts*, which has twenty-three categories, combined from CABI's *Forestry Abstracts* and *Forest Products Abstracts*,[30] and in *AGRICOLA*. Subject codes (e.g., 0F01, K130) are assigned at the time of indexing as an indication of the primary subject matter of the citation. In many cases, more than one category code is assigned to a citation when several subjects are involved. However, since primary category codes are not weighted and cannot be extracted singly from total code counts, all categories vary in degrees of duplication. This is especially a problem with subject codes pertaining to biological aspects of forestry, which overlap heavily with breeding, ecology, entomology, pathology, physiology, and plant taxonomy. Dates in these tables represent the date the journals or books were published.

In a previous study, upwards of 20% of *AGRICOLA* citations carried duplicate category codes.[31] Using only 1992 entries, several hundred *AGRICOLA* forestry citations were examined, and the rate of category code du-

27. Y. M. Rabkin and H. Inhaber, "Science on the Periphery: A Citation Study of Three Less Developed Countries," *Scientometrics* 1 (3) (1979): 261–274.

28. *Science Citation Index*, 1961+ (Philadelphia: Institute for Scientific Information).

29. *Index to Scientific Book Contents*, 1922+ (Philadelphia: Institute for Scientific Information).

30. *Forest Products Abstracts* was separated from *Forestry Abstracts* in 1978. Both are used in this analysis .

31. Wallace C. Olsen, "Literature Patterns and Trends," in *The Literature of Animal Science and Health*, ed. Wallace C. Olsen (Ithaca, N.Y.: Cornell University Press, 1993), chapter 2, 21p.

Table 3.1 Dates of publication totals of forestry citations in *AGRICOLA*

AGRICOLA		1980	1983	1986	1989	1980–89	1960–69	1970–79
K000	General forestry	481	300	154	116	3,414		
K510	Forest products: wood	1,140	919	467	321	8,689		
K530	Forest products: pulp and paper	521	360	233	228	4,059		
K590	Forest products: miscellaneous	96	82	63	34	1,018	Summary data only	
K500	Forest products: general	96	82	134	68	1,289		
K540	Forest products: chemicals, resins, tars	561	297	159	97	3,116	for these decades.	
K520	Forest products: composite	292	240	114	68	2,008		
K100	Forestry production	354	301	239	84	3,307		
K250	Mensuration	567	581	391	237	5,502		
K800	Forest pathology	218	301	251	150	2,698		
K200	Forest management	997	960	615	372	8,939		
K120 & F110	Nursery/forest tree production	16	20	7	3	144		
K810	Fire management	179	234	179	152	2,493		
K001 & F300	Forest ecology	653	556	477	371	5,247		
K001 & F130	Forest range management	38	37	32	15	285		
K130	Harvest and engineering	610	388	243	132	4,371		
K001 & F200	Forest tree breeding/genetics	54	392	303	186	2,850		
K110	Forestry: natural regeneration	290	404	141	91	2,615		

K120	Forestry: artificial regeneration	916	1,027	498	362	8,230		
K001 & F600	Forest tree physiology	636	592	770	592	6,366		
K001 & F500	Forest tree nutrition	71	160	135	107	1,100		
K001 & F400	Forest tree structure	166	170	159	130	1,523		
K001 & F700	Forest tree taxonomy	207	533	266	195	2,530		
K001 & F820	Forest pests of plants (animals)	15	15	36	30	249		
K001 & F821	Forest pests of plants (insects)	688	757	524	410	6,205		
K001 & F822	Forest pests of plants (nematodes)	21	24	20	12	225		
K001 & F830	Forest plant diseases (general)	39	30	24	21	285		
K001 & F831	Forest plant diseases (fungal)	316	356	298	139	2,749		
K001 & F832	Forest plant diseases (bacterial)	8	12	6	3	112		
K001 & F833	Forest plant diseases (viral)	9	3	6	2	45		
K001 & F840	Forest plant diseases (physiological)	7	37	14	9	180		
K001 & F900	Forest weeds	84	91	114	67	971		
K001 & J000	Forest soil sciences	9	3	0	3			
K001 & J100	Forest soil biology	89	226	172	77			
K001 & J200	Forest soil chemistry and physics	125	123	171	97	4,282		
K001 & J500	Forest soil fertility	121	148	171	90			
K001 & J700	Forest soil cultivation and cropping	5	26	53	37			
K001 & J800	Forest soil conservation	30	63	33	21			
	Totals	10,925	10,850	7,672	5,129	97,096	75,910	73,165

Table 3.2. Totals of forestry citations in *CAB Abstracts*

OF = *Forestry Abstracts* IF = *Forest Products Abstracts*	1980	1983	1986	1989	1980–89	1960–69	1970–79
0F01				795			
0F02 General forestry	1,100	1,464	1,360	495	15,846		
1F07 Wood-based materials	415	409	384	375	4,019		
1F08 Pulp industries	199	253	223	243	2,334	Summary	
1F09 Minor forest products	242	299	308	423	3,069		
1F02 Forest products and industry	478	420	350	343	4,048	data only	
1F03 Wood properties	706	807	733	577	6,785		
1F06 Utilization of wood	332	382	350	240	3,390	for these	
1F10 Marketing, trade, economics	180	240	332	185	2,386		
0F04 Mensuration	787	1,332	1,226	869	13,104	decades.	
0F14 Protection by forests/watersheds	277	357	323	269	3,839		
0F09 Mycology/pathology	630	704	653	593	7,495		
0F10 Insects and other invertebrates	711	737	862	837	8,785		
1F05 Timber protection	377	405	332	266	3,244		
0F03 Silviculture	1,511	2,070	2,237	1,789	23,640		
0F06 Fire	154	210	175	121	2,003		
0F05 Physical environment	869	1,270	1,549	1,527	15,547		
0F15 Nature conservation	452	684	637	757	7,691		
0F11 Range	83	80	63	57	874		
1F04 Timber extraction	794	914	855	553	7,749		
0F08 Genetics/breeding	305	362	383	412	4,354		
0F07 Plant biology	1,079	1,725	1,885	2,415	22,421		
0F16 Dendrochronology	39	45	45	50	587		
1F01 General publications and techniques	381	550	482	391	4,565		
Totals	12,101	15,719	15,747	14,543	167,775	56,221	70,370

plication within "K" (forestry) was found to be between 15% and 35%; the highest overlap was with tree breeding and genetics, the lowest with wood products. *CAB Abstracts* uses similar coding and also has nearly the same duplication of entries with its forestry categories.

One clear difference becomes apparent in the counts from CABI's two abstracting journals and *AGRICOLA* during the 1980–89 decade. The sharp drop in *AGRICOLA* reflects a decision by the creators of the database to concentrate on the indexing and cataloging of U.S. imprints. The three decade totals for both are still 48,000 apart. Another difference is the heavier emphasis of *Forestry Abstracts* on higher plant biology and to a lesser extent subjects like ecology (code 0F07) and nature conservation (code 0F15). Many citations in forestry have limited relevance to tree biology or directly to silviculture practice. For example, under the headings "Insects and Other Invertebrates" (code 0F10) in *CAB Abstracts*, insect ecology, behavior, systematics and breeding are covered. These are not subjects generally indexed in the *AGRICOLA* K file, whose Forest Pests of Plants (Insects) (code K001 & F821 combined) inclines toward indexing material on insect damage directly related to forests and nurseries, and omits invertebrate zoology and ecology. The higher count for CABI is therefore partly explained by its selective inclusion of allied sciences such as plant biology, ecology, and invertebrate zoology as deemed appropriate for its users.

A study[32] conducted in 1980 reported that between 1974 and 1980, *AGRICOLA* indexed 316, and *CAB Abstracts* 516 articles out of a possible 532 from the bimonthly *Forest Science*, the most highly cited journal in the discipline.[33] The study revealed that *AGRICOLA* had 38.8% fewer indexed items. The same study showed that *AGRICOLA* did a more complete job of indexing *Journal of Forestry*, *Forestry Chronicle*, and *American Forests* by small margins. One reason for the lower numbers reflected in *AGRICOLA* is that recent administrative policies at NAL have allowed the numbers of citations entered in the database to decline over the last ten years.[34]

Table 3.3 is a synthesis of the subject divisions of forestry in *CAB Abstracts* and *AGRICOLA* into comparable categories. The numbers are decade totals for literature published from 1980 through 1989. Although the numbers differ, the same four subjects have the greatest number of citations in each case: forest pathology, silviculture, forest ecology and forest tree biology. These four range between 10.7% and 20.5% of all forestry citations in *AGRICOLA*, and 13.4% and 14.1% of *CAB Abstracts*.

In the 1980s, as with historical antecedents, a clear pattern emerges with *CAB Abstracts* emphasizing biological aspects of forestry in *Foresty Abstracts*, while *AGRICOLA* and *Forest Products Abstracts* stress production and industry aspects. By late 1994, CAB International had revised its subject coding system and introduced it online with the Dialog system in the United States. These changes had been underway for some time and bring their new CABICodes into closer alignment with the subject codes used by AGRICOLA and AGRIS.[35] The two index files online at Dialog were merged into one and recodified becoming available for full use under the new codes in early 1995. New and revised subject codes were introduced as well as some searching, storage, and format enhancements. The subject

32. Kristina Brooks, "A Comparison of the Coverage of Agriculture and Forestry Literature on *AGRICOLA*, BIOSIS, *CAB Abstracts* and *SciSearch*," *Database* 3 (1980:) 38–49.

33. Its ranking is explained in Chapter 12, "Primary Journals and Serials in Forestry and Agroforestry," along with the high impact factor derived from the *Science Citation Index*.

34. (a) Wallace C. Olsen, "Characteristics of Agricultural Engineering Literature," in *The Literature of Agricultural Engineering*, ed. Carl Hall and Wallace C. Olsen (Ithaca, N.Y.: Cornell University Press, 1993), p. 221. (b) Wendy Simmons, "The Development of *AGRIS*: A Review of the United States Response," *Quarterly Bulletin of the International Association of Agricultural Librarians and Documentalists* 31 (1) (1986): 11–18.

35. *CAB International Database News* (20) (June 1994):1.

Table 3.3. Comparisons of grouped code categories, 1980–1989, in *AGRICOLA* and *CAB Abstracts*

Subject	AGRICOLA N = 97,096	CAB ABSTRACTS N = 167,775
Wood products	9978	10,804
Pulp & paper	4059	2,334
Non-wood products	6142	3,069
Production, economics, & industry	3307	2,386
Mensuration	5502	13,104
Forest pathology & protection	13,719	23,363
Silviculture	19,928	23,640
Fire	2,493	2,003
Forest ecology	10,629	23,238
Range management	285	874
Timber extraction	4,371	7,749
Tree breeding & genetics	2,850	4,354
Forest tree biology	10,419	22,421
General forestry	3,414	15,846

(Note: Codes in *AGRICOLA* and *CAB Abstracts* were matched where clear subject comparisons were available such as FIRE. Where there was no clear match, related subjects were combined for the sake of comparison as in the case of FOREST TREE BIOLOGY. For *AGRICOLA* this meant combining code K001 with F400 + F600 + F700, and comparing this total against *CAB Abstracts'* codes 0F07 "plant biology" with 0F054 and 0F055. For FOREST PATHOLOGY AND PROTECTION this meant combining K001 with F8?? in *AGRICOLA* with *CAB Abstracts'* 0F09 + 0F10 + 1F05. Similar comparisons were made for all the codes using logical subject groupings.)

coding changes should make comparisons of subjects in the three major agricultural databases more compatible, and reduce duplicate entry counts.

Therefore, the numbers in Tables 3.2 and 3.3 were reexamined to see if the recoding had shifted subject emphases significantly. The following Table 3.4 makes these comparisons with the earlier CAB *Abstracts* numbers. The data were obtained from the Dialog file in August 1995. The new CABICodes for forestry are all coded KK.

Using the new CABICodes increases the numbers 7.1% for the same publication years. This is partially the result of late entries into the database since the earlier searching was done in 1993, only three years after the last date of publication, and by 1995 two more years of late entries had been added. However, the total of 11,912 additional citations published before 1990 is sizable. CABI personnel suggest that these entries were added when a backlog was cleared up. The new searching manual with details on coverage of the new CABICodes was not published at the time of the searching and this writing.

Table 3.4. Online search of CABICodes for forestry in 1995

		1980	1983	1986	1989	1980–1989
KK100	General forestry	5,591	6,405	6,657	8,100	65,044
KK110	Silviculture	2,535	2,806	3,572	4,832	32,691
KK120	Forest mensuration & management	1,110	1,498	1,502	1,670	14,241
KK130	Forest fire management	185	225	199	218	1,951
KK140	Protection forestry	380	385	360	442	3,952
KK150	Other land use	515	646	661	888	6,669
KK160	Aboriculture	122	160	140	205	1,507
KK500	Forest products (General)	972	1,151	1,154	1,153	11,566
KK510	Wood properties & utilization	1,408	1,548	1,453	1,447	14,898
KK515	Wood processing	929	1,062	1,005	971	10,052
KK520	Forest products (Wood-based materials)	442	492	508	512	4,796
KK530	Forest products (Pulping & chemical utilization of wood)	235	259	252	346	2,746
KK540	Forest Products (Misc., minor forest products)	250	306	324	559	3,591
KK600	Agroforestry	243	433	580	1,630	6,737
	Totals	13,917	17,376	18,367	22,973	180,441

The Core Agricultural Literature Project identified numerous old codes which had no apparent counterpart in the new coding system. Discussions with CABI clarified how the old coding system was translated into the new codes. CAB International kindly provided the following translations into CABICodes by personal letter (Gillian Petrokofsky, 6 September 1995): "In KK100, General Forestry, the following old codes were subsumed: 0F05 - physical environment; 0F07 - plant biology; 0F08 - genetic/breeding; 0F09 - mycology, pathology; 0F10 - insects/pests; 0F16 - dendrology. In KK150, Other Land Use, old code 0F11—range management, was included. In KK500, Forest Products, these codes were included: IF01—general publications, techniques; IF10—marketing, trade, economics. In KK510, Wood Products & Utilization, IF06 - timber protection was included. Finally, KK515, Wood Processing, included IF04 - timber extraction."

Table 3.5 details the forestry-related citations for the printed *Forestry Abstracts* and the National Agriculture Library's *Bibliography of Agriculture* for the years not covered by computer databases; these are the precursors of *CAB Abstracts* and *AGRICOLA*.

In 1988, CABI began another abstracting journal dealing with forestry: *Agroforestry Abstracts* (code 7Y) which is a subfile in *CAB Abstracts* done in collaboration with the International Council for Research in Agroforestry

Table 3.5. Decade totals for forestry citations in *Forestry Abstracts* and *Bibliography of Agriculture*

Year	Bibliography of Agriculture	Forestry Abstracts
1950–1959	63,851	49,446
1960–1969	75,910	56,221
1970–1979	73,165	75,170[a]

[a]This number includes items from *Forest Product Abstracts* begun in 1978; *FPA* published 6,000 references in 1978 and 1979, of which an estimated 20% were duplicated in *Forestry Abstracts*.

(ICRAF).[36] CABI subfiles such as *Agroforestry Abstracts* are almost entirely culled from other files, in this case mostly from *Forestry Abstracts* (71.4%).[37] Almost all records in subfiles are given multiple subject and abstract codes so the citation may be used in different journals. In *Agroforestry Abstracts* an average of 76.4% of the records are to journal articles; English predominates at 83.6% of the file, with Spanish second at 4.3%, and French at 3.1%. Other languages include German, Japanese, Chinese, Portuguese, and Russian which are evenly distributed among the remaining records. About 1,500 records per year are entered in this new abstracting journal. It is doubtful if enough time has lapsed to obtain valid citation patterns from *Agroforestry Abstracts*.

The author studied the serial publications indexed by *Agroforestry Abstracts*. Of the 317 titles abstracted in 1989, 94 titles or 29.6% were published in the Third World. This is an impressively high percentage reflecting major work being done in agroforestry and published in low-latitude countries. The subject coverage of the indexed journals is extensive with emphases on forestry, general agriculture, soil science, botany, hydrology, environmental sciences, ecology, and college and rural education. By 1992, portions of 594 journals were being indexed for *Agroforestry Abstracts*, a 46% expansion of coverage in three years. But as pointed out in the preceeding paragraph, nearly all of the citations have been extracted from indexing for other CABI abstracting journals. The analysis of *Agroforestry Abstracts* was not extensive, so other means were used to quantify subject concentrations. The leading journal for low-latitude forestry is *Agroforestry Systems* (begun in 1981) which is also the most highly cited tropical forestry journal in the Core Agricultural Literature Project analysis (see Chapter 12). In a study commissioned by *Agroforestry Systems* to inaugurate its

36. *Agroforestry Abstracts*, 1988+ (Wallingford, U.K.: CAB International).
37. CABI, Forestry Department, furnished this overlap data. Three other CABI abstracts have major overlaps with its *Agroforestry Abstracts: Herbage Abstracts* (14.3%); *Horticultural Abstracts* (11.6%), and *Soils and Fertilizers* (11.4%).

tenth volume, the number of articles on specific topics were analyzed for a nine year period, 1982–1990. These are some results: 55.3% of articles were on soil nutrient cycle, 12.7% on fruit trees, 10.6% on fodder, 8.5% on water and climate, with fuelwood and root research each 6.3%. The clearest feature to emerge is "the importance assigned to soil fertility and nutrient cycling" in agroforestry systems.[38]

A comparative analysis was also made of the *AGRIS* bibliographic database. Its coverage of Third World publications is today more extensive than *AGRICOLA*. In the 1970s, *AGRIS* and *AGRICOLA* reworked their coding systems so that they would be in agreement. These codes were readjusted in 1985–1986 and now the matches between the two files are not as exact. *AGRICOLA*'s coverage is more detailed and extensive, with thirty-eight forestry subject categories as opposed to seven in *AGRIS*. Yet these seven forestry categories include only 63.5% of all forest science coverage in *AGRIS*; the remaining forestry-related citations are dispersed among other *AGRIS* categories, notably, "plant breeding" which contains all forest biology and the category "economics, development, and rural sociology" which contains all pertinent social forestry citations including those on the international timber trade. There is currently no practicable way to extract purely forestry-related citations from these categories using the *AGRIS* schema.[39]

Table 3.6 provides decade category totals for forestry in the *AGRIS* database, which has two dissimilar versions, one on-line in the United States, the other on CD-ROM. The *AGRIS* online version in the United States, available from DIALOG Information Retrieval Service, does not contain the U.S. imprints downloaded into *AGRIS* from *AGRICOLA*. The compact disk product from SilverPlatter, the version most commonly found in libraries and research centers of the Third World, includes the U.S. imprints. Thus, counts from the online version in the United States provide a unique window on materials published outside the United States as well as the influence of U.S. imprints. A comparison of these two versions reveals that citations from *AGRICOLA* downloaded into the *AGRIS* compact disk account for 35.3% of the Forestry (K category) citations, 1980–1989, which means that roughly 65% of the *AGRIS* forestry citations are not of U.S. imprint. A study by Norbert Deselaers[40] of all subjects in the three

38. Anthony Young, "Change and Constancy: An Analysis of Publications" in *Agroforestry Systems,* Volumes 1—10," *Agroforestry Systems* 13 (1991): 198.

39. Abraham Lebowitz, "Forestry Information: The View from *AGRIS* and CARIS" Unpublished paper (Rome: FAO, 1992).

40. Norbert Deselaers, "The Necessity for Closer Cooperation among Secondary Agricultural Information Services: An Analysis of *AGRICOLA, AGRIS,* and *CAB Abstracts,*" *Quarterly Bulletin of the International Association of Agricultural Librarians and Documentalists* 31 (1) (1986): 19–32.

Table 3.6. AGRIS category code totals by decade

AGRIS (from CD-ROM with U.S. publications)		1980	1983	1986	1989	Decade total 1980–89
K00	Forestry	633	406	n/d	n/d	3,254
K01	General Aspects	n/d	n/d	468	575	1,998
K10	Forest Production	1,832	1,812	3,050	2,582	23,894
K11	Forest Engineering	n/d	n/d	84	45	263
K20	Forest Management	1,324	1,202	n/d	n/d	8,034
K50	Wood Technology	2,190	1,503	1,631	1,931	19,333
K70	Forest Protection	274	376	801	926	5,491
	Total by year	6,253	5,299	6,034	6,059	62,267

AGRIS (from CD-ROM without U.S. publications)		1980	1983	1986	1989	Decade total 1980–89
K00		332	338	n/d	n/d	1,553
K01		6	41	45	485	2,394
K10		1,393	1,408	1,786	1,880	16,311
K11		n/d	7	86	45	362
K20		803	795	n/d	n/d	3,853
K50		1,249	1,049	1,060	1,190	11,496
K70		138	251	706	717	4,529
	Total by year	3,912	3,889	3,683	4,317	40,498

n/d = No data for that year

major agricultural databases revealed that citation overlapping was as high as 34–46% in 1983. Another comparative study on tropical tree citations only, showed an overlap between *AGRICOLA* and *AGRIS* as 1.4% on average.[41] This study also found the overlap of the same subject between *AGRIS* and *CAB Abstracts* to be 11%.

For a better understanding, the Core Agricultural Literature Project did a random sample of 150 citations published in 1989 and listed in the K01 - Forestry (General Aspects) category of *AGRIS*. The sample revealed that the only titles indexed by both *AGRICOLA* and *AGRIS*, 15.7% of the sample, were those to the *Philosophical Transactions of the Royal Society of London, Series B, Biological Sciences*, along with two monographs. It appears that the forestry in *AGRIS* includes most internal FAO reports (*Series*

41. Chen Quibo, "A Check on Overlapping between *AGRICOLA, AGRIS*, and CAB for Tropical Agricultural Records," *Quarterly Bulletin of the International Association of Agricultural Librarians and Documentalists* 34 (2) (1989): 67–72.

FAO-FO), symposia from developing countries, and numerous non-English-language journals, notably German, Swedish, and Japanese not found in *AGRICOLA*. Using the same forestry sample, *CAB Abstracts* included 58.5% of the *AGRIS* citations, omitting some of the FAO report series and pamphlet-length material in non-European languages.

On average, the CD-ROM version of *AGRIS* indexes about 1,200 fewer forestry citations a year than *AGRICOLA* with the largest proportion of omissions being in Forest Engineering (code K11) and Forest Protection (code K70) where *AGRICOLA* had almost twice the number of citations (1980–1989). One study of *AGRIS* revealed that 69.8% of forestry citations were assigned primary category codes in forestry, while the remainder were secondary subject codes to citations from subjects secondary to forest science. The same study showed that forestry-related citations accounted for 7.1% of the total *AGRIS* database.[42] The *AGRICOLA* and *CAB Abstracts* forestry citations account for 8.2% and 12.6% respectively of their total databases (1980–1989).

The Core Agricultural Literature Project sought to identify the documents in *AGRIS* and *CAB Abstracts* dealing with specific species of trees. In 1988, *AGRIS*, using keyword searching for tree species and genera, found that 30.7% of all forestry citations had indexing entries for specific trees.[43] Table 3.7 ranks the most common species indexed by *AGRIS* and *CAB Abstracts*.

Both databases demonstrate that the primary literature indexed, which probably reflects the research emphasis, is of temperate-climate trees, primarily conifers, the softwood timbers of high latitudes. *Pinus* in both columns received almost twice the number of citations as the second place *Abies* or firs, and roughly fifty-five times the number of citations as the twentieth ranked teak. Conifers accounted for 51.2% of the tree species citations in *AGRIS* and 49.9% in *CAB Abstracts*. It appears that angiosperms, or hardwood deciduous species, are far less important in the literature. Two tropical trees, neem and teak, are at the bottom with fewer than two hundred citations each in *CAB Abstracts* for the three years analyzed. Hevea or rubber, really more a crop than a timber product, falls to 17th place in the CABI system with other tropical trees, whereas its importance in *AGRIS* is much higher. *Carya*, the hickory tree, is unique to the North American continent, which probably explains its rank at twentieth in the CABI scheme, for it is a hardwood not endemic to the British Isles or its former tropical colonies.

42. Lebowitz, "Forestry Information: The View from *AGRIS* and CARIS," p. 6.
43. Ibid., p. 8.

Table 3.7. Ranking of frequency of tree species or genera citations in *AGRIS* and
CAB Abstracts, 1986–1988

Genera	*AGRIS* Ranking	*CAB Abstracts* ranking
Pinus (all pine species)	1	1
Abies (firs) and *Pseudotsuga* (Douglas fir)	2	2
Picea (spruces)	3	3
Quercus (oaks)	4	4
Populus (poplars)	5	5
Fagus (beech)	6	8
Eucalyptus	7	6
Hevea (rubber)	8	17
Betula (birch)	9	7
Acer (maple)	10	10
Carya (hickory)	11	20
Larix (larches)	12	9
Juglans (walnut)	13	15
Acacia	14	11
Salix (willow)	15	12
Fraxinus (ash)	16	13
Ficus (fig)	17	14
Tsuga (hemlock)	18	16
Azadirachta (neem tree)	19	18
Tectona (teak)	20	19

A fourth major indexing tool in addition to the three major databases is *Social Sciences in Forestry* (SSIF) which began in 1963 as a joint venture between the U.S. Department of Agriculture and the College of Forestry at the University of Minnesota. It is published four times a year and currently includes ninety published issues, and adds almost 3,000 records annually. The focus of *SSIF* is the social impacts of forest policy worldwide, agroforestry, forest economic impacts on local communities, and environmental damage with little or no emphasis on forest biology, ecology, or pathology. An online version searchable via keyword is available on the Forestry Library Internet Gopher at the University of Minnesota (see Chapter 13). Searches using keywords with this online version reveal that the number of agroforestry citations has risen since 1985, the earliest searchable year in *SSIF* database. Using the keyword "agroforestry" the number of resulting entries has almost doubled between 1985 and 1992 reflecting the growing interest in the field by scientists and researchers and the increased use of the word "agroforestry."

TREECD

In 1992, CAB International inaugurated a compact disk product called TREECD. This is a major effort covering the literature abstracted during 1939–1992 with more than 300,000 citations on forestry subjects drawn from literature around the world. This is the first time that the Core Agricultural Literature Project had computer access to extensive journal and monograph literature data of a discipline back to 1939. Therefore TREECD was examined and analyzed extensively, revealing several problems in extracting data resulting from the category or subject coding.

The category coding in TREECD is cumbersome because of the three coding systems used in the original indexing. For 1939–1951, it uses the International Decimal Classification of Philipp Flury; for 1951–1977, the Oxford Decimal Classification; and from 1978 onward the same codes as CABI's *Forestry Abstracts* and *Forest Products Abstracts*. Trying to match subjects over this time period is difficult. This was partly overcome by grouping subjects as in Table 3.3 in order to create valid comparisons. Table 3.8 provides the subject counts for literature published from 1955 through 1975. The years 1980 onward are covered adequately in Table 3.3. In Table 3.8 the subject codes are those of the Oxford Decimal Classification.

Counts from TREECD show the number of records indexed over the last forty-five years to have increased threefold, from 2,536 in 1940 to 9,785 in 1985.

Table 3.8. TREECD subject totals[a]

Subject codes		1955	1960	1965	1970	1975
0	Forestry and general forest products	117	173	314	891	1049
1	Factors of the environment	1328	1861	2051	2723	2504
2	Silviculture	1099	1469	1442	1714	1443
3	Harvesting and forest engineering	291	459	564	560	503
4	Forest injuries and protection	9857	1457	1474	1744	1320
5	Forest mensuration	343	437	552	728	593
6	Forest management	129	208	308	348	316
7	Forest products marketing	76	99	127	180	129
8	Forest products	1536	1610	2001	2185	1868
9	Social economics and forest policy	278	312	385	476	609

[a]These numbers were checked or supplied by Forestry Department, CABI.

D. Language Concentrations

In 1955, *Forestry Abstracts* did an appraisal of forestry literature indexed in the journal to that time. Publications from nearly 100 countries in over thirty languages were analyzed by the Commonwealth Forestry Bureau. The focus of Table 3.9, recast from the study, is journals which were weighted for frequency. The low counts of German-language material during and immediately following World War II climbed to 12.4% by 1954. English clearly predominates in all samples; nearly 50% of the English-language citations were published in the United States.[44]

Table 3.9. Language concentrations in *Forestry Abstracts*, 1946–1954

Language	1946	1950	1954
English	73.5%	70.0%	61.2%
French	9.0	7.5	8.3
German	6.0	6.6	12.4
Spanish, Portuguese, Italian, & Romanian	5.0	6.3	6.0
Scandinavian and Dutch	3.2	3.3	5.3
Russian and other Slavic	3.2	4.5	4.5
Oriental and other	0.1	1.8	2.3
Total	100.0	100.0	100.0

Source: *Forestry Abstracts* 16 (2)(1955): 157–160.

The scope of the current agricultural databases, *CAB Abstracts*, *AGRIS*, and *AGRICOLA*, is worldwide, thus providing the most accessible source for information on the languages most heavily used. Table 3.10 contains these data for the three major databases. It should be noted that *AGRICOLA* attempts to include all documents from the United States, and since 1985 has reduced its coverage of non-English-language publications. For this reason, *CAB Abstracts* data on language distribution are more representative of worldwide publishing trends in forest science. *CAB Abstracts* for all subjects has steadily declined on non-English publications from 32.5% in 1984, to just under 25% today,[45] whereas from 1970 to the present the *CAB Abstracts* searching manual reports that 61% of all documents in the database are in English.[46]

44. Commonwealth Forestry Bureau, "The Literature of Forestry: An Appraisal," *Forestry Abstracts* 16 (2) (1955): 157–160.

45. Personal communication, June 1992. CABI gave these percentages when reviewing the chapter "Literature Trends and Patterns," by Wallace C. Olsen, in *The Literature of Animal Science and Health*.

46. *CAB Abstracts Online Manual, 1989 Edition* (Wallingford: CAB International, 1989), p. 8.

Table 3.10. Languages of forestry literature listed in *CAB Abstracts, AGRICOLA,* and *AGRIS*

Language	CAB Abstracts 1980–1990	AGRICOLA 1980–1990	AGRIS (online in U.S.) 1974–1992[a]	AGRIS (on CD-ROM) 1981–1988
English	65.2%	89.1%	29.0%	60.0%
German	8.2	3.2	39.7	5.8
French	3.2	2.1	8.2	7.7
Spanish	0.2	0.9	3.9	6.8
Japanese	0.1	0.1	6.1	5.9
Russian	6.1	3.2	0.1	0.1
Other	17.0	1.4	13.0	13.8

[a]*AGRIS* online in the U.S. does *not* include U.S. imprints.

The language concentrations for *AGRIS* CD-ROM, may more closely represent publishing patterns of forest science literature in the Third World, primarily of pamphlets and reports not in English. FAO documents, alone, account for 6.8% of the total *AGRIS* file.[47]

The language counts relate to subject specialties. Forestry production literature is clearly a North American phenomena. The bulk of the citations come from *AGRICOLA*, which has a heavy representation of literature on wood products and production, where English dominates at over 90%. *Forestry Abstracts* indexes more non-English journals and includes much more forest biology, an area in which Germans, Russians, and Scandinavians excel. Counts in these languages reflect these strengths. The paucity of Russian documents cited in *AGRIS* (about 1.5%) reflects the fact that the U.S.S.R. only recently agreed to submit its records to *AGRIS*.

Table 3.11 gives the concentrations of five major languages from 1940 through 1975 in TREECD. One of the useful features of this database is that it makes accessible computer-generated data sets of early trends in for-

Table 3.11. TREECD language counts, 1940–1975

	1940	1945	1950	1955	1960	1965	1970	1975
English	1,801	1,679	2,135	2,604	3,131	3,427	4,859	4,617
German	237	58	256	548	671	698	645	716
French	77	116	287	332	340	271	295	522
Spanish	33	49	208	147	673	755	904	933
Russian	29	69	77	79	153	172	126	157
Total	2,536	2,217	3,502	4,607	6,053	6,577	7,824	8,345

47. Lebowitz, "Forestry Information: The View From *AGRIS* and CARIS."

estry literature. The low German counts for 1945 reflect the effects of World War II.

The last six dates (1950–1975) have English concentrations ranging from 51.7% to 62.1%. Several other forestry literature databases have been created with extensive citation coverage. Ernahrungs Land- und Forstwirtschaftliches Information System (ELFIS), begun in 1984 in Germany, is a bibliographic database which covers forestry, the timber industry, and allied agricultural fields with a concentration on international and European literature. ELFIS has approximately 70,000 document entries since the mid-1980s with a current annual update near 15,000 records. A recent study revealed that 20.9% of the ELFIS citations were not in *AGRIS*, *AGRICOLA*, or *CAB Abstracts*.[48]

France has its own forestry database called VELLEDA, created by the École Nationale du Génie Rural, des Eaux et des Forêts, in Nancy.[49] The French and German databases, as well as the Information Systems and Terminology Subject Group (S6.03) of the International Union of Forestry Research Organizations (IUFRO), the largest international association in forestry, have expressed interest in wider implementation of FAO's multilingual macrothesaurus AGROVOC currently in use with the *AGRIS* database. It is expected that the implementation of this thesaurus would facilitate forest science indexing of non-English material.[50] The staff of *AGRIS* has worked closely with Centro de Informacion Bioagropecuria y Forestal de la Universidad del Nordeste Resistencia (CIBAGRO) in Argentina to bring AGROVOC up to date in Spanish terminology. It is reasoned that non-English language material may reach indexing parity with the use of such a multilingual thesaurus.[51]

Table 3.12 identifies the primary languages resulting from the earlier four agricultural subject areas analyzed by the Core Agricultural Literature Project. English heavily predominates in all cases. Analysis of source documents for the historical component of the Core Agricultural Literature Project in Chapter 14 shows that one hundred years ago it was not English but German which dominated emerging forestry science literature. German's preeminence has steadily declined, although it is still a strong second today.

48. E. Stage, "ELFIS—An Agricultural Bibliographic Database of Forestry-Related Information," in *Information Systems for Forestry-Related Subjects: Access, Search Techniques and User Needs*, ed. R. Louis (Birmensdorf, GDR: International Union of Forestry Research Organizations, 1988 [Subject Group S6.03]), pp. 63–69.

49. M. Lionnet, "A Documentation Network for French Foresters," in *Information Systems for Forestry-Related Subjects*, pp. 123–130.

50. Simmons, "The Development of *AGRIS*: A Review of the United States Response."

51. Lebowitz, "Forestry Information: The View From *AGRIS* and CARIS."

Table 3.12. Core Agricultural Literature Project language concentrations

Subject	English	Second most common language	
Agricultural Economics and Rural Sociology	87.6%[a]	French	2.5%[a]
	64.1[b]	German	8.0[b]
Agricultural Engineering	88.4[a]	Russian	5.0[a]
	70.9[b]	German	8.9[b]
Animal Science and Health	73.0[a]	German	4.9[a]
	67.0[b]	German	7.1[b]
Soil Science	86.8[a]	German	3.3[a]
	79.6[b]	German	5.6[b]
Forest Science and Agroforestry	89.1[a]	German	3.2[a]
	65.2[b]	German	8.2[b]

[a]Data from *AGRICOLA* online.
[b]Data from *CAB Abstracts* online.

E. Types of Publications

The literature of forest science is published in all formats: serials, mono-graphs, conference proceedings, reports, theses, standards, patents, maps, laws, and a variety of tools such as measurement tables and compact disks. While journals clearly dominate, forty years ago over a third of forestry-related serial literature was published in non-forestry publications.[52] The subjects where this is most pronounced are plant biology and to a lesser extent economics. This is corroborated by the analysis of forest science periodical literature in Chapter 12, where ten of the top twenty journals cited in the Core Agricultural Literature Project analyses were not forestry publications.

In 1955, a retrospective analysis of *Forestry Abstracts* revealed that serials, including report series and journals, accounted for 84% of the citations. The remaining 16% were to monographs, patents, and laws.[53] In the intervening years the percentage of serially published material has risen slightly in *CAB Abstracts* and *AGRICOLA*. All three major agricultural databases are encoded with the type of document from which each bibliographic entry emanates, although some annual conferences in numbered series and other special reports are not precisely encoded. Format codes differ from system to system, but all essentially divided their entries into

52. Commonwealth Forestry Bureau, "The Literature of Forestry."
53. Ibid., p. 160.

Table 3.13. Document types of forestry citations, 1984–1990

	CAB Abstracts	*AGRICOLA*	*AGRIS* (CD-ROM)	Core Agricultural Literature Citation Analysis[a]
Journals & serials	89.1%	88.5%	65.7%	58.2%
Monographs & diss.	10.9	11.5	34.3	41.8

[a]The Core Project included reports in series as monographs, which is the reason for this variation. *AGRIS* does the same.

monographs, serials and journals, and "unconventional formats" such as maps, dissertations, patents, standards, and laws. The latter are rarely numerous and generally fall into the monograph category. (For the definition of a "monograph" in this study, see Chapter 11.) Table 3.13 compares these sources as well as data from citation analysis of the Core Agricultural Literature Project in Chapter 11. The high count of monographs in *AGRIS* is due, primarily, to the great number of reports indexed in the database. The *AGRICOLA* and *CAB Abstracts* format percentages are almost identical. In order to better understand citation patterns, the author tallied the formats of citations in articles of six issues of the journal *Forest Science* for 1991.[54] This publication is the most highly cited journal in the discipline. Of the 3,301 citations counted, 63.3% were to journal literature, 35% to monographic literature, and 1.7% to university theses. The high number of monographs cited was surprising since *Forest Science* publishes cutting-edge research which tends to cite journal literature most heavily.[55] That authors writing for a journal dedicated to rigorous biometric and statistical research should rely so heavily upon monographs and reports stands in slight contradiction to the norm for general science journals which tend to divide 80% journals, 20% monographs.[56] This is largely explained by the fact that of the 1,156 citations to monographs in *Forest Science*, 321 were to USDA Forest Service series reports, which were counted as monographs.

A special proceedings issue of *Forest Ecology and Management* subtitled "Agroforestry: Principles and Practice" was also analyzed by the author for similar format data.[57] The count was evenly split with 49.5% to monographs

54. *Forest Science.*
55. Editorial, *Forest Science* 38 (4) (1993): 724–726.
56. Eugene Garfield, *Citation Indexing: Its Theory and Application in Science, Technology and Humanities* (Philadelphia, Pa.: ISI Press, 1983); see Chapter 9.
57. P. G. Jarvis, "Preface" in "Agroforestry: Principles and Practices. Proceedings . . . International Conference, 23–28 July 1989 at the University of Edinburgh, U.K.," *Forest Ecology and Management,* 45 (1–4) (1991): 1–3.

including reports, 49.1% to journals, and 1.2% to theses. Here, the heavy monographic counts were because of government-sponsored reports published in series cited from a variety of developing countries as well as a heavy reference of occasional or non-serial conference proceedings. In an emerging science such as agroforestry, reliance on conferences, feasibility studies, and government reports seems logical. Of the journal citations, 57% were to *Agroforestry Systems*, a case of journal self-citation.[58] (See Chapter 12.)

Report Series

Report series are clearly of great importance in the literature of forest science and deserve mention here, although a fuller examination is given in Chapter 11. Numerous government agencies, such as the U.S. Forest Service, and international organizations, both public and private, publish reports in series which are heavily cited in the literature. Generally these publications follow a pattern of examining a single topic in a brief paper. In the Third World, many are feasibility studies for new techniques and environmental impacts in forestry practice. In the Core Agricultural Literature Project analysis, 8.9% of all forest monographic citations were to reports.

Dissertations

Forest science dissertations appear to have a greater influence in Third World research literature than in that of developed countries, and in the Third World the more heavily cited subjects in dissertations relate more to the biological sciences than to economics, engineering, or forest management and policy. In the Core Agricultural Literature Project analysis, 2.3% of the citations in temperate forestry monographs were to dissertations, both masters and Ph.D.'s. For Third World or tropical forest research, these accounted for 4.1% of the monographic counts. The combined average is 3.1%. This matches exactly the dissertation citation counts in the Core Agricultural Literature Project analysis of the literature of agricultural engineering.[59]

In a study of forestry school degrees granted in the United States, 1971–

58. (a) *Agroforestry Systems; An International Journal*, International Council for Research in Agroforestry (Dordrecht, Netherlands: Kluwer Academic Publishers). (b) Anthony Young, "Change and Constancy: An Analysis of Publications in *Agroforestry Systems*, Volumes 1–10," *Agroforestry Systems* 13 (1991): 195–202.

59. Olsen, "Characteristics of Agricultural Engineering Literature," p. 223.

1981, the total for masters degrees was 7,398, and 1,621 were Ph.D.'s.[60] For the decade 1982–1991, the number of doctoral dissertations in forestry was 1,955.[61] A rough measure of their influence can be obtained from the Core Agricultural Literature Project analysis of the thirty source documents which were submitted to citation analysis to gather data on the literature of forest science (see Chapter 11). All the source documents analyzed were published between 1981 and 1992. Of the 8,979 citations to monographs, 283 were to dissertations of which 81% were written at United States universities. Based on the citation analysis data and the total of 1,955 dissertations produced during 1982–1991, no more than an estimated 15% of all dissertations will be cited in the literature within the first ten years after completion.

Patents

The patent literature in forest science is fairly comprehensive, but sparsely cited. Most patents are concerned with wood products, such as lamination techniques or particle boards, or other aspects of wood product technology. A few concern tree growth regulators.

Congressional Reports

The United States Congress publishes many hearings and reports concerned with forest policy. Recently, many of these hearings have dealt with forestry impacts on the environment and on wildlife. In the Core Agricultural Literature Project monograph analysis (see Chapter 11), 1.6% of the citations were to congressional hearings and reports. The U.S. Environmental Protection Agency accounted for 0.9% of the citations. Another 0.7% were to U.S. Government Accounting Office publications.

F. Some Observations

From studies of forest science literature databases, it is clear that a substantial percentage of local and site-specific literature is not included in the three international databases used extensively in this chapter. This must be kept in mind in analyzing the database numbers with the realization that up

---

60. Ronald Christensen, "Forestry School Enrollment and Degrees Granted, 1971–1981," *Journal of Forestry* 81 (10) (1983): 660–662.
61. Online search of *Dissertations Abstracts* in DIALOG File 35, using Descriptor Code 0478—Forestry.

to 20% of the world's literature concerned with forestry and its agricultural relationships are missing in the large databases. That which is covered, however, represents the most widely known and utilized forestry literature.

From the TREECD collection of literature provided by CAB International, the primary subject concentrations from 1955 through 1975 were consistently factors of the environment, silviculture, forest injuries and protection, and forest products (Table 3.8). These are corroborated by averaging decade totals from the *Bibliography of Agriculture* and *AGRICOLA* with those of the CABI for the same time periods (Tables 3.3 and 3.5). The 1950–1959 literature citations averaged for these two sources are 56,648 and those for 1980–1989 are 132,435. This is also a threefold increase, but the subject concentrations have not shifted appreciably during this forty-year period. There is a greater concentration of literature on subaspects of the larger subjects such as for pest management, intercropping and agroforestry, and ecological systems. The only major subject area appearing to be declining in literature is forest products, and this might be reversed with anticipated literature increases from the Third World and more complete reporting of its forest products literature.

The TREECD shows that near 60% of all literature was in English for the 1940–1975 period (Table 3.11) which is only slightly lower than the *CAB Abstracts* 1980–90 average of 65.2% (Table 3.10). The strong second position of German-language forestry literature continues.

The heavy use of report literature, particularly from governmental sources, provides a different pattern of formats than with recent agricultural and other scientific literature. Clearly, much of the short-term literature with limited applications goes into report series in forestry, whereas in many other related disciplines this literature is more often published in journals. The forestry pattern appears to be shifting to journals also.

The assistance of the Forestry Department, CAB International, which provided clarifications on subject categories, the text, and the citation numbers, is gratefully acknowledged. Similar assistance with *AGRICOLA* was provided by John Forbes and Caroline Early of the National Agricultural Library. Helga Schmid, of the *AGRIS* office, also provided clarifications.

4. Agroforestry Directions and Literature Trends

P. K. R. NAIR

University of Florida

Agroforestry is the purposeful growing of trees and crops in interacting combinations for a variety of objectives. Although such farming practices have been used throughout the world for a long time, agroforestry attained prominence as a land-use practice only during the late 1970s. Acting as an interface between agriculture and forestry, agroforestry today is considered to be a promising approach to land use especially in the developing countries of the tropics and subtropics. This article will review the developments in agroforestry literature during the relatively short span of its existence and examine the emerging trends and directions. It is important to review developments in the subject and its underlying concepts as a prelude to the discussion of its literature; this is particularly important in the case of a new and integrated discipline such as agroforestry.

A. History of Agroforestry

With the advent of industrialization, rapid advances were made in agriculture and forestry as commercial enterprises with emphasis on production of single commodities. The traditional, mixed production systems of raising food crops, trees, and animals together, as well as exploiting a multiple range of products from natural woodlots did not fit into the developmental framework of such monocultural systems. However, the strategy of commercial, monocultural production paid rich dividends, and it formed the basis of modern agriculture and forestry in the industrialized world. When the newly independent nations of the developing world were faced with the problem of feeding their millions, the successful model of modern agriculture used in the industrialized world was thought to be the best solution. Several technologies were developed according to this model and tried in the tropics, and some of them resulted in substantial increases in agricultural production. The most significant among these is the so-called green revolution of the 1970s.

However, serious doubts soon began to be expressed about the relevance of these policies and approaches. In particular, there was concern that the basic needs of the poorest farmers, especially those in the rural areas, were neither considered nor adequately addressed. Soon it became clear that many of the technologies that contributed to the green revolution, such as irrigation systems, fertilizers, and pesticides, were not affordable by the poor farmer. It also came to be recognized that most tropical soils, which are poorer and more easily degraded than temperate-zone soils, were unable to withstand the impact of high-input technology. At around the same time, the disastrous consequences of deforestation of the world's tropical regions, which increased at an alarming rate, were recognized. It was also soon recognized that a major cause of deforestation was the search for more land to provide food and fuelwood for rapidly increasing populations.

As King discusses in his review of the history of agroforestry, it was this emerging concern about the plight of the poor, and the inapplicability of the modern agricultural and forestry technologies for the majority of the environment and the poorer sections of the societies in developing countries that paved the way for new approaches to land management, which would later be called agroforestry.[1] Faced with these problems, land-use experts and institutions intensified their search for appropriate land-use approaches that would be socially acceptable, enhance the sustainability of the production base, and meet the need for production of multiple outputs. These collective efforts led to studies of age-old practices based on combinations involving trees, crops, and livestock on the same land unit. The inherent advantages of traditional land-use practices involving trees—sustained yield, environmental conservation, and multiple outputs—were recognized quickly. Agroforestry thus began to come of age in the late 1970s; the event was institutionalized with the establishment, in 1977, of the International Council (now Centre) for Research in Agroforestry (ICRAF) in Nairobi, Kenya.

B. Concepts

Basically, agroforestry involves the deliberate growing of trees and shrubs, collectively called woody perennials, on the same unit of land as

---

1. K. F. S. King, "The History of Agroforestry," in *Agroforestry: A Decade of Development*, ed. H. A. Steppler and P. K. R. Nair (Nairobi: ICRAF, 1987), pp. 1–11.

agricultural crops or animals, either in some form of spatial mixture or temporal sequence. In agroforestry systems there is significant interaction, ecological and economical, between the woody and the nonwoody components. Thus an agroforestry system normally involves two or more species of plants or plants and animals, at least one of which is a woody perennial; two or more outputs; a production cycle of more than one year; and both ecology and economics that are more complex than a monocropping system. Additionally, all agroforestry systems are characterized by three basic sets of attributes, namely, productivity (production of preferred commodities as well as productivity of the land's resources), sustainability (conservation of the production potential of the resource base), and adoptability (acceptance of the practice by the farming community). This implies that the merits of agroforestry systems are assessed not only in terms of quantities, but also on the extent to which the resource base is sustained and the practice is adopted by the local land users.

In addition to agroforestry, several other terms with "forestry" endings became prominent in the late 1970s and 1980s as a consequence of increasing global interest in tree planting activities. Notable among these are *Community Forestry, Social Forestry,* and *Farm Forestry.* Although these terms have also not been defined precisely, it is generally accepted that they emphasize the self-help aspect—people's participation—in tree planting activities, not necessarily in association with agricultural crop and/or animals as in agroforestry, but with social objectives ranking equally in importance with production objectives. Thus, social forestry is the practice of using trees and/or tree planting specifically to pursue social objectives, usually betterment of the poor, through delivery of the benefits or tree planting to the local people. Community forestry, a form of social forestry, refers to tree planting activities undertaken by a community on communal lands or the so-called common lands; it is based on the local people's direct participation in the process, either by growing trees themselves, or by processing the tree products locally. Farm forestry, a term commonly used mainly in Asia, indicates tree planting on farms.

The major distinction between agroforestry and these other terms is that while agroforestry emphasizes the interactive association between woody perennials and agricultural crops and/or animals for multiple products and services, the other terms refer to tree planting, often as woodlots. In practice, however, all these labels directly or indirectly refer to growing and using trees to provide food, fuelwood, fodder, medicines, building materials, and cash income. Only blurred lines, if any at all, separate them, and they all encompass agroforestry concepts and technologies. Therefore, in common land-use parlance as well as in the literature, these terms are often used synonymously, even out of context.

C. Developments in Agroforestry Literature

The basis for the development of agroforestry as an interface between agriculture and forestry was provided by the congruence of people, concepts, and institutional changes that occurred during the 1970s. It was only natural that the literature in agricultural, forestry, ecology, and the related land-use disciplines were drawn upon to build the foundation of this new discipline. These historical developments in agroforestry literature can be grouped into three major categories: biophysical, social, and institutional.

Biophysical Advances

One of the most important developments in agricultural research during the 1960s and 1970s that was instrumental in initiating organized efforts in agroforestry was the renewed and heightened interest in the concept of intercropping and integrated farming systems. It was demonstrated that intercropping may have several advantages over sole cropping. Research results in different parts of the world indicated that in intercropping systems more effective use was made of the natural resources of sunlight, land, and water. The research also indicated that intercropping systems might have beneficial effects on pest and disease problems; that there were advantages in growing legumes and non-legumes in mixture; and as a result, that higher yields could be obtained per unit area even when multi-cropping systems were compared to sole cropping systems.[2]

It also became obvious that although a great deal of experimentation was being carried out in the general field of intercropping, there were many gaps in the knowledge. In particular, there was a need for a more scientific approach to intercropping research, and that greater efforts were needed with respect to crop physiology, agronomy, yield stability, biological nitrogen fixation, and plant protection.[3] Concurrently, the International Institute of Tropical Agriculture (IITA) in Ibadan, Nigeria, extended its work to

2. R. I. Papendick, P. A. Sanchez, and G. B. Triplett, eds., *Multiple Cropping* (Madison, Wisc.: American Society of Agronomy, 1976 [Special Publication No. 27]).

3. P. K. R. Nair, *Intensive Multiple Cropping with Coconuts in India: Principles, Programmes and Prospects* (Berlin and Hamburg: Verlag Paul Parey, 1979).

include integration of trees and shrubs with crop production.[4] Other research organizations had initiated serious work on, for example, the integration of animals with plantation tree crops such as rubber, and the intercropping of coconuts.[5] Building upon the success of these studies, agricultural scientists began investigating the feasibility of intercropping in stands of other trees as well as studying the role of trees and shrubs in maintaining soil productivity and controlling soil erosion. Livestock management experts also began to recognize the importance of indigenous tree and shrub browse in mixed farming and pastoral production systems.[6]

Environmental concerns also became very conspicuous as these changes and developments were happening in the land-use of tropical forestry and agriculture. Deforestation of the world's tropical region attained the status of a hot topic on the agenda of almost all environment-related discussions at all levels during the 1980s, and was a major environmental issue even during the 1970s. Although there were various definitions and estimates of the rates of deforestation, there was no divergence of opinion of the seriousness of its consequences: soil-productivity decline, accelerated erosion, siltation of dams and reservoirs, destruction of wildlife habitats, and loss of plant genetic diversity.[7] It was also generally agreed that the main causes of this deforestation were population resettlement schemes, forest clearance for large-scale agriculture, forestry enterprises and animal production, and, in particular, shifting cultivation. A 1982 estimate by the Food and Agricultural Organization (FAO) of the United Nations showed that shifting cultivation was responsible for almost 70% of the deforestation in tropical Africa, and that forest fallows resulting from shifting cultivation occupied an area equivalent to 26.5% of the remaining closed forest in Africa, 16% in Latin America, and 22.7% in tropical Asia.[8] Faced with these challenges and maladies of deforestation, several studies and efforts were made to reduce the extent of deforestation and suggest alternative land-management strategies. Though the problem was not contained, several strategies were suggested by different researchers. For example, ecologists produced convincing evidence of positive influence of forests and trees on the stability of ecosystems leading to the call for measures to protect the remaining forests

4. B. T. Kang, G. F. Wilson, and L. Sipkens, "Alley Cropping Maize (*Zea mays* L.) and Leucaena (*Leucaena leucocephala* Lam.) in Southern Nigeria," *Plant and Soil* 63 (1981): 165–179.

5. (a) Nair, *Intensive Multiple Cropping*. (b) P. K. R. Nair, "Agroforestry With Coconuts and Other Plantation Crops," in *Plant Research and Agroforestry*, ed. P. A. Huxley (Nairobi: ICRAF, 1983), pp. 79–102.

6. H. N. Le Houérou, ed., *Browse in Africa* (Addis Ababa: *ILCA*, 1980).

7. World Bank, *The Forest Sector* (Washington, D.C.: World Bank, 1991).

8. Food and Agricultural Organization, *Tropical Forest Resources Project* (Rome: FAO, 1982).

to introduce more woody perennials into managed land-use systems, and to change farming attitudes. Studies by anthropologists and social scientists on farmer attitudes to improved land-use systems showed the importance of mixed systems in traditional cultures and highlighted the need to build upon these practices when developing new approaches. These efforts, although not adequately coordinated, provided important knowledge about the advantages of combined production systems involving crops, trees, and animals. The biophysical and environmental literature provided the scientific basis for the conceptualization of agroforestry and the early planning for agroforestry research.

Social Change

One of the basic premises of agroforestry is that it entails the inseparable mixture of biophysical principles and social objectives. Because of this special emphasis on social objectives, agroforestry literature had rather heavy underpinnings of social science applications from the very beginning. Following the experience of the green revolution of the 1960s and 1970s, there was a general recognition that social factors play a serious role in all biological and science applications in developing countries. Concerns were raised about the inappropriateness of the green-revolution-technology-development pathway for scientific development of agroforestry: it is not "a matter of discovering some new technologies—tissue culture, super trees, fertilizers, ... For some scientists, the green revolution may seem to be an appealing model for agroforestry because it is familiar; it seems modern; and it involves the same set of biophysical scientists doing the same kind of research."[9] It was repeatedly emphasized that agroforestry technologies were expected to be especially suitable and adoptable by small-scale land-users with low capital and low energy requirements, and to yield products and benefits directed to immediate human needs rather than commercial advantages. Besides, social issues such as land tenure (and tree tenure) were considered to be of special importance in agroforestry.[10]

A significant development was appropriate research methodologies with special emphasis on social sciences, especially the Diagnosis and Design (D

9. R. B. Burch, and J. K. Parker, eds., *Social Science Application in Asian Agroforestry* (Arlington, Va.: Winrock International, 1991), p. 63.

10. (a) J. B. Raintree, *D & D User's Manual: An Introduction to Agroforestry Diagnosis and Design* (Nairobi: ICRAF, 1987). (b) L. Fortmann, "Tree Tenure Factors in Agroforestry with Particular Reference to Africa," in *Whose Trees—Proprietary Dimensions of Forestry*, ed. L. Fortmann and J. W. Bruce (Boulder, Colo.: Westview Press, 1988), pp. 16–33. (c) P. A. Francis, "Land Tenure Systems and the Adoption of Alley Farming," in *Alley Farming in the Humid and Subhumid Tropics*, ed. B. T. Kang and L. Reynolds (Ottawa: IDRC, 1989), pp. 182–195.

& D). Modeled along the farming system methodology, developed around the mid 1970s with considerable social science input, a notable feature of the D & D is the emphasis on farmers' participation in technology development. This is quite different from the conventional models of agricultural technology development and adoption, where the roles of researchers, extensionists, and farmers have been rigidly defined: new technologies were developed by researchers, "taught" or demonstrated by extensionists, and adopted by farmers. There was thus a one-way "transfer of technology" from researchers to farmers.[11] The 1970s and 1980s witnessed a shift in this strategy based on the realization that this model as well as the technologies developed according to this model were inappropriate, especially with respect to small-scale farmers. Strong criticisms asserted that technology development was not the exclusive domain of the research scientist and that farmers and extension workers had an important role to play. It was argued that rural people may even possess an inherent advantage over research institutions when dealing with trials of complex, location-specific, land-use systems.[12] On-farm research (OFR) was a response to the realization of the importance of involving farmers in the technology-generation process (simply stated, the essence of OFR is to conduct research and test technologies on farms or in farmers' fields). Thus, early literature in agroforestry contains many reports on social science aspects and related methodologies and procedures.

Policy and Institutional Changes

The significant policy and institutional reforms that occurred in the land-use sector during the 1970s had a major role in the development of agroforestry as a distinct activity. However, this is one area where the developments in the subject are not adequately reflected in the conventional literature, especially during the early phase. The main reason is that such policy and institutional changes do not result in journal papers or book chapters to the extent that scientific studies do.

Against the backdrop of concern for the rural poor, the World Bank actively considered the possibility of supporting nationally oriented forestry programs. As a result, it formulated a Forestry Sector Policy paper in 1978, which has been used as the basis for much of its lending in the forestry

11. R. Chambers, A. Pacey, and L. A. Thrupp, eds., *Farmer First: Farmer Innovation and Agricultural Development* (London: Intermediate Technology Publications, 1989).

12. R. Chambers, "Farmers-First: A Practical Paradigm for the Third-World Agriculture," in *Agroecology and Small Farm Development*, ed. M. Altieri, and S. Hecht (Boca Raton, Fla.: CRC Press, 1989).

subsector in the 1980s.[13] Indeed, its social forestry program, which has been expanded considerably since the 1980s, not only contains many elements of agroforestry but is reportedly designed to assist the peasant and the ordinary farmer by increasing food production and conserving the environment as much as it helps the traditional forest services to produce and process wood.[14]

It was around the same time (early 1970s) that FAO made a serious assessment of the forestry projects which it was helping to implement in developing countries, as well as the policies which it had advised the Third World to follow. The assessment showed that there was notable success, but there were also areas of failure. FAO redirected its focus and assistance in the direction of the rural poor. Its new policies, while not abandoning the traditional areas of forestry development, emphasized the importance of forestry for rural development.[15] It also focused on the benefits that could accrue to both the farmer and the nation if greater attention were paid to the beneficial effects of trees and forests on food and agricultural production, and advised land managers in the tropics to incorporate both agriculture and forestry into their farming system, and "eschew the false dichotomy between agriculture and forestry."[16]

To these two strands of forest policy reforms, which evolved independently, one in an international funding agency and the other in the FAO, were added the simultaneous efforts of a large number of tropical land-use experts and institutions. Perhaps the most significant single initiative that contributed to the development of agroforestry came later to be known as the Bene Commission report that embodied the results and recommendations of a study commissioned by the International Development Research Centre (IDRC) of Canada.[17] The recommendations of this report led to the establishment of ICRAF in 1977 and the institutionalization of agroforestry.

D. Early Literature in Agroforestry

In the development of a new discipline, it is difficult to distinguish between literature that specifically pertains to that discipline and other disci-

13. The World Bank's Forestry Policy, which was further revised in 1991, gives even more emphasis to agroforestry and "trees outside the forest."

14. J. Spears, "Agroforestry: A Development-Bank Perspective," in *Agroforestry: A Decade of Development*, ed. H. A. Steppler and P. K. R. Nair (Nairobi, Kenya: ICRAF, 1987), pp. 53–66.

15. FAO, *Forests for Research and Development* (Rome: Food and Agriculture Organization, 1976).

16. K. F. S. King, "Agroforestry and the Utilization of Fragile Ecosystems," *Forest Ecology and Management* 2 (1979): 162.

17. J. G. Bene, H. W. Beall, and A. Côte, *Trees, Food, and People* (Ottawa: IDRC, 1977).

plinary literature that is drawn upon to form the basis of the new discipline. Therefore, there is some unavoidable ambiguity. As we have seen, agroforestry as a term and concept was first discussed in scientific literature in the mid-1970s; but many publications prior to that also are very relevant to agroforestry. All literature produced since the mid-1970s bearing the term agroforestry or variants term can be considered as agroforestry literature. These publications can be grouped as conferenc and workshop proceedings, journal articles, and reports.

ICRAF, CATIE (Centro Agronomico Tropical de Investigacion y Enseñanza Turrialba, Costa Rica), FAO, and a few other international agencies, and some national and regional institutions (e.g., Indian Council of Agricultural Research, ICAR) organized several agroforestry conferences and published their proceedings. Nine international seminars, workshops, and conferences organized by ICRAF and proceedings published during the first decade of its existence (1977–1987), were reviewed by Nair.[18] Each meeting focused on an agroforestry theme included soils,[19] international cooperation,[20] agroforestry in Kenya;[21] plant research;[22] agroforestry systems for small-scale farmers;[23] professional education;[24] multipurpose tree germplasm;[25] tenure issues;[26] and application of meteorology to agroforestry systems planning.[27] ICRAF also published a commemorative state-of-the-art volume on a decade of developments in agroforestry in 1987.[28] Other conference proceedings prior to 1987 include the proceedings of seminars in India,[29] Costa Rica,[30] and Nigeria.[31] Additionally, there are several other

18. P. K. R. Nair, "International Seminars, Workshops, and Conferences Organized by ICRAF," *Agroforestry Systems* 5 (1987): 375–381.

19. H. O. Mongi and P. A. Huxley, eds., *Soils Research in Agroforestry: Proceedings of an Expert Consultation* (Nairobi: ICRAF, 1979).

20. T. Chandler and D. Spurgeon, eds., "International Cooperation in Agroforestry," *Proceedings of ICRAF/DSE Conference* (Nairobi: ICRAF, 1979).

21. L. Buck, ed., *Proceedings of the Kenya National Seminar in Agroforestry November, 1980* (Nairobi: ICRAF; University of Nairobi, 1981).

22. P. A. Huxley, ed., *Plant Research and Agroforestry* (Nairobi: ICRAF, 1983).

23. D. A. Hoekstra and F. M. Kuguru, eds., *Agroforestry Systems for Small-Scale Farmers* (Nairobi: ICRAF, 1982).

24. E. Zulberti, ed., *Professional Education in Agroforestry* (Nairobi: ICRAF, 1987).

25. J. Burley and P. G. Carlowitz, eds., *Multipurpose Tree Germplasm: Proceedings of a Planning Workshop to Discuss International Cooperation* (Nairobi: ICRAF, 1984).

26. Raintree, *D & D User's Manual*.

27. W. S. Reifsnyder and T. O. Darnhofer, eds., *Meteorology and Agroforestry* (Nairobi: ICRAF, 1989).

28. H. A. Steppler and P. K. R. Nair, eds., *Agroforestry: A Decade of Development* (Nairobi: ICRAF, 1987).

29. Indian Council of Agricultural Research, *Proceedings of the National Seminar on Agroforestry, May, 1979* (New Delhi: ICAR, 1979).

30. G. de las Salas, ed., *Proceedings of the Workshop on Agroforestry Systems in Latin America* (Turrialba, Costa Rica: CATIE, 1979).

31. L. H. McDonald, ed., *Agroforestry in the African Humid Tropics* (Tokyo: United Nations University, 1982).

published conference proceedings containing papers on agroforestry issues in a number of places around the world. Such conference proceedings constitute a major part of the early literature in agroforestry.

The other major early literature is journal articles. Because of the newness of the agroforestry concept and the nonexistence of a specialized journal for agroforestry, the number of articles in early agroforestry is relatively low. The initiation of the journal *Agroforestry Systems* in 1982 and the increase in scientific activities caused an upsurge in the number of journal articles toward the later part of 1980s. Experimental articles were practically nonexistent in the early phase of agroforestry literature because of the rather long time needed for planning and conducting rigorous agroforestry experiments. Results of intercropping and other forerunners published in agricultural and other thematic research journals, though very relevant to agroforestry, are not included in this discussion. The concept of agroforestry was intellectually attractive to many scientists of land-use disciplines resulting in a bandwagon effect that was echoed in the "concept-and-potential-of-agroforestry" articles that dominated the early journal articles, especially in *Agroforestry Systems*.

A third category of early agroforestry literature consists of special reports and brief publications especially on agroforestry systems and components. Notable among these topics are firewood,[32] agricultural crops,[33] underexploited and other indigenous trees,[34] soils and soil productivity under agroforestry,[35] and agroforestry systems.[36]

In addition there is a considerable body of literature dealing with research on agroforestry prior to the mid-1970s. For example, shifting cultivation and taungya are two examples of tropical land management practices that have a very strong relevance to agroforestry, and this literature can be found in the libraries of research institutions of former colonial powers. However, much of this information is not readily available.[37] Information relevant to agroforestry also remains in the grey literature that is not easily accessible nor noted in computerized information systems.

32. National Academy of Sciences, *Firewood Crops: Shrub and Tree Species for Energy Production*, Vol. 1 (Washington, D.C.: National Academy of Sciences, 1980); Vol. 2 (Washington, D.C.: National Academy of Sciences, 1983).

33. P. K. R. Nair, *Soil Productivity Aspects of Agroforestry* (Nairobi: ICRAF, 1980).

34. (a) Le Houérou, *Browse in Africa*. (b) H. J. von Maydell, *Trees and Shrubs of the Sahel: Their Characteristics and Uses* (Eschborn, Germany: GTZ, 1986).

35. Nair, *Soil Productivity*.

36. Montagnini, *Systemas Agroforestales* (San José, Costa Rica: Organization for Tropical Studies, [OTS]/CATIE, 1986).

37. R. Labelle, "Ten Years of Work in Agroforestry Information and Documentation," *Agroforestry Systems* 5 (1987): 339–352.

E. Trends in Agroforestry Journal Articles

The trends and directions in agroforestry development and research have been reflected somewhat consistently in the pages of *Agroforestry Systems* since it started publication in late 1982.[38] A. Young undertook a critical analysis of the nature of publications in the first ten volumes covering eight years (1982–1990).[39] Reviewing the 213 articles published during the period, he found that the main trends have been a decline in the number of methodological discussions and system descriptions, and an increase in experimental results. I extended this analysis to cover eight more volumes (11–18) that appeared through June 1992. Table 4.1 shows the total number of articles in the journal, grouped according to the major areas of subject matter. As the table shows, three subject areas have dominated: methodological, system and component descriptions, and experimental.

This analysis shows the following trends in journal articles in *Agroforestry Systems*:

1. The total number of articles per year has increased considerably;
2. Most publications deal with one of the three subject areas: methodological, system and component description, and experimental/quantitative;
3. The numbers of methodological papers used to be relatively high in the early 1980s but have gradually declined since the late 1980s;

Table 4.1. Main topics of articles in *Agroforestry Systems* during the first ten years of its publication (late 1982–June 1992; Vols. 1–18)

Subject matter	Number of articles
Methodology	70
System and component description	133
Experimental and quantitative	137
Economic	24
On-farm	24
Education[a]	14
Total	378[b] (402)

[a]Proceedings of an international workshop on the subject, published as a special issue of the journal.

[b]The total number of articles published in the journal during the period is 378; additionally there were 13 review articles in a special issue. Some articles have been considered under more than one category; therefore, the total of the table exceeds 378.

38. P. K. R. Nair, "Journal Articles in Agroforestry: Trends and Directions in the 1990s," *Agroforestry Systems*, editorial 13 (3) (1991): iii–v.

39. A. Young, "Change and Constancy: An Analysis of Publications in *Agroforestry Systems*, Volumes 1–10," *Agroforestry Systems* 13 (1991): 195–202.

4. The system and component descriptions that used to dominate the scene in the early 1980s have declined in number, but still continue to be a major type of study;
5. Experimental and quantitative publications have showed a steady increase since the late 1980s;
6. Publications dealing with economic and on-farm research are on the increase, though their numbers are still relatively lower than those of experimental papers.

This analysis has some limitations because it is based on just one journal, whereas agroforestry papers are also published in a number of other journals and publications. But *Agroforestry Systems* vividly reflects the general trends in agroforestry journal articles at large, and therefore the analysis can be taken as indicative of the general trend. It is clear that agroforestry research, and therefore journal articles, that started off as predominantly descriptive are becoming quantitative and experimental.

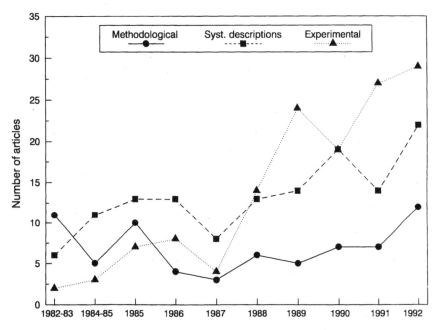

Figure 4.1. Trends in the nature of papers published in the first 20 volumes of *Agroforestry Systems* (1982–1992).

F. Nature of Agroforestry Publications Today

Publications in any subject can be classified according to several criteria: subject focus, year and country of publication, form and periodicity. These apply to agroforestry publications as well. In the previous sections, agroforestry publications were grouped as books, journals, and special reports. For reasons of simplicity and consistency, the same grouping can be retained in the remaining discussions.

Several books on agroforestry have been published during the past few years (1988–1992). These include conference proceedings,[40] system descriptions,[41] multi-author compendiums;[42] and several other specialized publications.[43] Several books have also been produced on specific components of agroforestry systems, especially the multipurpose trees/genera such as *Sesbania* spp.,[44] *Gliricidia sepium*,[45] and *Erythrina* spp. (an international conference which was held at CATIE in Costa Rica, in October 1992; proceedings in press).

In general, conference proceedings form the major category of books and reports. As opposed to the agroforestry conferences of the 1980s which covered many aspects of the subject in the same conference, conferences of the 1990s are becoming more specialized, focusing on specific topics (e.g., individual components, methodologies, and evaluations). A notable exception to this trend is the North American Agroforestry Conferences; two such

40. (a) W. W. Budd, I. Duchhart, L. H. Hardesty, and F. Steiner, eds., *Planning for Agroforestry* (Amsterdam: Elsevier, 1990). (b) K. Landauer and M. Brazil, eds., *Tropical Home Gardens* (Tokyo: United Nations, University Press, 1990). (c) P. G. Jarvis, ed., *Agroforestry: Principles and Practice* (Amsterdam: Elsevier, 1991). (d) G. M. Sullivan, S. M. Huke, and J. M. Fox, eds., *Financial and Economic Analyses of Agroforestry Systems: Proceedings of a Workshop held in Honolulu, Hawaii, USA., July 1991* (Paia, Hawaii: Nitrogen Fixing Tree Association, 1992).

41. (a) Montagnini, *Systemas Agroforestales*. (b) P. K. R. Nair, ed., *Agroforestry Systems in the Tropics* (Dordrecht, Netherlands: Kluwer Academic Publishers, 1989). (c) Z. Zhaohua, C. Mantang, W. Shiji, and J. Youxu, eds., *Agroforestry Systems in China* (Ottawa: Chinese Academy of Forestry, Beijing, China, and International Development Research Centre, 1991).

42. (a) K. G. MacDicken, G. V. Wolf, and C. B. Briscoe, *Standard Research Methods for Multipurpose Trees and Shrubs* (Arlington, Va.: Winrock International, 1991). (b) M. E. Avery, M. G. R. Cannell, and C. K. Ong, eds., *Biophysical Research for Asian Agroforestry* (New Delhi: Oxford University Press, 1990).

43. (a) M. Baumer, *Agroforesterie et Desertification* (Wageningen, Netherlands: Centre Technique de Cooperation Agricole et Rurale, 1987). (b) D. E. Rocheleau, F. Weber, and A. Field-Juma, *Agroforestry in Dryland Africa* (Nairobi: ICRAF, 1988). (c) A. Young, *Agroforestry for Soil Conservation* (Wallingford, U.K.: CAB International, 1989). (d) Burch, *Social Science Application*.

44. (a) D. O. Evans and P. P. Rotar, *Sesbania in Agriculture* (Boulder, Colo.: Westview Press, 1987). (b) B. Macklin and D. O. Evans, eds., *Perennial Sesbania Species in Agroforestry Systems* (Waimanalo, Hawaii: Nitrogen Fixing Tree Association, 1990).

45. D. Withington, N. Glover, and J. L. Brewbaker, eds., **Gliricidia sepium *(Jacq.) Walp: Management and Improvement** (Waimanalo, Hawaii: Nitrogen Fixing Tree Association, 1987).

conferences have been held at two-year intervals in 1989 and 1991, and their proceedings published; the third of the series was in August 1993. These conferences contain papers on a large number of agroforestry topics.[46]

Agroforestry Systems continues to be the single thematic journal for agroforestry. Its frequency of publication has increased to one issue every month, forming three volumes per year. Agroforestry research articles are also appearing in increasing numbers and frequency in other research journals such as *Forest Ecology and Management, Plant and Soil*, and *Experimental Agriculture*. A significant feature of these research articles is the clear shift in emphasis from descriptive and conceptual papers to primary research papers on biophysical aspects, and in-depth investigations on social, economic, and policy aspects.

Special reports are another notable type of publication in agroforestry. The various research methodologies such as the *D & D Manual*,[47] MPT measurement methodology,[48] reports on economic analysis of specific projects,[49] and the periodic reports of the Nitrogen Fixing Tree Association (NFTA) belong to this category. A significant publication in the area of agroforestry policy is the proceedings of a 1991 expert consultation.[50] The special publications of the National Academy of Sciences on several species/genera of multipurpose trees (e.g., *Leucaena, Casuarina, Acacia, Calliandra*, and so on) are examples of another set of such special reports.

Databases of various types now regularly index material on agroforestry. The most conspicuous of this category is the comprehensive multipurpose tree database compiled at ICRAF.[51]

Various magazines, newsletters, and technical reports on different aspects of agroforestry are also published. Notable among these are ICRAF's *Agroforestry Today* (4 issues a year), NFTA's *Leucaena Research Reports*, and other occasional publications. Newsletters of the Alley Farming Network for Africa (IITA, Nigeria), the International Society of Tropical For-

46. (a) P. A. Williams, ed., *Agroforestry in North America. Proceedings of the First Conference on Agroforestry in North America, August 13–16, 1989* (Ontario: University of Guelph, 1991). (b) H. E. Garrett, ed., *Proceedings of the Second Conference on Agroforestry in North America* (Columbia, Mo.: University of Missouri, 1991).

47. Raintree, *D & D User's Manual*.

48. K. G. MacDicken, G. V. Wolf, and C. B. Briscoe, *Standard Research Methods for Multipurpose Trees and Shrubs* (Arlington, Va.: Winrock International, 1991).

49. W. W. Dunn, A. M. Lynch, and P. Morgan, "Benefit-Cost Analysis of Fuelwood Management Using Native Alder in Ecuador," *Agroforestry Systems* 11 (1990): 125–139.

50. H. M. Gregersen, P. Oram, and J. Spears, eds., *Priorities for Forestry and Agroforestry Policy Research: Report of an International Workshop* (Washington, D.C.: International Food Policy Research Institute, 1992).

51. P. G. von Carlowitz, G. V. Wolf, and R. E. M. Kemperman, *Multipurpose Tree and Shrub Database: An Information and Decision Support System* (Nairobi: ICRAF, and Eschborn, Germany: GTZ, 1986).

esters (ISTF, Washington, D.C.), and *Farm Forestry News* (Winrock International, Arlington, Va.) now commonly cover agroforestry topics. A large number of magazines and newsletters on a wide range of other ecological and land-use subjects also contain regular news and information items on agroforestry.

It would be interesting to assess the geographical orientation of these various agroforestry publications and their impact in different geographical regions. Obviously such a task entails considerable resources; specialized libraries such as those of ICRAF, and abstract services (e.g., *CAB Abstracts*) may be encouraged to undertake such tasks periodically. An earlier attempt by ICRAF in this direction used the information on 6400 institutional and individual subscriptions to ICRAF's *Newsletter* which showed that about 30% of the addresses were from Africa, 20% from Asia, and 22% from Europe; Latin America and N. America accounted for about 8% each.[52] As regards the disciplinary backgrounds, 27% listed forestry, and 23% agronomy as their backgrounds; ecology, soil science, extension, farming systems, rural development, and information sciences were identified as background subjects by 5–6% each. In the absence of any updated information, it can only be conjectured that the present situation is similar.

The current status of the nature of agroforestry publications can be gauged by an analysis of a cross section of such publications. For the preparation of a textbook entitled *An Introduction to Agroforestry*, the author reviewed most of the available publications on the subject up to around August 1992.[53] The book covers, in twenty-five chapters, most if not all, aspects of agroforestry and lists 788 thematic references (excluding references to dictionaries and such other general publications). Figure 4.2 depicts the trend in these literature citations grouped according to their year and type of publication. This information along with previous tables gives a vivid picture of the current status of agroforestry publications. References to books and book chapters consistently outweigh journal articles as the major agroforestry literature source, and the numbers of publications has more than doubled during the past few years.

Education and Training Materials

Recognition of the need for organized research in agroforestry is a recent development; even more recent is the recognition of serious deficiencies in agroforestry education and training. Concern has been raised about a num-

---

52. Labelle, "The History of Agroforestry."
53. P. K. R. Nair, *An Introduction to Agroforestry* (Dordrecht, Netherlands: Kluwer, 1993).

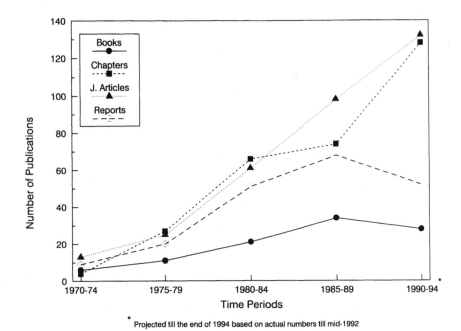

Figure 4.2. Nature and publication dates of 802 literature sources cited by the author in his *An Introduction to Agroforestry* (Dordrecht, Netherlands: Kluwer, 1993).

ber of related issues. For example, it is argued that existing educational programs are so specialized that they do not give serious attention to the study of indigenous land-use practice. There is a corresponding need for people to be trained at various levels to conduct research, implement development projects, and undertake extension work. Moreover, there is no uniform approach to educational program development in agroforestry. These issues have been addressed in two international workshops held at ICRAF (1982) and the University of Florida (1988), and their proceedings have been published.[54] A third international workshop on this subject was held at the Universidad Autonoma Chapingo, Mexico, in May 1993.

The current state of agroforestry education was aptly portrayed by J. P. Lassoie.[55] As a field of study, agroforestry is directly relevant to a wide variety of individuals and offers the opportunity to bring together broad

54. (a) Zulberti, *Professional Education in Agroforestry.* (b) P. K. R. Nair, *The Prospects for Agroforestry in the Tropics* (Washington, D.C.: World Bank, 1990 [*Technical Paper* No. 131]).

55. J. P. Lassoie, "Towards a Comprehensive Education and Training Program in Agroforestry," *Agroforestry Systems* 12 (1990): 121–131.

fields of forestry and agriculture to offer the scientific underpinnings for the development of a new, comprehensive, and integrative land-use strategy. There are a number of constraints that must be considered as this new field develops further. However, agroforestry education programs will only be successful if an integrative and comprehensive approach is adopted and maintained. Also, there can be no single education or training model that is universally applicable, because approaches to education and training in agroforestry must be country-specific, depending on ecological, socioeconomic, and cultural needs.[56] The current status of agroforestry training and education world-wide was reviewed by E. Zulberti. She noted that agroforestry had been incorporated in education and training programs at an unprecedented level since 1982; full undergraduate and postgraduate programs in agroforestry were being formulated and implementation started in several United States universities, and many students were choosing agroforestry-related research projects for their dissertations. Emerging trends indicate that traditional forestry programs are broadening the scope of the discipline (from forests to integrated land-use systems) while agriculturists are recognizing that trees play important roles as soil improvers and protectors, fodder, food, and fuel, and serve other domestic and commercial purposes. New institutional structures are evolving to allow for educational programs with coursework and research projects spanning many disciplines.[57]

Surveys conducted by ICRAF and by the U.S. Department of Agriculture show that a large number of institutions around the world offer specialized training and education programs in agroforestry.[58] At these institutions, agroforestry is taught as both a separate subject and as an integral part of other curricula at the undergraduate and postgraduate levels including several short-term training courses.

In spite of this increasing interest in education and training and the swelling literature in agroforestry, there is no textbook in the subject. The situation at present is such that for each training program or educational course, reading materials are assembled, usually by photocopying the relevant materials from various sources. Both the students and the instructors of agroforestry are often disappointed that they have to contend with such tempo-

56. Ibid.

57. (a) Training: Non-degree programs of relatively short durations; Education: Degree programs in universities or other academic institutions. (b) E. Zulberti, ed., *Professional Education in Agroforestry*.

58. (a) Ibid. (b) U.S. Dept. of Agriculture, *Training and Educational Opportunities in Agroforestry: A Directory of Institutions in the United States and Overseas* (Washington, D.C.: Forestry Support Program, U.S. Department of Agriculture, 1991).

rary arrangements. A recent book, *An Introduction to Agroforestry*, is expected to make up for this deficiency to some extent.[59]

Temperate-Zone Agroforestry: The Status of Its Literature

The developments in agroforestry have so far been more pronounced in the tropics and developing countries than the developed countries and the temperate zone. The main reason for this special focus is that agroforestry as an approach to land-use has more relevance and application in the tropics than in the temperate zone: traditional agroforestry systems are far more numerous and widespread in the tropics, and many land-use problems and constraints in the tropics are readily addressed by agroforestry. As in the tropics, however, there is a long tradition of meeting people's needs in the temperate zone through both purposeful combinations of trees, animals, and crops, and opportunistic use of natural ecosystems. Although not quite comparable to the extent of activities and developments in the tropical agroforestry scene, significant expansion in the scope of agroforestry is occurring in the temperate zone, with similar hope that the meshing of agriculture and forestry will bring new solutions to many land-use problems.

Agroforestry land use occurs throughout the range of temperate zone conditions, but unlike the great variety of systems and practices in the tropics, only a few agroforestry systems are practiced in these regions. The two most common systems have been agrisilvicultural use of windbreaks and shelterbelts to prevent soil erosion in the plains, and silvopastoral practices with livestock grazing in many different woodland and range ecosystems. Agrisilvicultural combinations of nut or fruit trees and other crops are an increasingly common third system. Socioeconomic conditions in the developed countries of the temperate zone have also strongly influenced land-use practices. Although small farms were dominant historically in the temperate zone, and still are in many regions, there have been significant trends in the twentieth century toward large, family, corporate or communal farms where production is largely in a few crops for local and distant markets. Agroforestry applications on such farms have often focused on one or two high-value crops and include high levels of mechanization. Combinations of trees and agriculture were viewed as an opportunistic approach to improving economic profitability.

The relatively slow progress of agroforestry in the temperate zone is also reflected in the rather slow development of its literature. We have seen that, in general, conference proceedings are a major form of agroforestry literature. Because of the tropical focus of agroforestry, most conferences have

59. Nair, *An Introduction to Agroforestry*.

dealt mainly with issues related to the tropics. But it has been customary in most conferences that a session or two were devoted to discussions on temperate-zone agroforestry. Additionally, several recent comprehensive reviews are available on agroforestry developments in the temperate zone and evaluations of their prospects.[60]

Given the contrasting socioeconomic and biophysical conditions of the tropics (developing countries) and the temperate zone (developed countries), and the special attributes of agroforestry, several questions are frequently asked in the land-use discussions. Is agroforestry necessary in temperate zone countries? How are low-input integrated practices such as agroforestry relevant to the commercialized, specialized, and modernized forestry and agricultural production enterprises of the developed countries and their largely urbanized societies? The role of, and opportunities in, agroforestry in the developed countries in this scenario have been reviewed by several authors, most notably Gold and Hanover,[61] Lassoie and Buck,[62] Lassoie et al.,[63] and Long and Nair.[64] Most of these reviews conclude that although the developments in agroforestry in the temperate zone have been rather slow, the possibilities and opportunities are certainly encouraging and multiple. However, there are also some formidable constraints to agroforestry development in these regions.

While great strides are being made in agroforestry development in the developing world, agroforestry development in the temperate zone seems poised to continue at a rather low-key level until there are major institutional changes. Such changes may depend on the unfolding of new land-use problems that stimulate significant interest in non-conventional efforts similar to the interest in tree crops following the 1930s Dust Bowl in the United States.

60. (a) M. A. Gold and J. W. Hanover, "Agroforestry Systems for the Temperate Zone," *Agroforestry Systems* 5 (1987): 109–121. (b) E. K. Byington, "Agroforestry in the Temperate Zone," in *Agroforestry Classification and Management*, ed. K. G. MacDicken, and N. T. Vergara (New York: John Wiley & Sons, 1990), pp. 228–289. (c) T. H. Bandolin and R. F. Fisher, "Agroforestry Systems in North America," *Agroforestry Systems* 16 (1991): 95–118. (d) A. J. Long and P. K. R. Nair, "Agroforestry Systems Design for the Temperate Zones: Lessons from the Tropics," in *Proceedings of the Second Conference on Agroforestry in North America*, pp. 133–139.

61. Gold and Hanover, "Agroforestry Systems for the Temperate Zone."

62. J. P. Lassoie, "Agroforestry in North America: New Challenges and Opportunities for Integrated Resource Management," in *Proceedings of the Second Conference on Agroforestry in North America*, pp. 1–19.

63. J. P. Lassoie, W. S. Teel, and K. M. Davies, Jr. "Agroforestry Research and Extension Needs for Northeastern North America," *Forestry Chronicle* 67 (1991): 219–226.

64. Long, "Agroforestry Systems Design for the Temperate Zones."

G. Conclusions: What to Expect?

In the 1990s the issues surrounding the young discipline of agroforestry and the direction in which it is going are becoming clearer. As Nair noted, the initial euphoria about agroforestry has died down and the rush to define it and provide it with a conceptual framework has abated.[65] Development agencies have accepted it as an important, fundable activity. Indeed the awareness of agroforestry as a potentially useful land-use approach has grown so dramatically over the past 10 or 15 years that there are now very few land-use related development projects that do not contain a significant agroforestry component. However, some agroforestry development enthusiasts see little need for research, and even scorn the methods of doing research. This unfortunate conflict between research and development, if allowed to continue, will be detrimental in the long run to the cause of agroforestry promotion—both development and research. With the increased emphasis being placed on agroforestry research by international bodies such as the Consultative Group on International Agricultural Research (which coordinates the activities of the many International Agricultural Research Centers), and many national research organizations, there is no doubt that research investments and contributions in agroforestry will increase in the coming years. These trends are perhaps not very different from those that many other established land-use disciplines had to undergo during their early stages of development.

Research scientists continue to express concerns about the lack of scientific data to support the widely held assumptions on the advantages of agroforestry as well as the inadequate methodologies currently being used and lack of trained personnel for agroforestry research. However, if the recent trends of journal articles in agroforestry are any indication, these concerns are being mitigated. Increasing numbers of scientists of various background are getting involved in agroforestry research, and agroforestry research is increasingly becoming experimental.

Multidisciplinary input is the key to the success of agroforestry. Scientific efforts in agroforestry have so far been dominated by topics such as management of multipurpose trees, and soil- and nutrient-related investigations especially under alley cropping and plantation-crop combinations. The main scientific foundation of agroforestry is multipurpose trees, and the success of agroforestry will depend upon the extent to which the productive, protective and service potentials of the multipurpose trees are understood, exploited, and realized. In order to accomplish that, we need the

65. Nair, *The Prospects for Agroforestry in the Tropics* (Washington, D.C.: World Bank, 1990).

collective and coordinated wisdom of multidisciplinary experts; scientists with different disciplinary backgrounds are being exposed to these challenges, and encouraged to publish their thoughts and results. Agroforestry research papers of the future will reflect this infusion of scientific ideas of a large number of scientists with varying backgrounds.

Sustainability is a key word in land-use parlance today. This is not at all a new term in agroforestry: sustainability is a cornerstone of the concept. The main problem in explaining the sustainability potential of agroforestry is the lack of a definition of sustainability and precise quantitative criteria to express it. With the current level of interest in sustainability among different groups of experts such as ecologists, economists, and biologists, and even politicians, we hear different explanations and criteria of sustainability. We hope to be given acceptable sustainability criteria, biological, economical, and ecological, that will weigh heavily in favor of agroforestry. Future research and therefore research publications in agroforestry will deal heavily with these sustainability parameters.

Biotechnology and its applications will also be applied to agroforestry. Ranging from biological nitrogen fixation to low-cost plant protection measures, from propagation of rare germ plasm to breeding of desirable plant ideotypes, the potential applications of biotechnology in agroforestry research are unlimited. Equally promising is the trend to use computers as an essential tool in research not only as an aid to store databases or analyze data, but also for development of predictive models and Expert Systems. Agroforestry research publications of the near future will thus be of a different genre from that of the 1980s.

The prognosis of developments in agroforestry education seems less promising than that of research. Developing separate programs in agroforestry at undergraduate and postgraduate levels will involve restructuring existing curricula or developing new ones; special courses need to emphasize interdisciplinary studies encompassing a wide range of subjects. The rigid institutional and disciplinary structures of most universities are not conducive to such new developments. The possibilities are rather remote for the establishment of new institutions for agroforestry similar to a large number of new agricultural universities that were established in some developing countries during the 1960s and 1970s along the model of the U.S. land-grant universities. It may be premature to talk about separate institutions for agroforestry education. In all probability, agroforestry education will remain within the existing educational framework of agriculture, forestry, and natural resources, and it is unlikely that there will be a proliferation of educational materials (textbooks) in agroforestry. However, with the current level of interest in short-term training in agroforestry, it is likely that more training materials of better quality will be produced in the future.

Agroforestry research seems to have gathered considerable momentum during the past few years. Research publications of various types (journal articles, book chapters, reports, etc.) will likely expand within the next few years. Agroforestry literature will be spread in a wider range of journals and other scientific publications than at present. Finally, the proceedings of agroforestry conferences of the future will become more thematic and focused than in the past.

5. Professional Societies and the Scientific Literature of Forestry

LARRY W. TOMBAUGH

North Carolina State University

This chapter deals with the role that professional forestry societies, particularly those in North America, have played in developing the scientific literature of forestry.

Forestry is an applied discipline as well as a process for managing forest ecosystems with roots in a large number of other applied disciplines as well as in the basic sciences. Most practitioners of the applied side of forestry are not scientists, and there are many research scientists who constantly expand the scientific base of forestry. Both call themselves foresters, although forest scientists may also be referred to by the scientific discipline of their specialty.

Forestry is significantly influenced by societal values and shifts in public opinion, dramatically evident in North America over the past 100 years. The dual nature of forestry, profession and applied science, combined with intense public interest in forest resources has profoundly shaped the evolution of forest science, forestry societies, and the literature they have produced.

A large number of forestry societies has emerged over the past 150 years. Some are membership organizations for people generally interested in trees and forests while others meet the need of people and groups with a commercial interest. Some organizations were formed to promote forestry from the view of professionals, developing codes of ethics, publishing professional journals and books, and forming societies.

The International Union of Societies of Foresters (IUSF) is composed of twenty-six national forestry associations. These national associations are the prime organizations and society publishers in the world. The IUSF members and their serial publications are listed in Table 5.2 at the end of this chapter. There are many more forestry societies operating throughout the world and many more forestry society publications. Their diverse nature and location make it difficult to capture them all.

This chapter concentrates on professional societies and their publications which have contributed to the science base of forestry. No attempt is made to be complete, or to identify all forestry societies or the major scientific publications in the field. Rather, the chapter provides an overview of the relationship between professional forestry societies and the ways their relationship has evolved with the scientific literature of forestry.

A. Early European Influences

Forestry literature and forestry societies originated in Europe. The first journal edited by a forester was Reitter's *Journal für Forst- und Jagdwesen*, published in Germany in 1790, which ran until 1797.[1] A variety of other forestry publications appeared in the early 1800s, and most of these were also short-lived. Interestingly, another journal entitled *Zeitschrift für Forst- und Jagdwesen* was published by Julius Springer from 1869 until 1943, and it played a significant role in the development of German forestry literature.

The Societät der Forst- und Jagdkunde may have been the first professional forestry society. It began a journal, *Diana*, in 1797 to record essays written by its members. *Diana* ceased with the third volume in 1805. Over the next few decades, several local professional forestry associations sprung up throughout Germany. The first forestry congress was held in Dresden in 1837, involving representatives from the local associations. All this activity lead to the Deutscher Forstverein (German Forestry Association) created in 1899.

Perhaps the oldest continuously published forestry journal is *Allgemeine Forst- und Jagdzeitung* founded in 1825 which continues to represent a significant contribution to the world's forestry literature. *Forstwissenschaftliches Centralblatt* was initiated in 1828 and it also continues to be one of the leading journals in the forestry literature. Another early association publishing in German was the Schweizerschen Forstvereins (Swiss Forestry Association) founded in 1843. Its journal was first published in 1850, *Schweizerische Zeitschrift für Forstwesen*.

Forestry and forestry societies were formalized later in the United Kingdom than on the European continent. The Scottish Arboricultural Society established in 1854 later became the Royal Scottish Forestry Society. Twenty-six years later, the English Arboricultural Society was founded, and it became the Royal Forestry Society of England, Wales, and Northern Ire-

---

1. B. E. Fernow, *A Brief History of Forestry in Europe, the United States, and Other Countries*, 2d ed. (Toronto: University of Toronto Press and *Forestry Quarterly*, 1911), p. 149.

land.[2] The latter organization began the *Quarterly Journal of Forestry* in 1907 which is still published. For details on early forestry activities in the United Kingdom, see Chapter 2.

Forestry and forestry societies were also highly developed by the late 1800s in India; *Indian Forester* began publication in 1875. Indian forestry was heavily influenced by German foresters and it, in turn, influenced forestry in the United Kingdom during the period of British colonial rule.

A particularly influential and enduring organization with strong European roots is the International Union of Forestry Research Organizations (IUFRO) which was designed to create a community of scholars among the foresters of seven European nations when created in 1892. Today, it is the leading international society for forest scientists, and maintains a headquarters staff in Vienna, Austria, to coordinate the activities of the 694 members from 106 countries. IUFRO provides an independent and unbiased forum for scientists to meet, exchange information, and cooperate on research. As a nongovernmental, nonpolitical body, the Union provides an opportunity for open debate on an enormous array of issues important to forestry. IUFRO is organized around six subject-matter divisions. Results of division meetings and their subdivisions are regularly published, as are the proceedings of each of its World Congresses held every four years. *IUFRO News* is issued periodically to report on recent and forthcoming IUFRO and other events and publications. Particularly useful is the calendar of meetings that is usually included. It is not uncommon for this calendar to list a hundred or more meetings to be sponsored by the various IUFRO research groups and scheduled for two or three years in advance at sites throughout the world. Some of these meetings and symposia result in publication.

B. Professional Societies in North America and Their Literature

Forestry began to assume importance as a profession in the United States during the first decade of the twentieth century. In 1900, there were only 123 people employed by the Division of Forestry in the U.S. Department of Agriculture, and sixty-one were student assistants from several prominent universities.[3]

Several events occurred within a few years that dramatically changed the position of forestry in the United States. Many of these were stimulated by the tireless and effective efforts of the first U.S. citizen formally educated

2. N. D. G. James, *A History of English Forestry* (Oxford: Basil Blackwell, 1981).
3. G. Pinchot, *Breaking New Ground* (New York: Harcourt, Brace, 1947).

in forestry, Gifford Pinchot. The Yale Forest School was established in 1900 through an endowment established by the Pinchot family. It played a significant role in providing technically trained personnel for the emerging forestry agencies. S. T. Dana and S. K. Fairfax indicate that professional forestry was born in the United States on November 30, 1900, the date Gifford Pinchot met with six other prominent foresters in his office to organize the Society of American Foresters (SAF).[4] An organization was needed to create a sense of professionalism and purpose among those entering the embryonic field of forestry. The purpose of the society was "to further the cause of Forestry in America by fostering a spirit of comradeship among foresters; by creating opportunities for a free interchange of views upon Forestry and allied subjects; and by disseminating a knowledge of the purpose and achievements of Forestry."[5] The demand for professionally-trained foresters and the need for a professional society increased substantially when the U.S. Forest Service was established in the Department of Agriculture in 1905, with Gifford Pinchot as its first chief.

It is difficult to overstate the importance of the interactions between the Society of American Foresters and the U.S. Forest Service in shaping the beginnings of professional forestry in the United States. The members of the society, according to Pinchot, became "the vital core of the Forest Service—vital in loyalty to all that the Service stood for, and with the highest morale to be found anywhere under the Government of the United States."[6] The two grew up together, and the bond continues to be strong today.

The fledgling profession clearly needed to develop a scientific and professional literature to guide forest practices under conditions that were unique to the United States. Most of the forestry literature referred to conditions and practices either in Europe or in European colonies throughout the world. In 1898, W.P. Cutter, Librarian of the U.S. Department of Agriculture, listed all publications related to forestry housed in the Department library. It consisted of 1,098 individual publications and 139 series, most of which had been written in Europe.[7]

American Journal of Forestry was an early and unsuccessful effort to develop a journal for people interested in forestry. This monthly periodical was devoted to "the interests of forest tree planting, the formation and care of woodlands and ornamental plantings generally, and to the various eco-

---

4. S. T. Dana, and S.K. Fairfax, *Forest and Range Policy* (New York: McGraw-Hill, 1980).

5. Pinchot, *Breaking New Ground*, p. 150.

6. Ibid., p. 151.

7. *List of Publications Relating to Forestry in the Department Library*, prepared under the direction of the Librarian (Washington, D.C.: Government Printing Office, 1898 [U. S. Department of Agriculture *Library Bulletin* 24]).

nomics therein concerned."[8] It was started in 1882 with Franklin B. Hough as its editor, but it lasted only one year.

The first scientific forestry journal in the United States was *Forestry Quarterly*, a student publication produced initially by the New York State College of Forestry at Cornell University.[9] The first volume was published in October, 1902, and when the college was closed in 1903, the *Quarterly* was published privately with Bernhard E. Fernow as its editor-in-chief.

The newly established Society of American Foresters began publishing a technical journal in 1905, *Proceedings of the Society of American Foresters*, whose primary purpose was to record papers read at the society meetings. The first issue of twenty-eight pages included an article by the president of the United States, Theodore Roosevelt, entitled "Forestry and Foresters." *Forestry Quarterly* and *Proceedings of the Society of American Foresters* were published separately until January, 1917, when they were merged to form the *Journal of Forestry*. This journal is still the flagship publication of SAF, and it makes a significant contribution to the world's literature on forestry.

In addition to *Journal of Forestry*, the Society of American Foresters has sponsored several other important contributions to the literature of forestry. It collaborated with the U.S. Department of Agriculture in 1905 to produce *Terms Used in Forestry and Logging*.[10] While a useful start to developing commonly accepted terminology, this document was severely criticized by the forestry experts of the day. The controversy spurred the creation of a committee to deal with the issue of forestry terminology. The results of the committee, chaired by Bernhard Fernow, were published in the *Journal of Forestry* in 1917. In 1944, the first edition of *Forestry Terminology* was published by the Society of American Foresters.[11] It is now in its third edition, and it provides a common language base for the profession.

Another major contribution of the SAF was the publication of *Forestry Handbook*.[12] This comprehensive volume deals with the major topics that comprise contemporary forestry. The first edition in 1955 was followed by a second in 1961, which continues as the standard reference volume for foresters in the United States.

8. J. L. Averell, "American Forest Literature From a Bibliography Point of View," *Journal of Forestry* 30 (2) (1932): 198.

9. E. L. Dennon, "*Forest Science*: A Quarterly Journal of Research and Technical Progress," *Forest Science* 1 (1) (1955): 3–5.

10. U. S. Forest Service, *Terms Used in Forestry and Logging* (Washington, D.C.: Government Printing Office, 1905 [*U. S. D. A. Bureau of Forestry Bulletin* 61]).

11. Society of American Foresters, *Forestry Terminology* (Washington, D.C.: Society of American Foresters, 1944).

12. R. D. Forbes, and A. B. Meyer, eds., *Forestry Handbook*, 2d ed. (New York: Ronald Press, 1961).

As the profession of forestry grew in importance and influence, its practitioners required an increasingly sophisticated scientific basis. A major focus of forestry during and after World War II was pushing the limits of productivity for more wood per acre. The forest industry grew rapidly during this period causing serious concerns about the future of commercial wood supplies. This led to the recognition of the need for more knowledge about genetics, physiology, soils, statistics, and economics.

The end of World War II spawned an enormous investment in domestic science. Federal funding for science became available in unprecedented amounts, and many of the nation's great universities became nationally recognized centers of basic research activity. By 1950, twenty-two universities were accredited by the Society of American Foresters to offer a professional degree in forestry.[13] The more competitive universities saw the need to expand the scientific basis of forestry, and scientific prowess became an important prerequisite for new faculty. The need for scientific forestry literature, in addition to a well-established professional literature, became apparent. Stephen Spurr, then a professor of forestry at the University of Minnesota, brought the issue to the attention of the profession in 1952, in which he recognized "that the *Journal of Forestry* is, and properly should be, basically a professional journal rather than a scientific one."[14]

Three years later, in 1955, the Society of American Foresters began *Forest Science* with Stephen Spurr the first editor. He even persuaded his new employer, the University of Michigan, to underwrite the first few issues through its Horace H. Rackham School of Graduate Studies. *Forest Science* has evolved into one of the leading scientific forestry journals in the world.

Another milestone in the continuing evolution of forestry literature was reached in 1977 when the Society of American Foresters published the first edition of the *Southern Journal of Applied Forestry*. In the lead article of the first volume, SAF president R. Keith Arnold stated that "science has three tasks—to replenish the reservoir of basic knowledge upon which future technological advances depend, to draw on the reservoir of basic knowledge to develop the applications of science to the problems of the day, and to get the results of research into usable form and into actual use."[15] *Forest Science* is designed to serve the first niche, the *Journal of*

13. C. H. Guise, "Statistics from Schools of Forestry for 1950: Degrees Granted and Enrollments," *Journal of Forestry* 49 (1) (1951): 7–12.

14. S. H. Spurr, "We Need a Scientific Journal in Forestry," *Journal of Forestry* 50 (5) (1952): 397.

15. R. K. Arnold, "Putting Forest Science to Work," *Southern Journal of Applied Forestry* 1 (1) (1977): 2.

Forestry contributes to the second, and the regional journals were meant to serve the third.

The *Northern Journal of Applied Forestry* appeared in 1984. In the first volume, the editors stated that their intended readership was 'green-collar' foresters and forestry-allied professionals, technicians, forestland owners, and other active forestry professionals. The *Western Journal of Applied Forestry* began publication in 1986.

Another important U.S. professional organization began contributing to the organized technical literature in 1951 with the publication of *Journal of the Forest Products Research Society*. In 1955, its name was changed to the *Forest Products Journal*. It, too, has evolved into an internationally recognized source of scientific literature.

Although not published by a professional society, *American Forests* magazine has played a significant role in the forestry literature. First appearing in 1895 as *New Jersey Forester*, it went through several name changes until 1931, when it took its present name. *American Forests* is a high-quality publication oriented to the interests of the general public and is published by the American Forestry Association, one of the oldest conservation organizations in the United States.

Forestry literature produced by the Canadian Institute of Forestry, Canada's professional forestry society, evolved a little differently than that of the United States. Its official professional publication, *Forestry Chronicle*, first appeared in 1925 published in both English and French. As in the United States, a need for a separate scientific publication to complement the professional literature was recognized. The result was *Canadian Journal of Forest Research* initiated in 1971 by the National Research Council of Canada. Thus, the professional organization has retained its responsibility for maintaining the professional literature while the federal science agency has assumed responsibility for the scientific literature in forestry.

C. Current Status of the Literature

How important are the various professional and scientific journals produced by forestry societies? Two measures address this question: the number of articles published, and a journal's impact factor as defined by the Institute for Scientific Information (ISI) in its *SCI Journal Citation Reports*. An impact factor attempts to measure a journal's influence based on the frequency with which it is cited by leading journals in the preceding year or two. In recent years, counts of the numbers of times other journals cite a journal have been used extensively because ISI indexes the citations at the ends of journal articles and makes this data available. A relative index has been established by ISI resulting by dividing the number of citations in a year of a journal by the number of times its earlier two volumes were cited by other journals that year. This is called an impact factor by ISI; if the impact factor is 1.0, it means that the number of citations in the current journal year is exactly the same as the times that journal was cited in other scientific journals in the two preceding years as indexed by ISI. This is a commendable impact. The impact factor tends to even out such factors as numbers of pages of a journal, publication frequency, and years of publication.[16]

Table 5.1 lists the seventeen forestry journals that are ranked in *SCI Journal Citation Reports* for 1990. It also identifies the number of articles that appeared in each from 1989–1992. Seven of the seventeen journals that had a high impact on the world's forestry literature in 1990 are official organs of a professional forestry or wood science society. Four of the seventeen are published by professional societies in the United States.

Differences in the ages of these journals span 150 years; but age does not guarantee impact. *Tree Physiology* first appeared in 1986 and was at the top of the list in 1990, a result of newness, a backlog of important literature,

Table 5.1. Leading forestry journals ranked by *Science Citation Index* 1990 impact factor

ISI impact factor rank	Journals	First published	Number of articles published 1989–1992
1	*Tree Physiology*	1986	230
2	*Canadian Journal of Forest Research*	1971	915
3	*Forest Science*	1955	319
4	*Wood Science & Technology*	1967	152
5	*Holzforschung*	1947	312
6	*European Journal of Forest Pathology*	1971	175
7	*Journal of Forestry*	1902	256
8	*Wood and Fiber Science*	1969	173
9	*Forest Ecology and Management*	1978	661
10	*Silvae Genetica*	1951	173
11	*Forest Products Journal*	1947	466
12	*Forestry*	1927	102
13	*Forestry Chronicle*	1925	265
14	*Agroforestry Systems*	1982	176
15	*Forstwissenschaftliches Centralblatt*	1828	108
16	*Allgemeine Forst und Jagdzeitung*	1825	70
17	*Annales des Sciences Forestieres*	1923	260

Sources: *Science Citation Index Journal Citation Reports* (1991); Table 4.1, this book.

16. *Science Citation Index Journal Citation Reports*, (Philadelphia, Pa.: Institute for Scientific Information, 1991), part in microfiche.

and an expanding topic. Rankings of journals by impact factor vary substantially from year to year, so years of publication is important.

The Core Agricultural Literature Project at Cornell University analyzed literature to arrive at a core list of current journals of greatest value to developed and Third World scholarship. These findings are detailed in Chapter 11.

D. Changes among Professional Societies

Throughout the world, forestry and forestry societies are currently undergoing enormous internal and external scrutiny. Public interest in forested ecosystems has never been more intense. Demands for the products of forests continue to grow, but the general public is also becoming increasingly aware of the environmental benefits of healthy, extensive, diverse forests.

In some parts of the world, public concern for wildlife, endangered plant and animal species, biological diversity, recreation, wilderness protection, and watershed management have overtaken concerns about timber supplies. Increasingly, forested ecosystems are being managed for their environmental benefits as well as for their commodity value.

These trends will likely continue. Professional forestry societies will find that their domains will increasingly overlap with those of other organizations whose members also have legitimate professional interests in forests. Hard lines between various natural resource management disciplines will likely continue to blur, and the drive will be toward more integrative approaches to resource management and more attention to social relevancy.

The literature stemming from, and supported by, professional forestry societies can be expected to respond to these trends. The *Journal of Forestry*, for example, has recently treated such issues as professional ethics, wilderness management, and racial and gender diversity in forestry. These articles are in direct response to the perceived interests of current and potential members of the organization.

Forest Science is also changing rapidly. Two clear needs in forest science—for more breadth and for greater depth—were identified in a recent report of the U. S. National Research Council entitled *Forestry Research: A Mandate for Change*.[17] According to this landmark document, "Forestry research must change radically if it is to help meet national and global needs. It must become broader in its clients, participants, and problems,

---

17. National Research Council, *Forestry Research: A Mandate for Change* (Washington, D.C.: National Academy Press, 1990).

Table 5.2. International Union of Societies of Foresters (IUSF) member societies and their serial publications

Society	Home of society	Publications
1. Association des Ingenieurs issus de la Faculté des Sciences Agronomiques de l'État Gembloux	Gembloux	*Annales de Gembloux*
2. Association Française des Eaux et Forêts	Paris	No serial publications
3. Association Nationale des Ingenieurs d'États des Eaux et Forêts	Rabat	No serial publications
4. Association of Consulting Foresters	Miami	*The Consultant*
5. Canadian Institute of Forestry	Toronto	*Annual Report; Forestry Chronicle*
6. Chinese Forestry Association	Taipei	*Quarterly Journal of Chinese Forestry; Taiwan's Forestry Monthly*
7. Estonian Society of Foresters	Tallinn	No serial publications
8. Forestry Association of Nigeria	Ibadan	*FAN Newsletter; Nigerian Journal of Forestry; Proceedings of the Annual Conference*
9. Forestry Association of Zimbabwe	Mutare	No serial publications
10. Indian Society of Tree Scientists	Solan	*Journal of Tree Sciences*
11. Institute of Chartered Foresters	London	*Forestry*
12. Institute of Foresters of Australia	Perth	*Australian Forestry*
13. Japan Overseas Forestry Consultants Association	Tokyo	No serial publications
14. Korean Forestry Society	Suweon	*Journal of the Korean Forestry Society*
15. Korean Society of Wood Science and Technology	Suweon	No serial publications
16. Nepal Forestry Association	Kathmandu	*Nepal Journal of Forestry*
17. New Zealand Institute of Forestry, Inc.	Wellington	*New Zealand Forestry*
18. Norwegian Union of Professional Foresters	Oslo	No serial publications
19. Professional Foresters Association of Cameroon	Yaounde	No serial publications
20. Society of American Foresters	Bethesda	*Forest Science; Journal of Forestry; Northern Journal of Applied Forestry; Southern Journal of Applied Forestry; Western Journal of Applied Forestry; Proceedings of the Society of American Foresters*
21. Society of Finnish Foresters	Helsinki	*Metsanhoitaja*
22. Society of Indian Foresters	Dehra Dun	*Van Vigyan*
23. Society of Irish Foresters	Wexford	*Irish Forestry*
24. South African Forestry Association	Pretoria	*South African Forestry Journal*
25. Tanzania Association of Foresters	Dar es Salaam	*Journal of the Tanzania Association of Foresters*
26. Unité des Eaux et Forêts	Louvain-la-Neuve	No serial publications

Sources: *International Union List of Agricultural Serials* (Wallingford, U.K.: CAB International, 1990).
The World of Learning 1993 (London: Europa Pubs., Ltd., 1992).
Ulrich's International Periodicals Directory, 1992–93 (New York: R. R. Bowker, 1992).

105

and at the same time it must both become more rigorous and be carried out in greater depth."[18]

To some extent, this mandate is beginning to be met. On the one hand, there seems to be a movement toward more interdisciplinary research directed toward significant environmental issues. This trend is partly, if not largely, driven by the need to address problems that are relevant to society. Most societal concerns do not map directly onto the array of scientific disciplines that have evolved over time. No one discipline can adequately address—or perhaps even define—such issues as biological diversity or ecosystem effects of global warming. Existing or yet to be defined professional societies can be expected to respond to this trend by creating publishing avenues for rigorous interdisciplinary research reporting.

The other trend in science, however, could result in a decreasing influence of professional forestry societies on the forest science literature. The National Research Council report stressed that forestry research must increasingly contribute to our understanding of fundamental processes affecting trees and forests, and this is already happening in some fields of forestry research. Scientists working in the area of forest biotechnology, for example, are practically indistinguishable from other types of plant molecular biologists.

As forestry research approaches fundamental areas of inquiry, forest scientists will increasingly find that they have more in common with scientists in related basic science disciplines than they do with their more applied colleagues in forestry schools. More and more manuscripts from forest scientists can be expected to find their way into the basic science literature. As this happens, forest scientists may well become less dependent on the scientific organs of professional forestry societies as outlets for the reporting of their research. Scientific forestry journals will continue to be vital to the orderly development of a body of scientific knowledge in forestry, but they will likely be competing with a wider array of publishing outlets.

Whether these speculations on the future will actually be realized or not is debatable. But it is clear that forestry and forest science are changing, and professional forestry societies will need clearly and regularly to evaluate the role they wish to play in expanding the body of forest science literature.

18. Ibid., p. 7.

6. Emergence and Role of Geographic Information Systems in Forestry

JOSEPH K. BERRY

Berry and Associates, Fort Collins, Colorado

WILLIAM J. RIPPLE

Oregon State University

Information is the cornerstone of effective decisionmaking. The modern office has evolved from typewriters, slide rules, and file cabinets to word processors, spreadsheets, and database management systems. The increasing complexity of current office needs has fueled a transition in other areas and disciplines. Natural resource information is particularly complex since it requires two descriptors, namely the precise location of what is being described, and a clear description of its physical features. For hundreds of years, explorers produced manually drafted maps which served to link these two descriptors. With their emphasis on accurate location of physical features, early maps helped explorers and navigators chart unexplored territory.

Today, these early attributes of maps have evolved from exploratory guides to physical space into management tools for physical properties. In short, analysis of mapped data has become the basis of resource planning. This new perspective marks a turning point in the use of maps, setting the stage for modern concepts in resource planning and management. It has extended information systems from an inventory focus to an analytical focus, emphasizing multiple interactive inputs to assist long-term planning strategies.

Modern resource information systems provide a means for quantitative modeling of spatial relationships. In one sense, this technology is similar to the conventional use of maps involving paper maps and drafting aids. What

This chapter, for which permission was granted, also appeared in *The GIS Applications Book: Examples in Natural Resources*, ed. William J. Ripple (Bethesda, Md.: American Society for Photogrammetry and Remote Sensing, 1994).

has changed is the purpose for which these new maps are used. Modern mapping systems provide a radically new analytic toolbox, driven by computer technology, for addressing complex issues in natural resource management, in which forests and forestry practice play a large part.

A. Historical Overview

Forestry and agroforestry are inherently spatial endeavors. In framing any forest plan, it is difficult to anticipate the potential long-term spatial effects of a plan's assumptions and policies. It is a formidable challenge to survey and incorporate a diverse set of opinions, in a manner consistent with the analysis process. These uncertainties and complexities create a need for more useful and timely information, expressed in spatial terms. The use of manual overlay mapping techniques to relate multiple spatial factors was popularized in the late 1960s.[1] The following excerpt depicts the process.

> The water systems map, vegetation map, soils map, and noise/visual impact map are placed over each other, taking care to see that the features of each map overlay one another exactly . . . . The entire bundle is then placed over a strong light source, such as a window . . . . Certain areas on the property will show through lighter than other areas. These lighter areas represent those portions of the property which have fewer restrictions on them . . . . A tracing of the outlines of the light areas is then made . . . ."[2]

Since the 1960s, decision making has become increasingly quantitative, and mathematical models have become commonplace. Two factors inhibited the full use of these techniques in natural resources. First, spatial analysis involves tremendous volumes of data. Manual cartographic techniques allowed manipulation of these data, but were inherently limited by their non-digital nature. Traditional statistics and mathematics enabled quantitative analysis, but the sheer magnitude and detail of the digital data sets were prohibitive for the geographic information systems of that time.

Early recognition of these limitations led to stratified sampling techniques developed in the early part of this century. These techniques treat spatial considerations at the onset of analysis by dividing geographic space into analysis units that are assumed to be homogenous. Most often, these parcels are manually delineated on an appropriate map, and the "typical" value for

---

1. I. L. McHarg, *Design with Nature* (Garden City: Doubleday/Natural History Press, 1969).
2. B. Hendler, *Subdividing the Wetlands of Maine* (Maine Land Use Regulation Commission, Augusta, Maine, August, 1973), p. 74.

each parcel is determined. Analysis results are assumed to be uniformally distributed throughout each parcel. The area-weighted average of several parcels statistically characterizes the typical response for an entire region. Mathematical modeling of forested systems followed a similar approach of spatially aggregating variation in model variables. Most ecosystem models, for example, identify level and flow variables presumed to be typical for vast geographic expanses.

The fundamental concepts of a comprehensive spatial statistics and mathematics have been described for many years by both theorists and practitioners.[3] The representation of spatial data as a statistical surface is indicated in most cartography texts.[4] A host of quantitative and analytical techniques are present in manual cartography. Muehrcke and Muehrcke, in an introductory text, outline a set of cartometric techniques allowing quantitative assessments of spatial phenomena.[5]

Many of these concepts were limited in their practical implementation until modern computers were able to manipulate data sets on digital maps. The computer has provided the means for both efficient handling of voluminous data and the numerical analysis capabilities that are required. From this perspective, the current revolution in Geographic Information Systems (GIS) is rooted in the digital nature of the computerized map. The complementary fields of systems analysis and GIS share this new perspective. Systems analysis for forest planning has been widely applied during the past two decades. Obvious needs in forest planning include assembling information for the variety of land resources found in a planning area, calculating the effect of various alternative treatments or practices on these land resources, and determining whether or not any combination of treatments could meet various output goals. The objective of such an exercise is generally stated as "meeting the output goals in a least-cost fashion." These output goals can be determined either by internal staff groups or by public interest groups. The resultant allocation of land treatments to various land units becomes a major part of the forest plan. The analytical problem is readily stated as a linear programming problem, and in fact linear program-

3. (a) L. A. Brown, *The Story of Maps* (Boston: Little, Brown and Company, 1949). (b) C. F. Steiitz, P. Parker, and L. Jordan, "Hand-Drawn Overlays: Their History and Perspective Uses," *Landscape Architecture* 66 (1976): 444–455. (c) R. L. Shelton and J. E. Estes, "Remote Sensing and Geographic Information Systems: An Unrealized Potential," *Geo-Processing* 1 (4) (1981): 395–420.

4. (a) A. Robertson, R. Sale, and J. Morrison, *Elements of Cartography*, 4th ed. (New York: John Wiley and Sons, 1978). (b) D. J. Cuff and M. T. Matson, *Thematic Maps* (New York: Methuen, 1982).

5. P. C. Muehrcke and J. O. Muehrcke, *Map Use: Reading, Analysis and Interpretation* (Madison, Wisc.: J. P. Publications, 1980), pp. 192–250.

ming is the most commonly used systems analysis technique for forest planning.[6]

B. Early Geographic Information Systems

The continued use of systems analysis for forest plans seems certain, and the laws that lead to the use of these techniques are not likely to change. The amount of information to be processed, the intensity of the calculations, and continued public interest in the world's forests require a well-designed and well-documented procedure. What is changing is how GIS's organize, handle, and analyze data.

In the early 1970s, GIS focused on computer mapping to automate the cartographic process. The points, lines and areas defining geographic features on a map are represented as an organized set of X and Y coordinates. These data form input to a pen plotter which can rapidly redraw the connections at a variety of scales and projections. The mapping programs of SYMAP and Odyssey developed at the Harvard Laboratory for Computer Graphics and Spatial Analysis are examples of this pioneering work. An obvious advantage of computer mapping is the ability to change a portion of the map and quickly redraft the entire area. Updates to resource maps which could take weeks, such as a forest fire burn, can be done in a few hours. The less obvious advantage is the radical change in the format of mapped data—from analog inked lines on paper to digital values stored in computers.

During the early 1980s, the change in format and computer environment of mapped data were exploited. Spatial database management systems were developed that linked computer mapping capabilities with traditional database management capabilities. In these systems, identification numbers are assigned to each geographic feature, such as a timber harvest or habitat parcel. These ID's are used as entry points into a database of information describing the composition and condition of each parcel. For example, a user is able to point at any location on a computer generated map and instantly retrieve information about the location. Alternatively, a user can specify a set of conditions, such as a specific forest and soil combination, and direct the result of the database search to be displayed as a map.

During the early development of GIS, two alternative data structures for encoding maps were debated. The vector data model closely mimics the

6. D. A. Jameson, "Forest and Forestry," in *McGraw-Hill Yearbook of Science and Technology* (New York: McGraw-Hill, 1993), pp. 150–51.

manual drafting process by representing map features as a set of lines which, in turn, are stored as an organized series of X and Y coordinates. An alternative structure, termed the raster data model, establishes an imaginary grid over a project area, then stores resource information for each cell in the grid. The early debate attempted to determine the universally best structure. The relative advantages and disadvantages of both were viewed in a competitive manner which failed to recognize the overall strengths of a GIS approach encompassing both.

By the mid-1980s, the general consensus within the GIS community was that the nature of the data and the processing desired determines the appropriate data structure. These considerations are summarized in Table 6.1.

The realization of the duality of mapped data structure had significant impact on geographic information systems. From one perspective, resource maps form sharp boundaries which are best represented as lines. Property ownerships, timber sale boundaries, and haul road networks are examples where the lines are real and the data are certain. Other maps, such as soils, site index, and slope are interpretations of terrain conditions. The placement of lines identifying these conditions are subject to judgement, statistical analysis of field data, and broad classification of continuous spatial distributions. From this perspective, the sharp boundary implied by a line is artificial and the data itself is based on probability.

The recognition of the need for both data structures focused attention on data exchange issues. Early programs incorporating both vector and raster data, such as MOSS(Map Overlay and Statistical System)/MAPS by the U.S. Fish and Wildlife Service, heralded the way for modern commercial systems such as GenaMap/GenaCell by Genesys and the GIS/ARC/INFO/ GRID by Environmental Systems Research Institute. These systems provide an integrated processing environment for a wide variety of mapped data. In forestry applications, it allows remotely sensed, digital elevation, roads, and vegetation maps to coexist in the same computing environment.

The increasing demands for mapped data focused attention on data availability, accuracy, and standards. Hardware vendors continued to improve

Table 6.1. Vector and raster data models

	Favoring vector	Favoring raster
Data nature	Lines real	Lines artificial
	Data certain	Data publication
Processing	Descriptive query	Prescriptive analysis
	Computer mapping	Spatial statistics
	Spatial DBMS	Spatial modeling

digitizing equipment, with manual digitizing tablets giving way to auto-mated scanners in many GIS facilities. A new industry for map encoding and database design emerged, as well as a marketplace for the sales of digital map products. Regional, national, and international organizations began addressing the necessary standards for digital maps to insure compat-ibility among systems.

C. Spatial Analysis and Modeling

As GIS continued its evolution, the emphasis turned from descriptive query to prescriptive analysis of maps. For the greater part, early GIS con-centrated on automating the cartographic process. If a forester had to over-lay several maps on a light-table repeatedly, an analogous procedure was developed within the GIS. Similarly, if repeated distance and bearing calcu-lations were needed, the GIS was programmed with a mathematical solu-tion. The result of this effort was functionality which mimicked the manual procedures in a forester's daily activities. The value of these systems was the savings in effort by automating tedious and repetitive operations.

By the mid-1980s, the bulk of descriptive query operations were avail-able in most GIS systems. A comprehensive theory of spatial analysis be-gan to emerge. The dominant feature of this theory is that spatial informa-tion is represented numerically, rather than in analog fashion as inked lines on a map. These digital maps are frequently conceptualized as a set of "floating maps" with common registration, allowing the computer to "look" down and across the stack of digital maps. The spatial relationships of the data can be summarized (database queries) or mathematically manipulated (analytic processing). Because of the analog nature of traditional map sheets, manual analytic techniques are limited in their quantitative process-ing. Digital representation, on the other hand, makes a wealth of quantita-tive as well as qualitative processing possible. The application of this new theory to forestry and agroforestry is revolutionary. Its application takes two forms—spatial statistics and spatial modeling.

Spatial statistics has been used by the geophysicist and statistician for many years.[7] This field of statistics seeks to characterize the geographic distribution, or pattern, of mapped data. They describe the spatial variation in the data, rather than assuming a typical response, or numeric average, to

7. (a) John C. Davis, *Statistics and Data Analysis in Geology* (New York: John Wiley and Sons, 1973). (b) Brian D. Ripley, *Spatial Statistics* (New York: John Wiley and Sons, 1981). (c) N. Cressie, *Statistics for Spatial Data* (New York: John Wiley and Sons, 1991).

be uniformally distributed in space. For example, field measurements of snow depth can be made at several sample plots within a watershed. Traditionally, these data are analyzed for the average depth (and standard deviation) to characterize the entire watershed. Spatial statistics, on the other hand, uses both the location and the measurement value at the sample plots to generate a map of relative snow depth throughout the watershed.

If the measurements are repeated a month later, change detection analysis (simple subtraction of the two maps) can be employed to determine a map of the change in snow depth. The full impact of this "map-ematical" treatment of maps on GIS is yet to be determined. The application of such concepts as spatial correlation, statistical filters, and map uncertainty await their translation from other fields, such as geophysics, environmental sciences, and quantitative geography. It is certain that spatial statistics has the potential to alter radically our current concepts and procedures in forest mensuration.

Spatial modeling, on the other hand, has many examples of assisting resource professionals in defining and evaluating spatial considerations in land management.[8] For example, forest managers can characterize timber supply by considering the relative skidding and log-hauling accessibility of harvesting parcels. Wildlife managers can consider such factors as proximity to roads and relative housing density in order to map human activity and incorporate this information into conventional wildlife habitat maps. Forest planners can assess the visual exposure of alternative sites for a facility or clearcut to sensitive viewing locations, such as roads and scenic overlooks.

Just as spatial statistics has been developed by extending concepts of conventional statistics, a spatial mathematics has evolved. This map algebra uses sequential processing of mathematical primitives to perform complex map analyses.[9] It is similar to traditional algebra in which primitive operations (e.g., add, subtract, exponentiate) are logically sequenced on variables to form equations. But in map algebra, entire maps composed of thousands or millions of numbers represent the variables.

Most of the traditional mathematical capabilities plus an extensive set of advanced map processing operations are available in modern GIS packages.[10] With spatial modeling, the spatial coincidence and juxtapositioning of

8. William J. Ripple, ed., *Geographic Information Systems for Resource Management: A Compendium* (Bethesda, Md.: American Society of Photogrammetry and Remote Sensing, 1987).

9. C. Dana Tomlin, *Geographic Information Systems and Cartographic Modeling* (Englewood Cliffs, N.J.: Prentice Hall, 1990).

10. Joseph K. Berry, *Beyond Mapping: Concepts, Algorithms, and Issues in GIS* (Fort Collins, Colo.: GIS World Publishers, 1993).

values among and within maps create new operations such as masking, proximity, and optimal paths. The set of map analysis operators can be flexibly combined through a processing structure similar to conventional mathematics. The logical sequence involves:

(a) retrieval of one or more maps from the database,
(b) processing that data as specified by the user,
(c) creation of a new map containing the processing results,
(d) and storage of the new map for subsequent processing.

This cyclical processing is similar to "evaluating nested parentheticals" in traditional algebra. For example, in the simple equation $A = (B - C)/B$, the variables B and C are first defined and then subtracted, with their difference stored as an intermediate solution. The intermediate value, in turn, is retrieved and divided by variable B to derive the value of the unknown variable A. This same structure provides the framework for spatial modeling, but the variables are represented as spatially registered maps. The numbers contained in the solution map (in effect, solving for map A) are a function of the input maps and the map analysis operations performed.

This mathematical structure forms a conceptual framework easily adapted to a variety of resource applications in a familiar and intuitive manner. For example, the equation discussed above is used in calculating percent change. If variables B and C are the snow depth maps derived by spatial statistics noted above, the resultant map would be the percent change in snow depth between the two periods. In a similar manner, a map of percent change in timber value for an entire ownership tract may be expressed using specified software packages now commercially available (e.g., Professional Map Analysis Package language, or pMAP, a PC-based system developed by Spatial Information Systems).

Within this model, data for current and past timber values for each forest parcel are derived by spatially evaluating tree growth and econometric models. The model might be extended to provide coincidence statistics, to create a table summarizing the spatial relationship between the wildlife habitat map and changes in timber values. Such a table shows which habitat parcels experienced the greatest increase in timber valuation. The basic model might be extended to include geographic searches which isolate those areas that experienced more than 20% change in timber value, assigning different weights to locations with big increases and to those with less.

Early map analysis efforts found foresters using GIS technology in the way that a city tax assessor would. Information about map parcels were retrieved and evaluated, and a decision was rendered based on the "repackaging" of existing mapped data similar to the above discussion. As systems matured, entirely new analytic capabilities became part of the GIS toolbox. For example, GIS was able to create a map of the visual connectivity of the road network in a resource area. The relative visual exposure to roads throughout a project area is mapped, with locations that are more frequently seen from roads receiving higher exposure values. In turn, this information may be summarized for each potential harvest unit by combining harvest data with its visual impact from nearby roads.

Similarly, a map of effective logging proximity to roads considering harvester operability can be derived by combining road data with barrier or key feature data. In this example, every location receives a value indicating how far it is from the nearest road (on average), considering intervening barriers, such as streams and excessively steep slopes, and how these, in turn, will impact harvesting. In effect, this new technique uses a "rubber ruler" that bends around actual barriers to give visual clues to harvester movement, resulting in a more realistic measurement of harvesting access. In short, a simple harvesting suitability model might combine visual exposure and effective proximity maps in order to identify those locations having low values, thus isolating the most suitable harvest units that have low exposure indices and are relatively close to roads.

The GIS "toolbox" contains many more advanced analytic concepts, such as optimal path routing and landscape diversity, shape, and pattern, which are radically changing how foresters characterize natural resources. These new modeling approaches to natural resource information combine and extend record-keeping systems and decision-making models into effective decision support systems.

D. GIS Applications in Forestry

There is currently extensive activity and support for GIS in forestry. For example, the British Columbia Ministry of Forestry and Lands has an operational GIS for managing their lands and making policy decisions on forest harvesting, reforestation, environmental assessment, and recreation planning. On the other hand, most other natural resource GIS activities are only at the database creation, or other basic inventory and query stages. We have not yet seen widespread use of GIS for making policy decisions based on simulating various future land management scenarios.

It seems that operational use of GIS in forestry has lagged behind other areas such as facilities management and urban applications. One reason for this is that GIS users in the natural resources have had problems in acquiring timely and accurate databases over very large geographic areas. For example, nearly all forestry applications need an accurate vegetation layer, but these data are typically very difficult to obtain. Advances in the field of remote sensing should help alleviate this problem in the future. Reviews are available on how remotely sensed data can be used as an information source for GIS in forest and natural resource management.[11] H. Lachowskie et al. described procedures and gave examples of how remote sensing and GIS are used to support all phases of resource management within the U.S.D.A. Forest Service. The authors illustrate an ever-widening use of aerial photographs, satellite imagery, airborne video, and global positioning systems (GPS). They also discuss how GIS has become important in the Forest Service for monitoring forest plans, vegetation mapping, and old-growth modeling, and range allotment monitoring. Also, recent advances in GPS in providing three-dimensional coordinates for any location should greatly assist the development of GIS for natural resource assessment.

Additional readings on GIS applications in forestry and natural resources can be obtained by referring to the two 1992 special issues on GIS in the *Journal of Forestry* (vol. 90, nos. 11 and 12), and several books including *GIS Applications in Natural Resources*, edited by M. Heit and A. Shorteid,[12] a volume on GIS applications in resource management,[13] a volume on fundamentals of GIS,[14] and a two-volume, sixty-chapter book on GIS principles and applications.[15] Other examples of GIS applications in forestry, as found in recent journal articles, can be divided into four categories including forest inventory, forest harvesting, forest fire modeling, and forest ecology.

Forest Inventory

The mapping of old growth forests in the Pacific Northwest is an important recent GIS application in forestry since the future management of these lands hinges upon accurate inventory data. R. G. Congalton et al. described the process involved in creating a GIS database for more than 30 million

---

11. (a) Craig M. Trotter, "Remotely-Sensed Data as an Information Source for Geographical Information Systems in Natural Resource Management: A Review," *International Journal of Geographical Information Systems* 5 (2) (1991): 225–39. (b) Henry Lachowskie, Paul Maus, and Bruce Platt, "Integrating Remote Sensing with GIS: Procedures and Examples from the Forest Service," *Journal of Forestry* 90 (12) (1992): 16–21.

12. M. Heit and A. Shortreid, eds., *GIS Applications in Natural Resources* (Fort Collins, Colo.: GIS World, 1991).

13. Ripple, William J. *Geographic Information Systems for Resource Management.*

14. William J. Ripple, ed., *Fundamentals of Geographic Information Systems: A Compendium* (Bethesda, Md.: American Society for Photogrammetry and Remote Sensing, 1989).

15. D. J. Maguire, M. F. Goodchild, and D. W. Rhind, eds., *Geographical Information Systems: Principles and Applications* (New York: John Wiley and Sons, 1991).

acres of national forest land in Oregon and Washington.[16] A. L. Maclean et al. demonstrated how remote sensing and GIS were used to analyze changes in timber stand conditions over time.[17] GIS has also been used to predict the distribution of tree species using digital terrain data with other environmental variables.[18]

Forest Harvesting

The wood products industry can benefit from innovative uses of GIS technology. For example, R. W. Brinker and B. D. Jackson used a GIS to study a wood procurement problem in Louisiana.[19] They assessed the current competition for pulpwood and located wood supply sources. E. Z. Baskent and G. A. Jordon used the geographic distribution of stand development types and changes over time to model timber harvesting, wood supply, and wildlife habitat in New Brunswick.[20] The Indonesian government has demonstrated how GIS can be applied to forest harvesting and management in the design of logging roads, forest monitoring, and forest land allocation.[21]

Forest Fire Modeling

M. P. Hamilton et al. have suggested that GIS can be used to provide a wildfire management probability model using data layers such as fuels, topography, weather, structures, water availability, and access routes.[22] The effects of wind during a forest fire were illustrated in a GIS by J. A. Zack and R. A. Minnich.[23] Differences in burn severity were related to post-fire vegetative cover in Michigan using GIS.[24]

16. R. G. Congalton, K. Green, and J. Teply, "Mapping Old Growth Forests on National Forest and Park Lands in the Pacific Northwest from Remotely Sensed Data," *Photogrammetric Engineering and Remote Sensing* 59 (4) (1993): 529–535.

17. Ann L. Maclean, David D. Reed, Glenn D. Mroz, Gary W. Lyon, and Thomas Edison, "Using GIS to Estimate Forest Resource Changes: A Case Study in Northern Michigan," *Journal of Forestry* 90 (12) (1992): 22–26.

18. F. W. Davis and S. Goetz, "Modelling Vegetation Patterns Using Digital Terrain Data," *Landscape Ecology* 4 (1) (1990): 69–80.

19. Richard W. Brinker, and Ben D. Jackson, "Using a Geographic Information System to Study a Regional Wood Procurement Problem," *Forest Science* 37 (6) (1991): 1614–31.

20. E. Z. Baskent, and G. A. Jordan, "Spatial Wood Supply Simulation Modelling," *Forestry Chronicle* 67 (6) (1991): 610–21.

21. S. Susilawati and J. C. Weir, "GIS Applications in Forest Land Management in Indonesia," *ITC Journal* 3 (1990): 236–44.

22. M. P. Hamilton, L. A. Salazar, and K. E. Palmer, "Geographic Information Systems: Providing Information for Wildland Fire Planning," *Fire Technology* 25 (1) (1989): 5–23.

23. J. A. Zack and R. A. Minnich, "Integration of GIS with a Diagnostic Wind Field Model for Fire Management," *Forest Science* 37 (2) (1991): 560–73.

24. M. E. Jakubauskas, P. L. Kamlesh, and P. W. Mausel, "Assessment of Vegetation Change in a Fire Altered Forest Landscape," *Photogrammetric Engineering and Remote Sensing* 56 (3) (1990): 371–77.

Forest Ecology

R. Iverson et al. reviewed how remote sensing can provide increase information on forested ecosystems in the areas of:[25]

(a) Mapping forest types
(b) Detection of forest changes
(c) Study of forest succession
(d) Assessment of stand structure
(e) Assessment of physiological parameters
(f) Assessment of forest productivity

Another indepth review, provided by L. B. Johnson, described a variety of ecological applications for analyzing spatial and temporal phenomena with GIS.[26] In the field of landscape ecology, both J. Paster and M. Broschort[27] and W. J. Ripple et al.[28] demonstrated how to use GIS to characterize forest landscape patterns to study the relationships between patterns and underlying ecological processes or wildlife habitat dynamics. Other wildlife applications with GIS include analyzing grizzly bear sightings,[29] predicting the presence or absence of red squirrels,[30] and studying long-term landscape alterations by beaver.[31]

The forestry GIS examples discussed above serve to illustrate an impressive potential for GIS in forestry. GIS applications in the future will involve a more functional and integrated system for spatial analysis. GIS will help provide a holistic view which will be required in order for us better to understand the forest system and how it functions.

25. R. Iverson, L. Graham, and A. Cook, "Applications of Satellite Remote Sensing to Forested Ecosystems," *Landscape Ecology* 3 (2) (1989): 131–43.

26. Lucinda B. Johnson, "Analyzing Spatial and Temporal Phenomena Using Geographical Information Systems: A Review of Ecological Applications," *Landscape Ecology* 4 (1) (1990): 31–43.

27. John Pastor and Michael Broschart, "The Spatial Pattern of a Northern Conifer-Hardwood Landscape," *Landscape Ecology* 4 (1) (1990): 55–68.

28. William J. Ripple, G. A. Bradshaw, and T. A. Spies, "Measuring Forest Landscape Patterns in the Cascade Range of Oregon," *Biological Conservation* 57 (1) (1991): 73–88.

29. J. K. Agee, S. C. F. Stit, M. Nyquist, and R. Root, "A Geographic Analysis of Historical Grizzly Bear Sightings in the North Cascades," *Photogrammetric Engineering and Remote Sensing* 55 (11) (1989): 1637–42.

30. Jose M. C. Pereira and Robert M. Itami, "GIS-Based Habitat Modeling Using Logistic Multiple Regression: A Study of the Mt. Graham Red Squirrel," *Photogrammetric Engineering & Remote Sensing* 57 (11) (1991): 1475–86.

31. Carol A. Johnston, and Robert J. Naiman, "The Use of a Geographic Information System to Analyze Long-Term Landscape Alteration by Beaver," *Landscape Ecology* 4 (1) (1990): 5–19.

E. GIS in the U.S. Forest Service

One of the most compelling challenges facing resource managers today is how to identify, measure, and monitor the cumulative effects of forest practices. In meeting these challenges, the U.S. Forest Service has outlined an ambitious plan to incorporate GIS and other computer technologies into a fully integrated resource information management system. The Forest Service has recognized the need for automating natural resources data for a long time. The earliest efforts were with Project INFORMS (Information Management System) in the 1970s which was developed on a large mainframe computer with limited availability to the field units. By the early 1980s, it was clear a readily available computer-based GIS capability was needed. Since 1984, under Project FLIPS (Forest Level Information Processing System), the Forest Service has automated everything in its office environment from words, to mail, to time and attendance reports. This microcomputer experience brought all their offices into the computer age and has greatly contributed to the agency's overall computer literacy. However, resource information management, and GIS in particular, were not part of this procurement.

In 1988, a plan was initiated for implementing a system combining office automation, database management, networks, and GIS technologies. The Project 615 procurement was awarded in 1994. The new workstation platform and software tools will allow for the storage, retrieval, analysis, and presentation of information about land-based resources, as well as management direction and activities. It will also provide all of the office automation and database capabilities of their current system in a fully integrated GIS environment. This means that a resource manager can begin a report in a word processing package, access tabular data from the DBMS, cut/paste a map from the GIS, and send the result by electronic mail to a colleague— all without leaving the word processing environment. It is estimated the GIS tools can serve fifty percent of the workload at ranger districts and forest science labs, thirty-three percent of the workload at forest supervisor's offices and research stations, and ten percent of the workload at regional offices and the Washington headquarters office.

These capabilities, however, do not come without a price, both literally and figuratively. The Forest Service anticipates an investment of US $1.2 billion over the eight-year life of the project, with $400–600 million going for hardware and software and the remainder for data conversion and support. During this period, thousands of resource and related maps will be digitized and entered into the GIS by Forest Service and contract personnel. The project recognizes that this electronic environment requires a new level of information organization and discipline in its management and use, as well as dramatic changes in the work patterns of resource professionals. The first year of the project will be a "pilot year" to capture user experiences and test the new system as they are implemented. The pilot year will address areas of concern such as configuration management planning, system sizing, network optimizing, information management standards development, interoffice information transfer, data communications with existing systems, conversion of existing data and applications, and development of "standard" GIS products. These findings will likely define the GIS of the future.

F. Future Trends

The 1990s will build on the cognitive basis, as well as the databases, of current geographic information systems. The system analysis models of the 1970s documented how forest actions are determined and how plans are developed. They focused attention on the analysis process and translated the constraints and objectives of forest management into mathematical terms. These models can be executed for a variety of scenarios which provide the forester with an understanding of the impacts of alternatives. However, a major limitation of this approach is the lack of spatial specificity, as landscape descriptors must be aggregated over large geographic areas. These aggregated statistics lose the unique character of individual forest parcels, thereby confounding the actual on-site implementation of forest plans.

In the 1980s, GIS technology captured this unique spatial character in modern natural resource information systems. Initially, these systems simply addressed drafting and management of inventory data. With the recognition of the digital nature of the mapped data, applications of GIS became increasingly quantitative. Complex spatial models, from wildfire behavior to harvest scheduling and wildlife habitat, were developed using an approach analogous to traditional math and statistics. These applications are constructed by logically sequencing basic map analysis operations (analytical toolbox) on spatial data (maps) to solve specific application equations (spatial models). However, the unknowns of these equations are represented as entire maps defined by thousands of numbers. This 'map-ematics' forms a conceptual framework for spatial modeling and the next step in GIS.

The technology is at a threshold that is pushing beyond mapping, management, and modeling to spatial reasoning and dialogue.[32] In the past,

---

32. Joseph K. Berry, "GIS in Island Resources Planning: A Case Study in Map Analysis," in *Geographical Information Systems*, ed. D. Maguire, M. Goodchild, and D. Rhind (New York: John Wiley, 1991), 2:285–95.

analysis models have focused on management options that are technically optimal—the scientific solution. Yet in reality, there is another set of perspectives that must be considered—the social solution. It is this final sieve of management alternatives that most often confounds resource decision making. It uses elusive measures, such as human values, attitudes, beliefs, judgment, trust, and understanding. These are not the usual quantitative measures amenable to computer algorithms and traditional decision-making models.

The step from technically feasible to socially acceptable options is not so much scientific and econometric modeling, as it is communication. Basic to effective communication is involvement of interested parties throughout the decision-making process. This new participatory environment surrounding GIS has two main elements—consensus building and conflict resolution. Consensus building involves technically driven communication and occurs during the alternative formulation phase. It involves the resource specialist's translation of the various considerations raised by a decision team into a spatial model. Once completed, the model is executed under a wide variety of conditions and the differences in outcome are noted.

From this perspective, a single map of a forest plan is not the objective. It is how maps change as the different scenarios are tried that becomes information. "What if avoidance of visual exposure is more important than avoidance of steep slopes in siting a new haul road? Where does the proposed route change, if at all?" Answers to these analytic queries focus attention on the effects of differing perspectives. Often, seemingly divergent philosophical views result in only slightly different map views. This realization, coupled with active involvement in the decision-making process, can lead to group consensus.

If consensus is not obtained, conflict resolution is necessary. This socially driven communication occurs during the decision formulation phase. It involves the creation of a conflicts map comparing the outcomes from two or more competing uses. Each management parcel is assigned a numeric code describing the actual conflict over the location. A parcel might be identified as ideal as a wildlife preservation, a campground, and a timber harvest. As these alternatives are mutually exclusive, a single use must be assigned. This assignment, however, involves a holistic perspective which simultaneously considers the potential assignments of all other locations in a project area.

Traditional scientific approaches are rarely effective in addressing the holistic problem of conflict resolution. Most are linear models, requiring a succession, or cascade, of individual parcel assignments. The final result is strongly biased by the ordering of parcel consideration. Even if a scientific solution is reached, it is viewed with suspicion by the layperson. Modern geographic information systems provide an alternative approach involving human rationalization and tradeoffs. This process involves statements like, "If you let me harvest this parcel, I will let you set aside that one for wildlife preservation." The statement is followed by a persuasive argument and group discussion. The dialogue is far from a mathematical optimization, but often closer to an effective solution. It uses the information system to focus discussion away from broad philosophical positions to a specific project area and its unique distribution of conditions and potential uses.

Planning and management have always required information as the cornerstone. Early geographic information systems relied on physical data storage and manual processing. With the advent of the computer, most of these data and procedures have been automated during the past two decades. Concurrently, the focus of these systems has extended from descriptive (inventory) to prescriptive (analysis) applications. As a result, resource information processing has increasingly become more quantitative. Systems analysis techniques developed links between descriptive data of the forest landscape to the mix of management actions which maximizes a set of objectives. This mathematical approach to forest management has been both stimulated and facilitated by modern information systems technology. The digital nature of mapped data in these systems provide a wealth of new analysis operations and a comprehensive ability to model complex resource issues spatially. The full impact of the new data form and analytical capabilities is yet to be determined. Resource professionals are challenged to understand this new environment and formulate new applications. It is clear that geographic information systems have and will continue to revolutionize the resource decision-making environment.

Selected Additional Readings on the Historical Evolution of GIS

Boyle, A. R. "Concerns About the Present Applications of Computer-Assisted Cartography," *Cartographica* 18 (1981): 31–33.

Boyle, A. R. "Developement of Equipment and Techniques," pp. 39–58, in *The Computer in Contemporary Cartography*. Edited by D. R. F. Taylor. (New York: John Wiley & Sons, 1980).

Burrough, P. A. "The Development of a Landscape Information System in The Netherlands; Based on a Turn-Key Graphics System," *GeoProcessing* 1 (1980): 257–274.

Burrough, P. A. *Principles of Geographical Information Systems for Earth Resources Assessment*. (Oxford: Clarendon Press, 1986).

Fisher, H. T. *Thematic Cartography—What it is and What is Different About It.*

(Cambridge, Mass.: Harvard Laboratory for Computer Graphics and Spatial Analysis, 1978). [Harvard Papers in Theoretical Cartography, no. 1.]

Hills, G. A. *The Ecological Basis for Land Use Planning.* (Ontario, Canada: Ontario Department of Lands and Forests, 1961). [Ontario Dept. of Land and Forests Research Report no. 46]

Hopkins, L. D. "Methods for Generating Land Suitability Maps: A Comparative Evaluation," *American Institute Planning Journal* 10 (1977): 386–400.

Kao, R. C. "The Use of Computers in the Processing and Analysis of Geographic Information," *Geographical Review* 53 (1962): 530–547.

McHarg, I. L. *Design with Nature.* (New York: Doubleday/Natural History Press, 1969).

Parent, P. J. *Geographic Information Systems: Evolution, Academic Involvement and Issues Arising from the Proliferation of Information.* Master's Thesis. University of California, Santa Barbara, 1988.

Parent, P. J., and R. Church. *Evolution of Geographic Information Systems as Decision Making Tools.* (San Francisco; GIS, 1988).

Rhind D. "Computer Aided Cartography," *Transactions of the Institute of British Geographers* 2 (1977): 71–97.

Tomlinson, R. F. "Geographic Information Systems—A New Frontier," in *Proceedings of the International Symposium on Spatial Data Handling, Aug. 1984.* (Zurich: Geographisches Institut, 1984).

Tomlinson, R. F. "The Impact of the Transition from Digital Cartographic Representation," *American Cartographer* 15 (1988): 252, 257.

Tomlinson, R. F., H. W. Calkins, and D. F. Marble. *Computer Handling of Geographic Data.* (Geneva: UNESCO, 1976).

Editor's Bibliographic Appendix

As delineated in this chapter, the value of GIS to forestry and agroforestry is increasing rapidly. This topic and its potential application prompted a reviewer of this book to suggest that additional bibliographic assistance would be worthwhile for the readers of this chapter. To this end, an analysis of the literature was made and parameters established in order to provide a concise and relatively short listing of the pertinent literature. There are two areas of concentration: application of computers and digitizing to mapping in forestry and agroforestry, and more general information about large databases, their relationship and use in GIS. Of particular interest is a recent two-volume work edited by David J. Maguire, Michael F. Goodchild, and David W. Rind entitled *Geographical Information Systems: Principles and Applications* (Longman Scientific & Technical). It covers many topics important to GIS and provides extensive bibliographic leads. Another new volume of specific interest to forestry is *The GIS Applications Book: Examples in Natural Resources,* edited by William J. Ripple (Bethesda, Md.:

American Society for Photogrammetry and Remote Sensing, 1994. 380p.) References appearing as page footnotes or as "Selected Additional Readings" in this chapter are not duplicated in this bibliographic appendix. Primary editor of this volume, Peter McDonald, assisted in the preparation of this listing.

Wallace C. Olsen
Series Editor

Aangeenbrug, R. T. and Y. M. Schiffman, eds. *International Geographic Information Systems (IGIS) Symposium: The Research Agenda*, November 15–18, 1987, Arlington, Virginia: Proceedings. Washington, D. C.: National Air and Space Adminstration, 1988. 3 vols.

Anderson, David L. and J. Dumanski. "Present-Day Soil Information Systems." In *The Literature of Soil Science*, edited by Peter McDonald. Pp. 143–160. Ithaca, N.Y.: Cornell University Press, 1994.

Anderson, J. R., et al. *A Land Use and Land Cover Classification System for Use with Remote Sensor Data*. Washington, D.C.: U.S. Geological Survey, 1976. (US Geological Survey Professional Paper, 964.)

Andersson, S. "The Swedish Land Data Bank." *International Journal of Geographical Information Systems* 1 (3) (1987): 253–263.

Aronoff, Stanley. *Geographic Information Systems: A Management Perspective*. Ottawa: WDL Publications, 1989. 294p.

Asrar, G., ed. *Theory and Applications of Optical Remote Sensing*. New York: Wiley, 1989.

Band, L. E. "A Terrain-Based Watershed Information System." *Hydrological Processes* 3 (1989): 151–62.

Barrett, E. C. and K. A. Brown, eds. *Remote Sensing for Operational Applications*. Nottingham, England: The Remote Sensing Society, 1989.

Bracken, I. and G. Higgen. "The Role of GIS in Data Integration for Rural Environments." *Mapping Awareness* 4 (8) (1990): 51–56.

Bracken, I. and C. Webster. *Information Technology in Geography and Planning: Including Principles of GIS*. London:Routledge, 1990.

Bracken, I. and C. Webster. "Towards a Typology of Geographical Information Systems." *International Journal of Geographical Information Systems* 3 (2) (1989): 137–152.

Burrough, P. A. "Matching Spatial Databases and Quantitative Models in Land Resource Assessment." *Soil Use and Management* 5 (1989): 3–8.

Burrough, P. A. and S. W. Bie, eds. *Soil Information Systems Technology*. Wageningen: PUDOC, 1984. 178p.

Campbell, W. G., et al. "The Role for a Geographical Information System in a Large Environmental Project." *International Journal of Geographical Information Systems* 3 (4) (1989): 349–362.

Chorley, R. S., ed. *Spatial Analysis in Geomorphology*. New York: Harper & Row, 1972.

Cliff, A. D. and J. K. Ord. *Spatial Processes: Models and Applications*. London: Pion, 1981.

Cocks, K. D., P. A. Walker and C. A. Parvey. "Evolution of a Continental-Sale Geographical Information System." *International Journal of Geographical Information Systems* 2 (3) (1988): 263–280.

Colwell, R. N., ed. *Manual of Remote Sensing*. 2d ed. Falls Church, Va.: American Society of Photogrammetry, 1983. 2 vols.

Coppock, J. T. and Rhind D. W. "The History of GIS." In *Geographical Information Systems: Principles and Applications*, edited by D. J. Maguire, M. F. Goodchild and Rhind D. W. Vol. 1. Pp. 21–43. London: Longman Scientific and Technical, 1991.

Croswell, P. L. and S. R. Clark. "Trends in Automated Mapping and Geographic Information Systems." *Photogrammetric Engineering and Remote Sensing* 54 (1988): 1571–1576.

Dale, P. F. and J. D. McLaughlin. *Land Information Management—An Introduction with Special Reference to Cadastral Problems in Third World Countries*. Oxford: Oxford University Press, 1988.

Damen, M. C. J., G. S. Smit and H. Th. Verstappen, eds. *International Symposium on Remote Sensing for Resources Development and Environmental Management: Proceedings of the Seventh International Symposium on Remote Sensing for Resources Development and Environmental Management ISPRS Commission VII, Enschede, 25–29 August 1986*. Rotterdam: Balkema, 1986–1988. 3 vols.

Dangermond, J., C. Freedman and D. Chambers. "Tongass National Forest Natural Resource Management Information Study—A Description of Project Methodology and Recent Findings." *GeoProcessing* 3 (1986): 51–75.

Davis, F. W. and J. Dozier. "Information Analysis of a Spatial Database for Ecological Land Classification." *Photogrammetric Engineering and Remote Sensing* 56 (5) (1990): 605–613.

Davis, J. R., P. A. Whigman and I. W. Grant. "Representing and Applying Knowledge About Spatial Processes in Environmental Management." *AI Applications in Natural Resource Management* 2 (4) (1988): 17–25.

Dent, B. D. *Principles of Thematic Map Design*. Reading, Mass.: Addison-Wesley, 1990.

Devine, H. A. and R. C. Field. "The Gist of GIS." *Journal of Forestry* August (1986): 17–22.

Dozier, J. "Spectral Signature of Alpine Snow Cover from the Landsat Thematic Mapper." *Remote Sensing of Environment* 28 (1989): 9–22.

Foley, J. D., et al. *Computer Graphics: Principles and Practice*. 2d. ed. Reading, Mass.: Addison-Wesley, 1990.

Franklin, J., et al. "Coniferous Forest Classification and Inventory Using Landsat and Digital Terrain Data." *IEEE Transactions on Geoscience and Remote Sensing* GE-24 (1986): 139–46.

Fuller, R. M., ed. *Ecological Mapping from Ground, Air and Space*. Amsterdam: Rudolf Muller, International Booksellers, 1983.

Gahegan, M. N. and S. A. Roberts. "An Intelligent, Object-Oriented Geographical

Information System." *International Journal of Geographical Information Systems* 2 (1988): 101–110.

Goldberg, M., M. Alvo and G. Karam. "The Analysis of LANDSAT Imagery Using an Expert System: Forestry Applications." *Proceedings of AUTOCART 6.* Pp. 493–503. Falls Church, Va.: American Congress on Surveying and Mapping/ American Society of Photogrammetry and Remote Sensing, 1984.

Goodchild, M. F. and S. Gopal, eds. *Accuracy of Spatial Databases.* London: Taylor & Francis, 1989.

Graetz, R. D., et al. "The Application of Landsat Image Data to Rangeland Assessment and Monitoring: The Development and Demonstration of a Land Image-Based Resource Information System (LIBRIS)." *Journal of Arid Environments* 10 (1986): 53–80.

Hall, F. G., D. E. Strebel and P. J. Sellers. "Linking Knowledge among Spatial and Temporal Scales: Vegetation, Atmosphere, Climate and Remote Sensing." *Landscape Ecology* 2 (1988): 3–22.

Jackson, M. J. and D. C. Mason. "The Development of Integrated Geo-Information Systems." *International Journal of Remote Sensing* 7 (1986): 723–740.

Jasinksi, M. F. and P. S. Eagelson. "The Structure of Red-Infrared Scattergrams of Semivegetated Landscapes." *IEEE Transactions on Geoscience and Remote Sensing* 27 (1989): 441–451.

Jensen, J. R. *Introductory Digital Image Processing.* Englewood Cliffs, N.J.: Prentice-Hall, 1986.

Jupp, D. L. B., J. Walker and L. K. Penridge. "Interpretation of Vegetation Structure in Landsat MSS Imagery: A Case Study in Disturbed Semi-Arid Eucalypt Woodlands. Part 2. Model-Based Analysis." *Journal of Environmental Management* 23 (1986):35–57.

Justice, C. O., et al. "Analysis of the Phenology of Global Vegetation Using Meteorological Satellite Data." *International Journal of Remote Sensing* 6 (1986): 1271–1318.

Kimes, D. S. "Remote Sensing of Temperature Profiles in Vegetation Canopies Using Multiple View Angles and Inversion Techniques." *IEEE Transactions on Geoscience and Remote Sensing* GE-19 (1981): 85–90.

Lesslie, R. G., B. G. Mackey and K. M. Preece. "A Computer-Based Method of Wilderness Evaluation." *Environmental Conservation* 15 (3) (1988): 225–232.

Lillesand, T. M. and R. W. Kiefer. *Remote Sensing and Image Interpretation.* 2d ed. New York: Wiley, 1987.

Maguire, D. J., M. F. Goodchild and D. W. Rhind, eds. *Geographical Information Systems: Principles and Applications.* London: Longman Scientific and Technical, 1991. 2 vols.

Masser, I. and M. J. Blakemore, eds. *Geographic Information Management: Methodology and Applications.* London: Longman Scientific and Technical, 1991.

Minnesota Department of Natural Resources, Division of Forestry and Office of Planning. *Modelling Direct Economic Returns to Timber Management as a Component of a Comprehensive, Multiple-use Forest Management Model.* St. Paul, Minn.: State of Minnesota, 1984.

Mounsey, H. M. and R. F. Tomlinson, eds. *Building Databases for Global Science.* London: Taylor & Francis, 1988.

Olsson, L. "Integrated Resource Monitoring by Means of Remote Sensing; GIS and Spatial Modelling in Arid Environments." *Soil Use and Management* 5 (1989): 30–37.

Parvey, C. and K. Grainger, eds. *A National Geographic Information System—An Achievable Objective?* Eastwood, New South Wales: AURISA, 1990. (AURISA Monograph, 4.)

Peuquet, D. and O'Callaghan J., eds. *Design and Implementation of Computer-Based Geographic Information Systems.* New York:International Geographical Union, 1983.

Proceedings of Autocarto 7. Falls Church, Va.: American Society for Photogrammetry and Remote Sensing, 1985.

Proceedings of the 3rd International Symposium on Spatial Data Handling. Columbus, Ohio: International Geographical Union, 1988.

Proceedings of the 4th International Symposium on Spatial Data Handling; July 23–27, 1990, Zurich, Switzerland. Columbus, Ohio: International Geographical Union IGU, Commission on Geographic Information Systems, Dept. of Geography, Ohio State University, 1990. 2 vols.

Proceedings of the GIS '87 Symposium. Falls Church, Va.: American Society for Photogrammetry and Remote Sensing, 1987.

Proceedings of the GIS/LIS'88 Symposium held in November 1988 in San Antonio, Texas. Falls Church, Va.: American Society of Photogrammetry and Remote Sensing, 1988.

Proceedings of the GIS/LIS '90. Bethesda, Md.: Association of American Geographers/American Congress on Surveying and Mapping/AMFM/American Society for Photogrammetry and Remote Sensing/Urban and Regional Information Systems Association, 1990.

Raper, J. F., ed. *Three Dimensional Applications in Geographical Information Systems.* London: Taylor & Francis, 1989.

Ripple, W. J., ed. *The GIS Applications Book: Examples in Natural Resources.* Bethesda, Md.: American Society for Photogrammetry and Remote Sensing, 1994. 380p.

Proceedings of the Sixth International Symposium on Automoted Cartography. Ottawa, Ontario: University of Ottawa, 1983.

Robinove, C. J. *Principles of Logic and the Use of Digital Geographic Information Systems.* Reston, Va.: U. S. Geological Survey, 1986. (U.S. Geological Survey Circular, 977).

Robinson, A. H., et al. *Elements of Cartography.* 5th ed. New York: Wiley, 1984. (New edition planned, 1995).

Running, S. W., et al. "Mapping Regional Forest Evapotranspiration and Photosynthesis by Coupling Satellite Data with Ecosystem Simulation." *Ecology* 70 (1989): 1090–1101.

Samet, H. *The Design and Analysis of Spatial Data Structures.* Reading, Mass.: Addison-Wesley, 1990.

Shumway, C. "Summary of the US Forest Service Geographical Information Systems Activities." In *Proceedings of Geographical Information Systems Workshop.* Pp. 49–52. Falls Church, Va.: American Society for Photogrammetry and Remote Sensing, 1986.

Star, J. and J. E. Estes. *Geographic Information Systems: An Introduction.* Englewood Cliffs, N.J.: Prentice-Hall, 1990.

Strahler, A. H. "Stratification of Natural Vegetation for Forest and Rangeland Inventory Using Landsat Digital Imagery and Collateral Data." *International Journal of Remote Sensing* 2 (1981): 15–41.

Turner, A. K., ed. *Three Dimensional Modelling with Geoscientific Information Systems.* Dordrecht: Kluwer, 1991.

U.S. Dept. of Interior. *Managing Our Land Information Resources.* Washington, D.C.: U.S. Department of the Interior, 1989.

U.S. Soil Conservation Service. *Technical Specifications for Line Segment Digitizing of Detailed Soil Survey Maps.* Washington, D.C.: Government Printing Office, 1984.

Ventura, S. J., B. J. Niemann, Jr. and D. D. Moyer. "A Multipurpose Land Information System Approach to Conservation and Rural Resource Planning." *Journal of Soil and Water Conservation* 43 (3) (1988): 226–229.

Wakeley, R. R. "G.I.S. and Weyerhaeuser—20 Years Experience." In *Proceedings of the GIS '87 Symposium.* Pp. 446–455. Falls Church, Va.: American Society for Photogrammetry and Remote Sensing, 1987.

Worrell, L., ed. *Geographic Information Systems: Developments and Applications.* London: Belhaven, 1990.

7. The Research Publishing Influence of the U.S. Department of Agriculture Forest Service

JOHN S. SPENCER, JR.

U.S. Forest Service, North Central Forest Experiment Station, St. Paul, Minnesota

The United States Forest Service, an agency of the U.S. Department of Agriculture, conducts the most extensive and productive program of integrated forestry research in the world. The research branch of the Forest Service is one of six; the other branches are administration, national forest system, state and private forestry, programs and legislation, and international forestry. Forest Service scientists produced 2,215 research publications in all formats on a wide variety of subjects in 1991.[1] Forest Service research publications account for only an estimated third of the total number of publications produced annually by the agency. The remaining two thirds of Forest Service publications are not research papers, and are published by the other branches of the Forest Service as brochures, manuals, handbooks, administrative documents, maps, policy and program guides, and periodicals on a wide range of subjects. *Tree Planters Notes* and *Fire Management Notes* are examples of two periodically issued titles. A recent exemplary brochure is entitled *How to Manage Eastern White Pine to Minimize Damage from Blister Rust and White Pine Weevil.*

This popular literature emanates from all levels of the Forest Service, nationally, regionally, and within small units with specific geographic responsibilities. Much of this material is informational and promotional. Public issue and policy statements mostly appear from the national offices although regions may deal with similar issues also. The popular literature includes information directories, topographical maps, recruiting material, interpretive pamphlets often on environmental features, and brochures on a

1. US Forest Service, *Profiles of the Research Program* (Washington, D.C.: USDA Forest Service, October 1992).

variety of topics including those of economic, scientific, or recreational interest. These publications, usually of few pages, constitute about two thirds of all publications of the Forest Service. However, in this chapter we will examine the research and scientific publications, which are a major factor in the academic and research literature of forestry.

The mission of Forest Service research is to develop and disseminate scientific and technical knowledge to enhance and protect economic productivity and environmental quality on all of the Nation's 1.6 billion acres of forests and associated rangelands. The Forest Service warrants investigation in this book because it is the largest publisher of forest science literature in the world, and makes a substantial impact on the conduct of forest research and forest management, both nationally and internationally.

A. Beginnings of Forest Service Research[2]

In 1876, Congress appointed Franklin B. Hough, a physician with interests in botany, history, meteorology, and statistics, as the first government forestry agent. His duties in the Department of Agriculture were "To determine the annual amount of consumption, importation, and exportation of timber and other forest products, the probable supply of future wants, the means best adapted to their preservation and renewal, the influence of forests upon climates, and the measures that have been successfully applied in foreign countries, or that may be deemed applicable in this country, for the preservation and restoration or planting of forests."[3] It can be said that the Forest Service was born as a research agency.

When the Division of Forestry was established in the Department of Agriculture in 1881 with Hough as its first chief, a body of knowledge of European forestry existed but little applied to American tree species or conditions. Forestry in America essentially had to start with very limited knowledge. Bernhard E. Fernow became chief of the division in 1886, the first professional forester in that position, and began investigations to "establish the principles upon which the forestry we advocate is to be carried on."[4] To this end, he inaugurated studies and published the life histories of important American forest trees and the conditions of their development in

2. The author acknowledges the help of Terry L. West, USDA Forest Service, History Unit, Washington, D.C. in preparing this section.

3. Robert K. Winters, *Fifty Years of Forestry in the U.S.A.* (Washington, D.C.: Society of American Foresters, 1950). p. 166.

4. Herbert C. Storey, *History of Forest Service Research: Development of a National Program* (Washington, D.C.: USDA Forest Service, History Unit, 1975), unpublished manuscript, p. 6.

the forest, such as *The White Pine*, authored by V. M. Spalding in 1899.[5] In addition, he initiated studies dealing with the physical, mechanical, and chemical properties of wood, which resulted in publications such as *Timber Physics, Part I*, produced by the division in 1892.[6]

Gifford Pinchot replaced Fernow as chief in 1898 and directed the division on a more activist, land management course. He established a Section of Special Investigations to provide answers to questions about the scientific management of forests, which was a cornerstone of his vision for the agency. By 1902 the Division of Forest Investigations within the Bureau of Forestry employed 55 persons and accounted for a third of the agency's budget.[7] The Bureau of Forestry was renamed the U.S. Forest Service in 1905, and by 1915 a Branch of Research, under the leadership of Earle H. Clapp, was established replacing a trio of investigative committees. This action brought together under a single administrative head all the disparate elements of research being conducted around the country, and greatly strengthened the role of research as a basic function of the Forest Service. It also meant that research was equal to and independent of the administrative branch of the Forest Service.

The first forest experiment station in the United States was established in 1908 at Fort Valley, Arizona, on the Coconino National Forest. This and similar stations created in Colorado, Idaho, California, Washington, and Utah were comparable to present-day experimental forests, and were viewed originally as serving the needs of national forests. Later, research efforts broadened from making empirical observations to testing hypotheses, and from studying problems associated with managing the national forests to studying problems on all forests and rangelands. In 1910 the Forest Service established the Forest Products Laboratory in Madison, Wisconsin, to research practical means to achieve greater utilization and less waste of wood, a major problem of the time, and to disseminate the findings to users, primarily the wood-using industries.

In a 1921 paper on the role of forest experiment stations, Clapp listed "What forest experiment stations need to do" as: "1) To determine what kinds of trees can and should be grown . . . , 2) To find out how to plant . . . , 3) To determine best methods of cutting and slash disposal . . . , 4) To answer many questions concerning forest management . . . , 5) To ascertain

5. V. M. Spalding, *The White Pine* (Washington, D. C.: USDA Division of Forestry, 1899 [*USDA Division of Forestry Bulletin* no. 22]).

6. USDA Division of Forestry, *Timber Physics, Part I* (Washington, D. C.: USDA Division of Forestry, 1892 [*USDA Division of Forestry Bulletin* no. 6]).

7. Harold K. Steen, *The U.S. Forest Service: A History* (Seattle: University of Washington Press, 1976).

growth and yields . . . , 6) To establish the basis for improved methods of fire protection . . . , 7) To cooperate in forest management problems closely related to wood utilization . . . , and 8) To furnish the scientific basis for forestry."[8]

By the 1920s, there were twelve regional experiment stations, which were later consolidated into eight, as well as branch field stations. The McSweeney-McNary Research Act of 1928 legitimatized the experiment stations, authorized forest research in all phases of forestry and wood utilization, provided funding, and gave congressional recognition to research as an important part of forestry.[9] William M. Jardine, then Secretary of Agriculture, said the act "authorized the inauguration of a far-reaching program of federal research in forestry, which for the first time affords the prospect that the basic knowledge essential for skillful timber growing will be progressively obtained at a rate commensurate with its importance."[10]

However, restricted funding limited research activities and staffing until after World War II, when the ensuing economic boom generated significantly increased appropriations, permitting an expanded research and publishing effort. During this expansion, basic research was given more emphasis than previously, until the 1970s when legislation passed by Congress shifted the emphasis back toward applied research. The Forest and Rangeland Renewable Resources Planning Act of 1974 and the Forest and Rangeland Renewable Resources Research Act of 1978 both redefined the charter of Forest Service research. Reduced federal spending in the 1980s resulted in staff reductions in Forest Service research from 964 scientists in fiscal year 1980 to 720 in fiscal year 1991.[11]

International Forestry (IF) was part of the research arm of the Forest Service before it became a separate and equal branch. From 1961 to 1971, part of the duties of the IF staff were to produce a library of publications and statistics on forest resources and industries for most of the nations of the world.[12]

8. Earl H. Clapp, *Forest Experiment Stations* (Washington, D.C.: USDA, 1921 [*USDA Circular* no. 183]), pp. 11–15.

9. Herbert C. Storey, *History of Forest Service Research: Development of a National Program*.

10. Gladys L. Baker, Wayne D. Rasmussen, Vivian Wiser, and Jane M. Porter, *Century of Service: The First 100 Years of the United States Department of Agriculture* (Washington, D.C.: USDA Economic Research Service, Agricultural History Branch, 1963), pp. 129–130.

11. US Forest Service, *Profiles of the Research Program*.

12. Terry L. West, "USDA Forest Service Involvement in Post World War II International Forestry" in *Changing Tropical Forests: Historical Perspectives on Today's Challenges in Central and South America*, ed. Harold K. Steen and Richard P. Tucker, Proceedings of a conference sponsored by the Forest History Society and IUFRO Forest History Group (Durham, N.C.: Forest History Society; Vienna, Austria: IUFRO Forest History Group, 1992).

B. Implementation of Research

Forest Service research today is partially shaped by environmental and political interests, such as the need to manage forests with sensitivity to endangered wildlife species, biodiversity, forest health, and old-growth forest issues.

Presently, the eight regional forest and range experiment stations of the Forest Service, along with the Forest Products Laboratory and the International Institute of Tropical Forestry in Rio Piedras, Puerto Rico (see Fig. 7.1), conduct studies that range geographically from the tropics to the Arctic, and from Hawaii and Pacific territories to Puerto Rico in the Atlantic Ocean. More than 2,800 studies are in progress at any one time, at seventy-four locations around the United States and the world where Forest Service scientists conduct or participate in research. Many studies focus on multifunctional or long-term natural resource issues of national or international scope.[13]

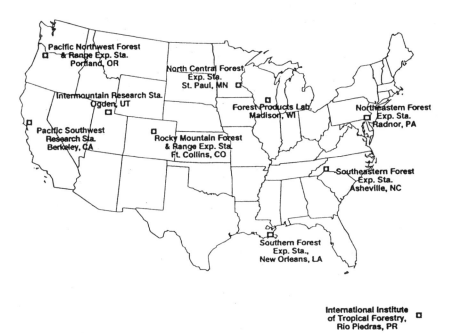

Figure 7.1. USDA Forest Service forest and range experiment stations, the Forest Products Laboratory, and the International Institute of Tropical Forestry.

13. USDA Forest Service, *Report of the Forest Service, Fiscal Year 1991* (Washington, D.C.: USDA Forest Service, June 1992).

Forest Service research is grouped into six broad research areas:[14] (1) Forest protection research develops better methods for preventing, controlling, and reducing the effects of wildfires, insects, and diseases. (2) Resource analysis research provides a scientific basis for assessing the current condition and outlook for forest land resources, forest product investments, and markets; as well as develops methods for improving management of outdoor recreation, wilderness, and urban forest resources. (3) Forest management research seeks ways to achieve higher levels of forest health, quality, and productivity through environmentally, biologically, and economically sound forest management practices. (4) Forest environment research seeks to develop the knowledge and techniques needed to manage forest, rangeland, and associated aquatic ecosystems to sustain biodiversity, and water, wildlife, and fish resources. (5) Forest products and harvesting research searches for better ways to harvest, produce, and use wood products. (6) Forestry inventory, economics and recreation.

Forest Service research is planned and coordinated with related efforts at the sixty-one forestry schools in the United States and the agricultural experiment stations of land-grant institutions throughout the country. Forest Service scientists also work closely with researchers and land managers from other public agencies, the forest industry, and, increasingly, from other countries.

In 1991, research appropriations were $168 million, approximately 11% of which supported cooperative studies with colleges, universities, industry, and domestic and international research organizations. The Forest Service also received supplemental research support from other government agencies and various private sector institutions totaling $3 million.[15]

The results of research findings are disseminated by publications, symposia, workshops, videos, newspaper articles, and direct public contact. Users include forestry and natural resource professionals, private landowners, resource-dependent industries, educators and students, legislators, citizens groups, homeowners, and the general public. In 1981, Callaham estimated that more than 48,000 professionals working in forty-two broad disciplines and employed by more than forty United States agencies constituted the federal clients for forestry research.[16] From this, he further estimated that a total of about 200,000 natural resource professionals in the United States use Forest Service research. These numbers, which account for only a portion of the total users, probably have risen since 1981.

14. Ibid.
15. Ibid.
16. Robert Z. Callaham, *Criteria for Deciding about Forestry Research Programs* (Washington, D.C.: USDA Forest Service, 1981 [*USDA Forest Service General Technical Report* WO-29]).

Table 7.1. Percent of research publications mailed from Forest Service experiment stations and the Forest Products Laboratory, by region of the country served by the station and recipient, 1977–1979

Recipients	Pacific Coast	Rocky Mtns & Great Plains	North	South	Forest Prod. Lab.	Average
Libraries and academia	20.0%	26.3%	15.3%	14.7%	15.1%	16.5%
U.S. Forest Service	8.8	11.0	3.3	2.6	9.0	6.0
Other federal and international agencies	6.6	15.2	2.7	3.2	6.6	5.4
State, county, and municipal agencies	4.4	8.0	14.1	4.6	—	5.4
Industry, trades, and media	9.5	8.2	18.1	19.0	41.9	23.0
Individuals in U.S.	35.8	30.0	29.7	46.8	27.4	35.4
Individuals outside U.S.	14.9	1.3	16.8	9.1	—	8.3
Total	100.0	100.0	100.0	100.0	100.0	100.0

Callaham also estimated that the largest portion of research publications, 35%, mailed from Forest Service experiment stations during 1977–1979 were sent to individuals in the United States (Table 7.1). These figures probably have not changed significantly. Forest Service research is unique in having separate appropriations, a broad mission ranging from applied to basic research, an extremely diverse clientele ranging from laymen to practitioners to scientists, and a system in which it competes for funds with nonresearch activities in the agency.

C. Emphasis of Forest Service Research and Publishing

Early research efforts were aimed at solving pressing problems, which often were related to commodity production. As new and different concerns surfaced, studies were added and completed ones were phased out. Current research emphasis is on building knowledge of the holistic nature of forest ecosystems, as well as knowledge of the social and economic processes that affect forests. Following are a few examples of the evolution of research in several selected fields.

One early research interest examined whether forests significantly changed the climate, increased rainfall, influenced streamflow, and reduced or ameliorated floods, a debate begun as early as 1864 by George Marsh in his landmark book, *Man and Nature*.[17] These were basic considerations in

17. George Perkins Marsh, *Man and Nature: Or, Physical Geography as Modified by Human Action* (New York: Charles Scribner, 1864).

the management of Western forests where water is scarce. Also, as a practical matter, it was important to determine the role of forests in floods because efforts begun in 1899 for legislation to empower the federal government to buy forest land for parks or forest reserves were blocked by the question of the practical value as well as the constitutionality of the purchases. The House Judiciary Committee ruled that the commerce clause of the Constitution authorized such purchases in watersheds of navigable streams, if it could be shown that forests prevented floods.[18] Beginning in 1910 on the Rio Grande National Forest in Colorado, Carlos Bates conducted the first controlled experiments on the hydrology of mountain watersheds, in cooperation with the U.S. Weather Bureau, and published the final report of the study in 1928.[19] Today, watershed research is being conducted at many grassland and forest locations around the country to learn how use or disturbance of the vegetation on different ecosystems impacts water quality and yield.

Range management research was also undertaken early because grazing, not timber use, was the primary concern of many of the first western forest managers. Frederick Coville undertook the first scientific range investigation in the country in 1897 at the Cascade Mountain Forest Reserve in Oregon and published his classic *Forest Growth and Sheep Grazing in the Cascade Mountains of Oregon* (1898).[20] Range research in 1907 on the Wallowa National Forest in Oregon determined grazing capacity and represented the evolution from rule-of-thumb management of rangelands by the Forest Service to scientific range management.[21] Succeeding research, seeking the best techniques for artificial and natural range reseeding, as well as methods for achieving the greatest grazing efficiency per unit of area, yielded results such as Selar Hutchings and George Stewart's *Increasing Forage Yields and Sheep Production on Intermountain Winter Ranges*, published in 1953.[22]

The Forest Products Laboratory (FPL) has conducted research on many diverse subjects, but only a few are shown here to suggest the changing

18. Richard S. Sartz, "Carlos G. Bates: Maverick Forest Service Scientist," *Journal of Forest History* 21 (1) (1977): 31–39.

19. Carlos G. Bates and A. S. Henry, "Forest and Streamflow Experiment at Wagon Wheel Gap, Colorado: Final Report on Completion of the Second Phase of the Experiment," *U.S. Monthly Weather Review*, Supplement 30 (1928): 1–79.

20. Frederick V. Coville, *Forest Growth and Sheep Grazing in the Cascade Mountains of Oregon* (Washington, D.C.: USDA Division of Forestry, 1898 [*USDA Division of Forestry Bulletin* no. 15]).

21. J. T. Jardine, *The Pasturage System for Handling Range Sheep* (Washington, D.C.: USDA Forest Service, 1910 [*USDA Forest Service Circular* no. 178]).

22. Selar S. Hutchings and George Stewart, *Increasing Forage Yields and Sheep Production on Intermountain Winter Ranges* (Washington, D.C.: USDA, 1953 [*USDA Circular* no. 925]).

priorities of demand. Studies to test the effect of preservatives on the life and durability of railroad ties, mine timbers, and poles were published by Weiss and Teesdale as *Tests of Wood Preservatives* in 1915.[23] Other early studies included tests to determine the mechanical properties (e.g., strength, hardness, and stiffness) of American woods. During World War I, scientists studied wood, plywood, and glues as components of airplanes, as well as boxing and crating design, and kiln drying techniques for lumber. In the 1920s, scientists at the FPL developed a technique for using southern pines to produce strong, white paper (rather than brown Kraft paper), which permitted the development of a southern pulp and paper industry in the 1930s.[24] Studies of prefabricated housing systems, wood chemistry, laminated arch construction, and methods of using poor quality trees followed. Some recent FPL research involves construction and use of timber bridges, as described in William McCutcheon's *Mormon Creek Bridge: Performance after Three Years,* published in 1992.[25] Other recent studies examine lignin-degrading fungi to soften wood chips before pulping, expanded uses of recycled fiber, and the combination of wood, plastic, and other material into composites useful for products.

In 1991, the Forest Service identified six priority national research programs for emphasis: global change, water quality, threatened and endangered species, declining forest-based economies in rural America, southern forest productivity, and catastrophic forest fires.[26]

D. Types of Research Publications

Forest Service scientists traditionally have disseminated their research findings in a broad array of publications, and those actually published by the Forest Service represent only a small portion of the total. Scientists author papers published by other government agencies, as well as those published in scientific journals, conference proceedings, books, trade journals, popular magazines, and other media not produced by the government.

The first scientific papers concerned with forestry and produced by the

23. Howard F. Weiss and C. H. Teesdale, *Tests of Wood Preservatives* (Washington, D.C.: USDA, 1915 [*USDA Bulletin* no. 145]).

24. C. E. Curran and M. W. Bray, *White Papers from Southern Pines, 1: Pulping Loblolly Pine for Strong, Easy-Bleaching Sulphate Pulp* (Madison, Wisc.: USDA Forest Service, Forest Products Laboratory, 1931 [*Forest Products Laboratory Report* no. 930]).

25. William J. McCutcheon, *The Mormon Creek Bridge: Performance after Three Years* (Madison, Wisc.: USDA Forest Service, Forest Products Laboratory, 1992 [*Research Paper* FPL-RP-509]).

26. USDA Forest Service, *Report of the Forest Service, Fiscal Year 1991.*

USDA Division of Forestry (followed by the Bureau of Forestry and later the Forest Service) were published in Washington, D.C., as bulletins. The first of the Division of Forestry *Bulletin* series was M.G. Kern's *Relation of Railroads to Forest Supplies and Forestry*, published in 1887.[27] Such *Bulletins* averaged fifty to sixty pages, and covered a wide range of subjects, such as Filibert Roth's *Timber: An Elementary Discussion of the Characteristics and Properties of Wood*,[28] and J.W. Toumey's *Practical Tree Planting in Operation*.[29]

The USDA Division of Forestry also published early research findings in *Circulars*, the first of which is dated 1886.[30] Like *Bulletins*, *Circulars* also covered diverse subjects, but were more general and shorter, averaging ten to twenty pages. Early *Circulars* were H.M. Suter's *Forest Fires in the Adirondacks in 1903*,[31] and Charles Herty's *Practical Results of the Cup and Gutter System of Turpentining* (1905).[32]

Early forest scientists also published in *USDA Farmers' Bulletins*, a popularized document to disseminate practical information to farmers and landowners. Bernhard Fernow published *Forestry for Farmers*[33] in this series in 1895, as did Perley Spaulding with *The White Pine Blister Rust*, (1916).[34]

The Division of Forestry *Bulletins* and *Circulars* were replaced by *USDA Bulletins* and *Circulars*, respectively, also printed in Washington, D.C., as the Department of Agriculture sought to reorganize and centralize its information program. *USDA Bulletins* (later called *USDA Department Bulletins*), which bore only the department name, were also long, and reported results of agricultural, and forestry, research. Therefore, *Bulletin* numbers of forest-related publications are not continuous. Carl Hartley's *Blights of Coniferous Nursery Stock*, published in 1913, is an example of an early *USDA*

27. M. G. Kern, *Relation of Railroads to Forest Supplies and Forestry* (Washington, D.C.: USDA Division of Forestry, 1887 [*USDA Division of Forestry Bulletin* no. 1]).

28. Filibert Roth, *Timber: An Elementary Discussion of the Characteristics and Properties of Wood* (Washington, D.C.: USDA Division of Forestry, 1895 [*USDA Division of Forestry Bulletin* no. 10]).

29. J. W. Toumey, *Practical Tree Planting in Operation* (Washington, D.C.: USDA Division of Forestry, 1900 [*USDA Division of Forestry Bulletin* no. 27]).

30. USDA Division of Forestry, *Request to Educators for Cooperation* (Washington, D.C.: USDA Division of Forestry, 1886 [*USDA Division of Forestry Circular* no. 1]).

31. H. M. Suter, *Forest Fires in the Adirondacks in 1903* (Washington, D.C.: USDA Bureau of Forestry, 1904 [*USDA Bureau of Forestry Circular* no. 26]).

32. Charles H. Herty, *Practical Results of the Cup and Gutter System of Turpentining* (Washington, D.C.: USDA Bureau of Forestry, 1905 [*USDA Bureau of Forestry Circular* no. 34]).

33. B. E. Fernow, *Forestry for Farmers* (Washington, D.C.: USDA, 1895 [*USDA Farmers' Bulletin* no. 67]).

34. Perley Spaulding, *The White-Pine Blister Rust* (Washington, D.C.: USDA, 1916 [*USDA Farmers' Bulletin* no. 742]).

Bulletin.[35] The final forest-related *USDA Bulletin* was T. R. Truax's *The Gluing of Wood*, published in 1929.[36]

The *USDA Technical Bulletin* series, also published in Washington, D.C., overlapped and then supplanted the *USDA Bulletins*. *Technical Bulletins* also addressed agricultural research, causing forest-related bulletin numbers to be discontinuous. The first forest-related *Technical Bulletin* was *The Relation of Highway Slash to Infestation by the Western Pine Beetle in Standing Timber*, by J. E. Patterson, published in 1927.[37] Some of the most noteworthy Forest Service research during this period was published in *USDA Technical Bulletins*, such as *The Yield of Douglas Fir in the Pacific Northwest*, by McArdle, Meyer, and Bruce, published in 1949.[38] From about 1913 on, USDA employees had approval to offer scientific and technical material for publication in media outside the department.[39] However, early *Technical Bulletins* provided the major outlet for government-published forestry research until experiment stations began publishing their own *Station Papers* or *Research Papers* (old series) in the late 1930s and early 1940s.

Yet another early outlet for publication was the *USDA Circular*, which replaced the Division of Forestry *Circulars* in the late 1920s. Although somewhat shorter than *USDA Bulletins*, the *Circulars* also covered a wide range of subjects, such as *Artificial Reseeding on Western Mountain Range Lands* by C. L. Forsling and W. A. Dayton, (1931).[40]

Individual experiment stations began publishing their own research findings as *Miscellaneous Publications, Technical Notes,* and *Research Notes* (old series) in the late 1920s or early 1930s. By the late 1930s, most stations were publishing *Station Papers* (sometimes called *Research Papers* of the old series) on a variety of subjects; and by the late 1930s to the mid-1940s, stations were publishing *Forest Survey Releases*, which announced the results of statewide forest inventories. In the 1950s, a few stations published *Technical Papers*.

In 1963, Forest Service research adopted the numbered series of publica-

35. Carl Hartley, *The Blights of Coniferous Nursery Stock* (Washington, D.C.: USDA, 1913 [*USDA Bulletin* no. 44]).

36. T. R. Truax, *The Gluing of Wood* (Washington, D.C.: USDA, 1929 [*USDA Department Bulletin* no. 1500]).

37. J. E. Patterson, *The Relation of Highway Slash to Infestation by the Western Pine Beetle in Standing Timber* (Washington, D.C.: USDA, 1927 [*USDA Technical Bulletin* no. 3]).

38. Richard E. McArdle, Walter H. Meyer, and Donald Bruce, *The Yield of Douglas Fir in the Pacific Northwest* (Washington, D.C.: USDA, 1949 [*USDA Technical Bulletin* no. 201, revised]).

39. Gladys Baker, et al., *Century of Service*.

40. C. L. Forsling and William A. Dayton, *Artificial Reseeding on Western Mountain Range Lands* (Washington, D.C.: USDA, 1931 [*USDA Circular* no. 178]).

tions in use at each station today: *Research Notes* (RN), *General Technical Reports* (GTR), *Research Papers* (RP), and *Resource Bulletins* (RB). *Research Notes* average only four to six pages, and serve to get new information out quickly, including preliminary findings, descriptions of new methods and devices, preliminary progress, data, and observations. *GTR's* report information of a technical nature, but not necessarily the product of a specific piece of original research. Included are symposium proceedings, annotated bibliographies, and descriptions of computer or statistical programs. *Research Papers* present results, analyses, and conclusions arising from formal studies. *Resource Bulletins* report findings, analyses, and data on forest resources and their use, primarily the results of Forest Inventory and Analysis (formerly Forest Survey) statewide forest inventories. Stations may publish miscellaneous unnumbered items, such as publication lists, informational brochures, and leaflets. Each numbered publication is identified by series, station, and number. For example, RP PNW-37 denotes *Research Paper* number 37 of the Pacific Northwest Forest and Range Experiment Station, in Portland, Oregon.

E. Publishing Outlets

During fiscal year 1991, Forest Service scientists produced a total of 2,215 research publications, in all publishing outlets.[41] A summary by specific outlet is not available for that year, but a study by P. J. Jakes and A. S. VanDyne[42] of the 4,523 publications (including 365 abstracts) produced by Forest Service researchers, exclusive of those at the Forest Products Laboratory, during fiscal years 1980 and 1981 showed that 24% of the total was published by the Forest Service (Table 7.2). These findings probably apply generally to more current years as well.

The study also found that the 1,112 articles published in scientific journals appeared in 240 different journals. The six journals in which articles were most frequently published were *Phytopathology, Journal of Forestry, Forest Science, Southern Journal of Applied Forestry, Canadian Journal of Forest Research,* and the *Forest Products Journal.* However, scientists working in more specific disciplines tended to publish in more specific journals; e.g. forest geneticists usually prefer to publish in *Silvae Genetica.*

---

41. U.S. Forest Service, *Profiles of the Research Program.*
42. Pamela J. Jakes and Andra Slimak VanDyne, "Forestry Literature: Who's Publishing What Where?" *Journal of Forestry* 85(9) (1987):33–36.

Table 7.2. Source or type of publication of U.S. Forest Service researchers, FY 1980 and 1981

Outlet		Percent
Forest Service publications		24
Conference proceedings (N = 1,150)		25
National conferences	42%	
Regional conferences	31	
International conferences	20	
Local conferences	7	
Scientific journals		24
Others (books, theses, trade journals, university staff papers)		27
	Total	100

Source: Pamela J. Jakes and Andra Slimak VanDyne, "Forestry Literature: Who's Publishing What Where?" *Journal of Forestry*, 85(9) (1987):33–36.

Among the 1,091 publications by the Forest Service in the study, 25% were *Research Papers*, as shown here:

Research Papers	25%
General Technical Reports	24%
Research Notes	24%
Others (*Resource Bulletins*, "How To" leaflets, and misc.)	27%

Publishing by Forest Products Laboratory (FPL) scientists was excluded from the cited study above, because their clientele differs substantially from that of the rest of the research organization, being diverse and specialized, including the wood-using industries, builders, recycling specialists, toxic waste specialists, paint chemists, and others. A sample of the 148 publications produced by the FPL during calendar year 1992 showed that seventy-one (48%) appeared in scientific journals.[43] Thirty different scientific journals were represented, most frequently *Forest Products Journal, Wood and Fiber Science, Holzforschung, Journal of Structural Engineering,* and *Applied and Environmental Microbiology*. Another fifty-three of the publications (36%) were published in conference proceedings, and only eleven (7%) were printed by the Forest Service, all but one in the *Research Paper* series. The remaining thirteen (9%) were published in books, trade journals, miscellaneous reports, abstracts, and newsletters.

43. USDA Forest Service, Forest Products Laboratory, *Dividends from Wood Research—Recent Publications, January-June 1992, and July-December 1992* (Madison, Wisc.: USDA Forest Service, Forest Products Laboratory, 1992 and 1993).

F. The Influence of Forest Service Research Publishing

One way of estimating the influence of scientific publishing is by analyzing the citations in certain source documents, selected as being international or national landmarks in their field, to determine how often a title or series is cited. A high frequency of citation suggests that prominent authors and scientists in the field find the work timely, important, and useable. The Albert R. Mann Library at Cornell University conducted such a citation analysis of fourteen recent international source documents in the field of forestry in developed countries of the world, selected as landmarks by a steering committee for the Library's Core Agricultural Literature Project (see Chapters 3 and 11). Analysts counted every citation and gathered data for each, including type of publication, place of publication, publisher, date of publication, and the number of times cited. Of the 13,830 citations in all fourteen source documents, 965 (7%) were from USDA publications of which 583 (4%) of the citations were clearly from Forest Service publications. This is a conservative estimate of the publishing influence of Forest Service scientists, however, because as noted earlier, Jakes and VanDyne found that only 24% of the papers attributed to agency scientists during fiscal years 1980 and 1981 were published by the Forest Service. Papers published by Forest Service scientists in scientific journals, conference proceedings, or other non-Forest Service outlets, 76% of their output, were not included in the Forest Service counts in the citation analysis. Also, some of the 965 USDA publications not cited as Forest Service publications probably should have been, but the specific agency could not be determined from the citation.

According to the Core Agricultural Literature Project citation analysis of the Forest Service publications since the 1963 adoption of the current numbered series, *Research Papers* and *General Technical Reports* were cited most often in the source documents, as shown in Figure 7.2.

R. Z. Callaham used another approach to grasp the significance of outputs and benefits from Forest Service research; he analyzed eighty-one innovations resulting from agency research during fiscal years 1977–1979.[44] Illustrations of the innovations from research are: development of fusiform rust-resistant southern pines, methods for sewage sludge disposal on forests, biocontrol of larch casebearer, differential campsite pricing methodology, methods of bark chip separation and segregation, methods of mechanized thinning of hardwoods, and methods for making wide boards from small logs. All benefits were categorized into sixteen general areas:

44. Robert Z. Callaham, *Criteria for Deciding about Forestry Research Programs*.

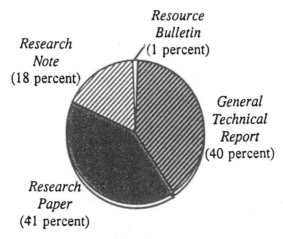

Figure 7.2. Percent of Forest Service research publications by series title from Core Agricultural Literature Project citation analysis.

1. Generated income/employment in forest industry or regional economy.
2. Increased utilization of natural resources.
3. Improved quality of physical/biological environment.
4. Lowered prices or costs to consumers.
5. Reduced costs for managing resources.
6. Improved methods for planning and evaluating alternative investments.
7. Resulted in new and improved products.
8. Improved visual environment and related amenities.
9. Increased resource productivity.
10. Reduced costs through improved processes.
11. Improved scientific methods and theory.
12. Improved social environment.
13. Raised quality and lowered cost of housing.
14. Enhanced health and safety.
15. Improved cultural/historical/geological environment.
16. Enhanced public involvement in decisionmaking.

Callaham found that monetary benefits could be shown for twenty-two of the eighty-one innovations, and that benefits from the remaining innovations, although equally important, were qualitative in nature and not easily quantified. He estimated that the monetary benefits, $2.6 billion, from just the twenty-two innovations would in the first year alone, exceed the cost of all prior Forest Service research, or would pay for the next twenty-four years of research of the agency at 1979 costs. Although he did not estimate an internal rate of return (the rate of interest earned from an investment) from forest research, he concluded that investments in such research have yielded a high rate of return, possibly parallelling those from agricultural research which range from 30% to 40%.

Perhaps the best measure of the influence of research and publishing efforts is the acceptance and use of the findings by forest practitioners and fellow researchers. Following are several examples of Forest Service research results currently being used intensively.

Research at the Southeastern Forest Experiment Station, Asheville, North Carolina, pioneered the introduction of selected mycorrhizal fungi on the roots of conifer seedlings in tree nurseries, permitting the trees to survive when planted on highly acidic, coal mine reclamation sites.[45] This technology has been widely adopted in the United States and is being demonstrated extensively in many other countries.

Scientists at the North Central Forest Experiment Station, St. Paul, Minnesota, have demonstrated a direct relationship between emissions from fossil fuel combustion and the acidity of precipitation, as evidenced by near background levels of acidic deposition in northwestern Minnesota, increasing to high levels in southeastern Michigan.[46] Studies of lakes along this deposition gradient found evidence of water chemistry changes related to acidic deposition. Soil and tree tissue along the gradient showed that the amount of sulfate deposited in acid precipitation is related to the amount of sulfur in soil and trees, suggesting that effects of acid deposition may be cumulative and that continued monitoring is advisable. These results helped set pollution emission/deposition limits in Minnesota and Wisconsin, and helped develop guidelines for evaluating permits for new emission sources that may affect air quality in wilderness areas in the Northeast.

Fire behavior models developed by Intermountain Research Station scientists, with administrative offices in Ogden, Utah, can be applied universally after the fuel model appropriate for a specific site is selected. The fire model will predict spread rates, intensity, flame dimensions, and expected size of fires in forests or rangelands. These predictions are based on such variables as type of fuel, fuel moisture, slope, and wind velocity.[47] The expected behavior of wildfires or prescribed fires can be assessed with these models for real-time field situations or for fire management plans. The pro-

45. D. H. Marx, *The Practical Significance of Ectomycorrhizae in Forest Establishment* (Falun, Sweden) Marcus Wallenberg Foundation Symposia Proceedings, 7 (1991): 54–90.

46. L. F. Ohmann and D. F. Grigal, "Spatial and Temporal Patterns of Sulfur and Nitrogen in Wood of Trees Across the North Central United States," *Canadian Journal of Forest Research* 20 (1988): 508–513.

47. R. C. Rothermel, *How to Predict the Spread and Intensity of Forest and Range Fires* (Ogden, Utah: USDA Forest Service, 1983 [*General Technical Report* INT-143]).

cedures are routinely applied by fire managers in the United States and in other countries.

Researchers at the Southern Forest Experiment Station, New Orleans, Louisiana, developed laboratory screening techniques for prospective insecticides to control termites, and, in addition, developed borate wood preservation treatments to control termites and other wood-destroying insects. They also devised methods for field testing the safety and effectiveness of the insecticides.[48] Any prospective termiticide must pass the two-year screening and five-year (minimum) field testing to be registered by the Environmental Protection Agency for wider use in the United States. These screening and field testing techniques are used broadly throughout the world where termites are a problem, and any termiticides used in the United States must have passed these tests.

Forest Products Laboratory scientists in Madison, Wisconsin, have developed techniques for biopulping by harnessing white-rot fungi to degrade lignin in wood chips to soften them before they are used to produce pulp.[49] Biopulping is superior to other methods because it saves 25% to 50% of the energy needed to make pulp by mechanical methods, paper produced from it is stronger by 25% to 200% than conventionally made paper, and pollution is reduced. The intense interest in this work is shown by the fact that a consortium of nineteen pulp and paper companies funded the feasibility research, and a second consortium is funding the research for commercializing the process.

Scientists at the Pacific Southwest Research Station, Berkeley, California, along with academic scientists funded by the Forest Service, studied the California spotted owl and recommended changes in forest management in the Sierra Nevada that would protect this sensitive species.[50] These recommendations were accepted by Forest Service land managers, resulting in extensive amendments to the Land Management Plans of all national forests in the Sierra Nevada range where the owl exists. The Forest Service will draw heavily from these studies as the agency develops a formal Environ-

48. B. M. Kard and C. A. McDaniel, "Field Evaluation of the Persistence and Efficacy of Pesticides Used for Termite Control," in *Pesticides in Urban Environments: Fate and Significance*, ed. K. D. Racke and A. R. Leslie (Washington, D. C.: American Chemical Society, 1993 [American Chemical Society Symposium Series 522]).

49. Kent T. Kirk, Richard R. Burgess, and John W. Koning, Jr., "Use of Fungi in Pulping Wood: An Overview of Biopulping Research," in *Frontiers in Industrial Mycology. Proceedings of Industrial Mycology Symposium; June 25–26, 1990*, ed. Gary F. Leatham (Madison, Wisc. and New York: Routledge, Chapman, and Hall, 1992), pp. 99–111.

50. Jared Verner, Kevin S. McKelvey, Barry R. Noon, R. J. Gutierrez, Gordon I. Gould, Jr., and Thomas W. Beck, *The California Spotted Owl: A Technical Assessment of its Current Status* (Albany, Calif: USDA Forest Service, Pacific Southwest Research Station, 1992 [*General Technical Report* PSW-133]).

mental Impact Statement to establish long-term guidelines for protection of the owl.

At the Rocky Mountain Forest and Range Experiment Station, Ft. Collins, Colorado, researchers are developing techniques for applying traditional agroforestry (growing trees in combination with agricultural crops to increase land productivity) and conservation agroforestry (working trees into agricultural and community ecosystems for human, economic, and environmental benefits) on semiarid sites.[51] Examples of semiarid agroforestry practices include windbreaks, living snowfences, livestock havens, riparian buffer strips, and wildlife habitat plantings. Through publications, education, demonstration, and application programs, these techniques are becoming widely known and are used by landowners and land managers of semiarid lands in the United States and internationally.

Results of research by scientists at the Northeastern Forest Experiment Station, Radnor, Pennsylvania, indicate that calcium is being depleted from eastern forest soils by leaching (probably induced by acid deposition) and by frequent whole-tree clearcutting, at a rate that could remove 50% of total soil and biomass calcium within 120 years.[52] This level of loss would result in serious calcium deficiencies and subsequent reduced tree growth. As a result of these findings, timber harvest practices in the eastern United States now leave more large, woody residue on the site. These research findings have also influenced national policy on air pollution regulations.

Scientists at the Pacific Northwest Forest and Range Experiment Station, Portland, Oregon, have demonstrated the importance of leaving large woody debris in streams, for the salmon and sea-going trout fisheries in the western United States.[53] This material, which was routinely cleared from streams in the past as a habitat improvement technique, was found to provide cover for fish; to protect fish from high water velocities and turbulence; to create pools, backwaters, and undercut banks; to aid fish migration during periods of low water; and to trap sediment that might otherwise smother spawning beds. As a result of these findings, public land management agencies have changed their policies about large instream debris.

51. W. J. Rietveld, J. R. Brandle, and G. A. Kuhn, *The Role of Trees in Sustainable Agricultural Systems*, Proceedings Great Plains Agricultural Council Forestry Committee, June 18–20, 1990, Colorado Springs, Colo., 1990), pp. 104–107.

52. C. A. Federer, J. W. Hornbeck, L. M. Tritton, C. W. Martin, R. S. Pierce, and C. T. Smith, "Long-Term Depletion of Calcium and Other Nutrients in Eastern U. S. Forests," *Environmental Management* 13 (1989): 593–601.

53. William R. Meehan, ed., *Influences of Forest and Rangeland Management on Salmonid Fishes and Their Habitats* (Bethesda, Md.: American Fisheries Society, 1991 [Special Publication no. 19]).

8. The Emergence of Sustainable Forestry

JAMES COUFAL

College of Environmental Science and Forestry,
State University of New York, Syracuse

DONALD WEBSTER

College of Environmental Science and Forestry,
State University of New York, Syracuse

Although sustainable forestry appears to be a worldwide topic of concern, this chapter focuses on the United States with some emphasis on Canada. Without implying complete transference of experience, this investigation can serve as an overview of the broad issues in sustainable forestry. Coverage will concentrate on the last twenty-five years. These changes underway will greatly alter the scholarship and literature of forestry.

Sustainable forestry must be approached with common grounds and understandings. The Society of American Foresters(SAF), says that forestry is "the science, the art and the practice of managing and using for human benefit the natural resources that occur on and in association with forest lands."[1] While this definition is broadly conceived, the practice of forestry has tended to focus on certain aspects, characterized by "the continuing use of their (forest) resources, material or other," "the husbandry of tree crops," and "the profitable exploitation of the resources intrinsic to forest land."[2] The discussion which follows will attempt to indicate how the practice of forestry science has overshadowed art, commodities have overshadowed amenities, and production has overshadowed the production system; in short, that the latter item in each of these cases is what is emerging as part of sustainable forestry. There must be understanding and some consensus on what is to be sustained before we can determine how to do it.

Sustainable forestry (taken here to be the same as environmental forestry) is but one part of the larger issue of sustainable development about which

1. F. C. Ford-Robertson, ed., *Terminology of Forest Science, Technology, Practice, and Production* 2d printing (Washington, D.C.: Society of American Foresters, 1983), p. 109.
2. Ibid., p. 9

147

there is controversy. The United Nations' World Commission on Environment and Development in their report, *Our Common Future,* defines sustainable development as a form of development or progress that "meet[s] the needs and aspirations of the present without compromising the ability to meet those of the future."[3] This definition has formed the basis of discussion of sustainable forestry as in Canada where, in the words of one study, "while meeting the needs of the present generation we must not place at risk the ability of future generations to meet their own needs. This means that development in the forest ecosystem must occur within environmental limits and that forest-related development must comply with nature."[4] A similar theme is offered from Sweden, where, as one commentator wrote, "forestry shall be performed in such a way that the total welfare benefits of forestry activities never decreases in the future. These welfare effects include benefits from volume yield and wood quality as well as from environmental factors."[5]

Even with some consensus on these rather loose characterizations of sustainable forestry, as much ethically as scientifically based, there is a need to move toward rationally-based, long-term strategies to make sustainable forestry a reality. Here the complexities begin. Sustainable forestry must operate at multiple scales, over long time periods, and balance land health with human needs.[6] Thus it is that R. P. Gale and S. M. Cordray ask "What should forests sustain?" and offer eight approaches that "imply distinctly different political consequences."[7] G. L. Baskerville offers sustainability as a "flexible" concept that needs to determine the land base at issue and answer the questions "What is being allowed and what is being sustained" if we are to match expectations with reality.[8]

The concept of sustainable forestry must consider context, space, and time, and be fundamentally concerned with the method of production as

3. FAO World Commissions on Environment and Development, *Our Common Future* (the Bruntland Commission Report) (New York: Oxford University Press, 1987), p. 40.

4. J. P. Hall, L. W. Carlson, and D. E. Dube, "A Forestry Canada Approach to Environmental Forestry," *Forestry Chronicle* 66 (1990): 138.

5. B. Hagglund, "Sustained-Yield Forest Management: The View from Sweden," *Forestry Chronicle* 66 (1990): 29.

6. H. Salwasser, "Some Perspectives on People, Wood, and Ecological Thinking in Forest Conservation: Why All the Fuss about Forests?" in *Southern Mosaic: Proceedings of the Southwestern Region New Perspectives University Colloquium* (Fort Collins, Colo.: Rocky Mountain Forest and Range Experiment Station, 1991 [*USDA Forest Service General Technical Report* RM216]).

7. R. P. Gale and S. M. Cordray, "What Should Forests Sustain? Eight Answers," *Journal of Forestry* 89 (1991): 36.

8. G. L. Baskerville, "Sustained Yield Management-Expectations and Realities," *Forestry Chronicle* 66 (1990): 26.

well as with the end product. Baskerville offers a succinct description of a linear progression:

> It is common to characterize social desires associated with the sustainable use of public forests as a form of succession. In the early stages of this succession, the social need is to clear the forests making way for agriculture. In the secondary stages of social succession, forest acquire value as a basis for industry, jobs, and taxes. In later stages there emerges a social desire for stability and economic security. In the climax stages of this succession, society becomes physically and intellectually separated from the forest, and social desires center on environmental well-being.[9]

It is likely that this climax stage will change as it faces natural disturbances. Forestry is moving from a "gospel of efficiency"[10] to a new environmentalism that lives by a "gospel of ecology," to provide sustaining options for the next generation.[11]

A. Historical Antecedents

The divine sanction to exploit found in Genesis 1:28, and later implied in the catch phrase Manifest Destiny, fostered unfettered use of natural resources and rapid territorial expansion. John Muir, when speaking of the pioneers' view of trees, suggested that they considered them little more than pernicious weeds. The earliest settlers also came seeking paradise, and this duality of relation of the puritans to the wild forest is summed by R. Nash: "Paradoxically, their sanctuary and their enemy were one and the same."[12] The first settlers came in search of spiritual values or the dollar, and this duality recurs throughout the history of conservation in the United States. To the religious the new world represented a wilderness to be tamed in the name of God. To others it represented unlimited natural resources to be exploited for profit. Both views were incompatible with conservation.

Exceptions to both positions evolved as attitudes changed in the nineteenth century. Some began to see the wilderness as a place for the perception and worship of God. Ralph Waldo Emerson and the transcendentalists

9. Ibid., p. 28.
10. S. P. Hays, *Conservation and the Gospel of Efficiency: The Progressive Conservation Movement, 1890–1920* (Cambridge: Harvard University Press, 1959).
11. (a) R. Nash, ed., *The American Environment: Readings in the History of Conservation* (Reading, Mass.: Addison-Wesley, 1976). (b) S. Postel and J. C. Ryan, "Toward Sustainable Forestry Worldwide," *Journal of Soil & Water Conservation* 46 (1991): 119–122.
12. R. Nash, *Wilderness and The American Mind*, 3d ed. (New Haven: Yale University Press, 1967), p. 35.

espoused that "nature was the symbol of the spiritual world and the container of the moral truths which permeated the universe."[13] Henry Thoreau's love of nature and his belief that the wilderness was the source of creativity influenced many preservationists. But it was George Perkins Marsh who set the stage for ushering in scientific understanding of nature and natural resources. In *Man and Nature* he wrote: "Man has too long forgotten that the earth was given to him for usufruct[14] alone, not for consumption, still less for profligate waste."[15]

According to Thomas Cox, Marsh's *Man and Nature* helped create an intellectual climate favorable to scientific studies of the natural environment—studies such as those of forester Bernhard Edward Fernow, hydrologist John Wesley Powell, naturalist C. Merriam Hart, and others.[16] The 1865 *Report of the Commissioner of Agriculture* of the United States charged the reader to "Hear G.P. Marsh." In the same report, the Rev. Frederick Starr, Jr., wrote "American Forests; Their Destruction and Preservation" in which he called for "extensive, protracted, and scientific experiments in the propogation and cultivation of forest trees be established . . . we should study and learn for ourselves what our country can do for its native trees, and what our trees can do for our country."[17]

Starr's work along with others of his time is remarkable in a number of ways. For one, in pre-Mendelian times, he vividly depicts the problems of what we now call "high-grading," or "cut the best and leave the worst"; he makes recommendations on what to cut and how to cut it to prevent damage to residual trees; he recognizes potential climatic consequences of overcutting, and declares government involvement in forest management to be necessary because of the long-term nature of forest growth. His message is very similar to that being given today, in developed and developing countries. One statement in particular became a focal point of forestry concern in North America.

The nation has slept because the gnawing of want has not awakened her. She has had plenty to spare; but within thirty years she will be conscious that not

13. R. Nash, "The American Wilderness in Historical Perspective," in *From Conservation to Ecology: The Development of Rural Concern* (New York: Crowell, 1973), p. 9.
14. usufruct: The right of enjoying or using the fruits or profits of an estate or other things belonging to another, without impairing the substance.
15. G. P. Marsh, *Man and Nature; or, Physical Geography as Modified by Human Action* (New York: Scribners, 1864), p. 36.
16. T. R. Cox, R. S. Maxwell, P. D. Thomas, and J. J. Malone, *This Well-Wooded Land* (Lincoln: Nebraska University Press, 1985), p. 164.
17. F. Starr, Jr., "American Forests: Their Destruction and Preservation," in *Report of the Commissioner of Agriculture for the Year 1865* (Washington, D.C.: Government Printing Office, 1865), p. 218.

only individual want is present, but that it comes to each from permanent national *famine* of wood.[18]

The threat of timber famine has played a large part in the history of forest policy and governments' role in forestry since that time. Starr ended with the reminder that "God has given us a great and goodly heritage, a grand and broad and luxuriant country; but it is our forests that have made this country so salubrious, so fertile."[19]

As the social temper of the times changed, forests came to be seen as important because of their imminent threat of depletion. In 1873 the American Association for the Advancement of Science appointed a committee whose task was "to memorialize Congress and the several state legislatures upon the importance of promoting the cultivation of timber and the preservation of forests and to recommend proper legislation for securing these objects."[20] In 1875 the American Forestry Association (AFA) was organized to promote forestry and timber culture. In 1876, Congress appropriated $2,000 to the Department of Agriculture to begin investigations into the use and condition of the forests. In 1882 the AFA sponsored the first American Forestry Congress, in which governors of various states participated.[21] This all occurred before the advent of the first professional forester in the United States.

According to the *Encyclopedia of American Forests and Conservation History*, from 1876 to 1886, Franklin Hough and his successor Nathaniel Eggleston, Chief of the Division of Forestry "collected, analyzed and abstracted available information and published four voluminous reports and a number of short bulletins."[22] The article further states that "Although there are earlier incidences that could be considered, continuous forestry research began in the United States with these efforts."[23] The year 1886 saw the Division of Forestry formally recognized by Congress and headed by the first professional forester to practice in the United States, Bernard E. Fernow, who comprehended the need for and the importance of research to provide a scientific and economic as well as a practicing base for forestry. Fernow resigned in 1898 to found one of the first forestry schools at Cornell University. Gifford Pinchot, his successor, appreciated research "but was

18. Ibid, p. 219.
19. Ibid., p. 234.
20. Henry Clepper, ed., *Origins of American Conservation* (New York: Ronald Press, 1966), p. 43.
21. Ibid.
22. T. F. McLintock, "Forestry Research," in *Encyclopedia of American Forest and Conservation History*, Vol. 1, ed. R. C. Davis (New York: Macmillan, 1983), p. 240.
23. Ibid., p. 240.

impatient and too busy with administrative affairs to pay much attention to it."[24]

Although it did not happen quickly, scientific forestry by the early twentieth century began to secure its place on the American scene. The early years of the twentieth century saw science and forestry melding into a force of considerable influence. The conservation movement could now turn to science to begin eliminating "profligate waste" and to rein in exploitation. But there was conflict. As some depended on science to make forests more efficient producers of wood and forest products; others, followed the transcendental example using aesthetics to make a case for preservation.

Gifford Pinchot who was not opposed to aesthetically pleasing forests nevertheless put his "wise use" policy first to ensure the economic well being of the nation. John Muir, who believed that the wilderness was God's temple, opposed "wise use" with "right use." The utilitarian position of Pinchot clashed with the "right or higher uses" of Muir. Later, Aldo Leopold with his land ethic elaborated on the usufruct theme earlier enunciated by Marsh. Man was part of a community, not a possessor of a commodity, in nature with all of its life forms and natural resources.

B. From Protection to Management

Forestry before the 1940s has been characterized as being "generally custodial or focused on meeting demands for resources in the surrounding area."[25] But the demand for timber increased after World War II affecting a shift from private forests to national forests. Such a shift created renewed need for scientific management of forests to ensure continuous flow of goods and services. Two concepts emerged from this shift and lead directly to sustainable forestry: sustained yield and community stability. The focus now centered on national forests because they were the lands where these concepts developed which in turn influenced landowners public and private.

In 1932, H. S. Graves addressed the annual meeting of the SAF on the current critical issues in forestry.[26] One of his major concerns was the rapid liquidation of private timber resources, almost exclusively timber, land ownership, land use regulations, land taxes, competition from the sale of publicly owned timber, and community stability. Regarding the latter, he noted:

24. Ibid., p. 241.

25. D. W. MacCleery, *American Forests: A History of Resiliency and Recovery* (Durham, N.C.: Forest History Society, 1992 [USDA Forest Service FS-540]), p. 43.

26. H. S. Graves, "Some Critical Issues in Forestry," *Journal of Forestry* 31 (1933): 131–140.

Of outstanding importance is the fact that the forest furnishes the basis for a great variety of industrial activities, through which people are employed, communities and institutions built up, taxable wealth created, and a contribution made by the region, where the forests are located, to the organized economic and social structure of the state and nation. These services cannot be rendered if, through a weak and uneconomic type of ownership, there are periodic interruptions of employment, recessions in values, delinquency in tax support of public enterprises, and constant danger of collapse of individual industries. Under such circumstances, the communities dependent on the forest are built on a shell that may crack at almost any time.[27]

What community stability meant and how it was to be obtained was an issue under debate, but that it was to be obtained seemed to become a tenet of forestry. The White Pine Blister Rust Protection Act of 1940 apparently was the first time that maintaining community stability was noted as the purpose of an act of the federal government.[28] Hinged to the ability of a forest to contribute to community stability was the necessity for it to produce sustained yield which also underwent a metamorphosis of definition.

T. Parry et al. trace the changing conceptions of sustained yield in two ways: first the U.S. Forest Service's sustained yield policy (an operational definition), and then by its linkage to broader societal concerns (a social meaning).[29] The "Organic Act" of 1897 (30 Stat. 34) stated that a major purpose of the forest reserves was "to furnish a continuous supply of timber for the use and necessities of citizens of the United States."[30] The "Multiple Use-Sustained Yield Act" (74 Stat. 215) defined sustained yield as "the achievement and maintenance in perpetuity of a high-level annual or regular periodic output of the various renewable resources of the national forests without impairment of the productivity of the land."[31] Sustained yield has been defined operationally in a number of ways. Parry et al. described sustained yield as:

1. harvest equal to growth; based on a regulated forest and brought by Gifford Pinchot from his European experience.
2. harvest based on regulating the growing stock rather than on regulating the yield; based on increasing data and recognition of the surplus of old timber in the U.S. and the non-regulated condition of the forest.

27. Ibid., p. 133.

28. Society of American Foresters, *Report of the Society of American Foresters National Task Force on Community Stability* (Bethesda, Md.: Society of American Foresters, 1989), p. 42.

29. T. Parry, H. J. Vaux, and N. Dennis, "Changing Conceptions of Sustained-Yield Policy on The National Forests," *Journal of Forestry* 81 (1983): 150–151.

30. *U.S. Statutes at Large* (Washington, D.C.: U.S. Government Printing Office) (30): 34.

31. *U.S. Statutes at Large* (Washington, D.C.: U.S. Government Printing Office) (74): 215.

3. harvest based on limiting the average annual cut to the continuous production capacity of the forest; largely influenced by David T. Mason and effectuated during the depression when community stability became a major selling point.[32]

Mason influenced the passage of the Sustained-Yield Forest Management Act of 1944 (58 Stat. 132),[33] which set up public/private cooperative sustained yield units that expanded and guaranteed timber supply for participating companies without requiring them to pay carrying charges, implied that the Forest Service could stabilize communities simply by regulating timber harvests, and in local areas gave monopoly control over public stumpage to specific companies. The critical point is that: "During this time, sustained yield (as distinct from forestry) became equated in the public mind with community stability (and) . . . the social meanings and operation definition became closely lined. Sustained yield was no longer just a technical matter, it was a social issue as well."[34]

The definition of sustained yield continued to shift such that it became:

4. an even flow harvest that shifted from regulation of growing stock; occurring in the late 1940s and early 1950s as the Forest Service shifted from a minor to a major supplier of timber (especially in the west) and as the timber industry became highly dependent on national forest timber.
5. harvest that featured nondeclining even flow where stabilization of communities and opportunities for employment were written into law (1963, *Code of Federal Regulations* 221.3(a)(3)); occurred during a time of pressure from a highly dependent timber industry and an awakening environmental movement.[35]

The Forest Service's concept of nondeclining, even flow was put into the 1976 *National Forest Management Act* (88 Stat. 476)[36] despite at least fifty years of history where "the social meaning of sustained yield has had an important effect on its operational definition, *despite a lack of evidence to show that desired social ends are, in fact, linked to their operational means.*"[37]

Review of the changing concepts of sustained-yield as they developed in relation to the idea of community stability provides several key elements of discussion in the emergence of sustainable forestry, including the following:

32. Parry, "Changing Conceptions," p. 150–154.
33. *U.S. Statutes at Large* (Washington, D.C.: U.S. Government Printing Office) (58): 132.
34. Parry, "Changing Conceptions," p. 152.
35. Ibid., p. 153.
36. *U.S. Statutes at Large* (Washington, D.C.: U.S. Government Printing Office) (88): 476.
37. Ibid., p. 154.

1. A dependable flow of resources may be necessary to community stability (however defined), but it is not sufficient. Consideration needs to be given to such items as world market conditions and demand shifts, increased efficiency and changing technology in the forest-products industry, demands for other uses of the forestland, and the diversity of local economies and the spirit of local peoples.[38] Changing traditional beliefs (e.g., community stability is directly and/or largely dependent on timber harvest) is not easy. There is a danger in focusing on single issues in sustainable forestry (e.g., biodiversity) such that we create new, simpleminded traditional beliefs.

2. Forestry (and sustainable forestry) is, at bottom line, as much a social phenomena as it is an ecological one. It is this recognition that, in part, is creating the move toward "ecological economics," whose goal is defined as seeking "to elaborate the connections between ecosystems and society's attempts to meet its goals—broadly defined as survival and a part of the ecological life-support system."[39] This will call into question such fundamental concepts as development and property rights.

3. The concurrent emergence of multicultural diversity and biodiversity are powerful concepts in sustainable forestry, and they start at the local (site-specific) level.[40]

4. As mediators between social needs and ecosystem realities, sustainable forestry has to involve foresters. At the beginning of the environmental era, Zivnuska warned that "If foresters are to have a responsible role in the face of unpredictable change, I think we must move up in our goals and accept our responsibility to be environmental managers rather than simply tree-growers or land managers."[41] To date, this issue has not been resolved and is often only implicit in the current debates about sustainable forestry, biodiversity, ecosystem management, and other current issues.

Because each of these and many other issues are still under debate, sustainable forestry will be learned by doing.

C. The Emergence of Sustainable Forestry

Aldo Leopold pondered the question of right use and expressed his understanding in practical terms: "I have read many definitions of what is a conservationist, and written not a few myself, but I suspect the best one is written not with a pen, but with an axe. *It is a matter of what a man thinks*

---

38. Society of American Foresters, *Report of the Society*.

39. D. P. Bradley and B. J. Lewis, "Ecological Economics: Integrating Natural and Social Dimensions," *Journal of Forestry* 90 (1992): 30.

40. Society of American Foresters, *Reports of the Society*, p. 37.

41. J. A. Zivnuska, "The Forester Prophet, Policy-Maker, or Professional?" *Journal of Forestry* 64 (1966): 819.

about while chopping, or while deciding what to chop. A conservationist is one who is humbly aware that with each stroke he is writing his signature on the face of his land. Signatures of course differ, whether written with an axe or pen, and this is as it should be" (italics added).[42]

The emergence of sustainable forestry does not mean that foresters are abandoning traditional forestry, but rather they are changing because of increased knowledge of ecosystems and changing societal values. Included in this change is a realization of an increased connection between economics and ecology; an increased awareness of the relationships of social justice and environmental justice, including the need for stockholders and stakeholders to participate in land use decisions that will impact them; and pressures from within the profession of forestry.

Sustainable forestry has been emerging for hundreds of years, but its day of arrival is closest to April 20, 1970, the first Earth Day. A. Backiel recently wrote that the difference between past changes and the new one is that "we have no more frontiers" to move to. We can no longer—either because of physical, political, or ethical realities—act as freemen and occupy and use our freeholds only to abandon them and move on to new land.[43] Backiel added that if the frontier thesis is finished, we need a new thesis to describe our goals. In comparing potentials for sustainable forestry in Europe and the United States, N. E. Koch claims the accelerating rates of change in forestry—the new goals—are explained by western societies evolving from rural agricultural states to urban, postindustrial societies. In forest science this is reflected in the way social values are communicated— from economic system dominance in communicating forest utilitarian values, to social and political systems which are more effective in communicating non-utilitarian and non-market forest values.[44]

There are other concerns and conflicts regarding the need for and potential implementation of sustainable forestry, but perhaps the most important one is a conflict of values. J. F. Franklin feels that two popular polarized views of forests have emerged: those of environmentalists and those of the traditional forester.[45] A. G. McQuillan believes that the advent of new forestry in the United States represents a traumatic shift in the philosophy of national forestry praxis, a broadening of values to include aesthetics and

---

42. A. Leopold, *A Sand County Almanac* (New York: Ballantine Books, 1966), p. 73.

43. A. Backiel, *Environmental Policy: Changing Attitudes, Expectations, and Aspirations* (Syracuse, N.Y.: SUNY College of Environmental Science and Forestry, 1993 [*Faculty of Forestry Miscellaneous Publication* no. 28]).

44. N. E. Koch, *Sustainable Forestry: Some Comparisons of Europe and the United States* (Corvallis, Ore.: Oregon State University Press, 1990).

45. J. F. Franklin, "An Ecologist's Perspective on Northwestern Forests in 2010," *Forest Watch* 10 (1989): 6–9.

sustainability of a natural ecological process; and coming closer to the "land ethic" of Aldo Leopold and away from the ethics of traditional forestry which are seen as "stoic utilitarian" and positivist.[46]

Stoic utilitarianism usually converts to a criticism that forestry has too long focused on the sustainability of timber and not the forest ecosystem.[47] This is partly due to the bureaucratic compartmentalization of renewable resources that result in a forestry and failed multiple-use, sustained-yield paradigm that seeks to perpetuate the physical supply of independent substances and services. A new paradigm of multiresource forest management would seek the simultaneous production of general interdependent substances and services.[48] H. Rolston and J. E. Coufal provide a list of instrumental and intrinsic values that they believe justify multi-value forest management,[49] which turn out to be similar to a list of values that go beyond economic justifications to serve the end of "the conservation of non-resources" as enunciated by Ehrenfeld.[50] In both cases, the values are largely anthropocentric and instrumental, but both also include intrinsic value of the land, with no consideration of human uses or values, as part of multi-value forest management.

Conflicts about sustainable forestry range from values to definitions to trust to technical matters. First, there is the question of what sustainable forestry really means, or how to define it operationally.[51] Perhaps at bottom there is agreement that forests present environmental opportunities but also have environmental limits, they degrade when these limits are exceeded, and sustainable forestry requires managing the impacts of human activities within these tolerance limits.[52] On the one hand, there is conflict regarding the sincerity of efforts to implement ecosystem management via "new for-

46. A. G. McQuillan, "New Perspectives: Forestry for a Post-Modern Age," *Western Wildlands* 17 (1992): 13–20.

47. L. S. Minkler, "Ecological Forestry for Renewable Forest Values," *Forest Watch* 11 (1990): 19–24.

48. R. W. Behan, "Multiresource Forest Management: A Paradigmatic Challenge to Professional Forestry," *Journal of Forestry* 88 (1990): 12–18.

49. H. Rolston III and J. E. Coufal, "A Forest Ethic and Multivalue Forest Management," *Journal of Forestry* 89 (1991): 35–40.

50. D. W. Ehrenfeld, "The Conservation of Non-Resources," *American Scientist* 64 (1976): 648–656.

51. (a) J. Mathews, "Forestry Word Games: Sustain," *Journal of Forestry* 89 (1991): 29–30. (b) G. R. Milne, "Sustainable Development: What Does It Really Mean?" *Forestry Chronicle* 67 (1991): 105. (c) T. O'Keefe, "Holistic (New) Forestry: Significantly Different or Just Another Gimmick?" *Journal of Forestry* 88 (1990): 23–24. (d) C. D. Rannard, "Sustainable Development (in Forestry): What Does It Really Mean?" *Forestry Chronicle* 67 1991: 109–112. (e) R. Shearman, "The Meaning and Ethics of Sustainability," *Environmental Management* 14 (1990): 1–8. (f) Gale, "What Should Forests Sustain?: Eight Answers."

52. J. S. Maini, "Practicing Sustainable Forest Sector Development in Canada: A Federal Perspective," *Forestry Chronicle* 67 (1991): 107–108.

estry" practices[53] even when new forestry has tenets rooted in biodiversity, ecological resilience, and long-term sustainability.[54] On the other, questions of whether the cumulative, worldwide effects of reduced timber supply caused by implementing "new forestry" (ecosystem management) have been considered sufficiently. Will there be trade-offs with environmental gains in one preserving region with off-setting losses in foreign regions?[55]

Do we know enough to be able to implement ecologically sustainable practices? For example, can we develop and have confidence in hybrid yield models that combine traditional mensurational models with a computer simulation of those growth-regulating processes (and do we understand them to begin with) that are significantly altered by changing management practices and/or by changing atmospheric chemistry and climate?[56] Can we get agreement between those interested in the productivity of a site—such as economists, silviculturists, and ecologists—on the definition of productivity and how to measure it and rates of change in it?[57] Or again, can we learn to measure and evaluate cumulative effects on ecosystem productivity that have not been explicitly addressed in the complexity of biological-physical interactions over large areas and time scales?[58]

The emergence of sustainable forestry hinges on the reexamination of two fundamental questions. What is the unique purpose of forestry that sets it apart from other resource professions? The second has to do with a proper environmental or land ethic for forestry which brings forth sub-questions such as: What are the ultimate values of forests that foresters cherish, profess, and act on with consistency? Operational definitions of sustainable forestry will be successful only if they are based on sound answers to such questions.

D. Sustainable Forestry Today

At least two themes seem to be common in the debate, despite the lack of agreement on what sustainable forestry or ecosystem management are.

53. (a) G. Aplet, "Forest Health: Ecological Crisis or Timber-Driven Hype?" *Forest Watch* 13 (1992): 19–22. (b) C. Master, *The Redesigned Forest* (San Pedro, Calif.: R&E Miles, 1988).

54. C. Fiedler, "New Forestry: Concepts and Applications," *Western Wildlands* 17 (1992): 2–7.

55. C. H. Schallau and A. Goetzl, "Effects of Constraining U.S. Timber Supplies," *Journal of Forestry* 90 (1992): 22–27.

56. J. P. Kimmins, "Future Shock in Forest Yield Forecasting: The Need for a New Approach," *The Forestry Chronicle* 61 (1985): 503–512.

57. D. E. Ford, *What Do We Need to Know about Forest Productivity and How Can We Measure It?* (Portland, Ore.: USDA Forest Service, 1983 [Report PNW-85]). p. 2–12.

58. G. E. Grant and F. Swanson, "Cumulative Effects of Forest Practices," *Forest Perspectives* 1 (1992): 9–11.

First, use of the land today must be done in a way that future generations will have equivalent, though not necessarily identical, options. Thus, sustainable forestry, as in the social forestry movement of the less developed countries, is more clearly tied to economic and social justice than traditional forestry has been, including the fundamental questions of who benefits, who pays, and who participates in land use allocation and management decisions.

Second, there is general agreement that to accomplish the first foresters must clearly think more in terms of systems. While this is often taken to mean that sustainable forestry means ecosystems management on the scale of watersheds and landscapes, it must also come to recognize the "holism" that is desired refers as much or more to completeness than to scale.[59] That is, that site specific prescriptions must be holistic in and for themselves, as well as being part of the larger scale of holistic ecosystem management as generally perceived.

Perhaps a third idea is also emerging as a common theme in the discussions of sustainable forestry and sustainable development: "Resource problems are not really environmental problems. They are human problems that we have created at many times and in many places, under a variety of political, social, and economic systems."[60] As such, we must include humans, in all their dimensions, as part of the ecosystem, and we must recognize that science and technology alone have not and likely will not prevent environmental exploitation or degradation.

In the context of disagreement of what is sustainable forestry and how is it accomplished on the ground, two models are summarized as representative of how sustainable forestry is being looked at. After tracing the social and political context and the forestry context, looking at inducements to change in forestry, including the political climate and the role of science, D. J. Brooks and G. E. Grant offer the following "significant tenets" or guiding principles that underlie the issues that are pushing for broad changes in the philosophy of forest management.[61] These are:

1. Forest management decisions must be based on an ecosystem perspective, and one which take into account uncertainty about our understanding of the system and about future conditions.
2. The effects of forest management need to be evaluated over a range of spatial

59. N. Richards, personal communication, June 1993.

60. D. Ludwig, R. Hilborn, and C. Walters, "Uncertainty, Resource Exploitation, and Conservation: Lessons from History," *Science* 260 (1993): 36.

61. D. J. Brooks and G. E. Grant, "New Approaches to Forest Management," *Journal of Forestry* 90 (1992) (Part I): 25–26; (Part II): 21–24.

scales, from microsite to forest stand to forest property to watershed to landscape and to region.

3. The effects of these decisions must be evaluated using ecologically relevant time scales, which bring into question such issues as cumulative effects, resilience, long-term site productivity and long-term population viability.

4. Because of uncertainty regarding the forest ecosystems and human social systems, today's decisions must be made to maintain as many future options as possible, recognizing that making a given decision may foreclose other future decisions.

5. Polarization must be overcome, and the full range of forest users must be encouraged to participate actively as equal partners in forest planning decisions.

Of interest, and closely related to #5 above, Brooks and Grant reject the idea that the current troubles faced by forest managers, and the current changes being suggested by many are the result of the recent emergence of "radical environmentalists." Rather, they believe that the changes are also the result of long-term forestry, and related to new scientific knowledge.[62]

The second model is that from the Society of American Foresters' 1993 Task Force report.[63] In 1991, based upon a grassroots request from the society's House of Society Delegates, SAF's president appointed a task force from the diverse societal membership to evaluate the ways to assure the long-term health and productivity of the nation's forest resources. A case can be made that ensuring long-term health and productivity of the nation's forests is strongly supported and representative of the SAF which is the only organization that can be said to represent forestry and foresters in the U.S. However, the task force's recommendation of ecosystem management as the model to use resulted in spirited discussion within SAF, and was not clearly accepted by all segments of the SAF.

Like Brooks and Grant, the task force looked at sustainable forest productivity within historical perspective, including the development of forest policy, and the social, scientific, and economic forces and issues arguing for change. They defined "sustaining the health and long term forest productivity to include all values and all forests, regardless of ownerships, but with attention to private property rights," and looked to managing these forests cooperatively across ownerships in large landscapes to ensure sustainable production of goods and services for humans in a multigenerational time frame.[64] They concluded that to reach this goal, three criteria will be

62. Ibid., p. 28.

63. Society of American Foresters, *Task Force Report on Sustaining Long-Term Forest Health and Productivity* (Bethesda, Md.: Society of American Foresters, 1993).

64. Ibid., p. 1.

necessary. Each strategy must: (1) maintain the structural and functional integrity of the forest as an ecosystem; (2) meet the diverse needs of the human community; and, (3) commit the technological, financial and human resources needed for implementation.

Based on the ideas encapsulated in Table 8.1, the task force concluded that "traditional sustained-yield management policies, strategies and practices are not fully sufficient for meeting contemporary social, political and technical goals"; they do not, in other words meet the criteria as listed above, and therefore proposed an ecosystem approach to forest management and research that would involve all ownerships at the landscape level.[65]

In proposing an ecosystem management approach which focuses on achieving a desired forest condition rather than on the production of a product or set of products, the task force made twenty-six specific recommendations under four major themes: (1) advocacy of ecosystem management; (2)

Table 8.1. A simplified comparison of management strategies

	Traditional sustained-yield	Ecosystem management
Objective processes	Sustained flow of specific products to meet human needs, constrained to minimize adverse effects	Maintains ecological and desired forest condition within which the sustained yield of products to meet human needs are achieved
Strategy for accomplishment	Resembles the agricultural model	Reflects patterns of natural disturbance
System character	Emphasizes production efficiency within environmental constraints	Retains complexity and processes, provides framework for the whole system
Unit of management	Stands and aggregations of stands within an ownership	Landscapes & aggregations of landscapes across ownerships
Time unit	Multi-rotations with rotation length determined by landowner objectives	Multi-rotations with length reflecting natural disturbance, although intensive management will cause some to be shorter
Current status	In transition, new knowledge is bringing in new values. Remains a valid strategy for portions of the landscape	Evolving, accepted for management on national forest lands

Source: Society of American Foresters, *Task Force Report on Sustaining Long-Term Forest Health and Productivity*, 1993; p. 36.

65. Ibid., p. 7–8.

the integration of ecosystem management into educational programs; (3) promotion of ecosystem management research; and (4) coordination of planning and management among landowners and with the public.[66] The compass of these four themes, and the twenty-six specific recommendations alone, imply a broad area for discussion, debate, and disagreement, which is further indicated by the SAF's Council actions regarding the task force report, which was accepted in December 1992. Such acceptance is usually followed by the adoption of an official SAF position paper on the topic. But the council agreed that the issue of ecosystem management was so important that the SAF members and others needed time to process and dialogue on the topic. How seriously they took this to be is indicated in the following two paragraphs, taken from the *Task Force Report* where the council added a "Foreword."

> The challenge presented by the Task Force is complex, and practical actions that can be taken in response are as uncertain as the consequences are unknown. The quantum changes recommended and the uncertainty of success raise worrisome questions. Are the necessary tools and technology available? (What) are the implications for the use and management of private lands? What mechanisms are available or may be needed to coordinate land and resource activities among landowners across landscapes that may involve many hundreds or thousands of acres? Does the cooperation implied necessarily mean more land-use regulation? How might the forestry profession itself be affected?
>
> Answers to these and other serious questions will not be found easily or quickly. Given the size of the challenge and its implications for society and the forestry profession, the SAF Council invites SAF members and other interested parties to begin a structured, contemplative study and discussion of this report before the Society adopts an official position. Through this approach, SAF members reaffirm their commitment to land stewardship as advocates and practioners of land management consistent with ecologically sound principles.[67]

One might reasonably conclude that the emergence of sustainable forestry will likely be a lengthy and difficult process.

Some simple examples show how this emergence is reflected in the literature. One student undertook a three-month effort aimed at comprehending the emergence, development, impact, and future direction of new perspectives and new forestry through the examination of publications on the sub-

66. Ibid.
67. Ibid., p. viii.

ject and related topics.[68] The resulting annotated bibliography, admittedly incomplete, cited 194 articles, by 146 authors, from twenty-four journals. The breakdown of these cited works by date of publication is as follows:

Table 8.2. Sustainable forestry citations in the annotated bibliography

1990>	=	124
1985–1989	=	36
1980–1984	=	10
1975–1979	=	6
1970–1974	=	12
1965–1969	=	4
1960–1964	=	0
1959<	=	2

Sources: *AGRICOLA; Bibliography of Agriculture.*

A brief computer search of the literature dealing with "sustainable development and forestry" found forty-five references, twenty of which were written in the 1980s and twenty-five in the 1990s. Of these, sixteen focused on worldwide sustainable development of forests with emphasis on developing countries; eleven dealt with Asia; nine with Africa; five with South America; four with the United States and Canada; and none with Europe. What is called "social" or "community" forestry in the developing countries might seem to go under the name of "new forestry" and/or "new perspectives" in the United States but they are all sustainable forestry.

In a similar pattern relating to the sudden growth of interest in sustainable forestry, the SAF task force on sustaining long-term forest health and productivity lists thirty-seven articles on or related to this topic that appeared in the *Journal Of Forestry*, saying that they "show the extent to which the debate, central to our report, has been prominent in the forestry literature in 1990, 1991, and 1992."[69] Sustainable development, and sustainable forestry as part of it, will continue to be prominent in the literature as values, processes and techniques are debated.

---

68. J. E. Rapant, *New Perspectives, New Forestry, Ecosystem Management, and Related Topics: An Annotated Bibliography* (Syracuse, N.Y.: SUNY College of Forestry & Environmental Science, unpublished).

69. Society of American Foresters, *Task Force Report on Sustaining Long-Term*, p. 13.

E. Future of Sustainable Forestry

In June 1992, after two years of a new perspectives program and considerable discussion of new forestry, the Chief of the U.S. Forest Service, Dale Robertson, formalized the adoption of an ecosystem approach to forest management as a policy of the U.S. Forest Service, and a similar policy was adopted for the U.S. Department of Interior's Bureau of Land Management. Thus, in the stroke of a pen, well over 200 million acres of forest land in the United States has by policy been officially put under ecosystem management.

Philosophical themes of the past were revisited in the New Perspectives including maintenance of nature's balance, waste-conscious resource development, and aesthetics. These themes have practical implications for national forest managers who now must be more cognizant of society's expectations for their national forests. These expectations include biological diversity, ecological function and balance, product yields, social values, and the beauty and integrity of natural environments.[70] Balance is sought between protection of natural environments while at the same time supplying products and services needed in society. The *Multiple Use Sustained Yield Act of 1960* directed the Forest Service to apply a multiple-use philosophy although it had previously applied such a philosophy informally prior to the 1960 act.[71] Multiple use has two main thrusts: (1) for the production of valuable forest products for human consumption, and (2) for recreational, educational, and aesthetic experiences. While the public still supports such uses, it has come to view the uses and values of wildlands in a larger context. According to Kessler, "the natural forests and grasslands are also appreciated as integral parts of the environmental security of the planet."[72]

New perspectives invoked a "back to the future" vision for the Forest Service by calling for a return to a land ethic and conservation philosophy of yore. A conservation philosophy according to George P. Marsh's *Man and Nature* was a view which included cultural and historical perspectives.[73] *Forestry Research: A Mandate for Change* calls for an alternate to the mul-

70. W. B. Kessler et al., "New Perspectives for Sustainable Natural Resources Management," *Ecological Applications* 2 (3) (1992): 221–225.

71. C. F. Wilkins and H. M. Anderson, *Land and Resource Planning in the National Forests* (Washington, D.C.: Island Press, 1987).

72. Kessler et al., "New Perspectives for Sustainable Natural Resources Management," p. 222.

73. C. J. Glacken, "The Origins of the Conservation Philosophy," *Journal of Soil and Water Conservation* 11 (1956): 63–66.

tiple-use paradigm.[74] It proposed a paradigm that uses a ecosystems approach to the management of forest lands. Soils, plants, animals, minerals, climate, water, topography, and linking ecological processes are all included. These are viewed as living systems important in their own right not just as commodities or amenities. The Forest Service's *Strategy for the 90's* outlines the approach to research in the current decade: understanding ecosystems; understanding people and natural resource relationships; and understanding expanding resource options.[75] The research agenda changes from individual use to a system-perspective approach that will require interdisciplinary efforts by the Forest Service.

Such swift change, even if only in policy as opposed to what actually transpires on the ground, is not likely to come soon on private forest lands, yet there are signs of change. The American Paper Institute has adopted a set of forest management principles which its members must agree to and abide by. These principles say, in part, that members will

Practice a land stewardship ethic which integrates the growing, nurturing, and harvesting of trees for useful products with the conservation of soil, air, and water quality, wildlife and fish habitat, and aesthetics.

Manage its forests in biologically, geologically, and historically significant areas in a manner that takes into account their special qualities.[76]

This is a formalized statement that takes into account multiple use, sustained yield and multiple values, and which is binding for membership.

The majority of forest land in the United States is owned by non-industrial private forest landowners. The 1989 *U.S. Farm Bill* included provision for a "Forest Stewardship Program," and "Stewardship Incentives Program." In general, the goal of these efforts to increase substantially the amount of non-industrial private forest landowners under "forest stewardship management plans," and to provide monetary incentives to them for a wide range of management practices, including water, wildlife, aesthetics, timber, fish, greenspace, and other enhancements.

In the area of research, three key documents have recommended expanding forestry's perspectives concerning people, natural resources, and eco-

74. National Research Council, Committee on Forest Research, *Forestry Research: A Mandate for Change* (Washington, D. C.: National Academy Press, 1990).

75. USDA Forest Service, *Strategy for the 90's for USDA Forest Service Research* (Washington, D.C.: U. S. Dept. of Agriculture, 1990).

76. J. P. McMahon, "Forest Industry's Commitment to the Public," *Journal of Forestry* 90 (1992): 38–40.

system sustainability, and they put increased emphasis on wildlife ecology and conservation biology as being important to managing natural resources. These are the National Research Council's report, *Forestry Research: A Mandate for Change*,[77] and the U.S. Forest Service/National Association of Professional Forestry Schools and Colleges' report, *Forests for America's Future*.[78] The Ecological Society of America's Sustainable Biosphere Initiative puts forestry research in a wider context of ecological needs.[79] All recommend natural resources research and management that is done in a worldwide context, stressing ecosystem health and sustainability relative to the quality of human life, and expanded to be large-scale and interdisciplinary, thereby, leading to adaptive management. The shift called for is from a utilitarian, production model to one based on a modern environmental paradigm based on both science and values.

The level of interest in forestry's changing role is perhaps best represented by *The Duluth Manifesto*, originally the work of three concerned, individual foresters, who expanded their effort so that the signatures numbered twelve. The objective of the manifesto is "To articulate and champion principles to ensure the vitality of forests to help sustain the global environment and economy."[80] The initial effort received support from the Institute for Forest Analysis, Planning, and Policy, a nonprofit, interdisciplinary research organization. In making its way through the country, it received the signature and support of fifteen additional prominent foresters, and more will have signed the next discussion draft.

In a more formal approach, twenty-eight invited panelists attended a seminar on land stewardship at Grey Towers National Historic Landmark, the family home of Gifford Pinchot, in November 1990. One result was the publication of *Land Stewardship in the Next Era of Conservation* which includes "The Grey Towers Protocol," a set of four principles meant to stimulate further discussion on future forest land management.[81] Stewardship was emphasized earlier in the discussion of the emergence of sustainable forestry is because, as in "The Grey Towers Protocol," it seems to be

77. National Research Council, *Forestry Research*.

78. USDA Forest Service, *Forests For America's Future: A Research Program for the 1990's* (Washington, D.C.: National Association of Professional Forestry Schools & Colleges & USDA Cooperative State Research Service, 1989).

79. J. Lubchenko et al., "The Sustainable Initiative: An Ecological Research Agenda," *Ecology* 72 (1991): 371–412.

80. P. R. Hagenstein et al., *The Duluth Manifesto: Principles to Guide Decisions on Forests, Discussion Draft 3.2* (Baltimore, Md.: Institute for Forest Analysis, Planning, and Policy, 1993), p. 1.

81. V. A. Sample, ed., *Land Stewardship in the Next Era of Conservation* (Milford, Pa.: Pinchot Institute for Conservation, Grey Towers Press, 1991).

the common theme or philosophy emerging as the basis for sustainable forestry. The fourth principle of "The Grey Towers Protocol" says that "Land stewardship must be more than good 'scientific management'; it must be a moral imperative" linking ethics with day-to-day forest management decisions and actions.[82]

Putting this into a broader context, the Society of American Foresters, after two years of advocacy and debate, approved by a three-to-one margin a new "land ethic canon" as an addition to their *Code of Ethics*, making it the first canon. It says that, "A member will advocate and practice land management consistent with ecologically sound principles." But perhaps even more important was the approval of changes to the *Preamble of the Code of Ethics*, wherein the first sentence now states that "Stewardship of the land is the cornerstone of the forestry profession."[83] This is a bold proclamation of intention and a commitment to pass the land and all of its resources, including intact, functioning forest ecosystems, to the next generation in as good or better a condition as when received. It is a commitment to sustainable forestry and ecosystem management, even if these have yet to be fully defined and even if they are never defined in the fullest sense.

It seems fitting to close this discussion of the emergence of sustainable forestry with a quote from Aldo Leopold. About achieving harmony with the land, a harmony more akin to ecosystem management and sustainable forestry than to traditional forestry, he said: "We shall never achieve harmony with land, any more than we shall achieve absolute justice or liberty for people. In these higher aspirations the important thing is not to achieve, but to strive."[84] If sustainable forestry is a higher aspiration, and if it is a form of stewardship with a moral imperative, the striving has begun and it is forestry's task to continue.

82. Ibid., p. ix.
83. Society of American Foresters, "Code of Ethics for Members of the Society of American Foresters" (revision of Nov. 2, 1992), *Journal of Forestry* 91 (1993): 4.
84. Leopold, *Sand County Almanac* (New York: Oxford Press, 1949), p. 210.

9. Forestry Education in
the United States

RICHARD A. SKOK

University of Minnesota

Formal forestry education in the United States will attain its centennial just before the arrival of the twenty-first century. The vision for the profession of forestry and forestry education has been profoundly influenced by many factors. Forces not from within the forestry community have dominated the changes that have occurred. Over the past century, forestry education has evolved in response to these changes. The purposes of this chapter are: (1) to examine the origin of forestry education in the United States, (2) to trace the nature of changes that have occurred, (3) to understand the goals and objectives these changes sought to attain, (4) to assess the current status of forestry education, and (5) to provide a vision of what lies ahead. Forestry education and research in the past twenty years have been the major forces behind the creation of scientific and applied forestry literature. This chapter demonstrates the diversity of forestry and forestry literature.

A. European Forestry Education: The Foundation for America

The European basis of forestry education in the United States had its beginning in the latter half of the eighteenth century. Prior to that time "foresters" were primarily trained in the elements of the hunt that took place in the context of feudal European land tenure.[1] Almost all forest lands were in the hands of the nobility or the crown. Formal education of foresters began in Germany at Wernigerode in 1763. "Master schools" were developed where the emphasis moved away from hunting to forestry and land management. Each master school was initially centered around a highly qualified forester who served as both instructor and director. By the early

---

1. Franz Heske, *German Forestry* (New Haven: Yale University Press, 1938).

1800s some master schools had expanded their staffs and were including instruction based on the emerging forest science. The early master schools in Germany were followed by state-operated forestry schools. Several private institutions were turned into state schools while others were dissolved. Universities which had previously sponsored lectures in forestry in some cases now established forestry training programs including the University of Berlin's Forestry College at Eberswalde (1821) and the Forestry Institute of the University of Giessen (1825).

France also began forestry education in the nineteenth century by establishing a forestry school at Nancy in 1825. In 1828 Sweden established a state forest institute in Stockholm, and Italy followed, in 1866, and then Great Britain in 1885.

Franz Heske suggests that Germany benefited by three types of forestry training: (1) the forestry program in a comprehensive university, (2) the separate stand-alone forestry college, and (3) the independent faculty of a polytechnical university.[2] Because of the multifaceted nature of forestry, Heske argued that this variety of training offered an advantage. While providing both a sound foundation in the basic sciences and a liberal education for the forester, it assured the integration of forest science with the larger science community. The technical foundations important to modern forestry education were first explored at these polytechnical institutions. The concept of a separate forestry school afforded a closer tie to the forest and optimized the students' opportunity for more practical training. This was the key to Germany's early dominance. By developing diversity through its institutional and programming approach, Germany was able to meet the variabilities of forestry training. This diversity provided a range of opportunities for the study of sustained management of forests, silvicultural and administrative systems, and the application of emerging technical developments to forestry needs.

B. Influence of the Forest Conservation Movement in the United States

The seemingly unending forest, which early American settlers cleared for town and pasture, masked an underlying shortcoming, that the forests of the nation were not inexhaustible. By the latter part of the 1800s, concern over a pending shortage of timber was exacerbated by the general abuse to forests. The waste resulting from forest fires and the associated dangers to human life and property were written about widely. By 1850, the forests of

2. Ibid.

· the eastern United States were thoroughly exploited. The public domain of the West was unmanaged, and laws to encourage settlement and development often caused exploitation of natural resources on public lands. Public concern was sufficiently aroused that a sequence of congressional actions resulted, culminating in the 1891 landmark legislation authorizing establishment of forest preserves from the public domain.

Initially, no agency had authority to administer these preserves. From 1897 to 1905 the Interior Department was given this responsibility. Then in 1905, the preserves were designated national forests and placed under the Forest Service in the U.S. Department of Agriculture for management and administration.[3] With the transfer came the recognition of a need to develop a cadre of personnel to administer, manage and protect these public lands.

Henry S. Graves noted that at least twenty-two land-grant colleges offered some instruction in forestry prior to 1897, mostly in the eastern United States.[4] The courses were of a general nature and not a concentrated program of a professional nature such as existed in Europe. They often reflected public concern directed at planting trees rather than the management of forests. Because these courses were generally offered through agricultural colleges, their educational orientation tended to reflect agronomics. But the need for formal programs to train and educate individuals for careers in forestry became evident to those needing to staff the national forests.

European and particularly German forestry provided the early basis for American forestry. Many who were prominent in efforts to establish the principles of forestry in the United States received their forestry education in Europe.[5] Among these were Bernhard Fernow, Gifford Pinchot, Henry Graves, Filbert Roth, and Carl Schenck. With these leaders, early forestry schools in America borrowed heavily from the formats and forestry experiences that had developed in Europe.

B. E. Fernow, a graduate of the Forest Academy at the University of Mundeen in Germany, was called upon in 1898 to organize and lead the first forestry school in the United States, the New York State College of Forestry at Cornell University. Fernow later performed a similar service at the University of Toronto in establishing the first forestry school in Canada. The Yale School of Forestry began in 1900 under the direction of Henry Graves who had studied forestry in France and Germany. The Biltmore

3. George A. Garrett, "Six Decades of Growth," in *American Forestry: Six Decades of Growth*, ed., H. Clepper and A. B. Meyer (Bethesda, Md.: Society of American Foresters, 1960).
4. Henry S. Graves and Cedric H. Guise, *Forest Education* (New Haven: Yale University Press, 1932).
5. Ibid.

Forest School, using another German model, was begun in 1898 by Carl Schenck a graduate of the University of Giessen.[6]

To respond to the rapid demand for foresters that emerged in the first decade of the 1900s, undergraduate work in forestry was given frequently through a fifth year of study to college graduates with degrees in other fields.[7] Yale was prominent in this endeavor. This fifth year became a quick means of providing academic training for the profession. At the same time, undergraduate programs in forestry were being developed and were offered in several land-grant and other state-supported universities encouraged by the Forest Service and the timber industry. To many young men this must have seemed like a crusade in the making.

Well-intentioned as these early efforts at forestry education were, the lack of forestry knowledge and teaching materials based on American forest characteristics and conditions was a handicap. The almost total absence of experienced teachers and facilities proved makeshift and unsatisfactory. Conflict became apparent between the breadth of education deemed needed by the forester and the essential professional forestry requirements of the day. Debate also emerged as to the proper balance between academic and practical training. These issues in one form or another have continued without resolution over the past ninety years.

By 1910, fourteen forestry schools were operating in the United States with three in Canada.[8] The U.S. Forest Service as a primary employer of the graduates of forestry schools had an interest in and influence on the type of education young foresters were given. On the other hand, the emerging forestry schools differed in the background and experience of faculty, their facilities, and the institutional administrative arrangement resulting in variations in the educational approach. Gifford Pinchot, first chief of the U.S. Forest Service, called for strengthening educational standards and, in 1909, organized a conference to address these and other issues in forestry education.

A committee established at that conference was chaired by Dean Graves of Yale and submitted a report two years later.[9] The committee affirmed the crucial role of the forestry schools in this formative stage of the United States forestry profession and in the science and practice of forestry. There were about twenty-four schools at this time offering a degree in forestry.

6. Samuel T. Dana and Evert W. Johnson, *Forestry Education in America Today and Tomorrow* (Bethesda, Md.: Society of American Foresters, 1963).

7. Hugh Baker, "The Forest School and the Education of the Forester," *Forestry Quarterly* 7 (1909): 15–22.

8. Graves and Guise, *Forest Education*.

9. Henry S. Graves et. al., "Standardization of Instruction in Forestry: Report of the Committee of the Conference of Forestry Schools," *Forestry Quarterly* 10 (3) (1912): 341–394.

The report identified the need for a set of standards to be prescribed for the curriculum. The committee proposing standardization recognized four different types of forestry training: (1) the core education for the profession of forestry that involved general education as well as technical forestry; (2) forest ranger training that was based on practical work and "common school" education comparable to forest technician training; (3) forestry instruction that supplemented agricultural education, essentially farm woodland management; and (4) general courses on conservation and forestry designed for non-forestry students. A Second Forest Schools Conference was called in 1911 to discuss the committee's proposal, and to focus on the standard of education for professional foresters as did the final Graves' Committee Report in 1912.

It was agreed that professional forestry education should be equal to that of other learned professions, should be of no less than four years of collegiate-grade undergraduate work, and that no post-graduate degree in forestry should be granted to anyone with less than two years of course work in technical forestry.[10] Specialization at the undergraduate level in fields such as forest engineering and forest utilization was rejected as incompatible with the forestry professional's needs during the four year course of study.

C. The Formative Years

Over the next decade employment opportunities for forestry graduates became more varied which led to more diverse expectations in the preparation of forestry students. Correct balance between practical versus academic training increasingly became the issue. As the profession matured, it generated a growing body of knowledge regarding American forestry with an accompanying urge to expand and proliferate courses. A trend developed toward specialization, conflicting with the earlier concepts of standardization of the curriculum.

Samuel Dana and Evert Johnson noted that logging engineering was the most prominent specialization to emerge among United States schools particularly in the West and to a lesser extent in the U.S. South.[11] Demand was strong for these graduates especially during World War I, which briefly revived the discussion as to the relative merits of the natural sciences versus engineering as the foundation for the education of a forester. With the advent of significant changes in logging and logging equipment in the 1920s,

10. Ibid.
11. Dana and Johnson, *Forestry Education in America Today and Tomorrow.*

engineering became less critical whereas the natural sciences supporting silviculture, protection, and utilization became the recognized knowledge base for general forestry education.

In 1920, the Yale School of Forestry called a conference to examine the direction of forestry education. The preliminary committee assignments prepared curricula for four- and five-year programs.[12] The conference debated the four- or five-year question and recommended the establishment of a permanent committee on forestry education by the Society of American Foresters, which was possibly the most significant result to emerge from this conference.[13]

During the 1920s more forestry schools were established which increasingly focused their curriculum on employment needs in their home state or region. One result was that the Forest Service's civil service examination used in hiring became less influential as a standard for the schools. This in turn led to greater diversification of programs among schools.

Articles critical of the deficiencies in forestry education were often published in the *Journal of Forestry*, the most influential U.S. forestry serial of its day.[14] In addition to concerns over curriculum content and sufficiency under the four-year degree constraints, writers were equally critical of the inadequate number of well qualified faculty at many schools, the emphasis schools were placing on research, and the inadequacies of facilities. Raphael Zon, in particular, was a strong advocate for a ranking system applied to forestry schools as an indication of the quality of its graduates. To address these growing concerns a special committee of the National Academy of Science appointed Dean Graves to prepare a report on forest education.[15] The report was based in part on a survey conducted by Graves.

In his report, Graves acknowledge the criticism being leveled at education in general and at forestry in particular. The report specified the purposes of forestry in the United States as a means of identifying the task of forestry education. Graves stated: "In forestry, we seek to make the forest lands of the country of the greatest possible service in the economic, industrial, and social development of the nation . . . a vast undertaking that involves the constructive utilization of from one-fourth to one-third of the

12. J. W. Toumey, "Second National Conference on Education in Forestry," *Journal of Forestry* 19 (1921): 167–172.

13. Dana and Johnson, *Forestry Education in America Today and Tomorrow*.

14. (a) Hugo Winkenwerder, "Problems in Forest Education: Report of Committee on Forest Education," *Journal of Forestry* 25 (3) (1927): 290–324. (b) Raphael Zon, "Our Forest Schools," *Journal of Forestry* 25 (3) (1927): 251–256. (c) R. D. Forbes, "The Next Generation," *Journal of Forestry* 25 (1927): 260–280.

15. Henry S. Graves, "Some Considerations of Policy in Forest Education," *Journal of Forestry* 26 (1928): 430–453.

land surface of the country."[16] As benchmarks for guiding college-level forestry education in meeting this mission, Graves noted: "forestry is essentially a science of forest production. The object of forestry is to obtain the greatest possible service from certain classes of land through the continuous growth of forests."[17] He further emphasized: "the success of our work will be judged by the character and condition of the forests we leave behind us."[18] This has proven to be a very prophetic observation.

Graves argued that this broad mission for forestry required such a diversity of capabilities that no single approach to training and education would be sufficient to supply the talents needed. This led him to support specialization by schools as well as greater development of institutions and programs that would provide the practical training of foresters. He discussed the differences among schools in giving primary emphasis to either the natural sciences or engineering in forestry education. He supported those who believed that the natural sciences were the essential disciplines serving the foundations of forestry education.

Central to Graves' report was the contention that the four-year time period for the curriculum leading to the baccalaureate degree in forestry was a shortcoming of United States forestry education. Specialization in the four-year program and graduate programs in forestry were seen as two approaches offering possible solutions. The elements by which forestry schools were to be evaluated emerged in this report. Graves recognized identification of the institution's objectives, the character of the faculty, the adequacy of the educational facilities, and to a lesser extent the curriculum as important measures in judging the strength of a forestry school.

The Graves Report recommended a more thorough study of forestry education. This was fulfilled through support of a grant from the Carnegie Corporation, and was conducted by Henry Graves under the auspices of the Society of American Foresters. This definitive study, referred to as the Forest Education Inquiry, resulted in the landmark book *Forest Education*.[19]

D. The Maturing Years: 1930–1945

Working with both a steering and an advisory committee, Graves and Cedric Hay Guise delved thoroughly into the education of those preparing for professional forestry work. There were twenty-six university level for-

16. Ibid., p. 431.
17. Ibid., p. 438.
18. Ibid., p. 439.
19. Graves and Guise, *Forest Education*.

estry programs in the United States in 1930 and four in Canada. The authors focused on issues regarding: (1) too many forestry schools, (2) too much variability in standards of educational performance among forestry schools, and (3) inadequate responsiveness of forestry education to the diverse demands that the practice of forestry required.

In approaching the task, Graves and Guise divided their report under five major headings: The Background of Forest Education; The Occupations of Foresters; The Education of Foresters; Problems of the Forest Schools; and Forest Education in Europe and in Other Foreign Countries. Graves and Guise noted:

> Education for professional work in forestry is exemplified by the collegiate forest school. The principles of education applicable to such schools have their analogy in schools of engineering, law, medicine, and other recognized professions. The professional forester needs a background both of general education and of principles and practices of forestry. Forest education in this country and in Canada is built on the principle that there is need for a body of professional men who have not only a thorough knowledge of the technique of forestry, but also a liberal education, and who are therefore competent to take the lead in creative and developmental work.[20]

Review of forestry occupations served as a basis in assessing the status of educational needs with the four-year program cited again as an underlying problem. Graves and Guise stated, "providing a reasonably thorough education in four years is one of the most difficult of all those (problems) confronting the undergraduate schools."[21] This undoubtedly contributed to the stinging criticism which they noted was frequently put forward by those they considered best qualified, that is employers, to make the assessment of the training of American foresters, "the lack of thoroughness in fundamentals. This applies to general education . . . and to the basic principles of forestry itself."[22]

The inquiry made an extensive review of four categories of forestry education, namely, general education, preparatory subjects, the pretechnical subjects, and technical subjects. Concern was expressed over the limited educational emphasis given wildlife, especially game management, despite the importance of forest habitats for game. Watershed and recreation management were also cited as being important, and their inclusion as part of courses such as forest economics, policy, and management was stressed.

20. Ibid., p. 99.
21. Ibid., p. 102.
22. Ibid., p. 104.

A suggested four year general forestry curriculum for the undergraduate degree was offered as a guide rather than as a call for standardization.[23] The inquiry recognized the variation among schools in their approach to curriculum. Two general patterns were noted in offering students elements of specialization at the undergraduate level. One was termed the parallel curricula in each field of concentration; the second presented a central core of courses, which all students were obliged to take, with electives providing the opportunity to meet individual interests. These patterns are discernible in forestry programs today. Field work was considered an essential element of forest education by the profession. The ionquiry concluded that most schools did not assure a program of adequate field experience either as part of the education program or in summer work opportunities. This illustrated again the ongoing struggle to resolve the need for both practical and academic work within a four-year degree. Criticism leveled at forestry school faculties in the Graves and Guise study identified the lack of professional forestry experience as a shortcoming that undoubtedly contributed to the status of field experience in a forestry curriculum.

Graves and Guise concluded that as a minimum, "forest school faculty should be composed of at least five men, experienced and competent to teach in a special field."[24] Specializations were silviculture, forest management, forest utilization, wood technology, forest economics, and administration. Thus, a strong emphasis on the timber aspects of forestry was explicit. Their review of faculty in forestry schools found them to be too few to cover the range and quantity of courses; almost half of the twenty-six schools fell short of a minimum faculty.

A Canadian view that similarly emphasized the timber orientation of forestry education at this time was expressed by C. D. Howe:

> In determining the method of educating for the profession of forestry, we should never lose sight of the ultimate objects of and the ultimate justification for the existence of the profession, namely, the continuous re-establishment of the new forests to replace those removed by any agency; the production of higher yields per acre and a better quality of wood so as to cheapen the cost of logging . . . ; in a word, so far as humanly possible, to maintain in continuous succession the supplies of raw materials for our wood using industries.[25]

Schools reported that financial limitations caused the lack of faculty, but Graves and Guise suggested that the cause was much more fundamental. "It

23. Ibid., pp. 182–183.
24. Ibid., p. 275.
25. C. D. Howe, "Forestry Education," *Forestry Chronicle* 8 (1) (1932): 22.

is due to lack of appreciation on the part of college authorities that forestry is not a subordinate branch of agriculture, but is as broad in scope and as diversified as the field of agriculture itself."[26] Eighteen of the twenty-six forestry schools were at land-grant universities making such a professional education problem pervasive. They linked this cause to the proliferation of forestry schools in observing; "The reason why there are so many forest schools today is that many of them are conducted on a basis of very small appropriations, wholly inadequate to build up a first class preparation in forestry of professional grade. Apparently the colleges with which the schools are associated have no realization of what is involved in the preparation for a profession so broad and diversified as forestry."[27]

In examining the research activities at the schools, Graves and Guise observed, "Many of the schools are feeling the pressure from the college or university authorities to conduct research that will result in scientific publications."[28] Research was acknowledged to be an important activity at many universities and colleges where forest schools were associated, and they believed that the schools had an important role to play in development of knowledge about forestry. However, they found that many forestry schools had no or poorly defined policy regarding research by faculty. Faculty were often unclear regarding institutional expectations in research and publication. Funding limitations as well as the small number of faculty with research training contributed to this problem; for the academic year 1929–1930 only 12% of the forestry faculty nationally held a doctoral degree.

Henry Schmitz reinforced these faculty concerns, and noted: "Probably the most urgent and pressing of all forest educational problems is the training and selection of competent teachers who not only can take their place among their scientific colleagues, but at the same time have not lost touch with the social and economic relations of forestry."[29]

As an outgrowth of the Graves and Guise report, the Society of American Foresters (SAF) authorized a study to classify the forest schools in the United States. The society had a long-standing involvement in the development and encouragement of forest education beginning in 1909.[30] SAF membership required training in the fundamental principles of professional forestry. Junior membership required graduation from a forestry school approved by the Council of the Society of American Foresters. The society

26. Graves and Guise, *Forest Education*, p. 262.

27. Ibid., p. 247.

28. Ibid., p. 282.

29. Henry Schmitz, "Some Problems in Forest Education," *Journal of Forestry* 30 (2) (1932): 204.

30. Henry Clepper, "The Society of American Foresters," in *American Forestry: Six Decades of Growth*.

had struggled with its membership criteria in an effort to establish forestry as a recognized profession. H. H. Chapman as president of the society, was outspoken on "the failure of foresters to grasp the distinction between a profession and a craft."[31] Now a system of identifying "approved schools" would be attempted as a means of assuring minimum standards.

Chapman prepared the *Professional Forestry Schools Report* that presented the grading of the twenty schools considered to be giving professional training in forestry. Of these, fourteen were given an approved rating and six were listed as offering professional forestry education but not fully meeting the standards. No others were considered to offer professional programs in the year of the study, 1934–1935. Graduates of the approved schools were eligible for junior membership in the Society of American Foresters with the endorsement of three present members. Thus began accreditation of forestry schools in the United States.

Several factors were considered in the grading process; these factors were weighted with personnel receiving the highest weight, and departmental status the lowest. The absence of specific curriculum content as a factor in the evaluation is noteworthy.[32] At this point evaluation was made of the institution, that is the "forestry school" and not the curricula. Based upon analyses used in the study to classify the forestry schools, Chapman sought to call attention to what he and others saw as the poor facilities and funding. Noted was the fact that of the twenty-four forestry schools offering professional education in forestry at that time, fifteen were part of agriculture at their university or college. While these forestry schools accounted, on the average, for about one-third of the students enrolled in agriculture in their institutions, they had slightly less then ten percent of the faculty and a little more than five percent of the budgets and facility space.[33]

Enrollments in forestry grew rapidly following the establishment in 1933 of the Civilian Conservation Corps (first called the Emergency Conservation Work program) which created a new and significant demand for foresters to direct work in forested areas.[34] Forestry was one of the few college-level programs that found favor with students during the depression of the 1930s, but support for forestry programs at their home universities had not responded to the increased demands. Chapman saw this as the result of

31. H. H. Chapman, "Society Affairs: Making the Society of American Foresters a Professional Organization," *Journal of Forestry* 32 (1934): 503.

32. H. H. Chapman, *Professional Forestry Schools Report* (Bethesda, Md.: Society of American Foresters, 1935), p. 174.

33. H. H. Chapman, "Forestry, the Cinderella of Agricultural Colleges," *Journal of Forestry* 34 (1936): 16–21.

34. Dana and Johnson, *Forestry Education in America Today and Tomorrow*.

two primary factors: first was the historically developed relationship that existed between the agricultural colleges and their forestry units. The federal support for agricultural research (Hatch funds), extension (Smith-Lever funds), and teaching favored agricultural plant and animal programs. At the same time the growth in forestry enrollments masked the depression-related declines in the numbers of agriculture students at these institutions. The second and more subtle factor Chapman identified was a long-standing perception among those in general agriculture that forestry was largely seen in the context of the woodlot associated with the farm enterprise. Further, forestry was seen as a less valuable land use competing with agriculture for public lands. Chapman conceded that the pervasive view of forestry as simply another department of agriculture was gradually changing among agricultural colleges but that the limited power of the forestry faculty required that the Society of American Foresters use acceptance to professional membership from approved schools as a means to assure the adequacy of the education of graduate foresters.[35] Accreditation was seen as a means of leverage to improve support for forestry schools. Chapman's conclusions about the limiting aspects of agricultural colleges on forestry programs echoed that reported by Graves and Guise. However, their conclusions predated the significant enrollment expansions of the mid-1930s.

These efforts had results. "The facilities and financial support of many American forestry schools have improved greatly in the past five years," stated Schmitz in an editorial in the *Journal of Forestry* in 1939.[36] Graves and Chapman were praised for their efforts which along with increased enrollments were believed to have been the principal factors in this achievement. Schmitz also expressed concern "that these improvements have not been more largely the result of the educational statesmanship and the scholarly activities of the forestry faculties."[37] The editorial implored its readers to recognize the need to improve the scholarly competencies of forestry faculties and to establish sound research programs in each forestry school, noting that less than 5% of the forestry research funds expended nationally were spent at forestry schools in 1934.

Enrollments in forestry peaked in the academic year 1937–1938 at 6,067, but with the advent of World War II enrollment declined precipitously.[38] Despite the growth period for forestry enrollments in the 1930s, only three

35. Chapman, "Forestry, the Cinderella of Agricultural Colleges."
36. Henry Schmitz, "Imminent Problems in Forestry Education," *Journal of Forestry* 37 (1939): 359.
37. Ibid.
38. Cedric H. Guise, "Statistics for Schools of Forestry for 1941: Degrees Granted and Enrollments," *Journal of Forestry* 40 (1942): 196–202.

new undergraduate forestry schools were formed; two were discontinued.[39] However, the Society of American Foresters' Division of Education, established in 1935, went on record the next year raising concern about the establishment of any new forestry schools.[40] The rationale included the observation that employment growth was limited and that existing schools would meet long-term needs.

A second round of accreditation was conducted by the Society of American Foresters in 1942 under the leadership of Chapman. His report reinforced the perception that on average the forestry schools had improved. He noted that accreditation by the society had "not interfered with or dictated the contents of curricula in any manner whatsoever" and further highlighted his belief that "if the Society of American Foresters . . . decides that it does no longer wish to represent a profession as such but become an association for administrators of 'wild' land including all its possible uses, then of course there will no further need for the Society to concern itself with what is taught in the schools."[41] Dana and Johnson noted,

> During the course of his 1934–1935 study of forestry schools, Professor Chapman broached a subject which later led to vigorous controversy—what is the body of knowledge of which a man should have command to qualify as a professional forester. Dealing specifically with the problem of accreditation and of admittance to membership in the Society of American Foresters, he called attention to the fact that approval of a school was based on its entire setup and included all of its curricula.[42]

In its simplest terms, the argument was over the narrow interpretation of forestry along the lines of timber management which Chapman favored, or a broader view. The latter would accept a diversity of curricula for preparing graduates in dealing with the varied uses of forested and related lands and a wider educational basis for admittance to the society. H. T. Gisborne succinctly expressed this broader view in a response to that argued by Chapman.[43] A group of executives from eight western forestry schools also took a different view recommending, "Education in forestry is education in multiple land use."[44]

39. Dana and Johnson, *Forestry Education in America Today and Tomorrow*.

40. R. P. Holdsworth, "Meeting of the Division of Forest Education," *Journal of Forestry* 34 (1936): 320–322.

41. H. H. Chapman, "Report of the Committee on Accrediting Schools of Forestry," *Journal of Forestry* 41 (1943): 229.

42. Dana and Johnson, *Forestry Education in America Today and Tomorrow*, p. 57.

43. H. T. Gisborne, "Is the Society Broad Enough?" *Journal of Forestry* 41 (1943): 543.

44. "Executives of the Western Forestry Schools Make Recommendations," *Journal of Forestry* 42 (1944): 603.

In a sense, the issue revolved around the continuing problem of meeting the perceived educational needs of the professional forester through a four-year degree. To accommodate courses addressing more than timber concerns meant the sacrifice of other forestry courses or the basic science and liberal education that everyone agreed were essential. Further evidence of the pervasive and divisive nature of this issue is in the report by C. F. Korstian which was neither approved nor disapproved at the 1939 annual meeting of the Division of Education which had commissioned the study.[45] Among the conclusions reached by the committee was its unanimous judgment that "a professional degree in forestry must rest upon more than four years of undergraduate work" and supported the exclusion of curricular matters from accrediting standards with, "The committee recognizes, and agrees with, the right and privilege of universities and colleges to establish their own requirements and designations for degrees in forestry as well as in other fields."[46]

E. The Watershed Years: 1945 to 1970

This period initially saw the flourishing of forestry education under the continuation of the forestry conservation paradigm championed by Gifford Pinchot under the wise use principle. Essential to this philosophy were "scientific forestry and rational planning as ways of using forests to raise living standards without destroying the land's ability to be used. . . . Foresters inherited Pinchot's 'tree farm' view of the forest and his belief in scientific forestry as the road to wise use of forest resources."[47]

With the perspective of time, R. Scott Wallinger summarized early forestry and forestry education as follows:

The men who launched American forestry and foresters were practical men with a conservation ethic. They were creating forests to grow wood and foresters to manage them for that purpose. . . . it was relatively straightforward to design an educational curriculum for them. . . . for the first half of the forestry's first century the objectives were fairly clear and widely endorsed. The roles of various professions (wildlife, water and range management) were distinct even though related, and the scientific base for each profession somewhat

45. C. F. Korstian, "Degrees in Forestry: Final Report of the Committee on Degrees in Forestry, Division of Education, Society of American Foresters," *Journal of Forestry* 38 (1940): 781–789.

46. Ibid., pp. 788–789.

47. National Research Council, *Forestry Research: A Mandate for Change* (Washington, D.C.; National Academy Press, 1990), p. 14.

limited. In that era and context, it was 'relatively' easy to define a forester . . .
forestry . . . a forestry school.[48]

Enrollments grew following World War II and expanding employment
opportunities accompanied this more active involvement in the handling of
forest resources. In the 1946–1947 school year, there were an unprecedented 7,010 undergraduates enrolled in forestry programs, many with GI
Bill benefits, which severely strained the teaching resources of schools.
W. F. McCulloch reported on the astounding growth of undergraduate de-
gree-granting forestry schools from twenty-one to thirty-two in the six year
period ending in 1948.[49] He expressed his concern noting, "Not all the new
schools are soundly premised, and there is possibility that in local areas the
number of trained foresters will exceed employment opportunities."[50]

Accreditation remained at the center of debate in the profession during
this period as the Society of American Foresters struggled once again with
efforts to agree on defining forestry and its scope as a basis for junior
membership. Institution-based accreditation initiated in 1935 led graduates
to specialized programs such as range and wildlife management, recreation
and forest products to become SAF members upon graduation. The resulting dilution of the profession concerned many members. Myron Krueger
discussed these changes and their implications for the society's membership
requirements and accreditation.[51] He had earlier cautioned the profession
that they should not rely too greatly on accreditation to meet the growing
and changing demands for technical forestry knowledge emerging in the
post-World War II period, and stated "accrediting cannot be of complete
help. . . it is a periodic process. . . . the major emphasis is on minimum
standards. . . . at best, the bases used for accrediting lag behind what is
actually being done by the stronger forestry schools."[52] He called for more
discussion of forestry education within the profession and for a thorough
study of the changes occurring in the profession in order to assess their
implications for forestry education. Wallinger noted that it was during this

48. R. Scott Wallinger, "Creating and Educating a 21st-Century Forest Resources Profession,"
Forest Resource Management in the 21st Century: Will Forestry Education Meet the Challenge?
Symposium Proceedings. (Bethesda, Md.: Society of American Foresters, 1991), p. 31.

49. W. F. McCulloch, "Current Developments in Forestry Education," *Proceedings: Society of
American Foresters Meeting 1948* (1949): 349–354.

50. Ibid., p. 354.

51. Myron Krueger, "Qualifications for Membership in the Society of American Foresters,"
Journal of Forestry 50 (1952): 409–415.

52. Myron Krueger, "The Society of American Foresters and Forestry Education," *Journal of
Forestry* 50 (1952): 6.

period that multiple use and nontimber values began to be a focal area of education.[53]

The Society of American Foresters undertook a study in 1957 to assess the progress and to establish future directions in forestry education. Samuel Dana assumed the leadership and the resultant book provides the most recent, comprehensive study of forestry education in the United States. The Society's committee on accreditation, chaired by Krueger, specified for the first time in 1948 the minimum credits that a forestry school must require for accreditation in the areas of silviculture, protection, management, economics, and utilization.[54] Heads of schools were given authority to decide if these requirements were met, but in time the SAF found it necessary to assume this responsibility. As a result, in 1950 accreditation became curricula rather than institution-based, and all forestry school curricula were reviewed that year. Thirteen of the twenty-five forestry schools were determined not to meet this new standard.[55] The line was clearly drawn indicating that the society saw timber management as the primary substance and scope of forestry. Dana and Johnson concluded,

> The effect of accreditation has been to emphasize timber management as the core of forestry. The management of other forest resources . . . is recognized as a forestry activity, but . . . courses in the management of these resources are . . . not required in the forestry curriculum. The growing recognition of nontimber values in recent years has resulted in little change in accreditation policy.[56]

A survey of a cross section of members of the profession conducted by Dana and Johnson found "a strong and growing belief that a forester should be a forest land manager—that his concern is not with timber alone but with all the products and services of the forest. Consequently, an understanding of the ecosystem (total environment) and its control is a prime requisite for the forest manager."[57]

Henry Vaux captured the meaning of this change accurately by noting that while the society had clarified the nature of the profession it had done so at a cost, and went on to state that the profession had given up an

---

53. Wallinger, "Creating and Educating a 21st-Century Forest Resources Profession." pp. 30–38.
54. Dana and Johnson, *Forestry Education in America Today and Tomorrow.*
55. Ibid.
56. Ibid., p. 360.
57. Ibid., p. 139.

important role on multiple-use management of forest lands which it had sought earlier to establish for itself.[58]

Dana and Johnson recommended that the society reexamine the curriculum and that accreditation replace specific courses and hours of study with fields that needed to be covered. They summarized thus:

> The goal of professional education can be achieved in forestry by establishing as "core requirements" certain subjects which must be taken by all students. . . .
> The forestry subjects suggested as core requirements are dendrology, forest ecology, silviculture, forest protection, forest measurements, forest economics, forest policy, forest administration, and forest management.[59]

This suggestion has many similarities to the curriculum outlined in 1932 by Graves and Guise with the notable exceptions of (1) a five-year designation instead of four for the program, (2) the exclusion of forest products and utilization, and (3) the designation of forest ecology.[60]

Additionally, they echoed the earlier observations that four years of education for the profession of forestry were insufficient and offered the proposal that five years should be the minimum required for admission as a professional member to the society. Considerable discussion aimed at countering the arguments against the five-year program was presented by Dana and Johnson, who concluded elsewhere that "the current standard four-year program results in many forestry graduates occupying positions for which they are either overtrained or undertrained."[61]

Forty-five forestry schools had been established by 1961 with twenty-seven of them accredited.[62] Dana and Johnson noted that this expansion had occurred despite repeated statements from the Society of American Foresters in 1936, 1946, 1949 and 1957 cautioning against the establishment of new professional forestry programs. However, states were clearly responding to the opportunities they saw in forestry education to serve local needs. A primary concern regarding these institutions reflected the observation that many of the newer schools were viewed as lacking funding, poorly staffed, and with inadequate facilities. They further noted many of these newer schools "failed to qualify for accreditation. . . . . This fact, however, appears to have had little effect either on attendance . . . or on the ability of

58. Henry Vaux, "Twenty Years of Population and Forestry Advances," *Proceedings of the Society of American Foresters 1959* (1960): 9–12.

59. Dana and Johnson, *Forestry Education in American Today and Tomorrow*, p. 330.

60. Graves and Guise, *Forest Education*.

61. Samuel T. Dana and Evert W. Johnson, "Forestry Education in America—Highlights of the SAF Study," *Journal of Forestry* 61 (1963): 484.

62. Dana and Johnson, *Forestry Education in America Today and Tomorrow*.

their graduates to find employment . . . This situation does not mean that there is no difference in the professional qualifications of graduates from accredited and unaccredited schools, but rather that most of the work for which 'foresters' are initially employed is technical rather than professional in character."[63] Clearly, accreditation was no impediment to the establishment of new forestry schools. Dana and Johnson concluded:

> Accreditation undoubtedly has resulted in the strengthening of some of the weaker schools . . . it has enabled the heads of these schools to induce their respective administrations to provide enough support to meet minimum standards. Unless these standards are at a high level, however, it is doubtful whether the profession gains substantially by increasing the number of schools which barely met the minimum. There is little evidence that accreditation has had much influence on the stronger schools, since the emphasis has been on keeping out the weak rather than building up the strong.[64]

Dana and Johnson recommended that curricula and faculty quality needed immediate attention whatever the age of the forestry school. They believed that existing requirements regarding coverage of subject matter deserved review to bring it up to date with the changing scope of forestry. They also called for improved means of evaluating the effectiveness of faculty particularly as teachers, and called for the society to accredit graduate programs and the training of technicians. They also emphasized the need for the society to shift its focus from accreditation to the improvement of education at each level, technical, professional, and graduate.

The growth of funded forestry research at both the state and federal level provided an opportunity for the expansion of forestry faculties and graduate education at publicly-supported forestry schools which directly strengthened undergraduate programs by broadening faculty expertise. As the role of research and associated graduate education became more prominent, the nature of forestry faculties began to change. Specialization demanded stronger scientific, discipline-oriented faculty whose principal academic achievements might only marginally be related to the practice of professional forestry. Dana and Johnson anticipated these changes. "Forestry research also will depend . . . to an increasing extent . . . on specialists without professional education in forestry."[65] The importance of research has grown accordingly with forestry faculties, particularly at the graduate-research universities. From the perspective of the forestry profession, faculty with

63. Ibid., p. 66–67.
64. Ibid., p. 361.
65. Ibid., p. 334.

limited forestry background were given equal sanction under university governance to participate in the shaping of future curricular requirements and to decide overall school policy. Indeed many faculty lacked any practical field experience while others had only marginal exposure to an academic program in forestry.

In a very real sense United States forestry schools responded to the culture of their individual academic environments and less to the professional demands which dominated earlier forest history. Academic rewards shifted increasingly to research and external funding sources. By the end of the 1960s, forestry education was changing in ways the accreditation process had not fully anticipated.

F. The Adaptive Years; 1970 to the Present

The last twenty-five years have been characterized by many forestry schools moving toward natural resource management and environmental sciences programming both in education and research. Such change has occurred, in part, at the expense of more traditional timber management-oriented, forestry education activities which now occupy a lessor role in the academic programming of many schools.

Accreditation in 1992 by the SAF recognized professional degree programs at forty-five schools as meeting established standards.[66] There was only one candidate program. The high accreditation rate among forestry schools results from a broadening of the definition of forestry within the profession, changes in SAF accreditation standards and the development by some schools to meet overall minimum standards. Accreditation has gradually moved toward changes suggested by Dana and Johnson by becoming less prescriptive through elimination of specific course hours as well as content requirements and by the substitution of more general subject matter.[67]

The broader definition for forestry adopted in 1989 establishes the current boundaries for accredited programs; it states:

Forestry is the science and art of attaining desired forest conditions and benefits. As professionals, foresters develop, use and communicate their knowledge for one purpose: to sustain and enhance forest resources for diverse benefits in perpetuity. To fulfill this purpose, foresters need to understand the many

66. "Curriculum: Accreditation and Recognition," *Journal of Forestry* 90 (3) (1992): 23–26.
67. Donald P. Duncan, Richard A. Skok, and Douglas P. Richards, "Forestry Education and the Profession's Future," *Journal of Forestry* 87 (9) (1989): 31–37.

demands that forests must satisfy and the potential for forest ecosystems to satisfy these demands now and in the future.[68]

Accredited programs are freer to display a diverse curriculum approach to meeting their own stated goals and mission. However, Wallinger argues, "until we agree what a forester is, we can't agree on what a forestry curriculum must contain. We have symbolically broadened the notion of a forester and forestry by using increasingly the term 'forest resources management' but this undefined extension doesn't resolve the issue."[69] He notes further with respect to efforts to define forestry and foresters:

> Their (committees and task forces) results to date generally conclude that foresters manage forest resources in conjunction with many other 'specialists'. . . . This means we have a traditional forestry profession that embraces related disciplines and certain related profess.ions that work on forest land. But there is no forest resource management profession today as such, and that les at the core of the problem.[70]

The emergence during the past several decades of widely expressed concerns over human impacts on the environment has become a major influence on educational programs in natural resource. By the mid-1970s enrollments in forestry schools grew to new levels spurred by student interest resulting from the first Earth Day in 1970. Many of the youth inspired by this movement saw university forestry education programs as one of the few available academic opportunities to express their interests. Not coincidentally, this corresponded with the opening of forestry education and the profession to women. It was only in the late 1960s that women in any numbers sought forestry education. Only in 1965 were more than one hundred women enrolled in forestry school undergraduate programs.[71] That number grew tenfold over the next six years and peaked at just under 6,000 in 1980.[72] These two developments expanded the pool of talented students entering forestry programs.

SAF survey data showed overall undergraduate enrollments peaked at the schools in the 1975–1976 academic year with a reported 22,345 students.

68. Society of American Foresters, *SAF Forest Policies* (1989).

69. Wallinger, "Creating and Educating a 21st Century Forest Resources Profession." p. 32.

70. Ibid., pp. 32–33.

71. Gordon D. Marckworth, "Statistics from Schools of Forestry for 1971" (Bethesda, Md.: Society of American Foresters, 1972), unpublished.

72. P. Gregory Smith, "SAF Enrollment and Degrees Survey" (Bethesda, Md.: Society of American Foresters, 1992), unpublished.

Forestry enrollment peaked at 11,746 in the 1977–1978 academic year. These levels stood at 18,988 and 7,155, respectively, in 1991 after having reached their corresponding lows of 13,684 in 1987 and 4,490 in 1988.[73]

Growth in the sciences underlying forests as well as their relationship to our overall physical and biological systems has crucially impacted the content of forestry curricula. This coupled with the changing and expanding values that society requires from our forest resources have steadily reshaped the form and content of forestry educational programs. The result is more diversity in curricular offerings at schools with accredited programs. The range of missions and goals which accredited schools aspire to in the natural resources and environmental sciences area has also significantly changed over this period. This is reflected by a number of schools offering more general curricula in natural resource management and environmental studies at the undergraduate level. These offerings generally supplement the accredited forestry programs and other resource-specific specialized programs such as range, wildlife, and recreation that have long existed.

Such newer academic offerings have appealed to a somewhat different group of university students thereby expanding the population from which students are attracted. This has helped a number of the schools recover from the low enrollments experienced in the mid-1980s. The proportion of students enrolled in accredited programs in the forty-five schools has steadily declined in recent years. As a result, SAF accreditation now applies to a minority of students (38% in 1991) in these schools.[74]

Research accounts for a more significant proportion of the budgets at most accredited forestry schools in 1990 than was the case in the 1950s and 1960s. Scientists-years in the sixty McIntire-Stennis forestry research program institutions doubled from approximately 420 in 1966 to 860 in 1984.[75] This has increased the base of faculty for teaching at both the undergraduate and graduate level at the forty-four public forestry schools in this program. Research has required greater specialization as the frontiers of knowledge have advanced in sciences basic to forestry. This contributed to changes in the profile of faculty at forestry schools. These included: (1) a significant increase in the number holding the doctorate, (2) an increase in the percentage of faculty with less than one year of professional forestry experience,

73. Ibid.
74. Ibid.
75. *A Quarter-Century of Progress: The McIntire-Stennis Cooperative Forestry Research Program, 1962–1987* (Washington, D.C.: National Association of Professional Forestry Schools and Cooperative State Research Service, USDA, 1986).

(3) an increase of faculty with no earned degree in forestry, and (4) a greater proportion of faculty time allocated to research.[76]

Over the past twenty years there has also been a shift toward more extramural-funded research relative to federal formula and state experiment station sources at forestry schools. Thus, the research objectives of granting agencies have increasingly impacted the direction of forestry schools.

Despite these changes substantial concern continues regarding the education of forestry professionals. Attracting high-quality students stands as the leading issue in a recent study of forestry education.[77] Other closely related issues that surfaced in this study included defining foresters' roles and the public image of forestry. Wallinger attributed the latter partially to "our currently inadequate concept of the profession that rests on a grossly low education base. That limitation is reflected in people who are scientifically trained—often in narrow specialties—but who lack the broader education in language, history, sociology and ethics to lead the profession and the public policy debate."[78]

Other high priority issues were improving teaching quality, strengthening curriculum, and meeting continuing education needs for professionals. This latter relates in an important way to the continuing issue of the adequacy of a four-year degree as the entry for professional forestry.[79] While generally conceding that the baccalaureate programs cannot adequately prepare the individual for professional forestery responsibilities, there seems to be growing acceptance that the realistic alternatives are post-baccalaureate education either through advanced degrees or informally through various professional continuing education avenues.

The SAF program of certified continuing education offerings and the move toward broadening the accreditation of master's degree programs supports this pressing need. During this period D. P. Duncan et al. noted, "Significant numbers of students with nonforestry degrees are enrolled at forestry schools. . . . These students bring to the profession a broader academic base but less technical education."[80] These students typically have been in pursuit of advanced degrees for entry either as professionals or as specialists in forestry. Interestingly, the first decade of forestry education in

76. Duncan, Skok, and Richards, "Forestry Education and the Profession's Future."

77. Ibid.

78. Wallinger, "Creating and Educating a 21st-Century Forest Resources Profession," p. 36.

79. (a) Duncan, Skok, and Richards, "Forestry Education and the Profession's Future." (b) Wallinger, "Creating and Educating a 21st-Century Forest Resources Profession."

80. Duncan, Skok, and Richards, "Forestry Education and the Profession's Future," p. 35.

the United States relied heavily on this graduate degree model of meeting early professional demands with Yale as the principal example.[81]

Changes in programs of forestry education and research since 1970 have gradually been reflected in name and organizational status of many schools. Of the thirty-three accredited forestry programs in 1970, thirty-one were reported to be in academic units with the term forestry in their name and of these, twenty-three were identified strictly as forestry academic units.[82] By 1992, thirty-five of the forty-five units with accredited programs carried forestry in their name and twenty-three had forestry as their exclusive identity.[83]

The needs of forestry programs became more distinct from those of agriculture over this period. The failure of academic agricultural administrators with programs in forestry to recognize this earlier had been previously acknowledged as a serious problem to forestry education development. A recent National Research Council study noted:

> Critical differences exist between forestry and agriculture. Forests must be managed over very long periods of time (long rotation), whereas many agricultural products are harvested annually (short rotation). Because of the vast size of forests and their natural state, management necessarily tends to be extensive rather than highly intensive, as agricultural production is. Forestry manages for an array of different products and values simultaneously; in agriculture, relatively few products are produced in a given area. Whereas agricultural fields are of human creation, most forests are essentially developed by natural systems. These natural processes are highly complex and can be very different from one forest area to the next.[84]

Overall, forestry schools developed greater autonomy from agriculture at a number of the land-grant colleges during the past twenty years. Presently, twenty-nine of the accredited programs are in colleges or schools where forestry or natural resource programs are dominant. This has programmatic importance for the broader educational and research response by universities to the needs of forest resource management.[85]

---

81. Graves and Guise, *Forest Education*.

82. "Institutions in the United States Offering Professional Education in Forestry," *Journal of Forestry* 68 (1970): 40–41.

83. "Curriculum: Accreditation and Recognition."

84. National Research Council, *Forestry Research: A Mandate for Change*, pp. 15–16.

85. Wallinger, "Creating and Educating a 21st-Century Forest Profession."

G. Graduate Education

Few comprehensive studies exist of graduate education in forestry. Graduate studies evolved slowly at forestry schools, and with few exceptions, advanced degrees were initiated as supplementary to the first professional degree, and growth has been closely linked to research programs. Opportunities beyond the baccalaureate were recognized from the beginning of forestry education in the United States. Post baccalaureate study for professional forestry degrees was often seen by forestry schools as a means of addressing the inadequacy of the four years of professional education.[86] The Yale School of Forestry, established in 1900, offered a two-year degree to students who had necessary requirements from their previous baccalaureate degree work in other fields. This was an expedient to address the immediate need for graduate foresters in the recently established forest reserves. Yale established high standards for programs that were to follow.[87]

At the outset, forestry in the United States had to borrow heavily from Europe for its knowledge base, but the need to establish the capability of generating knowledge about American forests and their management was immediately recognized. This meant forestry graduate degrees and capable scientists, but there were few qualified instructors available for advanced degree work.

For the first seventy years there was a strong belief that "men preparing for research in the field of forestry should have a knowledge such as is represented by a forestry degree. . . . most of our research work must be done by men who understand forestry in both its theoretical and practical aspects."[88] This was emphasized as a necessary credential for forestry school faculty as evidenced by frequent criticisms of their lack of experience in practice. Over the last two decades, the sophistication and breadth in forestry research and graduate programs coupled with the increase in the diversity of undergraduate curricular offerings at forestry schools has led to liberalization of such views. The result has been a significant and growing number of faculty as well as graduate students with little or no previous exposure to forestry.[89] This has been a needed change from the perspective of graduate study and research, and receives support in a recommendation of the National Research Council's Committee on Forestry Research: "Enhance the quality of forestry research by opening it to the broader scientific

86. (a) Graves, "Some Considerations of Policy in Forest Education." (b) Duncan, Skok, and Richards, "Forestry Education and the Profession's Future."
87. Graves and Guise, *Forest Education*.
88. Graves, "Some Considerations of Policy in Forest Education," p. 446.
89. Duncan, Skok, and Richards, "Forestry Education and the Profession's Future."

community and encourage increased participation by scientists currently in the community."[90]

The ninety years of forestry education has seen a significant shift toward graduate education in programming emphasis among the accredited forestry schools. Degrees granted provide an effective measure of this as shown in Table 9.1.

Table 9.1. Degrees granted

	Undergraduate	Master	Doctor
1910	61	48	N/A
1920	160	25	N/A
1930	308	69	34[a]
1940	1,072	105	9
1950	2,321	275	29
1960	1,473	296	36
1970	2,703	517	152
1980	4,899	840	160
1990	1,951	815	192

[a]A total of 34 doctoral degrees were granted by forestry schools through 1931. No record is available on an annual basis.

Sources: Gordon D. Marckworth, "Statistics from Schools of Forestry for 1971" (Bethesda, Md.: Society of American Foresters, 1972), unpublished; P. Gregory Smith, "SAF Enrollment and Degrees Survey" (Bethesda, Md., Society of American Foresters, 1992), unpublished.

Approximately 46% of the masters degrees and 61% of the doctoral degrees reported as granted in 1990 to students from accredited forestry schools were in fields with primary emphasis in forestry including forest products. A lack of historic data limits conclusions relative to long-term changes, but a reasonable speculation is that a much higher concentration of graduate degrees in these fields would have been noted in the first three quarters of this century.

The enactment and funding of the McIntire-Stennis Cooperative Forestry Research Program by the U.S. government in the early 1960s had a major influence. Among its objectives this legislation included enhancement of graduate education in the forest sciences along with forestry research. The increase in doctoral degrees subsequent to this is particularly noteworthy. The McIntire-Stennis legislation provided formula-funded federal research dollars to each state for distribution to the designated public institutions of higher education with the capacity to conduct forestry research.

The McIntire-Stennis Act provided a focus for federally-funded forestry

90. National Research Council, *Forestry Research: A Mandate for Change.* p. 54.

research that many forestry school administrators believed had not been possible through the Hatch Act funding coming to agricultural stations. R. H. Westveld stated, "Although Hatch Act funds could be and are being used for forestry research at the land-grant institutions, the directors of the state agricultural experiment stations have not allocated substantial funds to forestry research."[91] In support of this conclusion, he cited the fact that less than 2.5% of Hatch funds were allocated to forestry research nationally in fiscal year 1960. Forestry schools had long noted the reluctance of Congress to appropriate research funds for nonagricultural purposes under the Hatch Act. They also were impressed by the fact that Hatch funding led to significant increases in state agricultural research appropriations because of the state matching requirement. This also was required under the McIntire-Stennis Act.[92]

Graduate degree programs offered by forestry schools have been authorized by the graduate school and graduate faculty of the parent institution.[93] Professional graduate programs such as the Master of Forestry degree are often an exception when they are the direct responsibility of the forestry school and its faculty. The master degree category has represented both professional and scientific degree offerings usually designated as the "master of forestry," the "master of science in forestry," and the "master of science." The latter emphasizes specialization with strong science orientation. In actuality, all have contributed substantially to filling the ranks of professionals in forestry. Presently, over 25% of degrees granted annually by forestry schools are in the master's category.

The Doctor of Forestry degree has been considered another alternative for advanced professional forestry study. Duke and Yale Universities have offered such programs over the years although they have been little used. Dana and Johnson indicated that the degree was not sufficiently differentiated from the Doctor of Philosophy to meet the professional degree opportunities it presented. They recommended that the Doctor of Forestry be more widely developed and used to focus greater distinction on the professional character intended.[94] More recently Wallinger has revived the call, stressing "a broad and comprehensive, three-year Doctor of Forestry degree should be the prerequisite for full forest resource management leadership in forestry's second century."[95]

91. R. H. Westveld, "Opportunities for Research and Graduate Education in Forestry," *Journal of Forestry* 61 (6) (1963): 420.

92. Ibid., pp. 419–421.

93. Dana and Johnson, *Forestry Education in American Today and Tomorrow*.

94. Ibid.

95. Wallinger, "Creating and Educating a 21st-Century Forest Resources Profession."

H. Future Directions for Forestry Education

In reviewing forestry education it is clear that certain issues have persisted over the years. These include: (1) concern over the adequacy of a four-year degree program to meet professional needs; (2) the changing requirements of forestry and the impact this has on defining the profession of forestry, foresters, and therefore forestry education; (3) the importance of the human resource element in the quality of the education equation; (4) the dangers of ambivalence in dealing with needs for specialization in forest resource management; (5) the growing linkages between education and research; (6) the need for life-long learning to achieve professionalism in forestry; (7) the increased understanding of the role of forests in biological, physical, cultural, and social systems that advances in science provide; and (8) the need to be adaptive in education to the subtle and persistent changes in human values related to forest resources.

The recent National Research Council analysis of needed changes in forestry research concluded that the concepts that have been used for forest resource management which relied heavily on the commodity view of the forest were no longer appropriate. They called for adapting an environmental paradigm that "holds that human beings and nature are interrelated, that humans are not superior to the natural world, but depend on the biosphere for their existence. The biosphere's resources are finite, and human activities must not destroy the biosphere's intricate workings."[96] This is consistent in many respects with the summary findings of the recent Forestry Education Symposium which drew on the collective insights of more than two hundred faculty, researchers, and resource managers across a spectrum of employers.[97] It also represents the essentials of the "sustainability" approach whereby the natural resource heritage of future generations is assured and needs of present generations are addressed within that constraint.

This recent symposium sought to address what education needs would be under such a revised paradigm for resource managers of the future. The need for critical thinkers with analytic problem-solving capabilities, knowledgeable about ecosystems, able to be effective in interdisciplinary settings, aware of the social and political as well as the global context for their work, and capable of continuing professional growth and development were identified from these deliberations as critical benchmarks guiding educational programming in the years ahead. "Forestry" is used here in its broadest

96. National Research Council, *Forestry Research: A Mandate for Change*, pp. 14–15.
97. *Forest Resource Management in the 21st Century: Will Forestry Education Meet the Challenge?*

sense and relates to the education of all professional resource managers working in forestry.

Curricular approaches as well as student and faculty characteristics considered important to accomplish these outcomes through forestry education also were explored.[98] The "forestry education" baccalaureate curriculum of the future was viewed as one that would favor the generalists' approach, highlight basic competencies at the expense of traditional courses, have a more balanced resources perspective within a global context, develop abilities to participate in the socio-political process, and afford experimental learning experiences. An SAF Task Force on Sustaining Long-Term Forest Health and Productivity includes among its strategies for achieving such a goal, "the narrow disciplinary and functional orientation of professional education needs to be replaced by an integrative orientation throughout the curriculum."[99]

W. R. Bentley et al. emphasized that, "improved education, however, is more than curriculum content. . . . The best curriculum is not fully effective unless the best people are enticed into forestry programs . . . young leaders who come from the varied cultures and perspectives that create our pluralistic society."[100] The future importance of attracting academically well qualified and demographically representative faculty and students to forestry education was stressed. Students should have leadership and communication capabilities in addition to academic qualifications; faculty are the other part of the human resource equation and emphasis was given for them to be interested, enthusiastic, and effective as teachers. Individuals with significant resource management experience should be involved as instructors in forestry school programs.

Accommodating change will require that forestry schools be increasingly aware of who their clientele are and the significant issues of interest to them.[101] Forestry schools must continue to adapt to changing interests of both employers and students in order to remain viable. Conflicts in emphases will be more prevalent among relevant interest groups. This will present both new opportunities and challenges. There will be increased regional and institutional differences in educational programs. The "environmental paradigm" or "sustainability" will influence the educational direc-

98. Ibid.

99. "Sustaining Long-Term Forest Health and Productivity: Executive Summary of the Task Force Report," *Journal of Forestry* 91 (7) (1993): 33.

100. W. R. Bentley, B. C. Larson, and M. S. Ashton, "A Glimpse of the Resource Professional of 2022," *Forest Resource Management in the 21st Century: Will Forestry Education Meet the Challenge?*, p. 28.

101. K. Jane Coulter, "A Quantum Commitment to Quality Education," *Forest Resource Management in the 21st Century: Will Forestry Education Meet the Challenge?*

tions of many forestry schools into the twenty-first century. It would be a mistake to assume that a single model will serve all.

Institutional missions and priorities will require a school to move programmatically with its parent institution. Major graduate and research universities will most likely continue a trend that deemphasizes their traditional forestry programs while moving toward broader natural resource management and environmental studies programs, and a greater focus on graduate study. This commits an academic unit to significant changes in faculty expertise over time, and means less emphasis and, in some cases, eventual elimination of the traditional forestry degree and some departments. Some schools will be less identifiable as forestry schools as they make these changes. James G. Teer argues that a less homogeneous profession may be essential to break away from long-established professional constraints and to bring a more holistic approach to resource management. He suggests that accreditation and certification must provide for diversity to support such changes. His arguments extend to all traditional resource professions, not just to forestry.[102]

Almost unanimous agreement exists regarding the inadequacies of the four-year degree in preparing a forestry professional, but no agreement on what to do about this. The baccalaureate degree for resource management majors will be accepted as preparing the technicians of the future which generally has been the circumstance in recent years.[103]

For resource management professionals to progress to higher levels will require education beyond the four-year degree involving informal professional continuing education. One example that is suggested as a pervasive need in this area results from the emerging "environmental/sustainability" paradigm. "Self-directed study and continuing education that incorporate ecosystem management are needed by the forestry professionals to supplement the education received early in their careers."[104] This will require employers, professionals, and academic organizations to work more cooperatively to assure a quantity and quality of educational opportunities.

Formal advance degree offerings for both those with previous academic degrees in resource management and those with baccalaureates in other fields will need development and modification to meet emerging needs. Simply utilizing the research-oriented advanced degree will not meet the challenges to educating resource managers of the future.

Bentley et al. and Teer offer advice on advanced study for such program-

102. James G. Teer, "Educating Resource Managers for the Future," *Forest Resource Management in the 21st Century: Will Forestry Education Meet the Challenge?*, pp. 12–19.

103. Wallinger, "Creating and Educating a 21st-Century Forest Profession."

104. "Sustaining Long-Term Forest Health and Productivity: Executive Summary of the Task Force Report," p. 33.

ming needs. They acknowledge the importance of technical knowledge as well as its shortcomings as the entire or even principal basis for a resource management education. In their view it needs to be supplemented with emphasis on the humanistic dimensions of resource management, the development of abilities to analyze and synthesize, and a working understanding of mediation and conflict resolution.[105] Along with John F. Hosner, they agree on the advanced degree objective of such education as that of creating resource management generalists who will be the professionals. Hosner further proposes specialized degrees at the undergraduate level with an advanced degree that would educated specialists regarding the "many components of total resource management." He further proposes that accreditation occur at the first professional degree which would be at least one year beyond the baccalaureate.[106] Wallinger argues for the Doctor of Forestry as the true professional degree.[107]

This emerging emphasis of forestry schools in broader resource management and environmental studies presents the opportunity to reach an expanded population of motivated and qualified students. The challenge is to provide a scientifically sound, experiential based, and problem-solving learning experience that addresses issues beyond traditional programming at schools. This will require collaborating with many disciplines across the university. A premium will be placed on establishment of partnerships with private and public organizations programs. R. S. Whaley has presented both the need for, the barriers to, and the principles for success in such partnerships.[108]

Future directions must be dealt with in a period where higher education in the United States is facing critical issues of funding, redefining of missions and priorities, escalating instructional costs, emphasis on graduation rates, distrust of its commitment to quality instruction, and lowered public confidence in the institutions. Stakeholders with direct interests in resource management will be essential in guiding and supporting changes the forestry schools must make. The school must reach out to the key stakeholders and involve them as they plan, implement, and conduct their future programs. The importance of the stakeholders cannot be overemphasized. Forestry schools will need strategically to position themselves and to provide strong leadership in their dealings both within the university and with external constituents in order to succeed in their redirections.

105. (a) Bentley et al., "A Glimpse of the Resource Professional of 2022," pp. 20–29. (b) Teer, "Educating Resource Managers for the Future."

106. John F. Hosner, "Meeting Demands for Breadth and Depth: A Two-Tiered Approach," *Journal of Forestry* 91 (3) (1993): 15–16.

107. Wallinger, "Creating and Educating a 21st-Century Forest Profession."

108. R. S. Whaley, "Working Partnerships: Elements for Success," *Journal of Forestry* 91 (3) (1993): 10–11.

10. Reflections on Research and Editing

T. T. KOZLOWSKI

University of California, Berkeley

When I was invited to write an account of my experience as a researcher and editor and to make the record available to investigators particularly in Third World countries, I accepted with enthusiasm for several reasons. For a long time I had been concerned with the essentiality of forest products and complexity of forestry and agroforestry problems in developing countries. My interest was fostered by consultancies in Malaysia for the Food and Agriculture Organization of the United Nations, to the Federal Forestry Research Organization of EMBRAPA in Brazil, assignments at the University of Ibadan in Nigeria, and cooperation with the International Centre for Research in Agroforestry (ICRAF) in Nairobi, Kenya. I became familiar with forestry topics in the tropics through many years of reviewing research proposals for the International Foundation of Science. Supervision of the research programs of graduate students and postdoctoral investigators who came from developing countries to my laboratory at the University of Wisconsin in Madison made me aware of, and sympathetic to, many problems in tropical forestry.

The rationale for my choice of research projects always was deeply rooted in my fascination for the dynamics of developing forest stands and my curiosity about how stand development is controlled.[1] I knew that the species in a mixed forest stand vary greatly in growth rate, crown form, phenology, and longevity. I also knew that all forest stands show several common characteristics during their development. For example, they progress sequentially through (1) a regeneration stage in which certain species invade the spaces created by disturbances or harvesting, (2) a thinning stage

---

1. (a) T. T. Kozlowski, "The Challenges of Research in Forest Biology," *Forestry Chronicle* 47 (1971): 192–195. (b) T. T. Kozlowski, P. J. Kramer, and S. G. Pallardy, *The Physiological Ecology of Woody Plants* (San Diego, Calif.: Academic Press, 1991). (c) C. H. Winget, G. Cottam, and T. T. Kozlowski, "Species Association and Stand Structure of Yellow Birch in Wisconsin," *Forestry Science* 11 (1965): 369–383.

during which competition among trees is so intense that many die, (3) an understory initiation stage during which gaps form in the canopy and the understory is invaded by advance regeneration, and (4) an old growth stage.

Forest stands are subjected to many continuous and periodic environmental stresses that influence their growth and development. Important abiotic stresses include shading,[2] drought,[3] flooding,[4] temperature extremes,[5] wind,[6] low soil fertility,[7] and fire.[8] Major biotic stresses include attacks by insects and pathogens, herbivores, and activities of humans, including pollution. Trees growing in stands are stressed more than isolated trees because competition among plants reduces the amount of light, water, and mineral nutrients available to individual trees.[9] The effects of environmental stresses on tree growth and survival are mediated through changes in physiological processes, involving predominantly carbohydrate,[10] water,[11] mineral,[12] and hormone relations.[13]

2. Kozlowski, *The Physiological Ecology.*

3. T. T. Kozlowski, "Water Supply and Tree Growth, Part I, Water Deficits," *Forestry Abstracts* 43 (1982): 57–95.

4. T. T. Kozlowski, "Water Supply and Tree Growth, Part II, Flooding," *Forestry Abstracts* 43 (1982): 145–161.

5. (a) P. W. Owston and T. T. Kozlowski, "Effects of Temperature and Photoperiod on Growth of Western Hemlock," *Proceedings of the Western Hemlock Management Conference* (Seattle, Wash.: College of Forest Resources, University of Washington, 1976), pp. 108–117. (b) P. W. Owston and T. T. Kozlowski, "Growth and Cold Hardiness of Container-Grown Douglas-Fir, Noble Fir, and Sitka Spruce Seedlings in Simulated Greenhouse Regimes," *Canadian Journal of Forest Research* 11 (1981): 465–474. (c) A. R. Sena Gomes and T. T. Kozlowski, "Effect of Temperature on Growth and Water Relations of Cacao (*Theobroma Cacao* Var. *Comum*) Seedlings," *Plant and Soil* 103 (1987): 3–11.

6. A. R. Sena Gomes and T. T. Kozlowski, "Responses of Seedlings of Two Varieties of *Theobroma Cacao* to Wind," *Tropical Agriculture* 66 (1989): 137–141.

7. Kozlowski, *The Physiological Ecology.*

8. T. T. Kozlowski and C. E. Ahlgren, eds., *Fire and Ecosystems* (New York: Academic Press, 1974).

9. (a) Kozlowski, *The Physiological Ecology.* (b) C. H. Winget and T. T. Kozlowski, "Seasonal Basal Area Growth as in Expansion of Competition in Northern Hardwoods," *Ecology* 46 (1965): 786–793.

10. (a) T. T. Kozlowski, "Effects of Continuous High Light Intensity on Photosynthesis in Forest Trees," *Forestry Sciences* 3 (1957): 220–225. (b) T. T. Kozlowski and T. Keller, "Food Relations in Woody Plants," *Botanical Review* 32 (1966): 293–382. (c) S. Sasaki and T. T. Kozlowski, "Variable Photosynthetic Responses of *Pinus Resinosa* Seedlings to Herbicides," *Nature* 209 (1966): 1042–1044.

11. (a) T. T. Kozlowski, "Water Relations and Growth of Trees," *Journal of Forestry* 56 (1958): 498–502. (b) T. T. Kozlowski, *Water Metabolism in Plants* (New York: Hooper and Row, 1964). (c) T. T. Kozlowski, "Water Balance in Shade Trees," *Proceedings of the 44th International Shade Tree Conference* (1968), pp. 29–42. (d) T. T. Kozlowski, "Tree Physiology and Forest Pests," *Journal of Forestry* 67 (1969): 118–123. (e) Kozlowski, "The Challenges of Research" (f) Kozlowski, "Water Supply and Tree Growth, I" (g) T. T. Kozlowski, "Carbohydrate Sources and Sinks in Woody Plants," *Botanical Review* 58 (1992): 107–223. (h) T. T. Kozlowski and A. C. Gentile, "Respiration of White Pine Buds in Relation to Oxygen Availability and Moisture Content," *Forestry Sciences* 4 (1958): 147–152.

12. (a) T. T. Kozlowski and S. G. Pallardy, "Effects of Flooding on Water, Carbohydrate, and

Succession in disturbed forest stands progresses toward a community of species with a high capacity to tolerate the stresses of shading, drought, and mineral deficiency. Theoretically, a climax forest eventually may develop with a relatively stable species composition. However, even mature forests are not completely stable. Rather they are maintained in an oscillating steady state that is determined by recurrent disturbances. For example, many forest stands are burned periodically. As a result, succession to a climax forest is arrested in many regions and the stands are maintained in subfinal "fire climax" stages. An example is the subclimax pine forest of the southeastern United States.[14]

The importance of physiological control of tree growth can be illustrated by cambial growth responses following thinning of a forest stand to reduce competition among trees. The released trees respond to the greater availability of resources (light, water, mineral nutrients) by increasing their crown size. The increase in leaf area is accompanied by higher rates of photosynthesis and hormone synthesis, followed by transport of more carbohydrates and hormonal growth regulators to the lower stem, thus stimulating cambial growth to produce wood and bark. However, the increase in cambial growth in the lower stem may not be apparent for a few years after the stand a thinned, emphasizing that much time may elapse before the growth-controlling physiological changes take effect.

Although conducting field, greenhouse, and laboratory experiments, I

Mineral Relations," in *Flooding and Plant Growth*, ed. T. T. Kozlowski (New York: Academic Press, 1984), pp. 165–193. (b) Kozlowski, *The Physiological Ecology*. (c) P. E. Marshall and T. T. Kozlowski, "Changes in Mineral Contents of Cotyledons and Young Seedlings of Woody Angiosperms," *Canadian Journal of Botany* 53 (1975): 2026–2031.

13. (a) R. Domanski and T. T. Kozlowski, "Variations in Kinetin-Like Activity in Buds of *Betula* and *Populus* during Release from Dormancy," *Canadian Journal of Botany* 46 (1968): 397–403. (b) R. Domanski, T. T. Kozlowski, and S. Sasaki, "Interaction of Applied Growth Regulators and Temperature on Root Initiation in *Salix* Cuttings," *Journal of the American Society of Horticultural Science* 94 (1969): 39–41. (c) R. F. Evert, and T. T. Kozlowski, "Effect of Isolation of Bark on Cambial Activity and Development of Xylem and Phloem in Trembling Aspen," *American Journal of Botany* 54 (1967): 1045–1055. (d) R. F. Evert, T. T. Kozlowski, and J. D. Davis, "Influence of Phloem Blockage in Cambial Growth of *Acer Saccharum*," *American Journal of Botany* 49 (1972): 632–641. (e) T. T. Kozlowski and R. F. Evert, "Effect of Phloem Blocks on Cambial Activity of *Populus Tremuloides*," *American Journal of Botany* 53 (1966): 616. (f) F. Yamamoto and T. T. Kozlowski, "Effect of Ethrel on Stem Anatomy of *Ulmus Americana* Seedlings," *IAWA Bulletin*, n.s. 8 (1987): 3–9. (g) F. Yamamoto and T. T. Kozlowski, "Effects of Flooding, Tilting of Stems, and Ethrel Application on Growth, Stem Anatomy, and Ethylene Production of *Pinus Densiflora* Seedlings," *Journal of Experimental Botany* 38 (1987): 293–310. (h) F. Yamamoto and T. T. Kozlowski, "Effects of Flooding of Soil and Application of NPA and NAA to Stems on Growth and Stem Anatomy of *Acer Negundo* Seedlings," *Environmental and Experimental Botany* 27 (1987): 329–340. (i) F. Yamamoto and T. T. Kozlowski, "Experts of Flooding, Tilting of Stems, and Ethrel Application on Growth, Stem Anatomy, and Ethylene Production of *Acer Platanoides* Seedlings," *Scandinavian Journal of Forest Research* 2 (1987): 141–156.

14. Kozlowski, *The Physiological Ecology*.

recognized that some questions about effects of specific environmental stresses could not be adequately answered by field experiments because of the many environmental interactions involved. For example, light, temperature, and humidity are so interdependent that a change in one alters the others. During much of my career, I was fortunate in having access to the University of Wisconsin Biotron, a sophisticated controlled-environment laboratory in which plant responses could be quantified in a constant environment, or by changing one environmental factor and not others, and even under programmed diurnal and seasonal changes in climate.[15] The many carefully controlled environmental combinations that were available enabled quantitative studies of the action and interaction of environmental stresses on numerous forest tree species. Use of the Biotron decreased the variability of our data and increased its reproducibility.

A. Seed and Seedling Biology

Woody plants face their greatest mortality risk when they are in the ungerminated stage of the seed and in the cotyledon stage of seedling development. Many seeds fail to germinate because of embryo or seed coat dormancy[16] and unfavorable environmental conditions or seedbeds.[17] Many of the plants that survive the critical seed germination and cotyledon stages of development are killed during intense competition among seedlings for light, water, and mineral nutrients.[18] If a seed germinates the young seedling lives precariously at threshold levels of growth requirements, including carbohydrates, water, mineral nutrients, and growth hormones.[19] In order to

15. (a) T. T. Kozlowski, "Controlled Environment Research on Crop Production," *BioScience* 33 (1983): 677. (b) T. T. Kozlowski and P. A. Huxley, "The Role of Controlled Environments in Agroforestry Research," in *Plant Research and Agroforestry*, ed. P. A. Huxley (Edinburgh: Billans and Wilson, 1983), pp. 551–570.

16. (a) T. T. Kozlowski, *Tree Growth and Environmental Stresses* (Seattle, Wash.: University of Washington Press, 1979). (b) T. T. Kozlowski and A. C. Gentile, "Influence of the Seed Coat on Germination, Water Absorption and Oxygen Uptake of Eastern White Pine Seed," *Forest Science* 5 (1959): 389–395. (c) T. T. Kozlowski and J. H. Torrie, "Effect of Hydrogen Peroxide on Germination of Eastern White Pine Seed," *Advancing Frontiers in Plant Science* 9 (1964): 131–144.

17. C. H. Winget and T. T. Kozlowski, "Yellow Birch Germination and Seedling Growth," *Forest Science* 11 (1965): 386–392.

18. (a) D. I. Dickmann and T. T. Kozlowski, "Seasonal Changes in the Macro- and Micro-nutrient Composition of Ovulate Strobili and Seeds of *Pinus Resinosa* Ait.," *Canadian Journal of Botany* 47 (1969): 1547–1554. (b) T. T. Kozlowski, "Light and Water In Relation to Growth and Competition of Piedmont Forest Tree Species," *Ecological Monographs* 19 (1949): 207–231. (c) Kozlowski, *The Physiological Ecology*.

19. (a) T. T. Kozlowski, "Susceptibility of Young Tree Seedlings to Environmental Stresses," *American Nurseryman* 144 (1968): 12–13. (b) T. T. Kozlowski, *Growth and Development of*

gain insight into the physiological ecology of establishment of woody plants, we studied seed germination and early seedling development under both favorable and stressful environmental conditions.[20]

Environmental Stresses and Seedling Development

Growth and development of seedlings shortly after seeds germinated were very sensitive to many environmental stresses including shading,[21] day length,[22] drought,[23] and temperature.[24] Seedling development also was inhibited by allelopathic compounds[25] and a wide variety of chemicals, including herbicides,[26] air pollutants,[27] insecticides,[28] fungicides,[29] growth retardants,[30]

Trees, I: Seed Germination, Ontogeny, and Shoot Growth (New York: Academic Press, 1971). (c) Kozlowski, *Tree Growth and Environmental Stresses*. (d) P. J. Kramer and T. T. Kozlowski, *Physiology of Woody Plants* (New York: Academic Press, 1979).

20. (a) G. A. Borger and T. T. Kozlowski, "Early Periderm Ontogeny in *Fraxinus Pennsylvanica, Ailanthus Altissima, Robinia Pseudoacacia,* and *Pinus Resinosa* Seedlings," *Canadian Journal of Forest Research* 2 (1972): 135–143. (b) G. A. Borger and T. T. Kozlowski, "Wound Periderm Ontogeny in *Fraxinus Pennsylvanica* Seedlings," *New Phytologist* 71 (1972): 709–712. (c) Kozlowski, *Tree Growth and Environmental Stresses*. (d) S. G. Pallardy and T. T. Kozlowski, "Early Root and Shoot Growth of Rooted Tip Cuttings of *Populus*," *Slivae Genetica* 28 (1979): 153–156.

21. (a) G. A. Borger and T. T. Kozlowski, "Effects of Light Intensity on First Periderm and Xylem Development in *Pinus Resinosa, Fraxinus Pennsylvanica,* and *Robinia Pseudoacacia*," *Canadian Journal of Forest Research* 2 (1972): 190–197. (b) T. T. Kozlowski and G. A. Borger, "Effect of Temperature and Light Intensity Early in Ontogeny on Growth on *Pinus Resinosa* Seedlings," *Canadian Journal of Forest Research* 1 (1971): 57–65.

22. G. A. Borger and T. T. Kozlowski, "Effects of Photoperiod on Early Periderm and Xylem Development in *Fraxinus Pennsylvanica, Robinia Pseudoacacia,* and *Ailanthus Altissima* Seedlings," *New Phytologist* 71 (1972): 703–708.

23. G. A. Borger and T. T. Kozlowski, "Effects of Water Deficits on First Periderm and Xylem Development in *Fraxinus Pennsylvanica*," *Canadian Journal of Forest Research* 2 (1972): 144–151.

24. G. A. Borger and T. T. Kozlowski, "Effects of Temperature on First Periderm and Xylem Development in *Fraxinus pennsylvanica, Robinia pseudoacacia,* and *Ailanthus Altissima*" *Canadian Journal of Forest Research* 2 (1972): 198–205.

25. R. J. Norby and T. T. Kozlowski, "Allelopathic Potential of Ground Cover Species on *Pinus Resinosa* Seedlings," *Plant and Soil* 57 (1980): 363–371.

26. (a) G. A. Borger and T. T. Kozlowski, "Effect of Growth Regulators and Herbicides on Normal and Wound Periderm Ontogeny in *Fraxinus Pennsylvanica* Seedlings," *Weed Research* 12 (1972): 190–194. (b) T. T. Kozlowski, "Effects of 1,3,6-TBA on Seed Germination, Early Development, and Mortality of *Pinus Resinosa* Seedlings," *European Journal of Forest Pathology* 16 (1986): 385–390. (c) T. T. Kozlowski and J. E. Kuntz, "Effect of Simazine, Atrazine, Propazine, and Eptam on Growth of Pine Seedlings," *Soil Science* 95 (1963): 164–174. (d) T. T. Kozlowski and S. Sasaki, "Effects of Direct Contact of Pine Seeds or Young Seedlings with Commercial Formulations, Active Ingredients, or Inert Ingredients of Triazine Herbicides," *Canadian Journal of Plant Science* 48 (1968): 1–7. (e) T. T. Kozlowski and S. Sasaki, "Germination and Morphology of Red Pine Seeds and Seedlings in Contact with EPTC, CDEC, CDAA, 2,4–D and Picloram," *Proceedings of the American Society for Horticultural Science* 93 (1968): 655–662. (f) T. T. Kozlowski and S. Sasaki, "Effects of Herbicides on Seed Germination and Development of Young Pine Seedlings," in *Proceedings of the International Symposium on Seed Physiology and Woody Plants* (Poland: Institute of Dendrology and Kornik Arboretum, 1970), pp. 19–25. (g) T. T.

and antitranspirants.[31] Exposure of seedlings in the cotyledon stage of development to environmental stress and various agricultural chemicals often induced rapid mortality. The phytotoxicity of chemicals applied to the soil varied greatly with plant species, the compound applied and its dosage, environmental conditions, and manner of application of the chemical. Toxicity was low if the chemical was applied to the soil surface, intermediate if incorporated in the soil, and highest if maintained in direct contact in solution or suspension with plant tissues.[32] The high absolute toxicity of chemicals generally was masked because chemicals applied to the soil were lost by evaporation, leaching, microbial or chemical decomposition, and adsorption in the soil. Some chemicals killed young seedlings, others caused abnormal developmental changes.[33]

The mechanisms of herbicide toxicity were complex and involved effects on photosynthesis, respiration, and protein metabolism.[34] Herbicides also

Kozlowski and J. H. Torrie, "Effects of Soil Incorporation of Herbicides on Seed Germination and Growth of Pine Seedlings," *Soil Science* 100 (1965): 139–146. (h) S. Sasaki and T. T. Kozlowski, "Influences of Herbicides on Respiration of Young *Pinus* Seedlings," *Nature* 210 (1966): 439–440. (i) Sasaki, "Variable Photosynthetic Responses" (j) S. Sasaki and T. T. Kozlowski, "Effects of Herbicides on Respiration of Red Pine (*Pinus Resinosa* Ait.) Seedlings, II: Monuron, Diuron, DCPA, Dalapon, CDEC, CDAA, EPTC, and NPA," *Botanical Gazette* 129 (1968): 286–293. (k) S. Sasaki, T. T. Kozlowski, and J. H. Torrie, "Effect of Pretreatment of Pine Seeds with Herbicides on Seed Germination and Growth of Young Seedlings," *Canadian Journal of Botany* 46 (1968): 255–262. (l) C. H. Winget, T. T. Kozlowski, and J. E. Kuntz, "Effects of Herbicides on Red Pine Nursery Stock," *Weeds* 11 (1963): 87–90. (m) C. C. Wu, T. T. Kozlowski, R. F. Evert, and S. Sasaki, "Effects of Direct Contact of *Pinus Resinosa* Seeds and Young Seedlings with 2,4–D or Picloram on Seedling Development," *Canadian Journal of Botany* 49 (1971): 1737–1742.

27. H. Constantinidou, T. T. Kozlowski, and K. Jensen, "Effects of Sulfur Dioxide on *Pinus Resinosa* Seedlings in the Cotyledon Stage," *Journal of Environmental Quality* 5 (1976): 141–144.

28. M. O. Olofinboba and T. T. Kozlowski, "Effects of Three Systemic Insecticides on Seed Germination and Growth of *Pinus Halepensis*," *Plant and Soil* 64 (1982): 255–258.

29. T. T. Kozlowski, "Effects on Seedling Development of Direct Contact of *Pinus Resinosa* Seeds or Young Seedlings with Captan," *European Journal of Forest Pathology* 16 (1986): 87–90.

30. T. T. Kozlowski, "Effect of Direct Contact of *Pinus Resinosa* Seeds and Young Seedlings with N-Diethylamino Succinamic Acid, (2–Chloroethyl) Trimethylammonium Chloride, or Maleic Hydrazide," *Canadian Journal of Forest Research* 15 (1985): 1000–1004.

31. W. R. Chaney and T. T. Kozlowski, "Effects of Antitranspirants on Seed Germination, Growth, and Survival of Tree Seedlings," *Plant and Soil* 40 (1974): 225–229.

32. Kozlowski, "Effect of Soil Incorporation of Herbicides."

33. (a) Kozlowski, "Effects on Seedling Development . . . with Captan." (b) Kozlowski, "Germination and Morphology of Red Pine." (c) T. T. Kozlowski, S. Sasaki, and J. H. Torrie, "Influence of Temperature on Phytotoxicity of Triazine Herbicides to Pine Seedlings," *American Journal of Botany* 54 (1967): 790–796. (d) T. T. Kozlowski, S. Sasaki, and J. H. Torrie, "Influence of Temperature on Phytotoxicity of Monuron, Picloram, CDEC, EPTC, CDAA, and Sesone to Young Pine Seedlings," *Silva Fennica* 32 (1967): 13–28. (e) C. C. Wu and T. T. Kozlowski, "Some Histological Effects of Direct Contact of *Pinus Resinosa* Seeds and Young Seedlings with 2,4,5–T," *Weed Research* 12 (1972): 229–233. (f) Wu, "Effects of Direct Contact."

34. (a) Sasaki, "Influences of Herbicides." (b) Sasaki, "Variable Photosynthetic Responses." (c) S. Sasaki and T. T. Kozlowski, "Effects of Herbicides on Respiration of Red Pine (*Pinus Resinosa* Ait.) Seedlings. I. S-Triazine and Chlorophenoxy Acid Herbicides," *Advancing Frontiers in Plant Science* 22 (1968): 187–202. (d) Saski, "Effects of Herbicides, II."

caused direct injury to cells and tissues. The toxic action was traced to the active components of commercial herbicides, the "inert" ingredients, and synergistic effects of both.[35]

Cotyledon Physiology

Our experiments emphasized the importance of physiologically active cotyledons for seedling development.[36] Even temporary environmental stresses that inhibited physiological processes in cotyledons, particularly photosynthesis, decreased growth of seedlings and often led to their mortality.[37]

The cotyledons of hypogeous seedlings remained below ground and served primarily as storage organs.[38] The epigeous cotyledons, those emerging above ground, of some species also stored carbohydrates. Those of other species accumulated small amounts of carbohydrates but became photosynthetically active shortly after they emerged from the ground.[39] Species with seeds that lacked endosperm had cotyledons adapted for both storage and photosynthesis. The cotyledons of species with endosperm acted as storage organs and also as transfer organs in absorbing food reserves from the endosperm and transferring them to meristematic regions.[40]

Germination of pine seeds was followed by sequential development of cotyledons, primary needles, and secondary needles. Following seed germination, the rate of carbohydrate synthesis shifted from cotyledons to primary needles to secondary needles. Development of primary needles depended on photosynthate produced by the primary needles. The rate of cotyledon photosynthesis and development of *Pinus resinosa* needles were

35. (a) Kozlowski, "Effects of Direct Contact of Pine Seeds." (b) Kozlowski, "Effects of Herbicides on Seed Germination." (c) Sasaki, "Influences of Herbicides."

36. (a) G. A. Borger and T. T. Kozlowski, "Effect of Cotyledons, Leaves, and Stem Apex on Early Periderm Development in *Fraxinus Pennsylvanica* Seedlings," *New Phytologist* 71 (1972): 691–702. (b) Kozlowski, *Tree Growth and Environmental Stresses*. (c) P. E. Marshall and T. T. Kozlowski, "The Role of Cotyledons in Growth and Development of Woody Angiosperms," *Canadian Journal of Botany* 52 (1974): 239–245. (d) P. E. Marshall and T. T. Kozlowski, "Photosynthetic Activity of Cotyledons and Foliage Leaves of Young Angiosperm Seedlings," *Canadian Journal of Botany* 52 (1974): 2023–2032. (e) P. E. Marshall and T. T. Kozlowski, "Importance of Photosynthetic Cotyledons for Early Growth of Woody Angiosperms," *Physiologia Plantarum* 37 (1976): 336–340. (f) P. E. Marshall and T. T. Kozlowski, "Changes in Structure and Function of Epigeous Cotyledons of Woody Angiosperms During Early Seedling Growth," *Canadian Journal of Botany* 55 (1977): 208–215.

37. T. T. Kozlowski, "Susceptibility of Young Tree Seedlings to Environmental Stresses," *American Nurseryman* 144 (1976): 12–13, 55–59.

38. (a) Kozlowski, *The Physiological Ecology*. (b) Kramer, *Physiology of Woody Plants*.

39. (a) Kozlowski, *Growth and Development of Trees*. (b) Kozlowski, "Carbohydrate Sources and Sinks."

40. P. E. Marshall and T. T. Kozlowski, "Importance of Endosperm for Nutrition of *Fraxinus Pennsylvanica* Seedlings," *Journal of Experimental Botany* 27 (1976): 572–574.

closely correlated. Initiation and expansion of the first few primary needles depended on carbohydrates stored in the megagametophyte of the seed. However, development of most of the primary needles depended on carbohydrates synthesized by the cotyledons.[41] Partial reduction of cotyledon photosynthesis by herbicides was followed by decreased expansion of primary needles and reduced rate of dry weight increase of seedlings.[42] Exposure of seedlings to low temperature or low light intensity during the cotyledon stage prevented initiation of most of the normal complement of primary needles. However, when the environmentally stressed seedlings were returned to favorable temperature and light conditions, primordia of primary needles formed readily and expanded.[43]

The cotyledons of epigeous species progressed through storage, transition, photosynthetic, and senescent stages. In the storage phase the cotyledons contained variable amounts of carbohydrates, fats, proteins, and mineral nutrients, that were depleted during seed germination and used in early seedling growth. The amounts and types of stored products varied among species. Embryonic cotyledons of *Acer negundo* and *Robinia pseudoacacia* stored large amounts of carbohydrates; those of *Ailanthus altissima* and *Fraxinus pennsylvanica* stored little. The cotyledons of *Acer negundo, Ailanthus altissima*, and *Robinia pseudoacacia* stored large amounts of lipids.[44]

The cotyledons of the exalbuminous seeds, those lacking endosperm, of *Acer rubrum* and *Robinia pseudoacacia* stored large amounts of mineral nutrients that subsequently were transferred to meristematic tissues during early growth of seedlings. The small cotyledons of *Fraxinus pennsylvanica* stored only small amounts of mineral nutrients.[45]

In the transition stage the cotyledons changed from transfer organs to assimilatory organs. The transition stage was characterized by cotyledon expansion, depletion and use of cotyledon reserves in seedling growth, differentiation of guard cells, emergence of cotyledons, and synthesis of chlo-

41. (a) S. Sasaki and T. T. Kozlowski, "Utilization of Seed Reserves and Currently Produced Photosynthates of Embryonic Tissues of Pine Seedlings," *Annals of Botany* 33 (1969): 472–482. (b) S. Sasaki and T. T. Kozlowski, "Effects of Cotyledon and Hypocotyl Photosynthesis on Growth of Young Pine Seedlings," *New Phytologist* 69 (1970): 493–500.

42. (a) S. Sasaki and T. T. Kozlowski, "Effects of Herbicides on Seed Germination and Early Seedling Development of *Pinus Resinosa*," *Botanical Gazette* 129 (1968): 238–246. (b) S. Sasaki and T. T. Kozlowski, "The Role of Cotyledons in Early Development of Pine Seedlings," *Canadian Journal of Botany* 46 (1968): 1173–1183.

43. Kozlowski, "Effect of Temperature and Light Intensity."

44. P. E. Marshall and T. T. Kozlowski, "Compositional Changes in Cotyledons of Woody Angiosperms," *Canadian Journal of Botany* 54 (1976): 2473–2477.

45. Marshall, "Changes in Mineral Contents."

rophyll. The cotyledons emerged earlier in species with exalbuminous seeds than in species with cotyledons embedded in endosperm.[46]

The photosynthetic stage began with initial net carbon dioxide absorption and ended with senescence of cotyledons. Both the rate of cotyledon photosynthesis and the duration of the photosynthetic stage varied greatly among species.[47] The importance of cotyledon photosynthesis for growth and development of seedlings was emphasized by inhibitory effects on growth of excision of cotyledons or inhibition of cotyledon photosynthesis with herbicides.[48]

The beginning of the senescent stage was characterized by chlorosis of cotyledons which continued until the cotyledons were shed. The functional life span of cotyledons varied widely among species and was strongly influenced by environmental stresses.[49]

B. Plant and Water Relations

My strong interest in plant water relations began about a half century ago when I was a graduate student of P. J. Kramer, a splendid teacher and pioneer in the field. I learned early that the distribution, species composition, and growth of plants depend more on water supply than on any other environmental factor. The importance of water to plants is linked to its role as a primary constituent of protoplasm, reagent in photosynthesis and hydrolytic processes, solvent in which gases and salts move into and through plants, and maintainer of cell turgor.[50] Diurnal and seasonal changes in water balance of plants occur widely.[51] Because soil water supplies are almost never ideal, it seemed important to study in some detail how plants respond to too little and too much water.

Inhibition of tree growth by drought has been well documented.[52] Shoot

46. Marshall, "Changes in Structure and Function of Epigeous."

47. Ibid.

48. Ibid.

49. Ibid.

50. (a) P. J. Kramer and T. T. Kozlowski, *Physiology of Trees* (New York: McGraw-Hill, 1960). (b) Kramer, *Physiology of Woody Plants*.

51. (a) Kozlowski, "Water Supply and Tree Growth, I. " (b) J. S. Pereira and T. T. Kozlowski, "Diurnal and Seasonal Changes in Water Balance of *Abies Balsamea* and *Pinus Resinosa*," *Oecologis Plant Carum* 11 (1976): 413–428. (c) J. S. Pereira and T. T. Kozlowski, "Diurnal and Seasonal Changes in Water Balance of *Acer Saccharum* and *Betula Papyrifera*," *Physiologia Plantarum* 43 (1978): 19–30.

52. (a) W. J. Davies, T. T. Kozlowski, W. R. Chaney, and K. Lee, "Effects of Transplanting on Physiological Responses and Growth of Shade Trees," *Proceedings of the International Shade Tree Conference* 48 (1972): 22–30. (b) T. T. Kozlowski, "Tree Growth, Action and Interaction of Soil, and Other Factors," *Journal of Forestry* 53 (1955): 508–512. (c) Kozlowski, "Water Rela-

growth is decreased by inhibitory effects of water deficits on bud formation and on bud elongation into a shoot.[53] In some species the importance of water deficits during bud formation on shoot growth in the subsequent year is shown by high correlation between shoot length and the amount of rain in the previous year.[54] However, in species that exhibit recurrent flushes of shoot growth during the same growing season, the growth of shoots commonly is reduced during the year of drought.[55] Both cambial growth and root growth also are highly responsive to drought.[56] The effects of moisture deficits on growth of plants are mediated by changes in many physiological processes including absorption of water, seed germination, stomatal opening and closing, photosynthesis,[57] enzymatic activity,[58] leaf shedding,[59] mineral relations,[60] and hormone relations.[61]

tions and Growth of Trees." (d) Kozlowski, "Water Balance in Shade Trees." (e) T. T. Kozlowski, "Introduction," in *Water Deficits and Plant Growth*, Vol. 1 (New York, Academic Press, 1968), pp. 1–21. (f) T. T. Kozlowski, "Soil Water and Tree Growth," in *The Ecology of Southern Forests*, ed. N. E. Linnartz (Baton Rouge, La.: Louisiana State University Press, 1969), pp. 30–57. (g) T. T. Kozlowski, "Role of Environment in Plant Propagation: Water Relations," *Proceedings of the International Plant Propagation Society Annual Meeting* (1970): 123–139. (h) T. T. Kozlowski, "Physiology of Water Stress" USDA Forest Service, 1972 [General Technical Report INT-1], pp. 229–244. (i) Kozlowski, *Tree Growth and Environmental Stresses*. (j) Kozlowski, "Water Supply and Tree Growth, I." (k) T. T. Kozlowski, "Reduction in Yield of Forest and Fruit Trees by Water and Temperature Stress," in *Crop Reactions to Water and Temperature Stresses in Humid, Temperate Climates*, ed. C. D. Raper and P. J. Kramer (Boulder, Colo.: Westview Press, 1983), pp. 67–88.

53. (a) Kozlowski, "Physiology of Water Stress." (b) T. T. Kozlowski, "How Healthy Plants Grow," in *Plant Pathology—An Advanced Treatise*, ed. J. Horsfall and E. Cowling (New York: Academic Press, 1978), Vol. 3., pp. 19–51.

54. (a) Kozlowski, *Growth and Development of Trees*. (b) Kozlowski, "How Healthy Plants Grow." (c) T. T. Kozlowski, J. H. Torrie, and P. E. Marshall, "Predictability of Shoot Length from Bud Size in *Pinus Resinosa*," *Canadian Journal of Forest Research* 3 (1973): 34– 38.

55. (a) Kozlowski, *Growth and Development of Trees*. (b) Kozlowski, "How Healthy Plants Grow." (c) Kozlowski, *The Physiological Ecology*.

56. (a) Kozlowski, "Light and Water In Relation to Growth" (b) T. T. Kozlowski, *Growth and Development of Trees, Vol. 2: Cambial Growth, Root Growth, and Reproductive Growth* (New York: Academic Press, 1971).

57. (a) W. J. Davies, T. T. Kozlowski, and J. Pereira, "Effects of Wind on Transpiration and Stomatal Aperture of Woody Plants," in *Mechanisms of Regulation of Plant Growth*, ed. R. L. Bieleski, A. R. Ferguson, and M. M. Creswell (Wellington, N.Z.: Royal Society of New Zealand, 1974 [Royal Society of New Zealand Bulletin no. 12]), pp. 433–438. (b) Kozlowski, "Light and Water in Relation to Growth." (c) T. T. Kozlowski, "Photosynthesis, Climate and Growth of Trees," in *Tree Growth*, ed. T. T. Kozlowski (New York: Ronald Press, 1962), pp. 149–164. (d) M. Ogigirigi, T. T. Kozlowski, and S. Sasaki, "Effect of Soil Moisture Depletion on Stem Shrinkage and Photosynthesis of Tree Seedlings," *Plant and Soil* 32 (1970): 33–49.

58. Kramer, *Physiology of Woody Plants*.

59. (a) T. T. Kozlowski, "Extent and Significance of Shedding of Plant Parts," in *Shedding of Plant Parts*, ed. T. T. Kozlowski (New York: Academic Press, 1973), pp. 1–44. (b) T. T. Kozlowski, "Water Supply and Leaf Shedding," in *Water Deficits and Plant Growth*, ed. T. T. Kozlowski (New York: Academic Press, 1976), Vol. 4, pp. 191–231. (c) Kozlowski, *The Physiological Ecology* . . .

60. Kramer, *Physiology of Woody Plants*.

61. Kozlowski, *The Physiological Ecology*.

Water Transport

Over the years much interest has been shown in the path of upward water transport in trees, partly because of its implication in control of insects and diseases. Many systemic chemicals follow the path of water ascent in the xylem. Hence, the distribution of injected chemotherapeutants in tree crowns varies greatly among species with differences in their water transport patterns.[62]

In ring-porous angiosperms (e.g., *Quercus, Fraxinus*) water transport was limited to relatively few large vessels located in a thin cylinder of the outer sapwood and mostly to those in the outermost ring of xylem. In diffuse-porous angiosperms (e.g., *Populus, Acer*) water moved upward in many vessels located in several xylem rings in the sapwood.[63] The path of water transport in seedlings varied somewhat from that in large trees. For example, in three-year-old seedlings of *Fraxinus americana* (ring-porous) water moved primarily in the large earlywood vessels of the current annual ring and in the vessels located in two thirds of the annual ring produced in the previous year.[64]

In 1964 and 1965 I was at Oxford University where I worked with L. Leyton and J. F. Hughes who were interested in studying water turnover in forests. To gain some insight into this complex problem we investigated the paths of upward water transport in both seedlings and large gymnosperms. In pole-sized *Larix, Chamaecyparis lawsoniana, Picea abies*, and *Pinus sylvestris* water moved upward in many tracheids of several annual rings.[65]

The path of upward water transport may be vertical or may deviate from a vertical path depending on xylem structure. In many trees the ascent of water followed a spiral pathway more often than a vertical one. Variations in water uptake patterns often were associated with differences in spiral

62. T. T. Kozlowski, J. E. Kuntz, and C. H. Winget, "Effect of Oak Wilt on Cambial Activity," *Journal of Forestry* 60 (1962): 558–561.

63. (a) T. T. Kozlowski, "The Movement of Water in Trees," *Forest Science* 7 (1961): 177–192. (b) T. T. Kozlowski, "Water Movement in Trees," *Recent Advances in Botany* (1961): 1301–1305. (c) T. T. Kozlowski, "Water Transport in Trees," *Journal of the International Society of Tropical Ecology* 3 (1962): 84–100. (d) Kozlowski, *Water Metabolism in Plants*. (e) T. T. Kozlowski and C. H. Winget, "Patterns of Water Movement in Forest Trees," *Botanical Gazette* 124 (1963): 301–311.

64. W. R. Chaney and T. T. Kozlowski, "Patterns of Water Movement in Intact and Excised Stems of *Fraxinus Americana* and *Acer Saccharum* Seedlings," *Annals of Botany* 41 (1977): 1093–1100.

65. (a) T. T. Kozlowski, J. F. Hughes, and L. Leyton, "Patterns of Water Movement in Dormant Gymnosperm Seedlings," *Biorheology* 3 (1966): 77–85. (b) T. T. Kozlowski, J. F. Hughes, and L. Leyton, "Movement of Injected Dyes in Gymnosperm Stems in Relation to Tracheid Alignment," *Forestry* 40 (1967): 209–227.

grain. A spiral pathway was especially common in conifers and often there was more spiralling of the ascending sap stream than could be accounted for by the amount of spiral of the xylem resulting from the way in which the bordered pits were arranged in the xylem.[66]

In many species of conifers there was a marked, preponderantly clockwise, spiral pattern of water dye ascent.[67] This pattern varied somewhat in seedlings. The outermost ring of xylem often transported less water than the previous year's ring. Within an annual ring, the first-formed, large-diameter tracheids conducted the most water, with the last-formed, small-diameter tracheids conducting little or no water. However, lateral transport along the xylem rays was observed.[68]

Shrinking and Swelling of Plants

Variations in size of growing plants result from progressive accretion due to division and expansion of cells as well as superimposed changes in hydration of tissues and in thermal changes. As leaves, stems, roots, and reproductive structures dehydrate and rehydrate they shrink and swell.[69] Changes in the size of tissues and organs caused by hydration changes sometimes are small; at other times they may exceed those resulting from irreversible growth changes.[70] Both diurnal and seasonal shrinkage and swelling of plants are important because they influence stomatal aperture; wilting; flow of oleoresins and latex; dissemination of spores, pollen, and seeds; measurements of plant growth when determined by volume or linear dimensional changes; root functions; and cracking and splitting of plants. For these reasons we initiated studies of diurnal and seasonal shrinkage of

66. Ibid.

67. Kozlowski, "Movement of Injected Dyes."

68. (a) Kozlowski, "Patterns of Water Movement in Dormant." (b) T. T. Kozlowski, L. Leyton, and J. F. Hughes, "Pathways of Water Movement in Young Conifers," *Nature* 205 (1965): 830.

69. (a) F. H. Braekke and T. T. Kozlowski, "Shrinking and Swelling of Stems of *Pinus Resinosa* and *Betula Papyrifera* in Northern Wisconsin," *Plant and Soil* 43 (1975): 387–410. (b) W. R. Chaney and T. T. Kozlowski, "Seasonal and Diurnal Expansion and Contraction of Fruits of Forest Trees," *Canadian Journal of Botany* 47 (1969): 1033–1038. (c) W. R. Chaney and T. T. Kozlowski, "Seasonal and Diurnal Variations in Water Balance of Fruits, Cones, and Leaves of Forest Trees," *Canadian Journal of Botany* 47 (1969): 1407–1417. (d) W. R. Chaney and T. T. Kozlowski, "Seasonal and Diurnal Expansion and Contraction of *Pinus banksiana* and *Picea glauca* Cones," *New Phytologist* 68 (1969): 873–882. (e) T. T. Kozlowski, "Biological Implications of Shrinking and Swelling in Plants," *What's New in Plant Physiology* 4 (1) (1972): 1–6.

70. (a) Kozlowski, "Water Relations and Growth of Trees." (b) T. T. Kozlowski, "Expansion and Contraction of Plants," *Advancing Frontiers of Plant Sciences* 10 (1965): 63–77. (c) T. T. Kozlowski, "Diurnal Variations in Stem Diameters of Small Trees," *Botanical Gazette* 128 (1967): 60–68. (d) T. T. Kozlowski, "Continuous Recording of Diameter Changes in Tree Seedlings," *Forest Science* 13 (1967): 100–101. (e) Kozlowski, "Soil Water and Tree Growth." (f) Kozlowski, "Biological Implications of Shrinking and Swelling."

leaves, stems, roots, and reproductive structures of several species of woody plants.[71]

Leaves shrink and swell as their turgor changes.[72] Daily changes in thickness of leaves of *Prunus cerasus* grafted on *Prunus mahaleb* rootstock were highly correlated with environmental conditions affecting stomatal opening and transpiration. When the vapor pressure deficit (VPD) increased in the morning the leaves shrank and when VPD decreased at night the leaves expanded. After a prolonged drought, however, the leaves shrank progressively and their thickness was no longer correlated with diurnal changes in VPD.[73] The thickness of leaves of Calamondin orange trees began to decrease near sunrise when the VPD began to increase or slightly later, and they continued to shrink until mid or late afternoon, by which time the VPD was decreasing. The leaves expanded during the night. During a drought the amount and rate of daily leaf contraction and expansion decreased progressively. Irrigation following a drought resulted in rapid resumption of leaf expansion during the night and shrinkage during the day.[74]

Diurnal shrinkage of tree stems varied among species and at different times during the summer. Diurnal shrinkage was greater in *Pinus resinosa* than in *Populus tremuloides*, but greater in both than in *Quercus ellpsoidalis*. Stems shrank little early in the growing season when leaves were not fully expanded and the soil moisture content was high. As leaves expanded, temperature increased, and transpiration accelerated, the amount of stem shrinkage during the day and expansion at night increased. Late in the summer, as the soil dried and the stems dehydrated, daily stem shrinkage declined.[75] During prolonged droughts, especially late in the growing season, some tree stems shrank continuously for days, weeks, or months.[76] Following each rainfall an appreciable amount of radial increase in tree stems was traceable to a hydration component.[77] Following irrigation of severely

71. (a) Kozlowski, "Biological Implications of Shrinking and Swelling." (b) T. T. Kozlowski, "Shrinking and Swelling of Plant Tissues," in *Water Deficits and Plant Growth*, Vol. 3, pp. 1–64.

72. (a) Ibid. (b) A. R. Sena Gomes, T. T. Kozlowski, and P. B. Reich, "Some Physiological Responses of *Theobroma Cacao* var. *Catongo* Seedlings to Air Humidity," *New Phytologist* 107 (1987): 591–602.

73. W. R. Chaney and T. T. Kozlowski, "Diurnal Expansion and Contraction of Leaves and Fruits of English Morello Cherry," *Annals of Botany* 33 (1969): 991–999.

74. W. R. Chaney and T. T. Kozlowski, "Water Transport in Relation to Expansion and Contraction of Leaves and Fruits of Calamondin Orange," *Journal of Horticultural Science* 46 (1971): 71–81.

75. (a) Kozlowski, "Water Relations and Growth of Trees." (b) T. T. Kozlowski and C. H. Winget, "Diurnal and Seasonal Variation in Radii of Tree Stems," *Ecology* 45 (1964): 149–155.

76. (a) F. H. Bormann and T. T. Kozlowski, "Measurements of Tree Ring Growth with Dial-Gauge Dendrometers and Vernier Tree Ring Bands," *Ecology* 43 (1962): 289–291. (b) Braekke, "Shrinking and Swelling" (c) Kozlowski, "Effects on Seedling Development" (1986).

77. T. T. Kozlowski, C. H. Winget, and J. H. Torrie, "Daily Radial Growth of Oak in Relation to Maximum and Minimum Temperature," *Botanical Gazette* 124 (1962): 9–17.

droughted *Fraxinus americana* seedlings, most leaves regained turgor within an hour and increases in stem diameter were recorded within two hours after irrigation.[78]

Tree stems may also shrink during large temperature drops. In Wisconsin the amount of winter stem shrinkage usually was several times greater than the amount of diameter increase by cambial growth during the entire previous growing season. Winter stem shrinkage varied appreciably among species.[79] Most of the thermal expansion and contraction was localized in the bark.[80]

Superimposed on growth changes of fruits and cones (strobili) are changes in size resulting from hydration changes, with shrinkage during the day and expansion at night. Shrinkage of *Prunus cerasus* fruits in a mid-stage of development began in the morning and continued until late afternoon, at which time the fruits began to expand. Shrinkage and expansion were correlated with factors controlling transpiration.[81] In *Citrus mitis* a lag in fruit shrinkage behind leaf shrinkage indicated that water had been transported out of the fruits. VPD influenced fruit shrinkage through leaf transpiration.[82]

Early in their development, gymnosperm cones showed little diurnal shrinkage. For example, young cones of *Picea glauca* and *Pinus banksiana* showed only progressive increases in diameter. In contrast, cones in a midstage of development showed reversible daily shrinkage and expansion.[83] These changes were inversely correlated with VPD. During maturation the cones generally shrank continuously as they dehydrated.[84]

Drought Tolerance

A variety of drought-avoiding adaptations occur in leaves, stems, and roots of trees. These include shedding of leaves; production of small or only few leaves; small, few, and sunken stomata; rapid closure of stomata during drought; heavy deposition of leaf waxes; occlusion of stomata with wax;

78. Ogigirigi, "Effect of Soil Moisture Depletion."

79. (a) C. H. Winget and T. T. Kozlowski, "Winter Shrinkage in Stems of Forest Trees," *Journal of Forestry* 62 (1964): 335–337. (b) Winget, "Seasonal Basal Area Growth."

80. I. W. McCracken and T. T. Kozlowski, "Thermal Contraction in Twigs," *Nature* 208 (1965): 910–912.

81. T. T. Kozlowski, "Diurnal Changes in Diameters of Fruits and Tree Stems of Montmorency Cherry," *Journal of Horticultural Science* 43 (1968): 1–15.

82. Chaney, "Water Transport in Relation."

83. (a) D. I. Dickmann and T. T. Kozlowski, "Seasonal Growth Patterns of Ovulate Strobili of *Pinus Resinosa* in Central Wisconsin," *Canadian Journal of Botany* 47 (1969): 839–848. (b) Kozlowski, "Shrinking and Swelling of Plant Tissues."

84. Chaney, "Seasonal and Diurnal Expansion . . . *Picea Glauca* Cones."

and strong development of palisade tissues.[85] Resistance to water loss varies among species and genotypes[86] because of variations in stomatal size, stomatal frequency, and rapid stomatal closure during droughts. Often stomatal size and frequency (number per unit area) are negatively correlated. Stomatal size and frequency may vary greatly among different species of the same genus and among genotypes.[87] Wide variations occur among stomatal responses to atmospheric factors.[88]

We conducted experiments on stomatal resistance of species that occur along an ecological gradient from xerophytic to mesophytic; stomatal resistance to diffusion varied along the gradient.[89] *Quercus velutina* had the highest water use efficiency and was the most drought resistant species studied. It fixed carbon dioxide rapidly while losing little water and hence was successful on hot and dry sites.[90]

Control of Water Loss

Much attention has been given to maintaining hydration of plants by suppressing water loss with antitranspirants. Such compounds include: (1) film-type antitranspirants that form films on leaves and often occlude stomatal pores; these include waxes, wax-oil emulsions, high alcohols, silicones, plastics, and resins, and (2) metabolic antitranspirants that prevent stomatal

85. (a) Kozlowski, "Water Supply and Leaf Shedding." (b) T. T. Kozlowski, "Drought Resistance and Transplantability of Shade Trees," in *Better Trees for Metropolitan Landscapes*, ed. F. S. Santamour, H. Gerhold, and S. Little (1976 [USDA General Technical Report, NE-22]), pp. 77–90. (c) T. T. Kozlowski, "Water Relations and Tree Improvement," in M. Cannell and F. T. Last, eds., *Tree Physiology and Yield Improvement* (London: Academic Press, 1976), pp. 307–327. (d) S. G. Pallardy and T. T. Kozlowski, "Cuticle Development in the Stomatal Region of *Populus* Clones," *New Phytologist* 85 (1980): 363–368. (e) A. R. Sena Gomes and T. T. Kozlowski, "Stomatal Characteristics, Leaf Waxes, and Transpiration Rates of *Theobroma Cacao* and *Hevea Brasiliensis* Seedlings," *Annals of Botany* 61 (1988): 425–432.

86. (a) S. G. Pallardy and T. T. Kozlowski, "Water Relations of *Populus* Clones," *Ecology* 57 (1981): 367–373. (b) J. E. Wuenscher and T. T. Kozlowski, "The Response of Transpiration Resistance to Leaf Temperature as a Desiccation Resistance Mechanism in Tree Seedlings," *Physiologia Plantarum* 24 (1971): 254–259.

87. (a) S. G. Pallardy and T. T. Kozlowski, "Frequency and Length of Stomata of Twenty-One *Populus* Clones," *Canadian Journal of Botany* 57 (1979): 2519–2523. (b) R. Siwecki and T. T. Kozlowski, "Leaf Anatomy and Water Relations of Excised Leaves of Six *Populus* Clones," *Arboretum Kornickie* 18 (1973): 83–105.

88. (a) W. J. Davies, T. T. Kozlowski, and K. J. Lee, "Stomatal Characteristics of *Pinus Resinosa* and *Pinus Strobus* in Relation to Transpiration and Antitranspirant Efficiency," *Canadian Journal of Forest Research* 4 (1974): 571–577. (b) Davies, "Effects of Wind on Transpiration." (c) J. S. Pereira and T. T. Kozlowski, "Influence of Light Intensity, Temperature, and Leaf Area on Stomatal Aperture and Water Potential of Woody Plants," *Canadian Journal of Forest Research* 7 (1976): 145–153.

89. Wuenscher, "The Response of Transpiration."

90. J. E. Wuenscher and T. T. Kozlowski, "Relationship of Gas-Exchange Resistance to Tree Seedling Ecology," *Ecology* 52 (1971): 1016–1023.

opening in the light and induce closure of already open stomata. These compounds include succinic acids, phenylmercuric acetate,(PMA), hydroxylsulfonates, atrazine, and sodium azide. We evaluated the effects of many film-type and metabolic antitranspirants on transpiration, photosynthesis, growth, and injury to both angiosperms and gymnosperms. Experiments were conducted in the field, greenhouse, and environmentally-controlled growth chambers.[91] We also studied the effects of antitranspirants applied to the soil on seed germination.[92]

Although most film-type antitranspirants reduced water loss, many were phytotoxic. Toxicity, which varied with the antitranspirant and dosage, plant species, method of application, and environmental conditions, was evident in reduced photosynthesis, altered metabolism, leaf lesions, chlorosis, browning of leaves, reduced growth, and mortality of plants. Spraying of film-type antitranspirants was much less toxic than was dipping of leaves. The adverse effects of film-type antitranspirants were especially long-lived in *Pinus resinosa*, a species in which transpiration reduction was associated with the antitranspirant combining with wax in the stomatal pores and forming impermeable plugs.[93]

The effectiveness of metabolic antitranspirants varied with the specific compounds used (atrazine, PMA, decenylsuccinic acid, or abscisic acid). Atrazine was not effective in closing stomata. Although PMA closed stomata, it caused browning of leaves and a decrease in chlorophyll content.[94]

91. (a) T. T. Kozlowski and J. J. Clausen, "Effect of Decenylsuccinic Acid on Needle Moisture Content and Shoot Growth of *Pinus Resinosa*," *Canadian Journal of Plant Science* 50 (1970): 355–356. (b) T. T. Kozlowski and W. J. Davies, "Control of Water Balance in Transplanted Trees," *Journal of Arboriculture* 1 (1975): 1–10. (c) G. C. Marks and T. T. Kozlowski, eds., *Ectomycorrhizae: Their Ecology and Physiology* (New York: Academic Press, 1973). (d) M. O. Olofinboba, T. T. Kozlowski, and P. E. Marshall, "Effects of Antitranspirants on Carbohydrate Synthesis, Translocation, and Incorporation in *Pinus Resinosa*," *Plant and Soil* 40 (3) (1974): 619–635. (e) Y. Waisel, G. A. Borger, and T. T. Kozlowski, "Effects of Phenylmercuric Acetate on Stomatal Movement and Transpiration of Excised *Betula Papyrifera* Marsh Leaves," *Plant Physiology* 44 (1969): 685–690.

92. (a) W. R. Chaney and T. T. Kozlowski, "Effects of Antitranspirants on Seed Germination, Growth, and Survival of Tree Seedlings," *Plant and Soil* 40 (1973): 225–229. (b) W. R. Chaney and T. T. Kozlowski, "Growth and Transpiration of Black Locust Seedlings in Soil with Incorporated Cetyl Alcohol," *Canadian Journal of Forest Research* 3 (1973): 604–606.

93. (a) W. J. Davies and T. T. Kozlowski, "Short- and Long-Term Effects of Antitranspirants on Water Relations and Photosynthesis of Woody Plants," *Journal of the American Society for Horticultural Science* 99 (1974): 297–304. (b) W. J. Davies and T. T. Kozlowski, "Effects of Applied Abscisic Acid and Plant Water Stress on Transpiration of Woody Angiosperms," *Forest Science* 22 (1975): 191–95. (c) W. J. Davies and T. T. Kozlowski, "Effect of Applied Abscisic Acid and Silicone on Water Relations and Photosynthesis of Woody Plants," *Canadian Journal of Forest Research* 5 (1975): 90–96. (d) Kozlowski, "Control of Water Balance in Transplanted Trees." (e) K. J. Lee and T. T. Kozlowski, "Effects of Silicone Antitranspirants on Woody Plants," *Plant and Soil* 40 (1974): 493–506. (f) Olofinboba, "Effects of Antitranspirants on Carbohydrate Synthesis."

94. Waisel, "Effects of Phenylmercuric Acetate."

Decenylsuccinic acid applied to *Pinus resinosa* injured the needles and inhibited late-summer development of buds. Shoot growth in the year after treatment was greatly reduced.[95] By comparison, abscisic acid (ABA) was a very effective, non-toxic antitranspirant. It initially reduced transpiration and photosynthesis but maintained leaf turgor as the soil dried. Eventually the rate of photosynthesis of ABA-treated plants was higher than that of control plants.[96]

Flooding of Soil

Growth of plants usually is adversely affected when the soil is flooded and followed by rapid depletion of soil oxygen, replacement of aerobic organisms by anaerobic organisms (primarily bacteria), decrease in the rate of decomposition of organic matter, and accumulation of toxic compounds in the plants and soil.[97] Inundation of soil adversely affected growth of many species of angiosperms and gymnosperms. Variations in response to flooding among species were much greater than responses to drought. Sensitivity to flooding varied not only with species and genotype but also with the age of plants, condition of the flood water, and time and duration of flooding.[98] Seedlings of species considered very flood tolerant (e.g., *Taxodium distichum*) were very sensitive to flooding when inundated with stagnant water.[99]

Flooding of soil variously reduced growth of leaves, stems, and roots of many angiosperms[100] and gymnosperms.[101] The flood tolerance of closely

95. Kozlowski, "Effect of Decenylsuccinic Acid."
96. Davies, "Effects of Applied Abscisic Acid and Silicone."
97. (a) Kozlowski, "Water Supply and Tree Growth." (b) T. T. Kozlowski, "Extent, Causes and Impacts of Flooding," in *Flooding and Plant Growth*, pp. 1–7. (c) T. T. Kozlowski, "Responses of Woody Plants to Flooding," in *Flooding and Plant Growth*, pp. 129–164. (d) T. T. Kozlowski, "Plant Responses to Flooding of Soil," *BioScience* 34 (1984): 162–167. (e) Kozlowski, "Soil Aeration and Growth of Forest Trees."
98. (a) Kozlowski, "Water Supply and Tree Growth." (b) T. T. Kozlowski, "Responses of Woody Plants to Flooding." (c) T. T. Kozlowski, "Plant Responses to Flooding of Soil," *BioScience* 34 (1984): 162–167. (d) Kozlowski, "Soil Aeration and Growth of Forest Trees." (e) Kozlowski, *The Physiological Ecology*.
99. Shanklin, "Effect of Flooding of Soil on Growth."
100. (a) Kozlowski, "Water Supply and Tree Growth, II." (b) Kozlowski, "Extent, Causes and Impacts of Flooding." (c) Kozlowski, "Responses of Woody Plants to Flooding." (d) Kozlowski, "Plant Responses to Flooding of Soil." (e) Kozlowski, "Soil Aeration and Growth." (f) R. D. Newsome, T. T. Kozlowski, and Z. C. Tang, "Responses of *Ulmus Americana* Seedlings to Flooding of Soil," *Canadian Journal of Botany* 60 (1982): 1685–1695. (g) R. J. Norby and T. T. Kozlowski, "Growth Responses and Adaptations of *Fraxinus Pennsylvanica* Seedlings to Flooding," *Physiologia Plantarum* 49 (1980): 373–377. (h) R. J. Norby and T. T. Kozlowski, "Flooding and Sulfur Dioxide-Stress Interaction in *Betula Papyrifera* and *B. Nigra* Seedlings," *Forest Science* 29 (1983): 739–750. (i) J. S. Pereira and T. T. Kozlowski, "Variations Among Woody Angio-

related species often varied appreciably. For example, seedlings of *Eucalyptus camaldulensis* adapted better than those of *E. globulus*,[102] and *Betula nigra* tolerated flooding better than *B. papyrifera*.[103] Flooding reduced growth earlier and more drastically in *Pinus banksiana* than in *P. resinosa*.[104] Responses to flooding also varied among three families of *Hevea brasiliensis*.[105]

sperms in Response to Flooding," *Physiologia Plantarum* 41 (1977): 184–192. (j) A. R. Sena Gomes and T. T. Kozlowski, "Growth Responses and Adaptations of *Fraxinus Pennsylvanica* Seedlings to Flooding," *Plant Physiology* 66 (1980): 267–271. (k) A. R. Sena Gomes and T. T. Kozlowski, "Responses of *Melaleuca Quinquenervia* Seedlings to Flooding," *Physiologia Plantarum* 49 (1980): 373–377. (l) A. R. Sena Gomes and T. T. Kozlowski, "Effects of Flooding on Growth of *Eucalyptus Camaldulensis* and *E. Globulus* Seedlings, *Oecologia* 46 (1980): 139–142. (m) A. R. Sena Gomes and T. T. Kozlowski, "Responses of *Pinus Halepensis* Seedlings to Flooding," *Canadian Journal of Forest Research* 10 (1980): 308–311. (n) A. R. Sena Gomes and T. T. Kozlowski, "Effect of Flooding on Water Relations and Growth of *Theobroma Cacao* var. *Catongo* Seedlings," *Journal of Horticultural Science* 61 (1986): 265–276. (o) A. R. Sena Gomes and T. T. Kozlowski, "Physiological and Growth Responses to Flooding of Seedlings of *Hevea Brasiliensis*," *Biotropica* 20 (1988): 286–293. (p) Z. C. Tang and T. T. Kozlowski, "Some Physiological and Morphological Responses of *Quercus Macrocarpa* Seedlings to Flooding," *Canadian Journal of Forest Research* 12 (1982): 196–202. (q) Z. C. Tang and T. T. Kozlowski, "Physiological, Morphological, and Growth Responses of *Platanus Occidentalis* Seedlings to Flooding," *Plant and Soil* 66 (1982): 243–255. (r) Z. C. Tang and T. T. Kozlowski, "Further Studies in Flood Tolerance of *Betula Papyrifera* Seedlings," *Physiologia Plantarum* 59 (1983): 218–222. (s) Z. C. Tang and T. T. Kozlowski, "Water Relations, Ethylene Production, and Morphological Adaptations of *Fraxinus Pennsylvanica* Seedlings to Flooding," *Plant and Soil* 77 (1984): 183–192. (t) Z. C. Tang and T. T. Kozlowski, "Ethylene Production and Morphological Adaptations of Woody Plants to Flooding," *Canadian Journal of Botany* 62 (1981): 1659–1664. (u) H. Tsukahara and T. T. Kozlowski, "Importance of Adventitious Roots to Growth of Flooded *Platanus Occidentalis* Seedlings," *Plant and Soil* 88 (1985): 123–132. (v) H. Tsukahara and T. T. Kozlowski, "Effects of Flooding and Temperature Regime on Growth and Stomatal Aperture of *Betula Platyphylla* var. *Japonica* Seedlings," *Plant and Soil* 92 (1986): 103–112. (w) Yamamoto, "Effect of Ethrel on Stem Anatomy." (x) Yamamoto, "Effects of Flooding . . . *Acer Platanoides* Seedlings."

101. (a) Z. C. Tang and T. T. Kozlowski, "Responses of *Pinus Bankosiana* and *P. Resinosa* Seedlings to Flooding," *Canadian Journal of Forest Research* 13 (1983): 633–639. (b) H. Tsukahara and T. T. Kozlowski, "Effect of Flooding on Growth of *Larix Leptolepis* Seedlings," *Journal of the Japanese Forestry Society* 66 (1984): 33–66. (c) H. Tsukahara, T. T. Kozlowski, and J. Shanklin, "Tolerance of *Pinus Densiflora, Pinus Thunbergii,* and *Larix Leptolepis* Seedlings to Sulfur Dioxide," *Plant and Soil* 88 (1985): 385–397. (d) H. Tsukahara, T. T. Kozlowski, and J. Shanklin, "Effects of Sulfur Dioxide on Two Age Classes of *Chamaecyparis Obtusa* Seedlings," *Journal of the Japanese Forestry Society* 68 (1986): 349–353. (e) F. Yamamoto and T. T. Kozlowski, "Effect of Flooding of Soil on Growth, Stem Anatomy, and Ethylene Production of *Thuja Orientalis* Seedlings," *IAWA Bulletin New Series* 8 (1986): 21–29. (f) F. Yamamoto and T. T. Kozlowski, "Effect of Flooding on Growth and Stem Anatomy of *Pinus Halepensis* Seedlings," *IAWA Bulletin, New Series* 8 (1987): 11–19. (g) Yamamoto, "Effects of Flooding . . . *Pinus Densiflora* Seedlings." (h) F. Yamamoto and T. T. Kozlowski, "Effect of Flooding of Soil on Growth, Stem Anatomy, and Ethylene Production of *Cryptomeria Japonica* Seedlings," *Scandinavian Journal of Forest Research* 2 (1987): 45–58. (i) F. Yamamoto, T. T. Kozlowski, and K. E. Wolter, "Effect of Flooding on Growth, Stem Anatomy, and Ethylene Production of *Pinus Halepensis* Seedlings," *Canadian Journal of Forest Research* 17 (1987): 69–79.

102. Sena Gomes, "Effects of Flooding . . . *E. Globulus* Seedlings."
103. Norby, "Flooding and Sulfur Dioxide-Stress."
104. Tang, "Responses of *Pinus Banksiana*."
105. Sena Gomes, "Physiological and Growth Responses."

Flooding reduced leaf area by arresting leaf initiation and expansion and by inducing leaf abscission in *Betula papyrifera, Platanus occidentalis, Ulmus americana,* and *Pinus halepensis;*[106] soil inundation caused extensive leaf shedding in *Betula papyrifera.*[107] By contrast, flooding did not induce leaf shedding in seedlings of *Betula nigra.*[108]

The effects of flooding on cambial growth were complex and varied from decreases to increases in stem diameters of tree seedlings. In general, however, xylem production was reduced by prolonged flooding as in *Pinus banksiana, P. resinosa,*[109] *Larix leptolepis,*[110] and *Acer negundo.*[111] In some species flooding increased stem diameters by stimulating bark production rather than by increasing xylem increment.[112] In *Thuja orientalis,* however, diameters of stem increased as a result of increased growth of both xylem and bark tissues.[113]

Aerobic root respiration is essential for uptake of mineral nutrients by roots.[114] The amount of energy released by roots in flooded plants is inadequate to sustain uptake of minerals in sufficient amounts. Reduced growth of roots under anaerobic conditions as well as decay of the root system also decrease total absorption of mineral nutrients. The decreased absorption of minerals also is associated with suppression of mycorrhizal fungi. When the soil oxygen supply is limited, the permeability of root membranes is affected and the roots lose ions by leaching.[115]

The amounts and balances of growth regulating hormones are greatly altered in flooded plants. Ethylene, auxins, and abscisic acid are increased while gibberellins and cytokinius in stems and roots are decreased.[116]

Ethylene production was greatly stimulated by flooding of both angiosperms and gymnosperms, including in several tree species.[117] Because

106. (a) Newsome, "Responses of *Ulmus Americana* Seedlings." (b) Sena Gomes, "Responses of *Pinus Halepensis* Seedlings to Flooding." (c) Tang, "Physiological, Morphological." (d) Z. C. Tang and T. T. Kozlowski, "Some Physiological and Growth Responses of *Betula Papyrifera* Seedlings to Flooding," *Physiologia Plantarum* 55 (1982): 415–420.

107. Tang, "Some Physiological and Growth Responses."

108. Norby, "Flooding and Sulfur Dioxide-Stress."

109. Tang, "Responses of *Pinus Banksiana.*"

110. Tsukahara, "Effects of Flooding on Growth of *Larix Leptolepis* Seedlings."

111. Yamamoto, "Effects of Flooding of Soil . . . *Acer Negundo* Seedlings."

112. (a) Sena Gomes, "Physiological and Growth Responses." (b) Yamamoto, "Effect of Flooding . . . *Cryptomeria Japonica* Seedlings." (c) Yamamoto, "Effects of Flooding . . . *Acer Platanoides* Seedlings."

113. Yamamoto, "Effects of Flooding . . . *Thuja Orientalis* Seedlings."

114. Kozlowski, *The Physiological Ecology.*

115. Ibid.

116. (a) Ibid. (b) D. M. Reid and K. J. Bradford, "Effects of Flooding on Hormone Relations," in *Flooding and Plant Growth,* pp. 195–219.

117. (a) Tang, "Some Physiological and Morphological . . ." (b) Tang, "Water Relations, Ethylene Production." (c) Sena Gomes, "Responses of *Melalcuca Quinquenervia.*" (d) Yamamoto, "Ef-

many claims have been made that ethylene regulates several aspects of plant growth, we investigated its role in control of regulation of reaction wood (compression wood in gymnosperms and tension wood in angiosperms). Ethylene did not directly regulate formation of compression wood. Flooding of soil stimulated ethylene production in *Pinus halepensis* seedlings and induced formation of rounded, thick-walled tracheids. However, these tracheids developed an S_3 wall layer that is not present in well-developed compression wood.[118] Application of ethrel (which leads to ethylene production) to *Pinus halepensis* stems induced formation of abnormal tracheids which lacked the essential features of compression wood.[119] Tilting of *Pinus densiflora* seedlings was followed by formation of compression wood but ethrel application to upright stems was not. Also ethrel application to tilted stems inhibited formation of compression wood.[120]

Several lines of evidence showed that ethylene did not directly regulate formation of tension wood in *Acer platanoides*: (1) ethylene production was stimulated more by flooding than by tilting of stems, yet only tilting induced formation of tension wood; (2) increase in the amount of ethylene in tilted seedlings was equal to or greater on the lower than the upper side of the stem, but tension wood formed only on the upper side; (3) ethrel application to upright stems did not induce formation of tension wood; and (4) ethrel application to tilted stems inhibited formation of tension wood.[121] Applying ethrel to *Ulmus americana* seedlings stimulated ethylene production but did not induce formation of tension wood.[122]

Formation of adventitious roots has been attributed to the stimulatory effects of ethylene. Our experiments indicated, however, that ethylene did not have a primary direct effect on initiating adventitious roots in flooded plants. Often the formation of adventitious roots was not correlated with ethylene levels. More ethylene was produced by seedlings of some flooded tree species and some seedlings increased ethylene production more than five times.[123] All of these variant reactions were recorded in a set of papers.[124]

fects of Flooding . . . *Acer Negundo* Seedlings." (e) Sena Gomes, "Physiological and Growth Responses." (f) Yamamoto, "Effects of Flooding . . . *Thuja Orientalis* Seedlings." (g) Yamamoto, "Effect of Flooding . . . *Cryptomeria Japonica* Seedlings." (h) Yamamoto, "Effect of Flooding . . . *Pinus Halepensis* Seedlings."

118. Ibid.

119. Yamamoto, "Effects of Flooding . . . *Pinus Halepensis* Seedlings."

120. Yamamoto, "Effects of Flooding . . . *Pinus Densiflora* Seedlings."

121. Yamamoto, "Effects of Flooding . . . *Acer Platanoides* Seedlings."

122. Yamamoto, "Effects of Ethrel on Stem."

123. (a) Tang, "Ethylene Production and Morphological." (b) Tang, "Some Physiological and Morphological."

124. (a) Tang, "Some Physiological and Morphological." (b) Tang, "Water Relations, Ethylene Production." (c) Tang, "Ethylene Production and Morphological."

Important adaptations of woody plants to flooding include oxygen transport from shoots to roots and subsequent release of oxygen through lenticels and oxidation of reduced soil compounds, production of lenticels in roots and submerged stems, production of aerenchyma tissues, and production of adventitious roots.[125] Many flood-tolerant species initiated adventitious roots on submerged portions of stems and on their original roots.[126] When the original roots of flooded plants died back to major secondary or primary roots, new roots often emerged from these points. The origin of adventitious roots varied among species.

Although there has been spirited controversy about the adaptive significance of adventitious roots in flooded plants, our experiments indicated that such roots increased flood tolerance. This conclusion was based on these observations: (1) flood tolerance and production of adventitious roots were correlated. Such flood-intolerant species as *Pinus halepensis*, *P. banksiana*, *P. resinosa*, and *Betula papyrifera* produced few or no adventitious roots when flooded.[127] In contrast, flood-tolerant species, including *Fraxinus pennsylvanica* and *Betula nigra*,[128] *Platanus occidentalis*,[129] and *Melaleuca quinquenervia*[130] produced many adventitious roots when flooded; (2) absorption of water and presumably mineral nutrients was greatly increased by flood-induced adventitious roots. Production of adventitious roots also was correlated with reopening of stomata that had closed earlier following flooding of the soil;[131] (3) when flood-induced adventitious roots were excised from the submerged stems of *Platanus occidentalis*, subsequent growth of plants was inhibited.[132]

C. Carbohydrate Relations of Plants

Approximately two-thirds of the dry weight of woody plants consists of transformed sugars. Their growth depends on carbohydrate synthesis, trans-

125. (a) Kozlowski, "Extent, Causes and Impacts of Flooding." (b) Kozlowski, "Responses of Woody Plants to Flooding." (c) Kozlowski, *The Physiological Ecology.*

126. (a) Kozlowski, "Water Supply and Tree Growth, II." (b) Kozlowski, "Extent, Causes and Impacts of Flooding." (c) G. Angeles, R. F. Evert, and T. T. Kozlowski, "Development of Lenticels and Adventitious Roots in Flooded *Ulmus Americana* Seedlings," *Canadian Journal of Forest Research* 16 (1986): 585–590. (d) Yamamoto, "Effects of Flooding . . . *Cryptomeria Japonica* Seedlings."

127. (a) Sena Gomes, "Responses of *Pinus Halepensis*." (b) Tang, "Some Physiological and Growth Responses." (c) Tang, "Responses of *Pinus Banksiana*."

128. Kozlowski, "Responses of Woody Plants to Flooding."

129. T. T. Kozlowski, "Soil Aeration and Growth of Forest Trees." Review Article in *Scandinavian Journal of Forest Research* 1 (1986): 113–123.

130. Sena Gomes, "Responses to *Melaleuca Quinquenervia*."

131. Sena Gomes, "Growth Responses and Adaptations."

132. Tsukahara, "Importance of Adventitious Roots."

port to meristematic regions, and assimilation into new tissues.[133] It therefore seemed imperative to study carbohydrate sources and sinks, environmental impacts on availability of carbohydrates, and patterns of utilization of carbohydrates in vegetative and reproductive growth. Most carbohydrates are produced by photosynthesis in leaves but small and variable amounts are synthesized in other green tissues such as cotyledons, hypocotyls, buds, twigs, stems, flowers, fruits, and cones.[134]

The dry weight of a woody plant represents only a very small part of the photosynthate it produced. The difference is traceable to large aggregate losses of tissues by consumption by fungi and herbivores as well as carbohydrate losses by respiration, leaching, exudation, secretion, translocation to other plants through root grafts and mycorrhizae, and losses to mistletoes and other sap feeders.[135]

The rate of photosynthesis varies among species and genotypes and is influenced by interactions of many environmental factors[136] and various biocides such as antitranspirants, fungicides, herbicides, and insecticides.[137] Photosynthetic rates varied widely among species.[138] They were higher in *Quercus* than in *Pinus* seedlings but the amount of the difference varied with the unit on which carbon dioxide absorption was based (e.g., leaf area or leaf dry weight).[139] Photosynthetic rates also varied appreciably among *Populus* clones.[140] Photosynthesis at low light intensities was more efficient in shade-grown leaves than in sun-grown leaves.[141] The rate of photosynthesis was reduced by both stomatal closure (hence by decreased carbon dioxide uptake) and adverse effects on the photosynthetic mechanism. The importance of stomatal conductance was shown by variation in carbon diox-

133. (a) Kozlowski, "Carbohydrate Sources and Sinks." (b) Kozlowski, "Food Relations in Woody Plants." (c) Kramer, *Physiology of Woody Plants*.

134. (a) Kozlowski, "Carbohydrate Sources and Sinks." (b) Kozlowski, "Food Relations in Woody Plants."

135. Kozlowski, "Carbohydrate Sources and Sinks."

136. (a) Kozlowski, "Effects of Continuous High Light." (b) Kozlowski, "Photosynthesis, Climate and Growth of Trees." (c) Kozlowski, "Light and Water in Relation to Growth . . ." (d) Kozlowski, "Effects of Flooding on Water." (e) Davies, "Effect of Wind on Transpiration." (f) Sena Gomes, "Responses of Seedlings." (g) Kozlowski, *The Physiological Ecology*. (h) Sena Gomes, "Effect of Temperature on Growth." (i) T. T. Kozlowski and H. A. Constantinidou, "Responses of Woody Plants to Environmental Pollution, Part II, Factors Affecting Responses to Pollution," *Forestry Abstracts* 47 (1986): 105–32.

137. (a) Davies, "Short- and Long-Term Effects." (b) Davies, "Effect of Applied Abscisic Acid and Silicone." (c) Olofinboba, "Effects of Antitranspirants." (d) S. Sasaki and T. T. Kozlowski, "Effects of Herbicides on Carbon Dioxide Uptake of Pine Seedlings," *Canadian Journal of Botany* 45 (1967): 961–971.

138. Kozlowski, "Photosynthesis, Climate and Growth of Trees."

139. Kozlowski, "Light and Water in Relation to Growth."

140. O. Luukkanen and T. T. Kozlowski, "Gas Exchange in Six *Populus* Clones," *Silvae Genetica* 21 (1972): 220–229.

141. Kozlowski, "Light and Water in Relation to Growth."

ide uptake by plants with different-sized stomata. Correlation between stomatal conductance and tolerance of gaseous air pollutants also emphasized the importance of stomatal aperture to gas exchange.[142]

Shoot Growth

The amounts and sources of carbohydrates that were used in shoot growth varied with species and genotypes. In some species the winter bud contained a telescoped, preformed shoot. Such species completed shoot internode expansion in only a third to a half of the frost-free season.[143] Most internode elongation of *Pinus resinosa* shoots was completed by the beginning of July in Wisconsin, although the needles elongated beyond that date.[144] Hence the shoots were strong carbohydrate sinks for a short part of the growing season only.

Early in the growing season the opening terminal buds of *Pinus resinosa* imported carbohydrates largely from the one-year-old needles. By June 1, small additional amounts were imported from two- and three-year-old needles. By mid-August the supply of current photosynthate from old needles declined, with the fully expanded current-year needles replacing the old needles as the primary carbohydrate source.[145] Overall the one-year-old needles were the most important carbohydrate source for shoot growth of *Pinus resinosa*.[146]

In some species shoot growth consists of expansion of internodes and preformed leaves that were present in the winter bud as well as late (neoformed) leaves that form sequentially during the growing season. The latter group of species continued shoot growth late into the frost-free season and

142. (a) T. W. Kimmerer and T. T. Kozlowski, "Stomatal Conductance and Sulfur Dioxide Uptake of Five Clones of *Populus Tremuloides* Exposed to Sulfur Dioxide," *Plant Physiology* 67 (1981): 990–995. (b) R. J. Norby and T. T. Kozlowski, "The Role of Stomata in Sensitivity of *Betula Papyrifera* Marsh. Seedlings to Sulfur Dioxide at Different Humidities," *Oecologia* 53 (1982): 34–39.

143. (a) T. T. Kozlowski, "Characteristics and Improvement of Forest Growth," *Advancing Frontiers of Plant Science* 2 (1963): 73–136. (b) T. T. Kozlowski, "Growth Characteristics of Forest Trees," *Journal of Forestry* 61 (1963): 655–662. (c) T. T. Kozlowski, "Shoot Growth in Woody Plants," *Botanical Review* 30 (1964): 335–392. (d) Kozlowski, "How Healthy Plants Grow." (e) T. T. Kozlowski and R. C. Ward, "Seasonal Height Growth of Deciduous Trees," *Forest Science* 3 (1957): 168–171. (f) T. T. Kozlowski and R. C. Ward, "Effect of Continuous High Light Intensity on Photosynthesis in Forest Trees," *Forest Science* 3 (1957): 220–225.

144. T. T. Kozlowski and R. C. Ward, "Shoot Elongation Characteristics of Forest Trees," *Forest Science* 7 (1961): 357–368.

145. D. I. Dickmann and T. T. Kozlowski, "Mobilization by *Pinus Resinosa* Cones and Shoots of C^{14}-Photosynthate from Needles of Different Ages," *American Journal of Botany* 55 (1968): 900–906.

146. T. T. Kozlowski and C. H. Winget, "The Role of Reserves in Leaves, Branches, Stems, and Roots on Shoot Growth of Red Pine," *American Journal of Botany* 51 (1964): 522–529.

hence used carbohydrates for shoot growth for a longer time than did species with shoots that were wholly preformed in the winter bud.[147]

Growth of shoots of some temperate zone species and many tropical species occurs in a series of flushes. This growth pattern is characterized by expansion of the contents of a previously resting bud, followed by rapid formation and expansion of another bud at the shoot apex. Such repeated flushing often occurred several times and late into the growing season, with each growth flush depleting carbohydrates. Some tropical pines exhibited more or less continuous shoot growth and consumed carbohydrates throughout much of the year.[148]

Cambial Growth

The rate and seasonal duration of carbohydrate use in cambial growth varied among species and in different locations on stems and branches. Carbohydrate use in xylem production continued later into the summer in evergreen than in deciduous trees.[149] Dominant trees in forest stands used more carbohydrates in cambial growth, and for more of the growing season, than suppressed trees.[150] More carbohydrates were used for production of xylem than phloem increments. Early-season xylem production of *Pinus resinosa* depended largely on current photosynthate rather than on stored carbohydrates. Exposure of one-year-old needles of trees to $^{14}CO_2$ in mid-June was followed by rapid incorporation of ^{14}C into structural compounds of the newly formed tracheids. Thickening of tracheid walls to produce latewood also used large amounts of current photosynthate.[151]

Growing fruits and cones are strong carbohydrate sinks as shown by preferential transport of reserve and currently produced carbohydrates into rap-

147. (a) J. J. Clausen and T. T. Kozlowski, "Heterophyllous Shoots in *Betula Papyrifera*," *Nature* 205 (1965): 1030–1031. (b) J. J. Clausen and T. T. Kozlowski, "Seasonal Growth Characteristics of Long and Short Shoots of Tamarack," *Canadian Journal of Botany* 45 (1967): 1643–1651. (c) J. J. Clausen and T. T. Kozlowski, "Observations on Growth of Long Shoots of *Larix Laricina*," *Canadian Journal of Botany* 48 (1970): 1045–1048. (d) T. T. Kozlowski and J. J. Clausen, "Shoot Growth Characteristics of Heterophyllous Woody Plants," *Canadian Journal of Botany* 44 (1966): 827–843. (e) T. T. Kozlowski and J. J. Clausen, "Seasonal Development of Long- and Short-Shoot Components of Tamarack," *Bulletin of the Ecological Society of America* 47 (1966): 113–114.

148. T. T. Kozlowski and T. E. Greathouse, "Shoot Growth and Form of Tropical Pines," *Unasylva* 24 (4) (1970): 6–14.

149. Winget, "Seasonal Basal Area Growth."

150. T. T. Kozlowski and T. A. Peterson, "Seasonal Growth of Dominant, Intermediate, and Suppressed Red Pine Trees," *Botanical Gazette* 124 (1962): 146–154.

151. D. I. Dickmann and T. T. Kozlowski, "Mobilization and Incorporation of Photoassimilated ^{14}C by Growing Vegetative and Reproductive Tissues of Adult *Pinus Resinosa* Trees," *Plant Physiology* 45 (1970): 284–288.

idly growing reproductive structures. Use of the bulk of available carbohydrate pool is correlated with reduction in growth of vegetative tissues,
emphasizing the strong sink strength of reproductive organs.[152]

The carbohydrate needs of gymnosperm cones varied greatly during their
development. First-year cones were weak carbohydrate sinks whereas second-year cones were strong sinks. Early in the season carbohydrate reserves
were important for growth of second-year cones because their dry weight
increased in mid-April in Wisconsin when the rate of photosynthesis was
very low. By early May, however, the cones were importing current photosynthate. The needles that expanded during the previous growing season
were the most important source of photosynthate for early and mid-season
growth of cones. Transport of carbohydrates to the cones from current-year
needles became important only late in the growing season after these needles had matured. Photosynthesis by the green cones was relatively unim-
portant[153] and compensated only partly for respiratory losses of carbohydrates and did not contribute appreciably to their dry weight increase.[154]

D. Environmental Pollution

Plants are threatened by a wide diversity of chemicals that contaminate
the air and soil. The major pollutants include sulfur dioxide, ozone, fluorides, oxides of nitrogen, peroxyacetylnitrates, and particulates such as ce-
ment kiln and foundry dusts, soot, lead particles, magnesium oxide, iron
oxide, and sulfuric acid aerosols. Sulfur dioxide and ozone probably induce
more injury to plants than all other pollutants combined. However, near
point sources of pollution such as smelters injury to plants often is caused
by fluorides, dusts, and heavy metals—pollutants generally considered less
important than sulfur dioxide and ozone.[155]

152. (a) Kozlowski, "Carbohydrate Sources and Sinks." (b) Kozlowski, "Food Relations in
Woody Plants." (c) Kozlowski, *The Physiological Ecology.*
153. (a) Dickmann, "Seasonal Growth Patterns." (b) D. I. Dickmann and T. T. Kozlowski,
"Seasonal Variations in Reserve and Structural Components of *Pinus Resinosa* Ait. Cones," *American Journal of Botany* 56 (1969): 515–521. (c) D. I. Dickmann and T. T. Kozlowski, "Photosynthesis by Rapidly Expanding Green Strobili of *Pinus Resinosa*," *Life Sciences* 9 (Part 2) (10)
(1970): 549–552. (d) D. I. Dickmann and T. T. Kozlowski, "Water, Nutrient, and Carbohydrate
Relations in Growth of *Pinus Resinosa* Ovulate Strobili," in *Proceedings of the 1st All Union
Symposium on Sexual Reproduction in Conifers* (USSR: Novosibirsk, 1973), pp. 195–209.
154. (a) Kozlowski, "Carbohydrate Sources and Sinks." (b) Kozlowski, *The Physiological Ecology.*
155. (a) T. T. Kozlowski, "Impacts of Air Pollution on Forest Ecosystems," *BioScience* 30
(1980): 88–43. (b) T. T. Kozlowski, "Responses of Shade Trees to Pollution," *Journal of Arboriculture* 6 (1980): 29–41. (c) Kozlowski, "Effect of Direct Contact of *Pinus Resinosa* Seedlings."
(d) T. T. Kozlowski and H. A. Constantinidou, "Responses of Woody Plants to Environmental
Pollution, Part I, Sources, Types of Pollutants, and Plant Responses," *Forestry Abstracts* 47 (1986):
5–51.

Injury

Gaseous pollutants injured leaves after being absorbed through stomatal pores. Symptoms included necrotic mottling and lesions, which often preceded early leaf senescence and abscission. Acute injury was severe and followed absorption of enough pollutant to kill tissue. Chronic injury, induced by sublethal amounts of pollutants absorbed over a long time, was characterized by slow development of chlorosis. In leaves of broadleaved trees sulfur dioxide injury consisted of areas of injured tissues located between healthy tissues along the veins.[156] Symptoms of injury were less definite for evergreen than for deciduous trees. Several pollutants caused tipburn of pine needles. Tipburn also was caused by herbicides, deicing salts, excess fertilizers, and freezing temperature, often making it difficult to identify the specific cause.[157]

Growth

Air pollutants, alone and in combination, variously reduced shoot growth, cambial growth, root growth, as well as dry weight increment and relative growth rate of woody plants.[158] They also altered plant community structure.[159] The rates of leaf expansion of *Ulmus americana* seedlings exposed to sulfur dioxide were only about 40% of those of unfumigated plants.[160] Seedlings in the cotyledon stage of development were particularly sensitive to sulfur dioxide. For example fumigation of two-week-old *Pinus*

156. (a) H. A. Constantinidou and T. T. Kozlowski, "Effects of Sulfur Dioxide and Ozone on *Ulmus Americana* Seedlings, I, Visible Injury and Growth," *Canadian Journal of Botany* 57 (1979): 170–175. (b) R. J. Norby and T. T. Kozlowski, "Relative Sensitivity of Three Species of Woody Plants to Sulfur Dioxide at High or Low Response Temperature," *Oecologia* 51 (1981): 33–36. (c) W. Suwannapinunt and T. T. Kozlowski, "Effect of Sulfur Dioxide on Transpiration, Chlorophyll Content, Growth, and Injury in Young Seedlings of Woody Angiosperms," *Canadian Journal of Forest Research* 10 (1980): 78–81.

157. (a) Kozlowski, "Impacts of Air Pollution on Forest Ecosystems." (b) Kozlowski, "Responses of Shade Trees to Pollution." (c) Kozlowski, "Responses of Woody Plants to Environmental Pollution, I."

158. (a) Ibid. (b) T. T. Kozlowski, "Measurement of Effects on Environmental and industrial Chemicals on Terrestrial Plants," in V. B. Vouk, G. C. Butler, D. G. Hoel, and D. B. Peakall, eds., *Methods for Estimating Risk of Chemical Injury: Human and Non-Human Biota and Ecosystems* (Chichester: John Wiley & Sons, 1985), pp. 573–609. (c) T. T. Kozlowski, "Effects of Environmental Stresses on Deciduous Trees," in *Response of Plants to Multiple Stresses*, ed. H. A. Mooney, W. E. Winner, and E. J. Pell (San Diego, Calif.: Academic Press, 1991), pp. 391–409. (d) J. Shanklin and T. T. Kozlowski, "Effect of Temperature Regime on Growth and Subsequent Responses of *Sophora Japonica* Seedlings to Sulfur Dioxide," *Plant and Soil* 88 (1985): 399–405.

159. (a) Kozlowski, "Impacts of Air Pollution on Forest Ecosystems." (b) T. T. Kozlowski, "Effects of Sulfur Dioxide on Plant Community Structure," in *Sulfur Dioxide and Vegetation: Physiology, Ecology, and Policy Issues*, ed. W. E. Winner, H. A. Mooney, and R. Goldstein (Stanford, Calif.: Stanford University Press, 1985), pp. 431–451.

160. H. A. Constantinidou, "Effects of Sulfur Dioxides and Ozone . . . I."

resinosa seedlings with sulfur dioxide at 50 pphm for only fifteen minutes reduced dry weight increment and chlorophyll content of both cotyledons and primary needles.

Fumigation with 200 pphm sulfur dioxide for one or two hours inhibited height growth of *Robinia pseudoacacia*,[161] *Larix leptolepis, Pinus densiflora,* and *P. thunbergii*[162] but not *Taxodium distichum* seedlings.[163] Environmental pollutants variously inhibited dry weight increment of woody plants, with seedlings in the cotyledon stage especially sensitive.[164] In seedlings beyond the cotyledon stage of development, sulfur dioxide concentrations up to 2 μl l^{-1} for up to seventy-two hours reduced dry weight increment of seedlings and various plant parts of *Pinus, Larix,*[165] *Taxodium,*[166] *Ulmus,*[167] and *Fraxinus.*[168]

The relative growth rate (RGR) permits comparison of the effects of environmental factors on the rate of plant growth, independently of the size of the plants that are being compared.[169] The RGR of many species was reduced by exposure to sulfur dioxide over a range of concentrations. Both the amount of reduction in RGR and partitioning of dry weight between roots and shoots varied among species and with sulfur dioxide dosage.[170]

Combined air pollutants were more toxic than single pollutants. For example, the effects of sulfur dioxide-ozone mixtures were more severe and were evident earlier than those of sulfur dioxide or ozone alone.[171]

Metabolism

Exposure of plants to air pollutants often was followed by disturbances in metabolism especially in carbohydrate, protein, lipid, and enzyme relations.

161. Suwannapinunt, "Effect of Sulfur Dioxide on Transpiration."

162. Tsukahara, "Tolerance of *Pinus Densitlora*."

163. Shanklin, "Effect of Flooding . . . *Taxodium Distichum* Seedlings."

164. (a) Constantinidou, "Effects of Sulfur Dioxide on *Pinus Resinosa*" (b) Suwannapinunt, "Effect of Sulfur Dioxide."

165. (a) Tsukahara, "Tolerance of *Pinus Densiflora*." (b) R. J. Norby and T. T. Kozlowski, "Response of Sulfur Dioxide-Fumigated *Pinus Resinosa* Seedlings to Post Fumigation Temperature," *Canadian Journal of Botany* 59 (1981): 470–475.

166. Shanklin, "Effect of Flooding . . . *Taxodium Distichum* Seedlings."

167. Constantinidou, "Effects of Sulfur Dioxide and Ozone . . . I."

168. J. J. Shanklin and T. T. Kozlowski, "Effect of Temperature Preconditioning on Response of *Fraxinus Pennsylvanica* Seedlings to Sulfur Dioxide," *Environmental Pollution* (Series A) 36 (1984): 311–326.

169. (a) Kozlowski, *The Physiological Ecology.* (b) Kramer, *Physiology of Woody Plants.*

170. (a) Kozlowski, "Effect of Direct Contact of *Pinus Resinosa* Seedlings." (b) Norby, "Response of Sulfur Dioxide-Fumigated." (c) Norby, "Relative Sensitivity of Three Species." (d) R. J. Norby and T. T. Kozlowski, "Interaction of Sulfur Dioxide Concentration and Post-Fumigation Temperature on Growth of Woody Plants," *Environmental Pollution* 25 (1981): 27–39. (e) Norby, "Flooding and Sulfur Dioxide-Stress." (f) Shanklin, "Effect of Flooding . . . *Taxodium Distichum* Seedlings." (g) Tsukahara, "Importance of Adventitious Roots."

171. Constantinidou, "Effects of Sulfur Dioxide . . . I."

Such changes were associated with altered growth of plants.[172] Exposure to sulfur dioxide reduced the amounts of nonstructural carbohydrates and proteins in *Ulmus americana* seedlings but caused only small changes in lipid contents.[173] Exposure of *Pinus resinosa* and *Betula papyrifera* seedlings to sulfur dioxide increased production of ethylene and ethane and induced production of acetaldehyde.[174]

Variations in Responses to Pollution

The responses of woody plants to pollution varied with species and genotype,[175] pollutant dosage,[176] types and combinations of pollutants,[177] plant responses measured,[178] developmental stages of plants,[179] and environmental regimes.[180] Differences among species and genotype responses were traceable largely to variations in absorption of pollutants by leaves because of differences in stomatal conductance.[181] Variations in susceptibility of *Pop-*

172. (a) Constantinidou, "Effects of Sulfur Dioxide . . . I." (b) T. T. Kozlowski and J. B. Mudd, "Introduction," in *Responses of Plants to Air Pollution*, ed. J. B. Mudd and T. T. Kozlowski (New York: Academic Press, 1975), pp. 1–8. (c) Kozlowski, *The Physiological Ecology* . . . (d) J. B. Mudd and T. T. Kozlowski, eds. *Responses of Plants to Air Pollution*.

173. Constantinidou, "Effects of Sulfur Dioxide and Ozone . . . II."

174. T. W. Kimmerer and T. T. Kozlowski, "Ethylene, Ethane, Acetaldehyde and Ethanol Production by Plants Under Stress," *Plant Physiology* 69 (1982): 840–847.

175. (a) K. F. Jensen and T. T. Kozlowski, "Absorption and Translocation of Sulfur Dioxide by Seedlings of Four Forest Tree Species," *Journal of Environmental Quality* 4 (1975): 379–381. (b) Kimmerer, "Stomatal Conductance." (c) Norby, "Relative Sensitivity of Three Species." (d) Tsukahara, "Tolerance of *Pinus Densiflora*." (e) Tsukahara, "Responses of *Betula Papyrifera*."

176. (a) Constantinidou, Effects of Sulfur Dioxide and Ozone . . . I and II." (b) Constantinidou, "Effects of Sulfur Dioxide on *Pinus Resinosa*." (c) Kozlowski, "Impacts of Air Pollution on Forest Ecosystems." (d) Kozlowski, "Effects of Sulfur Dioxide on Plant Community Structure." (e) Kozlowski, "Responses of Woody Plants to Environmental Pollution, I." (f) Norby, "Interaction of Sulfur Dioxide Concentration."

177. (a) Constantinidou, "Effects of Sulfur Dioxide and Ozone . . . I and II." (b) Kozlowski, "Responses of Woody Plants to Environmental Pollution, I."

178. (a) Kozlowski, "Measurement of Effects of Environmental." (b) Kozlowski, "Responses of Woody Plants to Environmental Pollution, I." (c) Kozlowski, *The Physiological Ecology*.

179. (a) Constantinidou, "Effects of Sulfur Dioxide on *Pinus Resinosa*." (b) Tsukahara, "Effects of Sulfur Dioxide on Two Age."

180. (a) Kozlowski, "Effects of Environmental Stresses on Deciduous Trees." (b) Kozlowski, "Responses of Woody Plants to Environmental Pollution, II." (c) Kozlowski, *The Physiological Ecology*. (d) Norby, "Response of Sulfur Dioxide-Fumigated." (e) Norby, "Interaction of Sulfur Dioxide Concentration." (f) Norby, "The Role of Stomata in Sensitivity." (g) Norby, "Flooding and Sulfur Dioxide-Stress." (h) Shanklin, "Effect of Temperature Preconditioning." (i) Shanklin, "Effect of Flooding . . . *Taxodium Distichum* Seedlings." (j) Shanklin, "Effect of Temperature Regime."

181. (a) W. J. Davies and T. T. Kozlowski, "Stomatal Responses of Five Woody Angiosperms to Light Intensity and Humidity," *Canadian Journal of Botany* 52 (1974): 1525–1534. (b) Kramer, *Physiology of Woody Plants*. (c) T. L. Noland and T. T. Kozlowski, "Effects of Sulfur Dioxide on Stomatal Aperture and Sulfur Uptake of Woody Angiosperm Seedlings," *Canadian Journal of Forest Research* 9 (1979): 57–62. (d) S. G. Pallardy and T. T. Kozlowski, "Relationship of Leaf Diffusion Resistance of *Populus* Clones to Leaf Water Potential and Environment," *Oecologia* 40 (1979): 371–380. (e) Siwecki, "Leaf Anatomy and Water Relations of Excised Leaves."

ulus clones to sulfur dioxide were correlated with their stomatal conductances.[182] Stomatal openings of *Acer saccharinum* seedlings, associated with high potassium availability, resulted in increased absorption of ozone and more injury than in potassium-deficient plants.[183]

The effect of a given dosage of a pollutant was influenced by prevailing environmental regimes as well as those occurring before and after a pollution episode. Prevailing light intensity, temperature, water supply, and air humidity were especially important. Woody plants absorbed more gaseous air pollutants in the daytime when the stomata were open, than at night when the stomata were closed. Prevailing temperature regime also was important. The stomata of *Betula papyrifera* absorbed more sulfur dioxide at high than at low humidity because the stomata were more open when the air humidity was high.[184]

Exposing plants to prefumigation stresses (e.g., drought, flooding, low temperature) influenced responses to gaseous pollutants, largely by affecting stomatal aperture and hence absorption of pollutants. *Fraxinus pennsylvanica* seedlings that had been grown at 15°C had fewer leaves and more-closed stomata than those grown at 25°C. When both groups of seedlings were fumigated with sulfur dioxide at 22°C, the seedlings that had been grown at 15°C absorbed less of the pollutant and were less severely injured.[185] Flooding of soil before exposing plants to air pollutants induced stomatal closure and decreased pollutant uptake by leaves.

Environmental regimes following a pollution episode also influenced responses to air pollutants, with the effects mediated by changes in metabolism.[186] At high post-fumigation temperature (22 or 32°C), *Pinus resinosa* seedlings that previously were fumigated with sulfur dioxide formed new needles, providing an additional source of carbohydrates, hence reducing growth inhibition and lowering the proportion of sulfur dioxide-injured needles. By comparison, at a post-fumigation temperature of 12°C the shoots grew slowly and did not replace their injured needles.[187]

In concluding this account of my research there remains the pleasant task of expressing my deepest gratitude to the many individuals and agencies that made my work a success. I was very fortunate in having many excellent graduate students, postdoctoral research associates, and faculty col-

182. Kimmerer, "Stomatal Conductance."
183. T. L. Noland and T. T. Kozlowski, "Influence of Potassium Nutrition in Susceptibility of Silver Maple to Ozone," *Canadian Journal of Forest Research* 9 (1979): 501–503.
184. Norby, "The Role of Stomata in Sensitivity."
185. Shanklin, "Effect of Temperature Preconditioning."
186. Norby, "Interaction of Sulfur Dioxide Concentration."
187. Norby, "Relative Sensitivity of Three Species."

leagues, all of whom contributed significantly to my research program. I extend special thanks to the University of Wisconsin and the Wisconsin Alumni Research Foundation (WARF) for funding some of my research and appointing me to a WARF chair professorship. The "sifting and winnowing" philosophy championed by the University of Wisconsin provided a uniquely conducive climate for research.

E. Forestry Problems and Publication Needs

Many of the serious forestry problems that are likely to receive attention in the immediate and near future are associated with increasing population growth. It is anticipated that such problems will be well publicized in forthcoming books. A few examples of important current issues include the following:

Tree Improvement

To meet the growing demands for wood for timber, pulp, and energy, much attention will continue to be directed to planting genetically improved fast-growing trees under intensive culture at intervals of perhaps two to ten years. In addition to traditional genetic improvement by parent selection, clonal selection, and interspecific hybridization, much work is needed in the exciting realm of biotechnology. There is accelerating emphasis on new methods of producing trees with a fast turnover of biomass and with preservation of genetic variability. Through biotechnology we are approaching a capacity to produce trees that are free of, or at least resistant to, insects, diseases, and drought. Rapid advances are being made in technologies of cellular and subcellular engineering such as gene splicing and recombinant DNA, cloning, and plant tissue culture.

Agroforestry

There is much interest in developing systems for increasing sustainable production of food and fiber by planting various combinations of herbaceous and woody plants in the same site. Such agroforestry systems pose many problems that have not been adequately studied. Some complex agroforestry systems involve as many as thirty species of plants. Only recently have studies been directed to quantifying and characterizing the many interactions among the components of agroforestry systems. Much more is known about the effects of cultural practices on the herbaceous crop components than on the woody perennials. Hence much opportunity exists for research in the physiological ecology of tested and potential woody components of these systems. The obtainable data are needed to provide a rationale for use of species in mixtures and cultural practices that will maximize use of radiant energy, optimize water use, and minimize losses of mineral nutrients. A feature of agroforestry systems in close interaction, which may be competitive or complementary, between herbaceous and woody plants, or between members within each of these groups. Because the plants in agroforestry systems vary widely in size, form, growth rate, and physiological responses to environmental stress, agroforestry systems often are more complex than many forests and fruit orchards. Problems in agroforestry are discussed further in Chapter 4.

Maintaining Biological Diversity

Biological diversity encompasses variations in life forms, genetic composition, biological processes, and ecological niches. Both natural processes and activities of humans affect biological diversity. There is alarming concern about the effects of overcutting of forests, pollution, climatic change, soil compaction, and erosion in losses of species, populations, and even entire ecosystems. The consequences of loss of biodiversity are current and future losses of food, fiber, and medicinal plants as well as non-commodity values of forests.

Although forest resources are under great pressure throughout the world, the heaviest pressures are in the tropics where there are urgent needs for income, food, fuel, and fodder. Tropical forests are disappearing at an alarming rate because of clearing for pasture, logging, and slash and burn agriculture. It has been estimated that unless more is done to deter or stop destruction of tropical rain forests, most of them will disappear by the end of the present century (except for those in the western Amazon Basin and Central Africa).

Tropical plants comprise an enormous natural resource. They contain 40 to 50% of all living organisms, an important gene pool for modern agriculture. These forests are a source of many products useful to humans (e.g., food, lumber, gums, resins, rubber, pesticides). Modern medicine has advantageously used many tropical forest plants as medicine and as chemical blueprints for development of synthetic drugs.

Organized efforts are needed to stop degradation of many forested ecosystems and to rehabilitate them. Much attention should be given to conserving tracts of natural vegetation and establishing protected areas that are crucial to the preservation of wild genetic resources for human welfare.

Global Warming

There is much concern that the "greenhouse" gases that may cause global warming (e.g., carbon dioxide, nitrous oxide, methane, chlorofluoromethanes, and a few other gases such as ammonia and sulfur dioxide) may adversely affect growth and distribution of trees. The potential effects of global warming on woody plants are highly controversial and will depend on how much the temperature will rise. The responses of different species to climatic change may be expected to vary with the specific responses measured, plant longevity, seed production and dispersal rates, methods of propagation, genotypic diversity, phenotypic plasticity, and degrees of competition and disturbance. There is much uncertainty about the long-term effects of global warming on forests. The modifying effects of other environmental conditions on the influence of increasing carbon dioxide concentration of the air contribute appreciably to the uncertainty. Natural ecosystems are likely to vary in response to increasing carbon dioxide concentrations to the extent that their physiological processes are controlled by such factors as water supply, soil fertility, and temperature conditions.

Control of Tree Growth

In order to grow trees efficiently we need a better understanding of how trees grow and why they respond to the environmental stresses and cultural practices to which they are subjected. During the last decade we have learned that environmental and physiological controls of tree growth are far more complex than envisioned earlier. It is an enormous challenge to address the many interactive effects of environmental stresses on tree growth. Nevertheless, without a comprehensive understanding of tree responses to stress, we will not be able to develop the best management systems for forests and to predict tree responses to new environmental regimes.

The important forestry problems of seed production, seed germination, wood production, and bud dormancy all involve physiological controls. Furthermore growth inhibition and death of trees are preceded by abnormal physiological events. Hence we need an understanding of how tree growth is physiologically controlled, especially by hormonal balances and interactions.

F. Editing and Publishing Experiences

My editorial experience includes service on editorial boards of several journals including *Forest Science* (1950–1971), *American Midland Natu-*

ralist (1965–1971), *Ecology* (1968–1970), *Canadian Journal of Forest Research* (1970–1976), *BioScience* (1984–), *Tree Physiology* (1985–), and the *Journal of Arboriculture* (1988–1991). I also served on the editorial board of the *Encyclopedia of Environmental Biology* (1992–) and organized and edited the seven volumes of *Water Deficits and Plant Growth* (1968–1983) for Academic Press. In addition I was the founding editor of the book series on *Physiological Ecology* published by Academic Press.

A rapidly proliferating interest in water resources and worldwide concern with the effects of drought on plants indicated a need for a comprehensive analysis of the physical basis of development and control of water deficits in plants, measurement of soil and plant water deficits, and physiological and growth responses of plants to drought. Hence, I approached Academic Press with a proposal to achieve these objectives in two contributed volumes that I would organize and edit. The enthusiastic response to these two volumes provided the impetus for expanding this work into seven volumes.[188] These books covered a wide range of topics on the water relations of healthy and diseased plants, with chapters contributed by the foremost experts. Much emphasis was placed on characterization and measurement of the water status of soil and plants and on effects of plant water deficits on photosynthesis, enzymatic activity, nitrogen metabolism, hormone relations, stomatal function, flow of latex, shrinkage of plant tissues, germination, growth of vegetative and reproductive organs and tissues, and plant mortality. The final two volumes were devoted to water relations of plant communities, including coniferous forests, temperate hardwood forests, tropical and subtropical forests, citrus orchards, apple orchards, tea plantations, cotton, small fruits, grape vines, peach orchards, and closely related woody plants.

My most satisfying, but also most difficult, editorial experience was launching the physiological ecology book series of Academic Press. I was aided by the splendid support and much latitude the publisher gave me in recruiting books for the series which consisted of single-authored, joint-authored, and contributed (edited) volumes. Although I organized and recruited most of the authors for books in the series (but with Academic Press concurrence), a few book proposals were submitted directly to Academic Press by potential authors, and I was free to accept or reject them for the series.

Some of the thirty-two books that were published in the series on my watch were more ecological than physiological and others were the oppo-

---

188. T. T. Kozlowski, ed., *Water Deficits and Plant Growth*, Vols. 1–7 (New York: Academic Press, 1968–83).

site. Most of these books dealt with plants but several with animals, including subjects such as population dynamics,[189] physiological genetics,[190] animal migration,[191] adaptations of marine organisms,[192] hibernation,[193] and habitat selection.[194] The books on plants dealt with growth and development of trees,[195] soil water,[196] halophytes,[197] grasses,[198] seed biology,[199] mycorrhizae,[200] shedding of plant parts,[201] responses of plants to environmental stresses,[202] allelopathy,[203] fire and ecosystems,[204] air pollution,[205] plant geography,[206] the boreal ecosystem,[207] plant disease,[208] effects of flooding on plants,[209] nitrogen relations,[210] and physiological ecology of woody plants.[211] I began this series in 1968 and, after twenty years, I asked to be relieved as the series editor. The series continues to thrive with the current editor, H. A. Mooney, who has the support of a strong editorial board.

189. B. G. Murray, *Population Dynamics: Alternative Models* (New York: Academic Press, 1979).

190. J. G. Scandalios, ed., *Physiological Genetics* (New York: Academic Press, 1979).

191. S. A. Gauthreaux, ed., *Animal Migration, Orientation, and Navigation* (New York: Academic Press, 1981).

192. F. J. Vernberg and W. B. Vernberg, eds., *Functional Adaptations of Marine Organisms* (New York: Academic Press, 1981).

193. C. P. Lyman, J. S. Willis, A. Malon, and L. C. H. Wang, *Hibernation and Torpor in Mammals and Birds* (New York: Academic Press, 1982).

194. E. L. Cody, *Habitat Selection in Birds* (New York: Academic Press, 1985).

195. T. T. Kozlowski, *Growth and Development of Trees*, Vol. 1 and 2 (New York: Academic Press, 1971).

196. D. Hillel, *Soil and Water. Physical Principles and Processes* (New York: Academic Press, 1971).

197. Y. Waisel, *Biology of Halophytes* (New York: Academic Press, 1972).

198. V. B. Youngner and C. M. McKell, eds., *The Biology and Utilization of Grasses* (New York: Academic Press, 1972).

199. (a) T.T. Kozlowski, ed., *Seed Biology*, Vol. 1: *Importance, Development, and Germination*; Vol. 2: *Germination Control, Metabolism, and Pathology*; Vol. 3: *Insects and Seed Collection, Storage, Testing, and Certification* (New York: Academic Press, 1972).

200. Marks, *Ectomycorrhizae.*

201. T. T. Kozlowski, ed., *Shedding of Plant Parts* (New York: Academic Press, 1973).

202. (a) Kozlowski, *The Physiological Ecology.* (b) J. Levitt, *Responses of Plants to Environmental Stresses* (New York: Academic Press, 1972). (c) J. Levitt, *Responses of Plants to Environmental Stresses*, Vol. 1: *Chilling, Freezing, and High Temperature Stresses*; Vol. 2: *Water, Radiation, Salt and Other Stresses*, 2d ed. (New York: Academic Press, 1980). (d) H. A. Mooney, W. E. Winner, and E. J. Pell, eds., *Response of Plants to Multiple Stresses* (New York: Academic Press, 1991).

203. E. L. Rice, *Allelopathy*, 2d ed. (New York: Academic Press, 1984) (1st ed., 1974).

204. Kozlowski, *Fire and Ecosystems.*

205. Mudd, *Responses of Plants to Air Pollution.*

206. R. Daubenmire, *Plant Geography* (New York: Academic Press, 1978).

207. J. A. Larsen, *The Boreal Ecosystem* (New York: Academic Press, 1980).

208. R. D. Durbin, ed., *Toxins and Plant Disease* (New York: Academic Press, 1981).

209. Kozlowski, "Extent, Causes and Impacts of Flooding."

210. R. J. Haynes, K. C. Cameron, K. M. Goh, and R. R. Sherlock, *Mineral Nitrogen in the Plant-Soil System* (Orlando, Fla.: Academic Press, 1986).

211. Kozlowski, *The Physiological Ecology.*

Viewed broadly the publishing process first involves the acquisition of books directly by the publisher or by editors chosen by the publisher. Working with coauthors or coeditors who are not nearby poses some difficult communication problems. Eventually my coauthors and I found it useful to convene early, outline together the contents and format of a book, and decide which one of us would write the first draft of each chapter. These first drafts were exchanged, often several times for criticism and revision. We also sent certain chapters to knowledgeable colleagues for comment.

Working with coeditors also had some problems. If possible we first met to outline the subject matter and chapter titles for a book and decided on the most knowledgeable authors we could find to write each chapter. Once the authors accepted our invitation, we took the view that they had wide latitude and responsibility for subject matter development and that we exercised responsibility for consistency of format. The drafts of each chapter were viewed by each coeditor and returned to the author with suggestions for revision.

I have been asked to comment on the qualifications of a good editor. I suppose the most important attributes are dedication, infinite patience (or stubbornness), a sense of humor, a high boiling point, and capacity to cope with the different attitudes of authors toward the central mission of producing scholarly publications on time. Almost all authors were very cooperative and sensitive to the need for meeting pre-agreed-on deadlines. Regrettably, however, "almost all" is not good enough and 100% cooperation is essential for a book project to succeed. Unfortunately a few authors failed to meet their deadlines and jeopardized the work schedule. Another problem had to do with submission of chapters in a standardized format. Again, most authors submitted manuscripts in a consistent prescribed format. Yet this was not always the case. Still another problem was coping with the waning interests of some authors.

Editing of a book series is difficult and very time consuming. A conscientious editor spends much time in reading entire book manuscripts, some of which must be rejected. In selecting book titles and the most qualified authors, the editor needs to spend much time to research the literature to identify the active researchers and authorities. This is much more easily done for a subject rather close to the editor's field. In selecting book titles and authors for zoological topics I necessarily sought the advice of many people.

In retrospect I wonder why I spent so much time editing books. I suppose a good short answer is that I did it "for the good of the order," that is, I hoped I could make a significant and lasting contribution by orchestrating the organization and bringing to fruition a series of authoritative books on important topics in biology. I also thoroughly enjoyed working closely with many leading biologists. Finally, I must add my thanks to Academic Press and its people for many years of pleasant cooperation and high-quality production of the books I edited.

Both the quality of research papers and management of the publication process have changed progressively over the years and undoubtedly will continue to do so. The editing of journal papers has become more rigid than the editing of book material. Journal editors are necessarily guided by rapidly accelerating publication costs, limited budgets, backlogs of manuscripts, limitations on allocations of pages, and standards of format that are set by editorial boards. All the material in research manuscripts must be germane, well-documented, statistically justified, and as brief as possible. There can be little or no straying beyond the data presented. Editors of books have greater latitude in allowing authors to be more speculative and to inject their style and personality into their writings.

In both journals and books new subject areas have begun to predominate in accordance with growing public concern about such matters as environmental pollution, destruction of tropical forests, and controversial issues such as potential global warming.

An important trend has been a progressive improvement in the quality of research papers. Introduction of new and refined research techniques including, for example, the tools of molecular biology, electron microscopy, phytotronics, labelled isotopes, and modelling of plant growth have provided investigators with deeper understanding of the nature and control of plant growth. Precision instruments are now available to measure biological parameters in seconds, automatically programmed by computers.

A significant development has been the growing recognition that plant growth is very complex and regulated by interacting environmental and endogenous control systems. The view is now widely held that plants are highly integrated units that are influenced by many abiotic and biotic stresses. As a stress in imposed on plants they respond with multiple biochemical and morphological adjustments. By comparison, when auxins were discovered it became fashionable to "explain" most developmental processes by auxins and antiauxins. As research progressed very distinct functions were assigned to each of the major classes of endogenous hormones. Regulation of cell enlargement was attributed to auxin, of stem growth to gibberellins, and of cell division to cytokinins. However, it is now recognized that hormonal regulation of plant growth is more complex and involves interactions among hormones as well as between hormones and "second messengers" which may not be hormones. Plant responses are also modulated by temporal variations in cellular sensitivity to hormones. Furthermore, in addition to the major classes of endogenous hormones, other compounds, including phenolic compounds, polyamines, and brassinosteroids, appear to have a role in regulating plant growth. Hence, simplistic explanations of single-factor controls of plant growth are now considered inadequate.

The publishing process has changed dramatically over the years. Access by investigators to scientific literature has been increased by the establishment of data banks, and the writing of papers has been facilitated by word processors. These developments were accompanied by an explosive increase in the number of published papers and books, requiring increasingly more time of investigators just to keep abreast of the flow of relevant literature. A survey of *Biosis* showed that the number of publications on forest ecology approximately doubled between 1967 and 1984. Thereafter the rate of publication accelerated even faster and tripled in the six-year period after 1984. This cascade of scientific literature was accompanied by a large increase in new journals. These developments have necessarily complicated the management of scientific journals. For example, in its infancy, one biological journal was handled largely by one editor and a few assistants. By 1993 it had an editor-in-chief, and had added more than twenty editors, and an editorial advisory board of sixty-nine members. Because of the diversity and sophistication of research papers that are currently submitted to journals, it has become necessary to call on the expertise of many subject-matter specialists for competent reviews. The large volume of high-quality research papers and books will introduce questing scholars to many opportunities and challenges for truly exciting careers in forest biology.

Editors' Note

The monographs of Theodore T. Kozlowski, currently a Visiting Scholar in Environmental Science, Policy, and Management at the University of California, Berkeley, have consistently ranked highly in the analysis of the Core Agricultural Literature Project. Of all authors on the cumulative monographs list for all seven agricultural subjects, Kozlowski had the greatest number of titles ranked as core. In a career spanning fifty years, his numerous books and articles have been widely cited in the literature and for this reason the Core Agricultural Literature Project invited Kozlowski to reflect on his work as a researcher, author, and editor.

In the Core Agricultural Literature Project core monograph lists for the disciplines of Soil Science, Crop Science, and Forestry, one title edited by Kozlowski was first ranked in all three of the seven volumes: *Water Deficits and Plant Growth* (New York: Academic Press, 1968–1983).

Three works were first ranked in the two disciplines of Crop Science and Forestry: T. T. Kozlowski. *Seed Biology*. New York: Academic Press, 1972. 3 vols.; P. T. Alvim and T. T. Kozlowski, eds. *Ecophysiology of*

Tropical Crops. New York: Academic Press, 1977. 502p; and for Soil Science, and Forestry: T. T. Kozlowski and C. E. Ahlgren, eds. *Fire and Ecosystems*. New York: Academic Press, 1974. 542p.

It is in the area of woody plant physiology, trees in particular, that Kozlowski has done most of his work. On the Core List of Monographs for Forestry and Agroforestry, five additional titles are included:

T. T. Kozlowski, ed. *Flooding and Plant Growth*. Orlando, Fla.: Academic Press, 1984. 356p..

T. T. Kozlowski. *Tree Growth and Environmental Stresses*. Seattle, Wash.: University of Washington Press, 1979. 192p.

T. T. Kozlowski, P. J. Kramer, and S. G. Pallardy, eds. *The Physiological Ecology of Woody Plants*. San Diego, Calif.: Academic Press, 1991. 657p.

P. J. Kramer and T. T. Kozlowski. *Physiology of Woody Plants*. New York: Academic Press, 1978. 811p. (Revision of Kramer and Kozlowski's *Physiology of Trees*. New York: McGraw-Hill, 1960, and Kozlowski's *Growth and Development of Trees*. New York: Academic Press, 1971)

J. B. Mudd and T. T. Kozlowski, eds. *Responses of Plants to Air Pollution*. New York: Academic Press, 1975. 383p.

His service on the editorial boards of several prominent journals was extensive and four of these publications are top-ranked in this Project analysis. The overall rankings are in brackets, and the dates of Kozlowski's affiliation with these journals follow:

Forest Science [1]	1950–1971
Canadian Journal of Forest Research [2]	1970–1976
Ecology [6]	1968–1970
BioScience [41]	1984–1994

T. T. Kozlowski received his Ph.D. from Duke University in 1947, and has been on the faculties of several distinguished universities. He has traveled extensively abroad. He turned sixty-three in 1980, yet during this following decade his literature output was prodigious. A measure of his influence can be found in the number of times his writings were cited in the *Science Citation Index*: for 1980, eighty-two times; for 1984, ninety-six; and for 1988, eighty-three times, when he was seventy-one.

T. T. Kozlowski has added greatly to the literature of crops and forestry, and has left a remarkable legacy in world agricultural literature.

11. Primary Monographs in Forestry and Agroforestry

PETER McDONALD

New York State Agricultural Experiment Station, Geneva

A. Purpose and Methods

One aim of the Core Agricultural Literature Project was to examine the monographs of the past forty years to determine which are still valuable in academic teaching and research today. A twin process of citation analysis and scholarly review was used in order to arrive at a core monographic listing in forestry and agroforestry.

Since 1950, citation counting and analysis have been widely used in libraries and in scholarly disciplines to provide information pertaining to patterns and trends in literature and to identify the most cited literature. Citation analysis has proven useful in determining scholarly output and in measuring research and education productivity, and as an indicator of growth and overlap in a variety of disciplines.[1]

In this study, citations in source documents were subjected to analysis in order to tabulate select citation elements, including type of publisher, place of publication, concentrations in subject coverage and date. The aim was to identify prominent forestry literature of greatest value to advanced students and researchers worldwide representing a continuum from collegiate to advanced research literature. A further aim was to determine the relative rank and merits of titles for both the academic community of developed countries and developing nations. This focus required careful application of citation analysis methods used to determine the primary or core literature.

To insure guidance and counsel in these efforts, the Core Agricultural Literature Project created a Forestry and Agroforestry Steering Committee comprised of scholars from a variety of forest science disciplines. One of

---

1. Eugene Garfield, *Citation Indexing: Its Theory and Application in Science, Technology, and Humanities* (Philadelphia: ISI Press, 1979).

their first tasks was to aid in determining the source documents to be used for analysis. The Steering Committee members were:

James E. Coufal State University of New York, Syracuse	P. K. R. Nair University of Florida
James P. Lassoie Cornell University	Donald F. Webster State University of New York, Syracuse

Thirty source documents for citation analysis were identified by the steering committee and the Core Project staff. Care was taken to identify as source documents, those landmark works in forestry and agroforestry with the widest subject application. In surveying the monographic literature it became clear that forestry is a multifaceted, interdisciplinary field with ties to many allied disciplines. In order to cover all these attributes of forest science, monographs only were extracted from a further thirty-two documents deemed important by the Steering Committee to insure comprehensive coverage of the discipline. The caveats outlined in Chapter 3 on subject coverage pertain equally to the selection of source documents for analysis. No journals or literature reviews were analyzed for data on monographs. Selected journals were analyzed for data on serial publications. These are covered in Chapter 12. Numerous literature reviews and review articles were studied for possible citation analysis. Most were deemed too narrowly focused for the purposes of this study and were not included. All source documents are listed in here:

Source documents in forestry and agroforestry

    * Monographs only extracted.
    # Submitted to citation analysis.
    (D) = developed countries; (TW) = Third World.

Avery, Thomas. *Forest Measurements*. 3d ed. New York: McGraw-Hill, 1981. 324p. (D)
* Binkley, Dan. *Forest Nutrition Management*. New York: J. Wiley & Sons, 1986. 290p.
* Bonga, J., and D. Durzan. *Cell and Tissue Culture in Forestry*. Dordrecht, Neth.: Nijhoff, 1987. (Vol. 1 only.)
* Boungiorno, J., and K. Gilles. *Forest Management and Economics*. New York: Macmillan Co., 1987. 285p.
* Budd, W.W. et al. *Planning for Agroforestry*. Amsterdam: Elsevier, 1990. 338p.

\# Cannell, M.G.R., and J.E. Jackson, eds. *Attributes of Trees as Crop Plants.* Huntington, UK: Institute of Terrestrial Ecology, 1985. 591p. (D)

* Douglas, R. W. *Forest Recreation.* 3d ed. New York; Pergamon Press, 1982. 326p.

\# Duryea, M. L., and P. Dougherty, eds. *Forest Regeneration Manual.* Dordrecht and Boston: Kluwer Academic, 1991. 433p. (D)

\# Ellefson, P. V., ed. *Forest Resource Economics and Policy Research.* Boulder, Colo.: Westview Press, 1989. 403p. (D)

\# Evans, Julian. *Plantation Forestry in the Tropics.* Oxford: Clarendon Press, 1984. 468p. (TW)

\# Fanta, J., ed., *Forest Dynamics Research in Western & Central Europe.* Wageningen: Pudoc, 1986. 303p. (D)

\# Garner, Willa Y., and John Harvey. *Chemical and Biological Controls in Forestry.* Washington, D.C.: American Chemical Society, 1984. 406p. (D)

* Gregerson, H., Sydney Draper, and Dieter Elz, eds. *People and Trees: The Role of Social Forestry in Sustainable Development.* Washington, D.C.: World Bank, 1989. 274p.

* Gregory, G. Robinson. *Resource Economics for Foresters.* New York: J. Wiley, 1987. 477p.

\# Gupta, V.K., and N.K. Sharma. *Tree Protection.* New Delhi, India: Indian Society of Tree Scientists, 1988. 462. (TW)

\# Haygreen, J., and J Bowyer. *Forest Products and Wood Science: An Introduction.* Ames, Iowa: Iowa State University, 1982. 495p. (D)

* Hocker, Harold. *Introduction to Forest Biology.* New York: J. Wiley, 1979. 467p.

\# Holm-Nielsen, L.B., et al. *Tropical Forest: Botaonical Dynamics, Speciation and Diversity.* New York: Academic Press, 1989. 390p. (TW)

* Hoover, R., and D. Wills, eds. *Managing Forested Lands for Wildlife.* Denver, Colo.: Colorado Div. of Wildlife, 1987. 469p.

\# Hunter, M. L. *Wildlife, Forests and Forestry: Principles of Managing Forests for Biological Diversity.* Englewood Cliffs, N.J.; Prentice-Hall, 1990. 370p. (D)

* Husch, B., C. Miller, and T. Beers. *Forest Mensuration.* New York: J. Wiley, 1983. 415p.

\# Huxley, P. A., ed. *Plant Research and Agroforestry; Proceedings . . .* Nairobi, Kenya: International Council for Research in Agroforestry, 1983. 617p. (TW)

* Johansson, Per-Olov, and K.G. Lofgren. *The Economics of Forestry and Natural Resources.* Oxford: Blackwell, 1985. 292p.

* Johnson, W., and H. Lyon. *Insects That Feed On Trees and Shrubs.* 2nd ed. Ithaca, N.Y.: Cornell University Press, 1988. 555p.

\# Kallio, M., D. Dykstra, and C. Binkley, eds. *The Global Forest Sector: An Analytical Perspective.* New York: J. Wiley, 1987. 644p. (TW)

\# Khosla, P. K., ed. *Improvement of Forest Biomass; Proceedings of a Symposium . . .* Solan, India: Indian Society of Tree Scientists, 1982. 472p. (TW)

* Kimmins, J. P. *Forest Ecology.* New York and London: Macmillan & Collier, 1987. 531p.

* Kossuth, Susan, et al., eds. *Hormonal Control of Tree Growth; Proceedings of* *the Physiology Working Group, Birmingham, Ala., 1986.* Boston: Nijhoff Publishers, 1987. 243p.

\# Kozlowski, Theodore T. *The Physiological Ecology of Woody Plants.* New York: Academic Press, 1991. 576p. (D)

\# Kramer, P.J., and T. Kozlowski. *Physiology of Woody Plants.* New York: Academic Press, 1979. 811p. (D)

* Lamprecht, H. *Silviculture in the Tropics: Tropical Forest Ecosystems . . .* Eschborn, Germany: Gesellschaft für Technische Zusammenarbeit (GTZ), 1989. 301p.

\# Lassoie, J., and T. Hinckley. *Techniques and Approaches in Forest Tree Ecophysiology.* Boca Raton, Fla.: CRC Press, 1991. 599p. (D)

\#Lieth, H., and M. J. A. Werger, eds. *Tropical Rain Forest Ecosystems.* Amsterdam and New York: Elsevier, 1989. 636p. (TW)

\# Longman, Kenneth A. and J. Jenik. *Tropical Forest and Its Environment.* 2d ed. Harlow, U.K: Longman: New York: J. Wiley, 1987. 347p. (TW)

\# MacDicken, K. G., and N. T. Vergara, eds. *Agroforestry: Classification and Management.* New York: J. Wiley, 1990. 382p. (TW)

* Manion, Paul. *Tree Disease Concepts. 2nd ed.* Englewood Cliffs, N.J.: Prentice-Hall, 1981. 402p.

\# Mather, Alexander. *Global Forest Resources.* London: Belhaven Press, 1990. 349p. (D)

* Myers, N. *The Primary Source: Tropical Foirests and Our Future.* New York: Norton, 1992. 415p.

\# Nair, P. K. R., ed. *Agroforestry Systems in the Tropics.* Dordrecht: Kluwer Academic Publishers, 1989. 664p. (*Forestry Science Series* No. 31) (TW)

* Norse, Elliott, K. Rossenbaum, et al. *Conserving Biological Diversity in Our National Forests.* Washington, D.C.: The Wilderness Society, 1986. 326p.

* *ODI: Social Forestry Network.* London: Social Forestry Network, 1990. 100p. (Bibliography)

\# Panayotou, T., and P. Ashton. *Not By Timber Alone: Economics and Ecology For Sustaining Tropical Forests.* Washington, D.C.: Island Press, 1992. 277p. (TW)

* Pirone, P.P. *Tree Maintenance.* New York: Oxford University Press, 1978.

* Price, Colin. *Theory and Application of Forest Economics.* Oxford and New York: Blackwell, 1989. 402p.

* Pritchett, W., and R. Fisher. *Properties and Management of Forest Soils.* 2d ed. New York: Wiley, 1987. 494p.

* Reichle, D. E. *Dynamic Properties of Forest Ecosystems.* New York: Cambridge University Press, 1981. 683p.

* Reifsnyder, William E., and T. O. Darnhofer, eds. *Meteorology and Agroforestry; Proceedings of an International Workshop on the Application of Meteorology to Agroforestry Systems Planning and Management, Sponsored by ICRAF, WMO, and UNEP, Nairobi, Feb. 1987.* Nairobi, Kenya: ICRAF, 1989. 546p.

* Repetto, R., and M. Gillis. *Public Policies and the Misuse of Forest Resources.* Cambridge, U.K.: Cambridge University Press, 1988. 432p.

\# Salas, Gonzalo de las. *Suelos Ecosistemas Forestales: Con Enfasis America*

Tropical. San Jose, C.R.: Insitituto Interamericano de Cooperacion para la Agricultura, 1987. 447p. (TW)

* Sethuraj, M. R., and A. S. Raghavendra, eds. *Tree Crop Physiology.* Amsterdam: Elsevier, 1987. 361p.

\# Sharma, Neranda, ed. *Managing the World's Forests: Looking for Balance Between Conservation and Development.* Dubuque, Iowa: Kendall, Hunt, 1993. 605p. (TW)

* Shepherd, Kenneth. *Plantation Silviculture.* Boston; Nijhoff Publishers, 1986. 322p.

\# Sinclair, Steven. *Forest Products Marketing.* New York: McGraw-Hill, 1992. 403p. (D)

* Sinclair, W. et al. *Diseases of Trees and Shrubs.* Ithaca, N.Y.: Cornell University Press, 1987. 574p.

* Smith, D. M. *The Practice of Silviculture.* 8th ed. New York: Wiley, 1986. 527p.

\# Steppler, H. A., and P. K. R. Nair, eds. *Agroforestry: A Decade of Development.* Nairobi, Kenya: International Council for Research in Agroforestry, 1987. 335p. (TW)

* Stoddard, Charles H., and Glenn M. Stoddard. *Essentials of Forestry Practice.* 4th ed. New York: J. Wiley, 1987. 407p.

* U.S. Dept. of Agriculture. Forest Service. *Wood Handbook: Wood as an Engineering Material.* Washington, D.C.: U.S. Dept. Agriculture, 1987. 466p. (*USDA Agriculture Handbook* No. 72.)

* White, T., and G. Hodge. *Predicting Breeding Values with Application in Forest Tree Improvement.* Dordrecht, Neth.: Kluwer Academic, 1989. 367p.

\# Whitmore, T.C. *An Introduction to Tropical Rainforests.* Oxford, U.K.: Clarendon Press, 1990. 231p. (TW)

* Wright, J. *Introduction to Forest Genetics.* New York: Academic Press, 1976. 463p.

* Young, Anthony. *Agroforestry for Soil Conservation.* Wallingford, U.K.: CAB International, 1989. 276p.

\# Zobel, B., and J. Talbert. *Applied Forest Tree Improvement.* New York: J. Wiley, 1984. 505p. (D)

\# Zottl, H. and R. Huttl. *Management of Nutrition in Forests Under Stress.* Dordrecht, Neth.: Kluwer Academic, 1990. 668p. (D)

B. Compilation and Citation Analysis

The analysis of the source documents resulted in the following data being gathered from each citation.

Title of publication;
Date of publication;
Format of publication (e.g. report series, conference, theses, monograph or serial);

Category of publisher (e.g. commercial press, university, organization or government);
Country of publication.

As analysis proceeded, titles of monographs were noted and entered into computerized lists. A tally was made each time a monograph or chapter in a monograph was cited. The same count was made each time a journal was cited. Throughout the process, additional data were gathered providing the basis for evaluation in this and the following chapter.

Before examining the results of the Core Agricultural Literature Project analysis, the definitions of monograph must be understood. Monographs have variation in that some are serially published. The distinguishing charascteristics for monographs in this study included: that they were cited as distinct works with an author or editor, that the title itself was specific to the work, and that the item was complete in itself. Most reports in series fall into this category. The same held true for many conferences, workshops, and symposia since they had a separate and unique title; otherwise, as with societal deliberations, they were counted in the journal category. Chapters in books were counted as monographic titles. These definitions generally followed the citation patterns of forest science authors.

Certain materials were excluded:

1. Pamphlet-like material, fifty pages or fewer which were cited only once or twice.
2. Local and national government documents of brief pagination on specialized, site-specific or subject-specific topics.
3. Select titles on subjects well outside the discipline of forest science. (Some of these titles were placed in the monographic lists of other disciplines.)
4. Specialized geographic material only of use for a restricted site.
5. Publications prior to 1950.
6. Data on theses and dissertations were gathered, but individual titles were not placed on the core lists for evaluation by reviewers.

In all cases, citations to early editions were combined with the latest edition although data were kept on all printings, reprints, and revised editions for the monographs.

The monographic literature of forest science was divided into four broad subject categories. As in other scientific disciplines, scientists working in forestry are specialists, therefore a subject approach facilitates reaching the right evaluations. By creating categories, subject strengths and trends in the literature are more easily compared. Finally, by defining the subjects of forestry, complete coverage of the discipline is assured. These subject categories were:

1) Forest Biology, Breeding, Genetics and Ecology
2) Forest Pathology and Protection
3) Forest Management, Silviculture, Wood Products, Recreation, Wildlife Management, Mensuration
4) Forest Policy and Economics, General Forestry, International Development, Forestry Education, Social Issues

The Third World list of monographs for evaluation was not divided by subject, in part because forest science in these regions relies less rigidly on subject specialties. Source documents were chosen to reflect the full range of subject and geographical diversity in low latitude forestry, including agroforestry and tropical silviculture, deforestation, rainforest ecology, and the international timber trade.

Every effort was made in selecting the source documents for analysis to represent the broadest scope of forest science. Several subject areas require clarification, notably forest engineering and wood chemistry. In both cases, these disciplines fall predominantly outside the scope of forestry as described in Chapter 3 for the Core Agricultural Literature Project. Forest engineering is covered extensively in another volume of the *Literature of the Agricultural Sciences* series.[2] Wood chemistry, whose literature comprises only 1.3% of *CAB Abstracts* Forestry Subfile (Codes 0F and 1F) and 3.8% of *AGRICOLA's* forestry coverage, while heavily cited by chemists and material scientists, seems not to have a great impact in forest science literature. Consequently, no source documents on these specific topics were analyzed, although two titles covered wood science and technology.[3]

C. Results of Source Document Analysis

Analysis of the source documents resulted in 21,486 citations being examined, 8,979 or 41.8% were to monographs. This led logically to determine the reason why monographs should rank so prominently in forest science. What the analysis of the Core Project revealed is that the high count for monographs is partly explained by the importance of report series in forestry. Prominent report series are covered later in this chapter.

Analysis of the literature clearly revealed that the United States was the

2. Carl Hall and Wallace Olsen, ed., *The Literature of Agricultural Engineering* (Ithaca, N.Y.: Cornell University Press, 1992). See Chapter 8.
3. (a) J. Haygreen and J. Bowyer, *Forest Products and Wood Science: An Introduction* (Ames: Iowa State University, 1982). 495p. (b) U.S. Forest Service, *Wood Handbook: Wood as an Engineering Material* (Washington, D.C.: U.S. Dept. Agriculture, 1987 [*USDA Agriculture Handbook no. 72*]), 466p.

dominant country of publication at 50.4%. Europe as a whole had 25.2%. The combined countries of the Third World published 10.8%. Table 11.1 lists the major countries and their percentages of the citation counts. Sixty-nine countries were represented.

Table 11.1. Citation counts of publishing country

Country	% of Total	Third World Ranking[a]
United States	50.4%	1
United Kingdom	11.2	2
Germany	4.2	6
Italy	3.6	3
Netherlands	3.5	5
Kenya	2.7	4
Canada	1.9	11
Costa Rica	1.7	7
India	1.6	8
France	1.2	9
Other	18.0	n/a

[a]Country of publication ranking from Third World source documents only.

Further analysis of the monographic material published in Italy, Kenya, and Costa Rica revealed that report series were the predominant type of publication and that most of the monographic counts were from tropical forestry source documents. Almost half of Kenya's output were conferences hosted by the International Centre for Research in Agroforestry (ICRAF). The Food and Agriculture Organization accounted for 98% of Italy's monographic count. Colombia, Nigeria, the Philippines, Malaysia, and Brazil, in that order, were the succeeding top ranked publishing countries for the Third World.

The Core Project also sought to gather data on types of publishers. These were independent organizations such as societies and associations, commercial publishers, governments including the United Nations, and universities. Table 11.2 lists the publisher types and their percent of the total monographic literature.

Table 11.2. Publisher types of monographs by citation analysis of source data

Types	Developed	Third World
Commercial	35.9%	30.2%
Governmental	24.4	20.1
University	20.2	19.5
Organizational and societies	19.5	31.2

Commercial monographs were clearly the most commonly cited type of publication. Sixty-four separate commercial publishing houses were identified which were located in twenty-two different countries. The five top commercial houses by descending rank were: Academic, Springer-Verlag, John Wiley, McGraw-Hill, and W. Junk.

Academic Press had over twice as many cited monographs as second ranked Springer-Verlag. However, the two oldest commercial publishing houses are German, Springer-Verlag and Paul Parey, both of which started in the nineteenth century and are important forest science publishers today. Parey was ranked fifteenth.

The United States Department of Agriculture was the most frequently cited government publisher with 1,068 or 12.6% of the total monographic counts. Over 85% of the USDA publications were Forest Service Experiment Station report series. The Food and Agriculture Organization was second with 4.2% of the counts, again almost exclusively report series. PUDOC, UNESCO, and the Canadian Forest Service were closely tied for third.

With universities, Cambridge and Oxford, garnered 2.6% and 1.3% of the total monograph counts respectively. The University of Washington, the most cited U.S. university, was next. In rough parity, with half as many citations, were the universities of Minnesota, California at Berkeley, and Florida, followed by Johns Hopkins, Oregon State, and North Carolina State universities. Agriculture experiment station (AES) publications accounted for the high citation counts for the universities of Minnesota and Florida, and North Carolina State university. The monographic counts for Washington State and Oregon State universities were divided evenly between AES publications and sponsored conferences. Berkeley and Johns Hopkins were cited entirely for conferences and monographs. Outside the United States and and the United Kingdom, the University of Tokyo in Japan alone was ranked, at thirteenth.

By far the most commonly cited organization was ICRAF which had twice the number as the second ranked organization, Centro Agronomico Tropical de Investigacion y Ensenanza (CATIE) in Costa Rica, followed by the World Bank, the International Union of Forest Research Organizations (IUFRO) and the National Academy of Sciences (U.S.). Report series were the predominant type of publication for ICRAF, and in the case of the World Bank and CATIE, reports were over 96% of their cited output.

As noted, one of the commonest publication types in forestry and agroforestry are report series. The primary producers of these sorts of literature are national government agencies such as the U.S. Forest Service, CSIRO in Australia and the FAO in Rome, as well as research centers such as

ICRAF and the International Livestock Centre for Africa (ILCA) in Nairobi, Kenya. Report series need some clarification. Generally they fall into the category of serial monograph and for the purposes of this investigation have been designated as such for the sake of comparison and analysis. Most report series in forestry, though published in numbered sequence, are in reality pamphlet-length publications covering a single topic per issue, complete by themselves, and more nearly approximate a monographic title. For this reason, this study has counted them as monographs.

Since almost a quarter of all monographs cited were reports it is worth describing some of the more important series briefly. The most prominent are clearly those published by the U.S. Forest Service.

Reports

U.S. Forest Service publications were the most commonly cited report series in the analysis. Ten separate Forest Service offices publish report series. These offices include eight regional experiment stations, the Forest Products Laboratory in Madison, Wisc., and the central Washington, D.C., Office (see Chapter 7). Since 1963, uniformity of publication type has been imposed on these autonomous offices, which formerly published a variety of material on subjects of regional interest. They now publish only four types of report series. These include:

General Technical Reports—Lengthy research investigations broadly dealing with a variety of aspects of forest science. These reports are often the proceedings of special conferences.

Research Papers—Article length investigations into current research projects conducted by the regional offices.

Research Notes—Brief news service type publications, covering updates, works in progress, listings.

Resource Notes—Almost exclusively statistical dealing with operations management and budgets at the various regional offices.

The report series publications of the Forest Service are dealt with more fully in Chapter 7.

Food and Agriculture Organization of the United Nations in Rome publishes two report series of primary importance to forestry practitioners in developing countries.

FAO Forestry Papers—These began publication in 1977 and now number over a hundred. Often of substantial length, these reports cover a single topic per issue and are aimed at Third World managers and researchers to be used as manuals or handbooks.

FAO Conservation Guides—Like *Forestry Papers*, these publications cover a single topic per issue on a variety of subjects of interest to instructors and managers dealing with conservation issues in tropical regions. Forestry applications play a large part in the coverage.

The FAO also copublished a number of *Technical Reports* with the United Nations Environmental Programme (UNEP), commonly cited under a numbered system. A typical FAO-UNEP report number looks like: *Technical Report 3. UN 32/6 1301–78–04.*[4] Tropical forest resources were commonly the subject of this series. Finally, several *FAO Soils Bulletins* have direct application to agroforestry and were commonly cited in the Core Project analysis.

The World Bank publishes many reports and feasibility studies relating to forestry in the tropics.[5] Key report series are:

World Bank Discussion Papers
World Bank Occasional Papers
World Bank Policy Papers
World Bank Regional and Sectoral Studies
World Bank Technical Papers

The International Centre for Research in Agroforestry (ICRAF), a member of Consultative Group of International Agricultural Research (CGIAR), has several report series which were highly cited in the literature.[6] Many of the *AFRENA Report* (Agroforestry Research Networks for Africa) series are in French for distribution in Saharan and sub-Saharan Africa.

ICRAF Working Papers
AFRENA Reports

4. FAO/UNEP, "Tropical Forest Resources Assessment Project" *Forest Resources of Tropical Asia* (Rome, Italy: FAO/UNEP, 1981 [UN 32/6. 1301–78–04. *Technical Report* 3]).
5. Examples: Dennis Anderson. *The Economics of Afforestation: A Case Study for Africa* (Baltimore: Published for World Bank by Johns Hopkins University Press, 1987. *World Bank Occasional Paper*, No. 1); and World Bank, *Forestry: The World Bank's Experience* (Washington, D.C.: World Bank, 1991. *World Bank Operations Evaluation Report*).
6. Example: D.A. Hoekstra and J. Beniest., comps. *Summary Proceedings: East and Central Africa AFRENA Conference on Agroforestry* (Nairobi, Kenya: ICRAF, 1988.) (*AFRENA Report* No. 44)

The Centro Agronomico Tropical de Investigacion y Ensenanza (CATIE), which is located in Costa Rica, published: *Boletin Tecnico*; *Serie Materiales de Ensenanza*; and *Contribuciones de los Participantes*.

The Board on Science and Technology for International Development (BOSTID) is part of the National Research Council and publishes many reports under the imprint of the National Academy Press or BOSTID.[7] Through the aegis of the U.S. Agency for International Development, these reports are intended for free distribution primarily in developing countries. Major topics of importance to agroforestry are covered by the following BOSTID subject divisions: resource management, energy, technology options for developing countries, biological resources, innovations in tropical reforestation, managing tropical animal resources.

The International Crops Research Institute for the Semi-Arid Tropics (ICRISAT), a member of CGIAR, is located in Patancheru Province, India. It is a nonprofit scientific educational institute doing important work in agriculture. Its primary report series pertains to publications of proceedings of working group meetings held intermittently at the institute on topics pertaining to agriculture and on agroforestry in particular.[8]

The Central Arid Zone Research Institute (CAZRI) is a government institute located in Jodhpur, India, with close ties to the Indian Council of Agricultural Research. Its *Annual Progress Report* covers agroforestry research in India in some depth. Other reports on arid-zone agroforestry are frequently published.[9]

These are the major publishers of report series in forestry and agroforestry. Conservation societies such as Resources for the Future, the World Resources Institute, and the International Union for Conservation of Nature and Natural Resources, while not prominent as separate publishers, accounted in aggregate for 9.3% of the organizational total, of which almost their entire output were report series or special papers.

Clearly the importance of report series in forestry, and tropical forestry in particular, is due to the fact that as a renewable natural resource, forests are managed by a variety of private concerns, government agencies and nongovernmental agencies. Planning at regional, national, and international levels plays an important part in forest policy formulation. Part of this pol-

---

7. Example: Advisory Committee on the Sahel. *Agroforestry in the West African Sahel* (Washington, D.C.: Advisory Committee on the Sahel—BOSTID, 1984).

8. Example: Rick J. Van Den Beldt, ed. *Agroforestry Research in the Semi-Arid Tropics. A Report on the Working Group Meeting held at ICRISAT Center, India, 5–6 August 1985* (Patancheru, India: ICRISAT, 1986).

9. Example: Shankarnarayan, K.A., ed., *Agroforestry in Arid and Semi-Arid Zones* (Jodhpur, India: CAZRI, 1984).

icy process entails the publication of reports by interested bodies. Examples of players in forestry policy formulation are the World Bank and UNEP.

In the case of the U.S. Forest Service, the situation is more straightforward. Official results of research, special publications, and conferences, as well as statistics, are disseminated by federal mandate in the four types of report series. Since the Forest Service oversees some of the largest forest tracts in the world and carries out extensive research, their impact on the literature of forest science is strong. Forest Service personnel also publish extensively in journals and at conferences which may have no connection to the federal government. Each Forest Service experiment station publishes an annual bibliography listing these publications of research staff. These bibliographies are available free of charge by writing or calling experiment station publication offices.

Other Measurement Data

One of the commonest measures in citation analysis is citation half-life. Cited half-life is defined as the number of years going back from the current or publishing year which account for 50% of the total citations recorded in an analysis. The half-life of cited material provides a guide to the relevancy of the literature. Older articles or monographs are less likely to be cited than recent literature. In aggregate, the 11,800 monograph citations tallied from the source documents show the half-life for all items to be 7.2 years; for the Third World 7.1 and for developed countries 7.3. This rather short half-life may demonstrate a rapidly changing science with active monograph production which in this case includes a quantity of reports. It could also represent the nature of foresters and forest ecologists who have less need for earlier landmark literature related to their work, although a look at the citations tends to discount this possibility. Citations tapered off precipitously by the 1950s. In the 11,800 citations, only thirty-four citations were to monographs published prior to 1929.

The Core Agricultural Literature Project has analyzed six other disciplines in agriculture, but only in forestry is the half-life of monographs shorter than for journals. The usual literature pattern is that journals have a half-life three to five years shorter than monographs. From the source monographs, the half-life for cited journals was 8.6 years and for monographs 7.2, a difference of 1.4 years, in reverse order from most scientific literature. This discrepancy was so pronounced that further analysis was conducted. Citation analysis was done of three recent years of the journal *Forest Science*, one of the most frequently cited journals in the discipline.

Forest Science published in 1987, 1990, and 1993 with 5,312 citations were analyzed revealing that in this research journal the pattern was reversed to 8.2 years for cited journals and 9.2 years for cited monographs.

Several factors may explain the fact that the difference in half-life between journals and monographs is 50% shorter than with other scientific and technical literature. Report series are one of the key features of the forest science monographic literature. Indeed, much forestry research done in the United States and the Third World is first presented in reports. In the United States, the U.S. Forest Service regional experiment station publications predominate. These reports take the place of publication in traditional journals which is common in other disciplines. This lowers the years of monograph half-life. The Core Agricultural Literature Project monograph analyses in forestry showed that citations to almost 40% of all monographs were report series with a half-life near seven years. In the *Forest Science* analysis, a full 23% of the citations were to Forest Service and USDA report series. With such a high number of report monographs publishing current research that does not appear in journals, the difference in half-life is partly explained. Forestry is obviously in a period of great volatility and change. Seminal articles marking this changing course will likely have long currency in the literature, serving as the touchstone for future investigations as the discipline evolves.

Conferences and dissertations were counted as monographs, and data on publisher and date were kept, but not country of publication. Conferences require careful attention since it is not uncommon for a conference to be held in one country under the sponsorship of several institutional bodies, while the proceedings may be printed or published in another country. Other conferences first appear in serial publications, often as an entire issue of a particular journal, and are sometimes published separately at a later date in monographic form. Citations to papers from conferences such as these can therefore add to the counts in both the journal and monograph columns although these are few. Where clear sponsorship could be ascertained, the data revealed that 54.4% of the conferences were organizationally sponsored and published with many of importance sponsored by IUFRO; 36.6% were government sponsored, predominantly by the U.S. Forest Service or FAO/UNEP; 9% were under university sponsorship.

In the developed world ranking three conferences stand out.

Cannell, M.G.R., and J.E. Jackson, eds. *Attributes of Trees as Crop Plants; Proceedings of a Conference . . . Gorebridge, Midlothian, Scotland, July 1984.* Abbots Ripton, U.K.: Institute of Terrestrial Ecology, Natural Environment Research Council, 1985. 592p.

Cannell, M.G.R., and F.T. Last, eds. *Tree Physiology and Yield Improvement; Proceedings of a Conference on Physiological Genetics of Forest Tree Yield, Goreridge, Scotland, July 1975*. London and New York: Academic Press, 1976. 567p.

Cardillichio, Peter A., et al., eds., *Forest Sector and Trade Models: Theory and Applications; Proceedings of an International Symposium, Nov. 1987*. Seattle, Wash.:Center for International Trade in Forest Products, University of Washington, 1987. 298p.

In the Third World ranking two conferences deserve mention. Cannell's *Tree Physiology*, mentioned above, and this one sponsored by ICRAF: Huxley, Peter A., ed. *Plant Research and Agroforestry. Proceedings of a Consultative Meeting, Nairobi, April 1981*. Nairobi, Kenya: International Council for Research in Agroforestry, 1983. 617p.

Regarding the latter, it should be noted that in an emerging discipline such as agroforestry many of the early cited monographs will of necessity be reports and conferences, where research strategies and polices for the emerging discipline are put forward; hence the high citation rate for *Plant Research* which was a landmark in the early development of agroforestry.

Dissertations, of which 291 were cited, seem to have brief currency with a cited half-life of about 3.2 years. Most dissertations are cited in the journal literature. The number of dissertations cited dealing with temperate or with tropical subjects was about even, with tropical research predominating by 2%. Of the thirty source documents analyzed only one from the tropical list and three from the temperate did not cite a single dissertation. The most frequently cited dissertations had to do with aspects of agroforestry, which is logical since agroforestry systems are site-specific and lend themselves to narrowly focused analysis, a keystone of doctoral work in the sciences.

Another type of university publication cited in the literature were the bulletin series of the land-grant colleges' agricultural experiment stations. Since only 133 of these publications were cited in the Core Project analysis, it can be deduced that their influence is slight. All but three of these bulletins were cited in the source documents pertaining to developed countries. Five universities dominated the field, the universities of Washington, Minnesota, and Florida, as well as Oregon State and North Carolina State universities.

D. Weighting the Monograph Lists

Several elements were used in order to rank the titles.

(1) Every time a monograph or a chapter of a monograph was cited in one of the source documents, the title was given a "hit." This element was given the weight of one per citation.

(2) Reviewers' scores were totaled for each title and divided into two ranks, a high and a low. These were then given a weighted prominence in the total scoring. List #2, pathology and protection, received only half the number of responses from reviewers as lists #1, #3, and #4 so the evaluations had to be equalized.

(3) If a title went through more than one edition, it was given a score of two.

(4) If a title was reprinted or translated it was a given a score of one.

The equation for the scores for each title was:

(# of citation hits) + (top rank × 2) + (2nd rank × 1) + (edition × 2) + (reprint × 1) + (translation × 1) = total score

The same formula was used for the developed and the Third World lists, and all lists were sent separately to scholars with appropriate backgrounds to rank the monographs qualitatively. As noted, the scope of analysis reflected temperate zone forest science as well as that pertaining to the tropics, which translated into literature of the developed world and of the developing world. By this equation, peer evaluation accounted for 55.2% and 59.3% of the final scores for the developed and Third World lists respectively. Figure 11.1 shows the scoring curve for both lists.

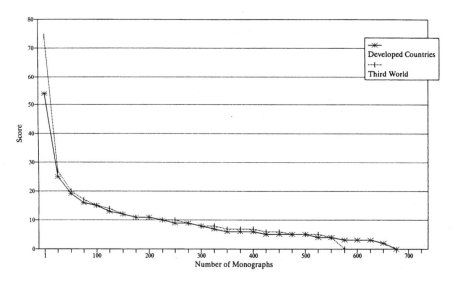

Figure 11.1. Cumulative scoring curve for monographs.

E. Comparison of the Core Monographic Lists.

The Third World and developed world lists combined have 1,035 distinct titles; 349 are unique to the Third World and 477 to the developed, with 209 common to both. Of the 209 titles common to both, the majority, 47.5%, dealt with aspects of forest biology or ecology; 30.2% dealt with policy; 19.2% with management; 7.1% with pathology and protection. As mentioned, the monograph list for developed countries was divided into four subject categories. The final percentages for the categories of the 686 core monographs were:

Forest biology and ecology	41.2%
Forest management	24.4
Policy and social aspects	22.2
Pathology and protection	12.2

As Gordon points out in Chapter 1, the bulk of scientific investigations in forest science are related to biology and ecology, with an exponential increase in the amount of research in tropical regions. Management and policy are about equally divided, and are closely tied for the simple reason that as new policies are implemented, management techniques tend to evolve to meet emerging trends. A clear example of this can be seen in the United States in the response of the federal government and the Forest Service toward the Endangered Species Act, which provides for the protection of individual species against adverse environmental harm, notably the much reported impact of clearcut logging on the spotted owl. Today, in the United States, forest management is beginning to focus less on preserving individual species as on protecting entire ecosystems which sustain endangered plants and animals. Foresters are beginning to change their thinking to incorporate these sorts of shifts in focus.

Forest pests and protection are of growing concern in managed tree monocultures and are a clear priority in U.S. Forest Service publications. Pests are less of a problem, however, with the intercropping techniques of agroforestry systems, which tend to lessen the impact of insect infestation, and may explain the small overlap of Third World titles in this subject category.

Care was taken to include on the monograph lists only those titles which would have wide application, either in subject scope or geographical range. Many titles which were locally published and site-specific, notably in the Third World, were dropped from the lists as being of marginal value when ranked by our reviewers. The majority of the these deletes were reports, both locally generated series and those published by international bodies such as the World Bank. A consistent characteristic of these reports was their short half-life, meaning they went out of date quickly. They also tended to be narrowly focused on a single geographical region.

Thirty-nine reviewers responded to rank the lists, twenty-two for developed countries and seventeen for developing. The return rate was 41.9%. Reviewers were instructed to evaluate the listings carefully and rank titles deemed important in forest science education and research. Evaluations were requested only for those titles or authors with which reviewers had some knowledge. Each reviewer was encouraged to recommend titles for inclusion to be reviewed by other scholars. The process of evaluation was iterative and four mailings were sent at two-month intervals. The names and affiliations of the reviewers are listed below.

REVIEWERS

Developed Countries Reviewers

P.F.J. Abeels
  Faculte des Sciences Agronomiques
  Universite Catholique de Louvain
  Louvain la Neuve
  Belgium

P.O. Backstrom
  Faculty of Forestry
  Sveriges Lantbruksunwersitet
  Uppsala, Sweden

David F. Batten
  CSIRO
  Highett, Victoria
  Australia

George Blakeslee
  University of Florida
  Gainesville

Clark S. Binkley
  School of Forestry & Environmental
  Studies
  Yale University
  New Haven, Conn.

J. M. Bonga
  Canadian Forestry Service-Maritimes
  Fredericton, Canada

Thomas M. Bonnicksen
  Dept. of Forestry
  Texas A&M University
  College Station

James Bowyer
  Dept. Head of Forest Products
  University of Minnesota
  St. Paul

Runar Brannlund
  Department of Forest Economics
  College of Forestry
  Umea, Sweden

Joseph Buongiorno
  University of Wisconsin, Madison

A. Cihal
  Faculty of Forestry
  Vysoka Skola Zemedelska
  Czech/Slovak Federated Republic

Paul Ellefson
  Forest Economics & Policy
  University of Minnesota
  St. Paul

Timothy J. Fahey
  Dept. of Natural Resources
  Cornell University
  Ithaca, New York

John Foltz
  University of Florida
  Gainesville

Jerry F. Franklin
  College of Forest Resources
  University of Washington
  Seattle

Piero Garoglio
  Università degli Studi Di Torino
  Turin, Italy

H.E. Garrett
  School of Forestry, Fisheries,
  and Wildlife
  University of Missouri
  Columbia

Fromard Gaupuelin
  Institut de la Carte Internationale
  de la Végétation
  Université Paul Sabatier
  Toulouse, France

Henry L. Gholz
  School of Forest Resources
  & Conservation
  University of Florida
  Gainesville

R. Grammel
  Faculty of Forestry
  Albert Ludwigs Universität
  Freiburg, Germany

Pierre Guillon
  Chambre Regionale de Commerce
  et d'Industrie
  Limoges, France

D. Guitard
  Université de Montpellier II
  Place E. Bataillon
  Montpellier, France

Jan-Erik Hallgren
  Department of Forest Genetics
  and Plant Physiology
  Swedish Universtiy of
  Agricultural Sciences
  Umea, Sweden

P.K. Hari
  Faculty of Agriculture & Forestry
  Helsingfors Universitet
  Helsinki, Finland

Richard Haynes
  Pacific Northwest Forest & Range
  Experiment Station
  U.S. Forest Service
  Portland, Oregon

John Hendee
  College of Forestry, Wildlife,
  & Range Science
  University of Idaho
  Moscow

Thomas C. Hennessey
  Department of Forestry
  Oklahoma State University
  Stillwater

Thomas M. Hinckley
  College of Forest Resources
  University of Washington
  Seattle

David Y. Hollinger
  Forest Research Institute
  Ministry of Forestry
  Christchurch, New Zealand

P. O. Johansson
  Dept. of Forest Economics, Swedish
  University of Agricultural Sciences
  Umea, Sweden

Jon Johnson
  University of Florida
  Gainesville

Walter G. Kauman
  Centre Technique du Bois
  et de l'Ameublement
  Paris, France

Dennis H. Knight
  Dept. of Botany
  University of Wyoming
  Laramie

M. Kosztka
  Erdeszeti Es Faipari Egyetem
  Sopron, Hungary

W. Liese
  Dept. of Biology
  Universität Hamburg
  Hamburg, Germany

William McKillop
  College of Natural Resources
  University of California-Berkeley

Tom Miller
  University of Florida
  Gainesville

E. K. S. Nambiar
  Division of Forestry &
  Forest Products
  Queen Victoria Terrace
  Canberra, Australia

V.I. Onegin
  S.M. Kirov Forestry Academy
  St. Petersburg, Russia

Stephen G. Pallardy
  School of Natural Resources
  University of Missouri
  Columbia

Hannu Paulapuro
  Dept. of Forest Products
  Helsinki University of Technology
  Espo, Finland

E.S. Petrenko
  V.N. Sukachev Inst. of Forest
  & Wood
  Siberian Branch Academy of Sciences
  Krasnoyarsk, Russia

Carl Reidel
  Director of Environmental Program
  University of Vermont
  Burlington

Ing. Sanchez-Palomares
  Departamento de Silvopascicultura
  E.T.S. Ingenieros de Montes
  Universidad Politechnica Madrid
  Madrid, Spain

F. Schmithusen
  Dept. Forest and Wood Science
  Eidgenossische Technische
  Hochschule
  Zürich, Switzerland

Roger A. Sedjo
  Forest Economics & Policy Program
  Resources for the Future
  Washington, D.C.

William K. Smith
  Department of Botany
  University of Wyoming
  Laramie

Douglas G. Sprugel
  Dept. of Forest Resource Management
  University of Washington
  Seattle

Robert O. Teskey
  School of Forest Resources
  University of Georgia
  Athens

Soren Wibe
  Department of Economics
  University of Umea
  Umea, Sweden

Jonathan Wright
  Dept. of Forestry
  Michigan State University
  East Lansing

Richard H. Waring
  School of Foresty
  Oregon State University
  Corvallis

Harold W. Wisdom
  College of Agriculture and
  Life Sciences
  Virginia Polytechnic Institute and
  State University
  Blacksburg

Third World Reviewers

H.K. Badola
  Forest Soils & Land
  Reclamation Div.
  Forest Research Institute
  Dehra Dun, India

F. E. Fasehun
  Dept. of Forest Resources
  Management
  University of Ibadan
  Ibadan, Nigeria

Hans Lamprecht
  Faculty of Forestry
  Georg-August-Universitat Gottingen
  Göttingen, Germany

James Lassoie
  Dept. of Natural Resources
  Cornell University
  Ithaca, New York

John Beer
  CATIE
  Turrialba, Costa Rica

Mark Follis
  Instituto de Estrategias Agropecuarias
  Quito, Ecuador

Douglas Boland
  International Council for Research
  in Agroforestry
  Nairobi, Kenya

I. Gupta
  Forest Soils & Land Reclamation Div.
  Foresty Research Institute
  Dehra Dun, India

Hans Loffler
  Department of Forestry
  Ludwig-Maximilians Universität
  Munich, Germany

Helvecia Bonilla
  Direccion de Desarrollo Forestal
  Instituto Nacional de Recursos
  Naturales
  Paraiso, Ancon, Panama

Wenyue Hsing
  Forest Ecology
  Nanjing Forestry University
  Nanjing, China

J. Maghembe
  Makoka Research Station
  Zomba, Malawi

Hugo Ramirez-Maldonado
  Chapingo, Mexico

M.N. Jha
  Forest Research Institute and Colleges
  Dehra Dun, India

James Coufal
  College of Envir. Science & Forestry
  State University of New York
  Syracuse

Norman Myers
  Consultant in Environment
  and Development
  Oxford, England

P. K. Khosla
  University of Horticulture
  and Forestry
  Himachel Pradesh, India

Laercio Cuoto
  Dep. de Engenharia Florestal
  Universidad Federal do Vicosa
  Vicossa, Brazil

L. C. Nwoboshi
  Forest Research Laboratory
  Kumasi, Ghana

Juan Carlos Kosarik
  Facultad de Ciencias Forestales
  Universidad Nacional de Misiones
  Misiones, Argentina

Gerard Deon
  Centre Technique Forestier Tropical
  Nogent-sur Marne, France

Marcos Pena-Franjul
  New York, New York

James Quashie-Sam
  Agroforestry and Wildlife
  University of Science and Technology
  Kumasi, Ghana

John B. Raintree
  Winrock International
  Bangkok, Thailand

M. Rao
  International Council for Research
  in Agroforestry
  Nairobi, Kenya

Jorge Roman Decano
  Zamorano
  Tegucigalpa, Honduras

Pedro A. Sanchez
  International Council for Research
  in Agroforestry
  Nairobi, Kenya

Arsenio Renda Sayous
  Instituto de Investigaciones Forestales
  Siboney
  Havana, Cuba

Kenneth Shepherd
  Dept. of Forestry
  Australian National University
  Canberra, Australia

Wang Shiji
  Research Institute of Forestry
  Chinese Academy of Forestry
  Beijing, P. R. of China

Stephen F. Siebert
  School of Forestry
  University of Montana
  Missoula

Toga Silitonga
  Ministry of Forestry
  Department Kehutanan
  Gedung Manggala Wanabhakti
  Jakarta, Indonesia

Komar Soemarna
  Forest Research and
  Development Centre
  Bogor, Indonesia

Douglas Southgate
  Instituto de Estrategias Agropecuarias
  Quito, Ecuador

Emmanuel Torquebian
  International Council for Research
  in Agroforestry
  Nairobi, Kenya

Mitsuo Uski
  Nature Conservation Bureau
  Environment Agency
  Tokyo, Japan

Carlos Gonzalez Vicente
  Instituto Nacional de Investigaciones
  Forestales y Agropecuarias
  Sonora, Mexico

H.J. von Maydell
  Institute of World Forestry
  and Ecosystems
  Hamburg, Germany

Sun Wensheng
  Forestry Research Institute
  Jungyuetan, Changchun
  P. R. of China

Ester Zulberti
  Training and Communications
  Division
  ICRAF
  Nairobi, Kenya

Instructions to reviewers stressed the aim of identifying the core monographs of scholarly and instructional value for advanced academic work. However, some titles clearly aimed at practitioners in the field were nevertheless deemed important for the Third World list and were retained. At the beginning of the process, the original size of the developed world list stood at 1,335, of which 649 or 48.6% were removed because of low evaluations creating a core listing of 686 titles. For developing countries, the original size of the list was 951 titles of which 393 or 41.3% of the original titles were dropped leaving 558 volumes as core. Deletions common to both lists were 160 titles. Obscure titles dropped out as the review process proceeded. Many dated titles cited in the literature also tended to drop out. Works from the 1950s and 1960s generally fell out unless of seminal importance. There was a clear trend among reviewers to exclude titles not published in English especially among reviewers of developing countries. Similarly, titles in non-Romance languages generally fell out. Still other titles were universally given low rankings and were rejected by reviewers.

Table 11.3. Comparisons of types of publishers of core monographs

Type	Developed	Third World
Commercial	53.9%	39.7%
Organizational and societies	12.6	27.3
University	18.9	15.8
Governmental	13.6	17.2

Comparison of the percentages in Table 11.3, reveals that commercial presses published the bulk of core titles. It would appear that organizations and governments have stronger publishing records in the Third World than in the developed.

Today, many publishing houses have offices around the globe. This makes analysis of country of publication problematic. In the comparative counts ranked in Table 11.4, the country of publication first on the title page was considered the place of publication. In both categories, the United States was the primary country of publication worldwide.

The column on the right shows that the scattering of country of publication for the Third World list is wider and the disparities between counts less, than on the developed world list where the U.S. dominates. The entire count for Italy on the Third World list are FAO publications.

Table 11.4. Country of publication of core monographs

Country	Developed	Third World
United States	65.5%	40.2%
United Kingdom	11.9	15.0
Netherlands	6.2	8.5
Germany	6.2	4.8
Canada	3.2	2.1
Italy	1.9	10.7
Costa Rica	—	2.1
Kenya	—	3.4
India	—	2.6
Other	5.1[a]	10.6[b]

[a]Developed world: Australia, Japan, France, Philippines, Sweden, India, Poland, Yugoslavia, Greece, China, Nigeria and Switzerland.

[b]Third World: Australia, Nigeria, Japan, Philippines, Malaysia, Colombia, Switzerland, Brazil, France, Ireland, China, Bangladesh, Singapore, Sweden and Greece.

The most prominent publishers of the core monographs are given in Table 11.5. All four types of publishers are combined in the same ranking system. For the developed countries' list, commercial publishers dominate.

Table 11.5. Top-ranked publishers of core monographs

Publisher	Developed	Third World
Academic Press	1	3–4
John Wiley	2	2
Springer-Verlag	3	5
USDA	4	—
McGraw-Hill	5	—
Cambridge University	6	7
FAO	7	1
Johns Hopkins University	8–9	—
Blackwell	8–9	10
Elsevier	10	8–9
ICRAF	—	3–4
Oxford University	—	6
National Academy of Sciences	—	8–9

The FAO in Rome was clearly the top publisher for the Third World with almost three times the publications (53) as third-fourth ranked Academic Press (18) and ICRAF (18) in Nairobi.

F. The Top Core Monographs

The twenty-five monographs for both developed countries and developing are displayed in Table 11.6. Forty-one separate titles are listed, twenty-six for the developed world, twenty-five for developing. Seven titles are common to both, which demonstrates only moderate agreement on which titles are equally valuable. T.T. Kozlowski, author of Chapter 10, with three top-ranked monographs, clearly stands out. That his *Water Deficits and Plant Growth* should be ranked as the second most important on the developed world list twenty-five years after publication is a tribute both to Kozlowski's scholarship and the volume's value.

Six of the titles for the Third World are specific to agroforestry; ten deal with tropical forest ecosystems, and nine titles cover general aspects of forest science primarily in the area of biology. Of the top-ranked titles on the developed list, seventeen deal with biological aspects, four with policy or economics, two with pathology or protection, one with wood technology, and two with management. The dearth of management titles in the top twenty-five belies the fact that almost 25% of the titles on the full core list are oriented toward forest management and silviculture. On the combined list, six titles are in new editions, and one was reprinted.

The top publisher in Table 11.6 is Academic Press with seven volumes,

Table 11.6. Top-ranked monographs in forestry and agroforestry

Third World rank	Developed World rank	
1	4	Cannell, M. G. R., and F. T. Last, eds. *Tree Physiology and Yield Improvement*; Proceedings of a Conference on Physiological Genetics of Forest Tree Yield, Gorebridge, Scotland, July 1975. London and New York; Academic Press, 1976. 567p.
2		Steppler, Howard A., and P. K. R. Nair, eds. *Agroforestry: A Decade of Development*. Nairobi, Kenya; International Council for Research in Agroforestry, 1987. 335p.
3		Huxley, Peter A., ed. *Plant Research and Agroforestry*; Proceedings of a Consultative Meeting, Nairobi, April 1981. Nairobi, Kenya; International Council for Research in Agroforestry, 1983. 617p.
	1	Kramer, Paul J., and Theodore T. Kozlowski. *Physiology of Woody Plants*. New York; Academic Press, 1979. 811p. (Rev., expanded and updated ed. of Kramer's Physiology of Trees and Kozlowski's Growth and Development of Trees.)
	2	Kozlowski, T. T., ed. *Water Deficits and Plant Growth*. New York; Academic Press, 1968. 7 vols.
4	3	Bowen, G. D., and E. K. S. Nambiar, eds. *Nutrition of Plantation Forests*. London and Orlando, Fla.; Academic Press, 1984. 516p.
5–7	16–17	Sutton, S. L., T. C. Whitmore, and A. C. Chadwick, eds. *Tropical Rain Forest: Ecology and Management*. Oxford and Boston; Blackwell Scientific; St. Louis, Mo.; Blackwell Mosby, 1983. 498p.
5–7		Tomlinson, P. B., and Martin H. Zimmermann, eds. *Tropical Trees as Living Systems*; Proceedings of the 4th Cabot Symposium, Harvard Forest, Petersham, Mass., April 1976. Cambridge and New York; Cambridge University Press, 1978. 675p.
5–7	9–10	Whitemore, T. C. *Tropical Rain Forests of the Far East*. 2d ed. Oxford, U.K.; Clarendon Press, 1984. 352p. (1st ed., 1975. 282p.)
	5	Cannell, M. G. R., and J. E. Jackson, eds. *Attributes of Trees as Crops Plants*; Proceedings of a Conference . . . Gorebridge, Midlothian, Scotland, July 1984. Abbots Ripton, Huntington, U.K.; Institute of Terrestrial Ecology, Natural Enviroment Research Council, 1985. 592p.
8	6	Bonga, J. M., and Don J. Durzan, eds. *Cell and Tissue Culture in Forestry*. Dordrecht and Boston; N. Nijhoff, 1987. 3 vols. (Rev. ed., 1982, of Tissue Culture in Forestry.)
	7	Nobel, Park S. *Biophysical Plant Physiology and Ecology*. San Francisco; W. H. Freeman, 1983. 488p.
9–10		Food and Agricultural Organization. *Tropical Forestry Action Plan*. Rome; FAO, 1987. 32p. (Earlier ed., 1985. 159p.)
9–10	19–24	Richards, Paul W. *The Tropical Rain Forest; An Ecological Study*. Cambridge, U.K.; Cambridge University Press, 1952. 450p. (Reprinted, 1979.)
20–23	8	Kallio, Markku, Dennis P. Dykstra, and Clark S. Binkley, eds. *The Global Forest Sector: An Analytical Perspective*. Chichester; Wiley, 1987. 717p.
11–13		Gregerson, Hans, Sydney Draper, and Dieter Elz, eds. *People and Trees: The Role of Social Forestry in Sustainable Development*. Washington, D.C.; World Bank, 1989. 272p.

Table 11.6. Continued

Third World rank	Developed World rank	
11–13		Raintree, John B., compiler and ed. *D & D User's Manual: An Introduction to Agroforestry Diagnosis and Design.* Nairobi, Kenya; International Council for Research in Agroforestry, 1987. 110p.
11–13	14	Repetto, Robert, and Malcolm Gillis, eds. *Public Policies and the Misuse of Forest Resources.* Cambridge and New York; Cambridge University Press, 1988. 432p.
14–16		Golley, Frank B., ed. *Tropical Rain Forest Ecosystems; Trends in Terrestrial and Aquatic Research.* New York; Springer-Verlag, 1975. 398p.
14–16		Jarvis, P. G., ed. *Agroforestry: Principles and Practices*; Proceedings of an International Conference, University of Edinburgh, July 1989. Amsterdam; Elsevier, 1991. 356p.
14–16		Sanchez, Pedro A. *Properties and Management of Soils in the Tropics.* New York; Wiley, 1976. 618p. (Available in Spanish as Suelos del Tropico: Caracteristicas y Manejo. San Jose, Costa Rica; Instituto Interamericano de Cooperacion para la Agricultura, 1981.)
	9–10	Levitt, J. *Responses of Plants to Environmental Stresses.* 2d ed. New York; Academic Press, 1980. 2 vols. (1st ed., 1972. 697p.)
18	11	Monteith, John L., ed. *Vegetation and the Atmosphere.* London and New York; Academic Press, 1975. 2 vols.
17		Prance, Ghillean T., ed. *Biological Diversification in the Tropics*; Proceedings of the 5th International Symposium of the Association for Tropical Biology, Macuto Beach, Caracas, Venezuela, Feb. 1979. New York; Columbia University Press, 1982. 714p.
	12–13	Landsberg, J. J. *Physiological Ecology of Forest Production.* London and Orlando, Fla.; Academic Press, 1986. 198p.
	12–13	Cardellichio, Peter A., Darius M. Adams, and Richard W. Haynes, eds. *Forest Sector and Trade Models: Theory and Applications*; Proceedings of an Interantional Symposium, Nov. 1987. Seattle; Center for International Trade in Forest Products, University of Washington, 1987. 298p.
19		Mergen, François, and Jeffrey R. Vincent, eds. *Natural Management of Tropical Moist Forests: Silvicultural and Management Prospects of Sustained Utilization.* New Haven, Conn.; Yale University, School of Forestry and Enivronmental Studies, 1987. 212p.
	15	Tennessee Valley Authority. *Forest Fertilization: Theory and Practice*; Proceedings of a Symposium . . . University of Florida, 1967. Muscle Shoals, Ala.; TVA, National Fertilizer Development Center, 1968. 306p.
	16–17	Panshin, Alexis J., and Carl de Zeeuw. *Textbook of Wood Technology: Structure, Identification, Properties, and Use of the Commercial Woods of the United States and Canada.* 4th ed. New York; McGraw-Hill, 1980. 722p. (1st ed., 1949–1952. 2 vols.)
20–23		MacDonald, L. H., ed. *Agro-Forestry in the African Humid Tropics*; Proceedings of a Workshop, Ibadan, Nigeria, April–May 1981. Shbuya-ku, Japan; United Nations University, 1982. 163p.
20–23		Mongi, H. O., and P. A. Huxely, eds. *Soils Research in Agroforestry*; Proceedings of an Expert Consultation held at the International Council for Research in Agroforestry, Nairobi, March 1979. Nairobi, Kenya; ICRAF, 1979. 584p.

Table 11.6. Continued

Third World rank	Developed World rank	
20–23		World Forestry Congress, 8th, Jakarta, Indonesia, 1978. *Forests for People.* Hamburg; Kommissionverlag Buch, 1979. 211p.
24–25		Halle, Francis, R. A. A. Oldeman, and P. B. Tomlinson. *Tropical Trees and Forests: An Architectural Analysis.* Berlin and New York; Springer-Verlag, 1978. 441p.
24–25		Ruthenberg, Hans. *Farming Systems in the Tropics.* 3d ed. Oxford, U.K.; Clarendon Press; New York; Oxford University Press, 1980. 424p. (1st ed., 1971. 313p.)
	18	Reichle, David E. *Dynamic Properties of Forest Ecosystems.* New York; Cambridge University Press, 1981. 683p.
	19–24	Davis, Lawrence S., and K. Norman Johnson. *Forest Management.* 3d ed. New York; McGraw-Hill, 1987. 790p. (Rev. ed. of Forest Management, by Kenneth P. Davis.)
	19–24	Esau, Katherine. *Plant Anatomy.* 2d ed. New York; Wiley, 1965. 767p. (1st ed., 1953. 735p.)
	19–24	Marks, G. C., and T. T. Kozlowski, eds. *Ectomycorrhizae.* New York; Academic Press, 1973. 444p.
	19–24	Royer, Jack P., and Christopher D. Risbrudt, eds. *Nonindustrial Private Forests: A Review of Economic and Policy Studies*; Proceedings of a Symposium, April 1983. Durham, N.C.; Duke University Press, School of Forestry and Environmental Studies, 1983. 398p.
	19–24	Sinclair, Wayne A., Howard H. Lyon, and Warren T. Johnson. *Diseases of Trees and Shrubs.* Ithaca, N.Y.; Comstock Pub. Associates, 1987. 574p.
	25–26	Likens, Gene E., et al. *Biogeochemistry of a Forested Ecosystem.* New York; Springer-Verlag, 1977. 146p.
	25–26	Gordon, J. C., and C. T. Wheeler, eds. *Biological Nitrogen Fixation in Forest Ecosystems: Foundations and Applications.* The Hague and Boston; M. Nijhoff/W. Junk, 1983. 342p.

followed by ICRAF and Cambridge University Press with four apiece. Of the total forty-one titles, twenty-two were commercially published, universities published ten, organizations seven, and governmental units, two.

G. Monograph List

Listed below are the core monographs for forestry and agroforestry. These 1,053 titles are divided into First Rank, titles which are highly cited in the literature and which reviewers considered of great value; Second

Rank, which generally received fewer citations but were still considered of importance by reviewers; and finally Third Rank, titles which had few counts but received mixed reviews distributed evenly between "important" to of "little value." Since there was little unanimity on these titles they were ranked Third. This list represents the core forestry monographs for research and instruction for use today. Table 11.7 shows how the monographs are divided by rank in each category.

Table 11.7. Percent of monographs in each rank

Developed countries		Third World
15.7%	FIRST rank	18.9%
25.1	SECOND rank	30.6
59.2	THIRD rank	50.5

Core List of Monographs for Developed and Third World Countries, 1950–1993, 1,053 titles

Developed Countries Ranking		Third World Ranking
First	Abeles, Frederick B., Page W. Morgan, and Mikal E. Saltveit. Ethylene in Plant Biology. 2d ed. San Diego, Calif.; Academic Press, 1992. (1st ed., 1973. 302p.)	
First	Adams, Darius M., and Richard W. Haynes. The 1980 Softwood Timber Assessment Market Model: Structure, Projections, and Policy Simulations. Washington, D.C.; Society of American Foresters, 1980. 64p. (Forest Science Monograph no. 22)	
Third	Adams, Norma R., and Foster B. Cady, eds. Modeling Growth and Yield of Multipurpose Tree Species. Arlington, Va.; Winrock International Institute for Agricultural Development, 1988. (Multipurpose Tree Species Network Technical Series no. 1)	Second
Second	Addicott, Fredrick T., ed. Abscisic Acid. New York; Praeger, 1983. 607p.	
Third	Addicott, Fredrick T. Abscission. Berkeley; University of California Press, 1982. 369p.	
	Agarwal, B. Cold Hearths and Barren Slopes: The Woodfuel Crisis in the Third World. New Delhi; Institute of Economic Growth, 1986. 209p.	Third
Third	Agee, James K., and Darryll R. Johnson, eds. Ecosystem Management for Parks and Wilderness. Seattle and London; University of Washington Press, 1988. 237p. (Institute of Forest Resources Contribution no. 65)	
	Agpaoa, A., et al. Manual of Reforestation and Erosion Control	Third

Developed Countries Ranking		Third World Ranking
	for the Philippines. Eschborn; German Agency for Technical Cooperation, 1976. 569p.	
	Agro-Forestry Seminar, Imphal, India, 1979. Proceedings . . . New Delhi, India; Indian Council for Agricultural Research, 1981. 267p.	Third
Second	Allen, George S., and John N. Owens. The Life History of Douglas-Fir. Ottawa; Information Canada, 1972. 139p.	
Third	Allen, Shirley W., and Grant W. Sharpe. An Introduction to American Forestry. 4th ed. New York; McGraw-Hill, 1976. 544p. (1st ed., 1938, by S. W. Allen. 402p.)	
	Almeda, Frank, and Catherine M. Pringle, eds. Tropical Rainforests: Diversity and Conservation. San Francisco, Calif; California Academy of Sciences, 1988. 306p.	Third
Third	Alston, Richard M. The Individual vs. the Public Interest, Political Ideology and National Forest Policy. Boulder, Colo.; Westview Press, 1983. 250p.	
	Alvim, Paulo de T., and T. T. Kozlowski, eds. Ecophysiology of Tropical Crops; Proceedings of an International Symposium . . . Manaus, Brazil, May 1975. New York; Academic Press, 1977. 502p.	First
	Amerena, Penelope. Agriculture and Natural Resources: A Manual for Development Workers. London; Voluntary Services Overseas, 1990. 117p.	Third
First	American Society of Agronomy. Methods of Soil Analysis. 2d ed. Madison, Wis.; American Society of Agronomy, Soil Science Society of America, 1982–1986. 2 vols.	
	Anderson, Anthony B., ed. Alternatives to Deforestation: Steps toward Sustainable Use of the Amazon Rain Forest; Proceedings of an International Conference, Belem, Brazil, Jan. 1988. New York; Columbia University Press, 1990. 281p.	Second
Third	Anderson, Dennis. The Economics of Afforestation: A Case Study in Africa. Baltimore, Md.; Published for the World Bank by Johns Hopkins University Press, 1987. 86p. (World Bank Occasional Papers, New Series no. 1)	Second
Third	Anderson, J. M., and J. S. I. Ingram. Tropical Soil Biology and Fertility: A Handbook of Methods. 2d ed. Wallingford, U.K.; CAB International, 1993. 221p. (1st ed., 1989. 171p.)	
	Anderson, Robert S., and Walter Huber. The Hour of the Fox: Tropical Forests, the World Bank, and Indigenous People in Central India. Seattle; University of Washington Press, 1988. 158p.	Third
Second	Anderson, Roger F. Forest and Shade Tree Entomology. New York; J. Wiley, 1960. 428p.	
	Aquino, Rosemary. Mounting a National Social Forestry Program: Lessons Learned from the Philippine Experience. Hon-	Second

Developed Countries Ranking		Third World Ranking
	olulu, Hawaii; Environment and Policy Institute, East-West Center, 1987. 72p.	
	Armitage, F. B. Irrigated Forestry in Arid and Semi-Arid Lands: A Synthesis. Ottawa; International Development Research Centre, 1985. 160p. (IDRC no. 234e)	Second
	Armitage, F. B., P. A. Joustra, and B. Ben Salem. Genetic Resources of Tree Species in Arid and Semi-Arid Areas: A Survey for the Improvement of Rural Living in Latin America, Africa, India and Southwest Asia. Rome; Food and Agriculture Organization, 1980. 118p.	Second
Second	Armson, K. A. Forest Soils: Properties and Processes. Toronto and Buffalo; University of Toronto Press, 1977. 390p.	Third
	Arnold, G. W., and C. T. de Wit, eds. Critical Evaluation of Systems Analysis in Ecosystems Research and Management; Papers presented at a Symposium during 1st International Congress of Ecology, The Hague, Sept. 1974. Wageningen, Netherlands; Centre for Agricultural Publishing and Documentation, 1976. 108p.	Third
	Arnold, J. E. Community Forestry: Ten Years in Review. Rome; Food and Agriculture Organization, 1991. 31p.	Third
	Arnold, J. E., preparer. An Introduction to Planning Forestry Development. Rome; Food and Agriculture Organization, 1974. 86p.	Third
Third	Arpan, Jeffrey S., et al. The United States Pulp and Paper Industry: Global Challenges and Strategies. Columbia; University of South Carolina Press, 1986. 1 vol.	
Third	Ashton, Floyd M., and Alden S. Crafts. Mode of Action of Herbicides. 2d ed. New York and Chichester, U.K.; J. Wiley, 1981. 525p. (1st ed., 1973. 504p.)	
Third	Asrar, Ghassem, ed. Theory and Applications of Optical Remote Sensing. New York; J. Wiley, 1989. 734p.	Third
Third	Association of Official Analytical Chemists. Official Methods of Analysis . . . edited by Kenneth Helrich. 15th ed. Arlington, Va.; AOAC, 1990. 2 vols. (7th ed., Washington, D.C.; AOAC, 1950. 910p.)	
First	Atkinson, D., et al., eds. Tree Root Systems and Their Mycorrhizas. The Hague and Boston; M. Nijhoff, 1983. 525p.	Second
Second	Attiwill, P. M., and G. W. Leeper. Forest Soils and Nutrient Cycles. Carlton, Victoria; Melbourne University Press; Beaverton, Or.; International Specialized Book Services, 1987. 202p.	Second
	Avery, M. E., M. G. R. Cannell, and C. Ong, eds. Biophysical Research for Asian Agroforestry. Arlington, Va.; Winrock International Institute for Agricultural Development, 1991. 292p.	Second

Developed Countries Ranking		Third World Ranking
Third	Avery, T. Eugene, and Graydon L. Berlin. Fundamentals of Remote Sensing and Airphoto Interpretation. 5th ed. New York; Macmillan; Toronto; Maxwell Macmillan Canada, 1991. (Earlier eds. as Interpretation of Aerial Photographs; An Introductory College Textbook and Self-Instruction Manual, Minneapolis; Burgess Pub. Co., 1962. 192p.)	Third
Second	Avery, T. Eugene, and Harold E. Burkhart. Forest Measurements. 3d ed. New York; McGraw-Hill, 1983. 331p. (1st ed., 1967. 290p.)	Second
	Ayanaba, A., and P. J. Dart. Biological Nitrogen Fixation in Farming Systems of the Tropics. Chichester, U.K. and New York; J. Wiley, 1977. 377p.	Third
Third	Baas, P., ed. New Perspectives in Wood Anatomy. The Hague, Netherlands; M. Nijhoff/W. Junk Pub., 1982. 252p.	
Third	Baas, P., A. J. Bolton, and D. M. Catling, eds. Wood Structure in Biological and Technological Research. Leiden, Germany; Leiden University Press, 1976. 280p. (Leiden Botanical Series no. 3)	
Third	Bailey, James A. Principles of Wildlife Management. New York; J. Wiley, 1984. 373p.	Third
Third	Bajaj, Y. P. S., ed. Trees I. Berlin and New York; Springer-Verlag, 1986. 515p. (Biotechnology in Agriculture and Forestry, 1)	
Third	Bajaj, Y. P. S., ed. Trees III. Berlin and New York; Springer-Verlag, 1991. 493p. (Biotechnology in Agriculture and Forestry, 16)	
Third	Baker, Andrew J., ed. Advances in Production of Forest Products. New York; American Institute of Chemical Engineers, 1983. 87p.	
Third	Baker, F. W. G., ed. Rapid Propagation of Fast-Growing Woody Species. Wallingford, U.K.; CAB International, 1992. 125p.	Third
First	Baker, Kenneth F., and R. James Cook. Biological Control of Plant Pathogens. San Francisco; W. H. Freeman, 1974. 433p. (Reprinted, St. Paul, Minn.; American Phytopathological Society, 1982.)	
Third	Baker, N. R., W. J. Davies, and C. K. Ong, eds. Control of Leaf Growth. Cambridge, U.K. and New York; Cambridge University Press, 1985. 350p. (Society for Experimental Biology, Seminar Series no. 27)	
	Bakshi, Bimal K. Forest Pathology: Principles and Practice in Forestry. Delhi, India; Controller of Publications, 1976. 400p.	Third
First	Ballard, Russell, and Stanley P. Gessel, technical eds. Forest Site and Continuous Productivity; Proceedings of an IUFRO Symposium, Seattle, Aug. 1982. Portland, Oreg.; Pacific	First

Developed Countries Ranking		Third World Ranking
	Northwest Forest and Range Experiment Station, 1983. 406p. (General Technical Report PNW no. 163)	
	Baltaxe, R. The Application of Landsat Data to Tropical Forest Surveys. Rome; Food and Agriculture Organization, 1980. 122p.	Third
	Bandhu, D., and R. K. Garg, eds. Social Forestry and Tribal Development. New Delhi; Indian Environmental Society, 1986. 162p.	Third
	Banerjee, Ajit K. Shrubs in Tropical Forest Ecosystems: Examples from India. Washington, D.C.; World Bank, 1989. 132p. (World Bank Technical Paper no. 103)	Third
Third	Barber, Stanley A. Soil Nutrient Bioavailability: A Mechanistic Approach. New York; J. Wiley, 1984. 398p.	Third
Second	Barbour, Michael G., and William D. Billings, eds. North American Terrestrial Vegetation. Cambridge, U.K.; Cambridge University Press, 1988. 434p.	
	Barnes, R. D., and G. L. Gibson, eds. Provenance and Genetic Improvement Strategies in Tropical Forest Trees; Proceedings of a Joint Work Conference . . . Mutare, Zimbabwe, April 1984. Oxford, U.K.; Dept. of Forestry, Commonwealth Forestry Institute, University of Oxford, 1984. 663p.	Second
Third	Barrett, E. C., and L. F. Curtis. Introduction to Environmental Remote Sensing. 2d ed. London and New York; Chapman and Hall, 1982. 352p. (1st ed., London: Chapman and Hall and New York: Wiley, 1976. 336p.)	
Second	Barrett, John W., ed. Regional Silviculture of the United States. 2d ed. New York; J. Wiley, 1980. 551p. (1st ed., New York; Ronald Press, 1962. 610p.)	
	Baumer, Michel. The Potential Role of Agroforestry in Combating Desertification and Environmental Degradation: With Special Reference to Africa. Wageningen, Netherlands; Technical Centre for Agriculture and Rural Cooperation, 1990. 250p. (Translation of Le Role possible de l'Agroforesterie dans la Lutte contre la Desertification et la Degradation de l'Environnement, 1987. 260p.)	Second
Second	Baumol, William J., and Wallace E. Oates. The Theory of Environmental Policy. 2d ed. Cambridge, U.K.; Cambridge University Press, 1988. 299p. (1st ed., Englewood Cliffs, N.J.; Prentice-Hall, 1975. 272p.)	Second
	Baur, George N. The Ecological Basis of Rainforest Management. Sydney, Australia; V.C.N. Blight, 1968. 499p.	Third
	Bawa, K. W., and M. Hadley, eds. Reproductive Ecology of Tropical Forest Plants. Paris; UNESCO, 1990. 421p.	Third
Second	Baxter, Dow V. Pathology in Forest Practice. 2d ed. New York; J. Wiley, 1952. 601p. (1st ed., London; Chapman & Hall, 1943. 618p.)	

Developed Countries Ranking		Third World Ranking
	Beer, J. W., H. W. Fassbender, and J. Heuveldop, eds. Advances in Agroforestry Research; Proceedings of a Seminar, CATIE, Turrialba, Costa Rica, Sept. 1985, sponsored by Centro Agronomico Tropical de Investigacion y Ensenanza (CATIE) and Deutsche Gesselschaft für Technische Zusammenarbeit (GTZ). Turrialba, Costa Rica; CATIE, 1987. 379p.	Third
	Bene, J. G., H. W. Beall, and Cote A. Trees, Food and People: Land Management in the Tropics. Ottawa; International Development Research Centre, 1977. 52p.	First
	Bengston, David N., et al. Forestry Research Capacity in the Asia-Pacific Region: An Evaluation Model and Preliminary Assessment. Honolulu; East-West Environment and Policy Institute, 1988. 96p. (EWEPI Occasional Paper no. 6)	Third
Second	Benyus, Janine M. Christmas Tree Pest Manual. Washington, D.C.; U.S. Supt. of Docs., 1983. 107p.	
Third	Berryman, Alan A., ed. Dynamics of Forest Insect Populations: Paterns, Causes, Implications. New York; Plenum Press, 1988. 603p.	
Second	Bewley, J. Derek, and M. Black. Physiology and Biochemistry of Seeds in Relation to Germination. Corrected printing of 1st ed. Berlin and New York; Springer-Verlag, 1983. 2 vols. (1st ed., 1978.)	
Third	Bingham, Gail. Resolving Environmental Disputes: A Decade of Experience. Washington, D.C.; Conservation Foundation, 1986. 284p.	Third
Third	Binkley, Dan. Forest Nutrition Management. New York; J. Wiley, 1986. 290p.	
Third	Bissonette, John A., ed. Is Good Forestry Good Wildlife Management?; Proceedings of a Joint Conference of New England Chapter of the Society of American Foresters; Maine Chapter of the Wildlife Society; Atlantic International Chapter of the American Fisheries Soceity, Portland, Maine, Mar. 1985. Orono; Maine Agricultural Experiment Station, University of Maine, 1986. 377p. (Maine AES Misc. Publ. no. 689)	
Third	Blaikie, Piers M. The Political Economy of Soil Erosion in Developing Countries. London; Longmans, 1985. 118p.	
Third	Blair, Harry W., and Porus D. Olpadwala. Forestry in Development Planning: Lessons from the Rural Experience. Boulder, Colo.; Westview Press, 1988. 205p.	Third
Third	Blanchard, Robert O., and Terry A. Tattar. Field and Laboratory Guide to Tree Pathology. New York; Academic Press, 1981. 285p.	
Third	Blandon, Peter. Soviet Forest Industries. Boulder, Colorado; Westview Press, 1983. 290p.	
Third	Blomquist, R. F., et al., eds. Adhesive Bonding of Wood and Other Structural Materials: A Compilation of Eduational Mod-	

Developed
Countries
Ranking

Third
World
Ranking

	ules Especially Prepared for Engineers and Architects at the 3d Clark C. Heritage Memorial Workshop on Wood, Madison, Wis., Aug. 1981. University Park, Md.; Educational Modules for Materials Science and Engineering (EMMSE) Project, Materials Research Laboratory, Pennsylvania State University, 1983. 436p.	
Third	Blyth, J., et al. Farm Woodland Management. 2d ed. Ipswich; Farming Press Books, 1991. 208p. (1st ed., 1987. 189p.)	
Third	Bodig, Jozef, and Benjamin A. Jayne. Mechanics of Wood and Wood Composites. Reprint ed. Malabar, Fla.; Krieger Pub., 1992. (1st ed., New York: Van Nostrand Reinhold, 1982. 712p.)	
	Boehnert, Joachim. Agroforestry in Agricultural Education: With a Focus on the Practical Implementation. Gaimersheim, Germany; J. Margraf, 1988. 182p. (Available in Spanish, 1990, as La agroforestria en la educacion agricola con enfasis en su aplicacion practica.)	Third
	Boland, D. J., ed. Trees for the Tropics: Growing Australian Multipurpose Trees and Shrubs in Developing Countries. Canberra; Australian Centre for International Agricultural Research, 1989. 247p. (ACIAR Monograph no. 10)	Second
Third	Bolin, B., and R. B. Cook, eds. The Major Biogeochemical Cycles and Their Interactions. Chichester U.K. and New York; Published for the Scientific Committee on Problems of the Environment of the International Council of Scientific Unions by J. Wiley, 1983. 532p. (SCOPE no. 21)	Third
First	Bonga, J. M., and Don J. Durzan, eds. Cell and Tissue Culture in Forestry. Dordrecht and Boston; N. Nijhoff, 1987. 3 vols. (Rev. ed., 1982, of Tissue Culture in Forestry.)	First
First	Bonner, Frank, ed. Flowering and Seed Development in Trees; Proceedings of a Symposium, Mississippi State University, May 1978. Starkville, Miss.; Published for the International Union of Forest Research Organizations by the Southern Forest Experiment Station, 1979. 380p.	
First	Bonner, James F., and Joseph E. Varner, ed. Plant Biochemistry. 3d ed. New York; Academic Press, 1976. 925p. (1st ed., 1950. 537p.)	
	Bormann, F. Herbert, and Graeme Berlyn, eds. Age and Growth Rage of Tropical Trees: New Directions for Research; Proceedings of a Workshop on Age and Growth Rate Determination for Tropical Trees, Harvard Forest, Petersham, Mass., Apr. 1980. New Haven, Conn.; Yale University Press, 1981. 137p.	Second
First	Bormann, F. Herbert, and Gene E. Likens. Pattern and Process in a Forested Ecosystem: Disturbance, Development, and the	Second

	Steady State Based on the Hubbard Brook Ecosystem Study. New York; Springer-Verlag, 1981. 253p. (1st printing, 1979.)	
	Borota, Jan. Tropical Forests: Some African and Asian Case Studies of Composition and Structure. Amsterdam and New York; Elsevier, 1991. 274p. (Translated from Slovak.)	Second
	Borota, Jan. Tropical Forests: Some African Case Studies of Composition and Structure. Amsterdam and New York; Elsevier, 1991. 274p. (Developments in Agricultural and Managed-Forest Ecology no. 22)	Third
Third	Bourke, Ian J. Trade in Forest Products: A Study of the Barriers Faced by the Developing Countries. Rome; Food and Agriculture Organization, 1988. 140p. (FAO Forestry Paper no. 83)	
First	Bowen, G. D., and E. K. S. Nambiar, eds. Nutrition of Plantation Forests. London and Orlando, Fla.; Academic Press, 1984. 516p.	First
Second	Bowes, Michael D., and John V. Krutilla. Multiple-use Management: The Economics of Public Forestlands. Washington, D.C.; Resources for the Future, 1989. 357p.	
First	Boyce, John S. Forest Pathology. 3d ed. New York; McGraw-Hill, 1961. 572p. (1st ed., 1938. 600p.)	
	Boyce, Stephen G., ed. Biological and Sociological Basis for a Rational Use of Forest Resources for Energy and Organics; Proceedings of an International Workshop, Michigan State University, East Lansing, May 1979. Asheville, N.C.; USDA Forest Service, Southeastern Forest Experiment Station, 1979. 193p.	Third
First	Brady, Nyle C. The Nature and Properties of Soils. 10th ed. New York; Macmillan; London; Collier Macmillan, 1990. 621p. (1st ed., by T. Lyttleton Lyon and Harry O. Buckman, 1922. 588p.)	Second
First	Brandle, J. R., D. L. Hintz, and J. W. Sturrock, eds. Windbreak Technology; Proceedings of an International Symposium . . . Lincoln, Nebr., June 1986. Amsterdam and New York; Elsevier, 1988. 598p.	First
Second	Branham, Susan J., and G. D. Hertel, eds. Integrated Forest Pest Management Symposium; Proceedings . . . Athens, Ga., 1984. Athens; University of Georgia, Center for Continuing Education, 1984. 281p.	Second
	Brewbaker, James L., et al. Leucaena: Forage Production and Use. Waimanalo, Hawaii; Nitrogen Fixing Tree Association, 1985. 39p.	Second
Third	Briscoe, C. Buford. Field Trials Manual for Multipurpose Tree Species. 2d ed. Arlington, Va.; Winrock International Institute for Agricultural Development, 1990. 141p. (Multipurpose	Second

Developed Countries Ranking		Third World Ranking
	Tree Species Network Research Series, Manual no. 3) (1st ed., 1989. 163p.)	
Third	Brockman, C. Frank, and Lawrence C. Merriam. Recreational Use of Wild Lands. 3d ed. New York; McGraw-Hill, 1979. 337p. (1st ed., 1959. 346p.)	
Third	Brokaw, Howard P., ed. Wildlife and America: Contributions to an Understanding of American Wildlife and Its Conservation. Washington, D.C.; Council on Environmental Quality, 1978. 532p.	
	Brokensha, David, and Alfonso Peter Castro. Fuelwood, Agro-Forestry, and Natural Resource Management: The Development Significance of Land Tenure and Other Resource Management Utilization Systems. Binghamton, N.Y.; Institute for Development Anthropology, 1984. 64p.	Second
	Browder, John O., ed. Fragile Lands of Latin America: Strategies for Sustainable Development. Boulder, Colo; Westview Press, 1989. 301p.	Third
Second	Brown, Arthur A., and Kenneth P. Davis. Forest Fire: Control and Use. 2d ed. New York; McGraw-Hill, 1973. 686p. (1st ed., by K. P. Davis, 1959. 584p.)	
Third	Brown, George W. Forestry and Water Quality. Corvallis, Oreg.; Oregon State University Book Stores, 1980. 124p.	
	Browne, Francis G. Pests and Diseases of Forest Plantation Trees: An Annotated List of the Principal Species Occurring in the British Commonwealth. Oxford, U.K.; Clarendon Press, 1968. 1330p.	Third
Third	Brubaker, Sterling, ed. Rethinking the Federal Lands. Washington, D.C.; Resources for the Future, 1984. 306p.	
Second	Bucher, Jurg B., and Inga Bucher-Wallin, eds. Air Pollution and Forest Decline; Proceedings of the 14th International Meeting for Specialists in Air Pollution Effects on Forest Ecosystems, IUFRO, Interlaken, Switzerland, Oct. 1988. Birmensdorf, Switzerland; Iedgenossische Anstalt fur das forstliche Versuchswesen, 1989. 564p.	
	Buck, Louise, ed. Kenya National Seminar on Agroforestry, Nairobi, Nov. 1980. Nairobi; International Council for Research in Agroforestry, 1981. 638p.	First
	Buck, Louise E. Agroforestry. Nairobi, Kenya; Cooperative for American Relief Everywhere International, 1980. 2 vols.	Third
	Budd, W., I. Duchhart, L. Hardesty, and F. Steiner, eds. Planning for Agroforestry; Papers presented at an International Symposium, Washington State University, Pullman, 1989. Amsterdam and New York; Elsevier, 1990. 338p. (Developments in Landscape Management and Urban Planning no. 6C)	First
	Budelman, Arnoud. Woody Species in Auxiliary Roles; Live	Third

Developed Countries Ranking		Third World Ranking
	Forest Service, 1986. 156p. (General Technical Report no. 111)	
First	Burns, Russell M., compiler. Silvicultural Systems for the Major Forest Types of the United States. Washington, D.C.; U.S. Forest Service, 1983. 191p. (USDA Agriculture Handbook no. 445) (Earlier ed., 1973. 114p.)	
Second	Burns, Russell M., and Barbara H. Honkala. Silvics of North America. Washington, D.C.; USDA Forest Service, 1990–1991. 2 vols. (USDA Agriculture Handbook no. 654) (Supersedes USDA Agriculture Handbook no. 271, Silvics of Forest Trees of the United States. Washington, D.C.; USDA, Forest Service, 1965. 762p.)	
Third	Busgen, Moritz. The Structure and Life of Forest Trees, English translation by Thomas Thomson (Bau und Leben unserer Waldbaume). 3d rev. and enl. ed. London; Chapman and Hall, 1929. 436p.	
Third	Butin, Heinz. Krankheiten der Wald- und Parkbaume: Leitfaden zum Bestimmen von Baumkrankheiten. Stuttgart and New York; Thieme, 1983. 172p.	
Third	Butterfield, Brian G., and B. A. Meylan. Three-Dimensional Structure of Wood: An Ultrastructural Approach. 2d ed. London and New York; Chapman & Hall, 1980. 103p. (Previous ed. by B. A. Meylan and B. G. Butterfield.)	
Third	Cannell, M. G. R., compiler. World Forest Biomass and Primary Production Data. London and New York; Academic Press, 1982. 391p.	First
First	Cannell, M. G. R., and J. E. Jackson, eds. Attributes of Trees as Crop Plants; Proceedings of a Conference . . . Gorebridge, Midlothian, Scotland, July 1984. Abbots Ripton, Huntingdon, U.K.; Institute of Terrestrial Ecology, Natural Environment Research Council, 1985. 592p.	First
First	Cannell, M. G. R., and F. T. Last, eds. Tree Physiology and Yield Improvement; Proceedings of a Conference on Physiological Genetics of Forest Tree Yield, Gorebridge, Scotland, July 1975. London and New York; Academic Press, 1976. 567p.	First
Second	Cannell, M. G. R., D. C. Malcolm, and P. A. Robertson, eds. The Ecology of Mixed-Species Stands of Trees. Oxford, U.K.; Blackwell Scientific, 1992. 312p.	Second
Second	Canny, M. J. Phloem Translocation. Cambridge, U.K.; Cambridge University Press, 1973. 301p.	
Third	Canny, M. J., et al. Transport in Plants. Berlin and New York; Springer-Verlag, 1975–1976. 3 vols.	
First	Cardellichio, Peter A., Darius M. Adams, and Richard W. Haynes, eds. Forest Sector and Trade Models: Theory and Ap-	

plications; Proceedings of an International Symposium, Nov. 1987. Seattle; Center for International Trade in Forest Products, University of Washington, 1987. 298p.

Third   Carlquist, Sherwin J. Ecological Strategies of Xylem Evolution. Berkeley; University of California Press, 1975. 259p.

Carlson, Les W., and Keith R. Shea, compilers. Increasing Productivity of Multipurpose Lands; Proceedings of an IUFRO Research Planning Workshop for Africa, Sahelian and North Sudanian Zones, Nairobi, Kenya, Jan. 1986. Hull, Ont.; International Union of Forestry Research Organization, 1986. 333p.   Second

Carpenter, Richard A., ed. Assessing Tropical Forest Lands: Their Suitability for Sustainable Uses; Proceedings of a Conference on Forest Land Assessment and Management for Sustainable Uses, Honolulu, June 1979. Dublin; Tycooly International, 1981. 337p.   Second

First   Carson, Rachel. Silent Spring. 1st Ballantine Books ed. New York; Fawcett Crest, 1982. 304p. (1st ed., Boston; Houghton Mifflin Co., 1962. 368p.) (Also available in Spanish as Primavera Silenciosa. Barcelona; Grijalbo.)

Third   Cellulose and Celluslose Derivatives2d comp. rev. and augm. ed. New York; Interscience Publishers, 1954–1971. 5 vols. (Vol. 1–3 edited by E. Ott, H. M. Spurlin, M. W. Grafflin; Vol. 4–5 edited by N. M. Bikales, L. Segal.)

Center for Research on Economic Development, University of Michigan. Agroforestry in Developing Countries. Ann Arbor, Mich.; CRED, 1984. 195p.   Third

Centro Agronomico Tropical de Investigacion y Enseanza. Sistemas Agroforestales: Principios y Aplicaciones en los Tropicos. San Jose, Costa Rica; Organizacion para Estudios Tropicales, CATIE, 1986. 818p.   Second

Cernea, Michael M., ed. Putting People First: Sociological Variables in Rural Development. 2d ed., rev. and expanded. New York; Published for the World Bank by Oxford University Press, 1991. 430p. (1st ed., 1985.)   Second

First   Chabot, Brian F., and Harold A. Mooney, eds. Physiological Ecology of North American Plant Communities. New York; Chapman & Hall, 1985. 351p.

Third   Chabreck, Robert H., and Robert H. Mills, eds. Integrating Timber and Wildlife Management in Southern Forests; 29th Forestry Symposium, Louisiana State University, 1980. Baton Rouge, La.; Div. of Continuing Education, Louisiana State University, 1980. 119p.

Third   Chandler, Craig, et al. Fire in Forestry. New York; J. Wiley & Sons, 1983. 2 vols.

Developed Countries Ranking		Third World Ranking
	Chandler, Trevor, and David Sprugeon, eds. International Cooperation in Agroforestry; Proceedings of an International Conference. Nairobi, Kenya; International Council for Research in Agroforestry, 1979. 469p.	Third
	Chapman, G. W., and T. G. Allan. Establishment Techniques for Forest Plantations. Rome; Food and Agriculture Organization, 1978. 183p. (FAO Forestry Paper no. 8)	Third
	Cherry, Joe H., ed. Environmental Stress in Plants: Biochemical and Physiological Mechanisms; Proceedings of a NATO Advanced Research Workshop, Norwich, U.K., 1987. Berlin and New York; Springer-Verlag, 1989. 369p. (NATO ASI Series G, Ecological Sciences no. 19)	Third
Third	Christiansen, M. N., and Charles F. Lewis, eds. Breeding Plants for Less Favorable Environments. New York; J. Wiley, 1982. 459p.	
Second	Chudnoff, Martin. Tropical Timbers of the World. Madison, Wis.; USDA Forest Products Laboratory, 1984. 464p.	Second
Third	Churchman, C. West, Albert H. Rosenthal, and Spencer H. Smith, eds. Natural Resource Administration: Introducing a New Methodology for Management Development. Boulder, Colo.; Westview Press, 1984. 228p.	Third
Third	Clark, Colin W. Mathematical Bioeconomics: The Optimal Management of Renewable Resources. 2d ed. New York; J. Wiley, 1990. 386p. (1st ed., 1976. 352p.)	
Third	Clark, F. E., and T. Rosswall, eds. Terrestrial Nitrogen Cycles: Processes, Ecosystem Strategies and Management Impacts; Proceedings of an International Workshop, Gysinge Vardshus, Osterfarnebo, Sweden, Sept. 1979. Stockholm; Swedish Natural Science Research Council, 1981. 714p.	
Third	Clark, J. R., and J. Benforado, eds. Wetlands of Bottomland Hardwood Forests; Proceedings of a Workshop on Bottomland Hardwood Forest Wetlands of the Southeastern United States, Lake Lanier, Ga., June 1980. Amsterdam and New York; Elsevier Scientific, 1981. 401p.	
Third	Clark, Roger N., and George H. Stankey. The Recreation Opportunity Spectrum: A Framework for Planning, Management, and Research. Portland, Oreg.; USDA Pacific Northwest Forest and Range Experiment Station, 1979. 32p. (PNW General Technical Report no. 98)	
Third	Clark, Thomas D. The Greening of the South: The Recovery of Land and Forest. Lexington; University Press of Kentucky, 1984. 168p.	
	Clark, William C., and R. E. Munn, eds. Sustainable Development of the Biosphere. Cambridge, U.K. and New York; Cambridge University Press, 1986. 491p.	Second

Developed Countries Ranking		Third World Ranking
Second	Clawson, Marion. The Economics of National Forest Management. Washington, D.C.; Resources for the Future, 1976. 117p. (RFF Working Paper no. EN-6)	
Third	Clawson, Marion. The Federal Lands Revisited. Washington, D.C.; Resources for the Future, 1983. 302p.	
Second	Clawson, Marion. Forests for Whom and for What? Baltimore; Published for Resources for the Future by Johns Hopkins University Press, 1975. 175p.	
Third	Clawson, Marion, and Carlton S. Van Doren, eds. Statistics on Outdoor Recreation. Washington, D.C.; Resources for the Future, 1984. 368p. (Earlier ed., 1958. 165p.)	
Second	Clawson, Marion, and Jack L. Knetsch. Economics of Outdoor Recreation. Baltimore; Published for Resources for the Future by Johns Hopkins Press, 1971. 328p. (1st ed., 1966.)	
Second	Cleve, K. Van, et al., eds. Forest Ecosystems in the Alaskan Taiga: A Synthesis of Structure and Function. New York; Springer-Verlag, 1986. 230p.	
Second	Clutter, Jerome L. et al. Timber Management: A Quantitative Approach. New York; J. Wiley, 1983. 333p.	Second
Third	Coggins, George C., and Charles F. Wilkinson. Federal Public Land and Resources Law. 2d ed. Mineola, N.Y.; Foundation Press, 1987. 215p. (1st ed., 1981. 849p.)	
Third	Cole, Dale W., and Stanley P. Gessel, eds. Forest Site Evaluation and Long-Term Productivity. Seattle; University of Washington Press, 1988. 196p. (Institute of Forest Resources no. 63)	
Second	Colwell, Robert N., editor. Manual of Remote Sensing. 2d ed. Falls Church, Va.; American Society of Photogrammetry, 1983. 2 vols. (1st ed., edited by Robert G. Reeves, 1975.)	First
Third	Conference on Genetic Manipulation of Woody Plants. Proceedings of a Conference . . . Michigan State University, 1987. New York; Plenum, 1988. 519p.	
	Cook, Cynthia C., and Mikael Grut. Agroforestry in Sub-Saharan Africa: A Farmer's Perspective. Washington, D.C.; World Bank, 1989. 94p. (World Bank Technical Paper no. 112)	Second
Third	Cooley, James L., and June H. Cooley, eds. Natural Diversity in Forest Ecoysytems; Proceedings of a Workshop, Athens, Nov.-Dec. 1982. Athens, Ga.; Institute of Ecology, University of Georgia, 1984. 290p.	Second
Second	Coombs, J., et al., eds. Techniques in Bioproductivity and Photosynthesis. 2d ed. Oxford, U.K. and New York; Pergamon Press, 1985. 298p. (1st ed., 1982. 171p.)	First
Second	Cooper, J. I. Virus Diseases of Trees and Shrubs. Cambridge, U.K.; Institute of Terrestrial Ecology, 1979. 74p.	

Developed Countries Ranking		Third World Ranking
Third	Cooper-Driver, Gillian A., Tony Swain, and Eric E. Conn, eds. Chemically Mediated Interactions between Plants and Other Organisms. New York; Plenum, 1985. 246p. (Recent Advances in Phytochemistry no. 19)	
First	Cordell, Charles E., et al. Forest Nursery Pests. Washington, D.C.; U.S. Dept. of Agriculture, Forest Service, 1989. 184p. (USDA Agriculture Handbook no. 680)	
Third	Core, H. A., W. A. Cote, and A. C. Day. Wood Structure and Identification. 2d ed. Syracuse, N.Y.; Syracuse University Press, 1979. 182p. (1st ed., 1976. 168p.)	
Third	Corner, E. J. H. The Seeds of Dicotyledons. Cambridge, U.K. and New York; Cambridge University Press, 1976. 2 vols.	Third
Second	Coulson, Robert N., and John A. Witter. Forest Entomology: Ecology and Management. New York and Chichester, U.K.; J. Wiley, 1984. 669p.	Third
Third	Countryman, David W., and Denise M. Sofranko, eds. Guiding Land Use Decisions: Planning and Management for Forests and Recreation. Baltimore; Johns Hopkins University Press, 1982. 251p.	Third
Third	Cox, Thomas R., et al. This Well-Wooded Land: Americans and Their Forests from Colonial Times to the Present. Lincoln; University of Nebraska Press, 1985. 325p.	
Second	Craswell, E. T., J. V. Remenyi, and L. G. Nallana. Soil Erosion Management; Proceedings of a Workshop held at PCARRD, Los Banos, Philippines, Dec. 1984. Canberra; Australian Center for International Agricultural Research, 1985. 132p. (ACIAR Proceedings Series no. 6)	First
Third	Cubbage, Frederick W., ed. Southern Forest Economics Workshop: Trends in Growing and Marketing Southern Timber, Athens, Ga., March 1985; Proceedings . . . Raleigh; North Carolina State University, 1985. 212p.	
Third	Cubbage, Frederick, Jay O'Laughlin, and Charles Bullock. Forest Resource Policy. New York; J. Willey, 1993. 562p.	
Second	Culhane, Paul J. Public Lands Politics: Interest Group Influence on the Forest Service and The Bureau of Land Management. Baltimore; Published for Resources for the Future by Johns Hopkins Univeristy Press, 1981. 398p.	
Second	Cummins, Goerge B., and Yasuyuki Hiratsuka. Illustrated Genera of Rust Fungi. Rev. ed. St. Paul, Minn.; American Phytopathological Society, 1983. 152p. (1st ed., Minneapolis; Burgess, 1959. 131p.)	
Second	Dallimore, William, and A. Bruce Jackson. A Handbook of Conifers and Ginkgoaceae, rev. by S. G. Harrison. 4th ed. London; E. Arnold, 1966. 729p. (1st ed., 1923. 570p.)	
First	Dana, Samuel T., and Sally K. Fairfax. Forest and Range Policy,	

Its Development in the United States. 2d ed. New York; McGraw-Hill, 1980. 458p. (1st ed., 1956. 455p.)

First   Daniel, Theodore, John A. Helms, and Frederick S. Baker. Principles of Silviculture. 2d ed. New York; McGraw-Hill, 1979. 500p. (1st ed., 1950, by Frederick S. Baker.) (Also available in Spanish as Principios de Silvicultura.)   First

Dargavel, John, Kay Dixon, and Noel Semple, eds. Changing Tropical Forests: Historical Perspectives on Today's Challenges in Asia, Australasia and Oceania; Workshop Meeting, Canberra, May 1988. Canberra; Centre for Resource and Environmental Studies, Australia National University, 1988. 446p.   Third

Dasgupta, P. S. The Control of Resources. Oxford, U.K.; Basil Blackwell, 1982. 123p.   Third

Third   Dasgupta, P. S., and G. M. Heal. Economic Theory and Exhaustible Resources. Cambridge, U.K.; Cambridge University Press, 1979. 501p.   Third

Third   Dasmann, Raymond F., John P. Milton, and Peter H. Freeman. Ecological Principles for Economic Development. London and New York; International Union for Conservation of Nature and Natural Resources by J. Wiley, 1973. 252p.   Second

Third   Daubenmire, Rexford F. Plant Geography: With Special Reference to North America. New York; Academic Press, 1978. 338p.

Daubenmire, Rexford F. Plants and Environment; A Textbook of Plant Autecology. 3d ed. New York; J. Wiley, 1974. 422p. (1st ed., 1947. 424p.)   Second

Davidson, J., Tho Yow Pong, and Maarten Bijleveld, eds. The Future of Tropical Rain Forests in South East Asia; Proceedings of a Symposium, Kepong, Malaysia, Sept. 1983, organized by the Forest Research Institute and the IUCN Commission on Ecology. Gland, Switzerland; International Union for Conservation of Nature and Natural Resources, 1985. 127p. (Reprint from the Environmentalist 5 (1985) Supplement.)   Third

First   Davis, Lawrence S., and K. Norman Johnson. Forest Management. 3d ed. New York; McGraw-Hill, 1987. 790p. (Rev. ed. of Forest Management, by Kenneth P. Davis.)   First

Third   Davis, Raymond E., et al. Surveying, Theory and Practice. 6th ed. New York; McGraw-Hill, 1981. 992p. (1st ed., 1928. 1016p.)   Third

Dawkins, Henry C. The Management of Natural Tropical Highforest with Special Reference to Uganda. Oxford, U.K.; Imperial Forestry Institute, 1958. 155p. (IFI Institute no. 34)   Third

Third   De Fossard, Ronald A. Tissue Culture for Plant Propagators. Rev. ed. Armidale, Australia; Dept. of Botany, University of New England, 1981. 409p. (1st ed., 1976.)

Developed Countries Ranking		Third World Ranking
	De Montalembert, M. R., and J. Clement. Fuelwood Supplies in the Developing Countries. Rome; Food and Agriculture Organization, 1983. 125p. (FAO Forestry Paper no. 42)	Second
Second	Deacon, Robert T., and M. Bruce Johnson, eds. Forestlands: Public and Private. 2d ed. San Francisco; Pacific Institute for Public Policy Research, 1985. 335p.	
	DeBach, Paul, ed. Biological Control of Insect Pests and Weeds. London; Chapman & Hall, 1973. 844p. (Earlier printing, 1964.)	Third
Third	Decker, Daniel J., and Gary R. Goff, eds. Valuing Wildlife: Economic and Social Perspectives. Boulder, Colo.; Westview Press, 1987. 424p.	Third
Third	DeGraaf, Richard M., and Keith E. Evans, compilers. Management of North Central and Northeastern Forests for Nongame Birds; Proceedings of a Workshop, Minneapolis, Jan. 1979. St. Paul, Minn.; USDA Forest Service North Central Forest Experiment Station, 1979. 268p. (General Technical Report NC no. 51)	
	Denslow, Julie S., and Christine Padoch, eds. People of the Tropical Rain Forest. Berkeley; University of California Press; Washington, D.C.; Smithsonian Institution Traveling Exhibition Service, 1988. 231p.	Second
	Dent, David. Soil Survey and Land Evaluation. London and Boston; Allen & Unwin, 1981. 278p.	Second
	Desai, Vasant. Issues in Agriculture and Forestry. Bombay, India; Himalaya Pub. House, 1984. 448p.	Third
Third	Desch, Harold E. Timber: Its Structure and Properties. 5th ed. New York; St. Martin's Press, 1973. 424p. (1st ed., London; Macmillan, 1938. 169p.)	
Third	Devall, Bill, and George Sessions. Deep Ecology. Salt Lake City, Utah; G. M. Smith, 1985. 267p.	
Third	Dickinson, C. H., and G. J. F. Pugh, eds. Biology of Plant Litter Decomposition. London and New York; Academic Press, 1974. 2 vols.	
Third	Dickmann, Donald I., and Katherine W. Stuart. The Culture of Poplars in Eastern North America. East Lansing, Mich.; Dept. of Forestry, Michigan State University, 1983. 168p.	
Third	Dietrich, William. The Final Forest. The Battle for the Last Great Trees of the Pacific Northwest. New York and London; Simon & Schuster, 303p.	
Third	Dilworth, John R., and J. F. Bell. Log Scaling and Timber Cruising. Rev. ed. Corvallis, Oreg.; Oregon State University Book Stores, 1984. 468p. (Earlier ed., 1949. 233p.)	
First	Dinus, Ronald J., and Robert A. Schmidt, eds. Management of Fusiform Rust in Southern Pines; Proceedings of a Symposium	

Developed Countries Ranking		Third World Ranking
	and Carmen Garcia-Dowining, eds. Development or Destruction: The Conservation of Tropical Forest to Pasture in Latin America. Boulder, Colo.; Westview Press, 1992. 405p.	
Third	Dreyer, E., et al., eds. Forest Tree Physiology. Paris; Editions Scientifiques Elsevier, 1989. 875p. (English and French. Supplement to Vol. 46 of Annales des Sciences Forestieres.)	
Third	Dropkin, Victor H. Introduction to Plant Nematology. 2d ed. New York; J. Wiley, 1989. 304p. (1st ed., 1980. 293p.)	
Third	Duerr, William A. Forestry Economics as Problem Solving. Blacksburg, Va., and Syracuse, N.Y.; W. A. Duerr, 1988. (Earlier ed., 1985.)	
Second	Duerr, William A. Fundamentals of Forestry Economics. New York; McGraw-Hill, 1960. 579p.	Second
Third	Duerr, William A., et al., eds. Forest Resource Management: Decision-Making Principles and Cases. Philadelphia; Saunders, 1979. 611p.	Third
Second	Duryea, Mary L., ed. Evaluating Seedling Quality: Principles, Procedures, and Predictive Abilities of Major Tests; Proceedings of a Workshop, Oct. 1984. Corvallis; Forest Research Laboratory, Oregon State University, 1985. 143p.	
Second	Duryea, Mary L., and Gregory N. Brown, eds. Seedling Physiology and Reforestation Success; Proceedings of the Physiology Working Group Technical Session, Society of American Foresters National Convention, Portland, Oreg., Oct. 1983. Dordrecht and Lancaster, Penn.; M. Nijhoff, 1984. 326p.	
Third	Duryea, Mary L., and Phillip M. Dougherty, eds. Forest Regeneration Manual. Dordrecht and Boston; Kluwer Academic, 1991. 433p. (FOSC no. 36)	Third
First	Duryea, Mary L., and Thomas D. Landis, eds. Forest Nursery Manual: Production of Bareroot Seedlings. The Hague and Boston; M. Nijhoff, 1984. 385p.	First
	Duvigneaud, P., ed. Productivity of Forest Ecosystems; Proceedings of a Symposium, Brussels, Oct. 1969. Paris; UNESCO, 1971. 707p.	First
First	Dykstra, Dennis P. Mathematical Programming for Natural Resource Management. New York; McGraw-Hill, 1984. 318p.	First
Third	Dyne, Goerge M. Van, ed. The Ecosystem Concept in Natural Resource Management. New York; Academic Press, 1969. 383p.	Second
	Earl, Derek E. Forest Energy and Economic Development. Oxford, U.K.; Clarendon Press, 1975. 128p.	Second
	Easter, K. William, John A. Dixon, and Maynard M. Hufschmidt, eds. Watershed Resource Management: An Integrated Framework with Studies from Asia and the Pacific. Boulder, Colo.; Westview Press, 1986. 236p.	Third
Second	Edmonds, Robert L., ed. Analysis of Coniferous Forest Ecosys-	

Developed Countries Ranking		Third World Ranking
	tems in the Western United States. Stroudsburg, Pa.; Hutchinson Ross Pub. Co., 1981. 419p. (US/IBP Synthesis Series no. 14)	
Third	Ek, Alan R., Stephen R. Shifley, and Thomas E. Burk, eds. Forest Growth Modelling and Prediction; Proceedings of the IUFRO Conference, Minneapolis, Minn., Aug. 1987. St. Paul, Minn.; U.S. Dept. of Agriculture, Forest Service, North Central Forest Experiment Station, 1988. 2 vols.	
	El-Swaify, S. A., W. C. Moldenhauer, and Andrew Lo, eds. Soil Erosion and Conservation. Ankeny, Iowa; Soil Conservation Society of America, 1985. 793p.	First
Third	Elfring, Chris. Sustaining Tropical Forest Resources: U.S. and International Institutions. Washington, D.C.; U.S. Office of Technology Assessment, 1983. 65p.	Third
Third	Ellefson, Paul V., ed. Forest Resource Economics and Policy Research: Strategic Directions for the Future. Boulder, Colo.; Westview Press, 1989. 403p.	Second
Third	Ellefson, Paul V. Forest Resource Policy: Process, Participants, and Programs. New York; McGraw-Hill, 1992. 554p.	Third
Second	Ellefson, Paul V., and Robert N. Stone. U.S. Wood-Based Industry: Organization of the Nation's Timber Enterprises. New York; Praeger, 1984. 479p.	
Third	Eltringham, Stewart K. Wildlife Resources and Economic Development. Chichester U.K. and New York; J. Wiley, 1984. 325p.	Third
First	Encyclopedia of Plant Physiology. New series. Berlin and New York; Springer-Verlag, 1975–1987. 19 vols.	Second
Third	Engel, J. Ronald, and Joan G. Engel, eds. Ethics of Environment and Development: Global Challenge, International Response. London; Belhaven, 1992. 256p. (1st ed., 1990. 264p.)	Third
	Epstein, Emanuel. Mineral Nutrition of Plants: Principles and Perspectives. New York; J. Wiley, 1972. 412p.	Third
First	Esau, Katherine. Plant Anatomy. 2d ed. New York; J. Wiley, 1965. 767p. (1st ed., 1953. 735p.)	
Third	Estrada, Alejandro, and Theodore H. Fleming, eds. Frugivores and Seed Dispersal. Dordrecht and Boston; W. Junk, 1986. 392p.	Third
Second	Evans, Julian. Plantation Forestry in the Tropics. 2d ed. Oxford, U.K.; Clarendon Press; New York; Oxford University Press, 1992. 400p. (1st ed., 1982. 472p.)	First
	Evans, L. T. Environmental Control of Plant Growth; Proceedings of a Symposium, Canberra, Australia, August 1962;. New York; Academic Press, 1963. 449p.	Third
Third	Everhart, William C. The National Park Service. Boulder, Colo.; Westview Press, 1983. 197p.	
	Ewing, Andrew J., and Raymond Chalk. The Forest Industries	Third

Developed Countries Ranking		Third World Ranking
	Sector: An Operational Strategy for Developing Countries. Washington, D.C.; World Bank, 1988. 83p. (WB Technical Paper no. 83)	
Second	Eyre, F. H., ed. Forest Cover Types of the United States and Canada. Washington, D.C.; Society of American Foresters, 1980. 148p. (Earlier ed., 1954, as Forest Cover Types of North America, Exclusive of Mexico. 67p.)	
Second	Faegri, Knut, and L. van der Pijl. The Principles of Pollination Ecology. 3d ed., rev. Oxford, U.K. and New York; Pergamon Press, 1979. 244p. (1st ed., 1966. 248p.)	
Third	Fairfax, Sally K. Coming of Age in the Bureau of Land Management: Range Management in Search of a Gospel. Boulder, Colo.; Westview Press, 1984. 1766p.	
Third	Fairfax, Sally K., and Carolyn E. Yale. Federal Lands: A Guide to Planning, Management, and State Revenues. Washington, D.C.; Island Press, 1987. 252p.	
Third	Fanta, J., editor. Forest Dynamics Research in Western and Central Europe; Proceedings of the Workshop held Sept. 1985 in Wageningen, the Netherlands. Wageningen, Netheralnds; Centre for Agricultural Publishing and Documentation, 1986. 320p.	
	FAO Conservation Guide. Watershed Management Field Manual: Road Design and Construction in Sensitive Watersheds. Rome; Food and Agriculture Organization, 1989. 196p. (FAO Conservative Guide	Third
	FAO World Symposium on Man-Made Forests and Their Industrial Importance. Proceedings . . . Canberra, Australia, April 1967. Rome; Food and Agriculture Organization, 1967. 3 vols. (FAO/MMF no. 67)	Second
Third	FAO/IUFRO Symposium on Internationally Dangerous Forest Diseases and Insects. Proceedings of a Symposium, Oxford, July 1964. Rome; Food and Agriculture Organization, 1966. 2 vols.	
	Fearnside, Philip M. Human Carrying Capacity of the Brazilian Rain-Forest. New York; Columbia University Press, 1986. 293p.	Second
	Fedden, Matthew. Forest Farm Husbandry. Kumasi, Ghana; Technology Consultancy Centre, 1988. 70p.	Third
	Felker, Peter, ed. Tree Plantings in Semi-Arid Regions; Proceedings of a Symposium . . . Texas A. & I. University, Kingsville, Tex., April-May 1985. Amsterdam and Oxford, U.K.; Elsevier, 1986. 444p.	Third
	Fernandes, Walter, and Sharad Kulkarni, eds. Towards a New Forest Policy: People's Rights and Environmental Needs. New Delhi; Indian Social Institute, 1983. 155p.	Third
Second	Figueroa Colon, Julio C., et al., eds. Management of the Forests	First

Developed Countries Ranking		Third World Ranking
	of Tropical America: Prospects and Technologies; Proceedings of a Conference, San Juan, Sept. 1986. Rio Piedras, Puerto Rico; Institute of Tropical Forestry, Southern Forest Experiment Station, 1987. 469p.	
	Findlay, W. P. K., ed. Preservation of Timber in the Tropics. Dordrecht and Boston; M. Nijhoff/W. Junk, 1985. 273p.	Third
Third	Finland/UNEP International Seminar on Sound Environmental Management in the Pulp and Paper Industry, Helsinki, May 1986. Proceedings . . . Paris; United Nationa Environment Programme, 1987. 246p.	
Third	Finley, James C., and Margaret C. Brittingham, eds. Timber Management and Its Effects on Wildlife; Proceedings 1989 Penn. State Resources Issues Conference. University Park, Penn.; Pennsylvania State University, 1989. 296p.	
	Finnish Forest Research Institute. Sexual Reproduction of Forest Trees; Preceedings of an International Union of Forestry Research Organizations Meeting, Varparanta, Finland, 1970. Helsinki; IUFRO, 1970. 2 vols.	Second
Third	Fins, Lauren, Sharon T. Friedman, and Janet V. Brotschol, eds. Handbook of Quantitative Forest Genetics. Dordrecht, etc.; Kluwer Academic, 1991. 404p. (Forestry Sciences Series no. 39)	
Third	Fisher, Anthony C. Resource and Environmental Economics. Cambridge, U.K. and New York; Cambridge University Press, 1981. 284p.	Third
Third	Fitter, A. H., et al., eds. Ecological Interactions in Soil: Plants, Microbes and Animals; Proceedings of a British Ecological Society Meeting, York, April 1984. Oxford, U.K. and Boston; Blackwell Scientific, 1985. 451p. (BES Special Publication no. 4)	Second
Third	Flather, Curtis H., and Thomas W. Hoekstra. An Analysis of the Wildlife and Fish Situation in the United States, 1989–2040. Fort Collins, Colo.; USDA Forest Service, Rocky Mountain Forest and Range Experiment Station, 1989. 146p. (General Technical Report RM no. 178)	
	Flor, Hildebrando de Miranda. Florestas tropicais: Como Intervir sem devastar. Sao Paulo; Icone Editora, 1985. 180p.	Third
	Foley, Gerald, and Geoffrey Barnard. Farm and Community Forestry. London; Earthscan, 1984. 236p.	Second
	Food and Agricultural Organization. Forestry Extension Organization. Rome; Food and Agriculture Organization, 1986. 167p. (FAO Forestry Paper no. 66)	Third
	Food and Agriculture Organization. Databook on Endangered Tree and Shrub Species and Provenances. Rome; Food and Agriculture Organization, 1986. 524p. (FAO Forestry Paper no. 77)	Third

Developed Countries Ranking		Third World Ranking
Third	Food and Agriculture Organization. Design Manual on Basic Wood Harvesting Technology. Rome; Food and Agriculture Organization, 1989. 1 vols. (FAO Training Series no. 18)	Third
	Food and Agriculture Organization. Food and Fruit-Bearing Forest Species. Rome; Food and Agriculture Organization, 1986. 308p. (FAO Forestry Paper no. 44/3) (1st pub., 1983, FAO Forestry Paper 44/1.)	Third
	Food and Agriculture Organization. Forest Products, World Outlook Projections: Projections of Consumption and Production of Wood-Based Products to 2000. Rome; Food and Agriculture Organization, 1988. 350p. (FAO Forestry Paper no. 84) (Earlier ed., 1986. 101p. Available in French and Spanish.)	Second
Second	Food and Agriculture Organization. Forestry for Local Community Development. Rome; Food and Agriculture Organization, 1978. 114p. (FAO Forestry Paper no. 7)	Third
Third	Food and Agriculture Organization. Forestry Policies in Europe. Rome; Food and Agriculture Organization, 1988. 283p. (FAO Forestry Paper no. 86)	
	Food and Agriculture Organization. Fruit-Bearing Forest Trees: Technical Notes. Rome; Food and Agriculture Organization, 1982. 177p. (FAO Forestry Paper no. 34)	Second
	Food and Agriculture Organization. The Fuelwood Situation in Developing Countries. Rome; Food and Agriculture Organization, 1981. (One map, scale ca. 1:25,000,000; 62 x 106 cm.)	Third
	Food and Agriculture Organization. Guidelines for Forestry Information Processing: With Particular Reference to Developing Countries, based on the work of J. W. van Roessel. Rome; Food and Agriculture Organization, 1986. 139p. (FAO Forestry Paper no. 74)	Third
	Food and Agriculture Organization. Handling Forest Tree Seed. Rome; Food and Agriculture Organization, 1955. 109p. (FAO Forestry Development Paper no. 4)	Third
	Food and Agriculture Organization. Household Food Security and Forestry: An Analysis of Socio-Economic Issues. Rome; Food and Agriculture Organization, 1989. 147p.	Third
	Food and Agriculture Organization. Improved Production Systems as an Alternative to Shifting Cultivation. Rome; Food and Agriculture Organization, 1984. 201p. (FAO Soils Bulletin no. 53)	First
	Food and Agriculture Organization. Institutional Aspects of Shifting Cultivation in Africa. Rome; Food and Agriculture Organization, 1984. 171p.	Third
	Food and Agriculture Organization. Land Evaluation for Forestry. Rome; Food and Agriculture Organization, 1984. 123p. (FAO Forestry Paper no. 48)	Second
	Food and Agriculture Organization. Manual of Forest Inventory:	Third

Developed Countries Ranking		Third World Ranking
	With Special Reference to Mixed Tropical Forests. Rome; Food and Agriculture Organization, 1981. 200p. (FAO Forestry Paper no. 27) (1st ed., 1973.)	
Third	Food and Agriculture Organization. Modeling World Trade, Tropical Timber. Rome; Food and Agriculture Organization, 1983. 105p.	
Third	Food and Agriculture Organization. Monitoring and Evaluation of Participatory Forestry Projects. Rome; Food and Agriculture Organization, 1985. 133p. (FAO Forestry Paper no. 60)	Second
Third	Food and Agriculture Organization. Pulping and Paper-Making Properties of Fast-Growing Plantation Wood Species. 2d ed. Rome; Food and Agriculture Organization, 1980. 2 vols. (FAO Forestry Paper no. 19) (1st ed., 1975. 466p.)	Third
	Food and Agriculture Organization. Report on the Agro-Ecological Zones Project. Rome; Food and Agriculture Organization, 1978–1981. 4 vols.	Third
	Food and Agriculture Organization. Simple Technologies for Charcoal Making. Rome; Food and Agriculture Organization, 1983. 154p. (FAO Forestry Paper no. 41)	Third
	Food and Agriculture Organization. Small and Medium Sawmills in Developing Countries: A Guide for Their Planning and Establishment. Rome; Food and Agriculture Organization, 1981. 149p. (FAO Forestry Paper no. 28)	Second
	Food and Agriculture Organization. Small-Scale Forest-Based Processing Enterprises. Rome; Food and Agriculture Organization, 1987. 246p. (FAO Forestry Paper no. 79)	Third
	Food and Agriculture Organization. Some Medicinal Forest Plants of Africa and Latin America. Rome; Food and Agriculture Organization, 1986. 252p. (FAO Forestry Paper no. 67)	Third
	Food and Agriculture Organization. Strategies, Approaches, and Systems in Integrated Watershed Management. Rome; Food and Agriculture Organization, 1986. 232p. (FAO Conservation Guide no. 14)	Second
	Food and Agriculture Organization. Tropical Forestry Action Plan. Rome; Food and Agriculture Organization, 1987. 32p. (Earlier ed., 1985. 159p.)	First
	Food and Agriculture Organization. Wood Gas as Engine Fuel. Rome; Food and Agriculture Organization, 1986. 173p. (FAO Forestry Paper no. 72)	Third
	Food and Agriculture Organization. Wood Resources and Their Use as Raw Material. Vienna, Austria; United Nations, 1983. 177p. (Sectoral Studies Series/UNIDO Division for Industrial Studies no. 3)	Third
	Food and Agriculture Organization. World Consultation on Forestry Tree Breeding, 2d, Washington, D.C., 1969. Rome; Food and Agriculture Organization, 1970. 3 vols.	Third

Developed Countries Ranking		Third World Ranking
Third	Food and Agriculture Organization. World Forest Products Demand and Supply 1990 and 2000. Rome; Food and Agriculture Organization, 1982. 346p. (FAO Forestry Paper no. 29)	Second
First	Ford, E. D., D. C. Malcolm, and J. Atterson, eds. The Ecology of Even-Aged Forest Plantations; Proceedings of the International Union of Forestry Research Organizations, Division I Meeting, Edinburgh, Sept. 1978. Cambridge, U.K.; Institute of Terrestrial Ecology, Natural Environment Research Council, 1979. 582p.	First
Third	Forman, Richard t. T., ed. Pine Barrens: Ecosystem and Landscape. New York; Academic Press, 1979. 601p.	
Second	Forman, Richard T. T., and Michel Godron. Landscape Ecology. New York; J. Wiley, 1986. 619p.	
	Fortmann, Louise, James Riddell, and Steve Brick. Trees and Tenure: An Annotated Bibliography for Agroforesters and others. Madison, Wis; Land Tenure Center, University of Wisconsin; and Nairobi, Kenya: International Council for Research in Agroforestry, 1985. 135p.	Second
Third	Fowells, H. A., ed. Silvics of Forest Trees of the United States. Washington, D.C.; USDA, Forest Service, 1965. 762p. (USDA Agriculture Handbook no. 271)	
Third	Francis, Charles A., ed. Multiple Cropping Systems. New York; Macmillan; London; Collier Macmillan, 1986. 383p.	First
Second	Franklin, E. Carlyle. Pollen Management Handbook. Washington, D.C.; Dept. of Agriculture, Forest Service, 1981. 98p. (USDA Agriculture Handbook no. 587)	
Second	Franklin, Jerry F., L. J. Dempster, and Richard H. Waring, eds. Research on Coniferous Forest Ecosystems; Proceedings of a Symposium, Northwest Scientific Association 45th Annual Meeting, Mar. 1972. Portland, Oreg.; USDA, Pacific Northwest Forest and Range Experiment Station, 1972. 322p.	
	Frederick, W. James, ed. New Process Alternatives in the Forest Products Industries. New York; American Institute of Chemical Engineers, 1980. 225p. (AIChE Symposium Series no. 200)	Third
Third	Freeman, A. Myrick. The Benefits of Environmental Improvement: Theory and Practice. Baltimore; Published for Resources for the Future by Johns Hopkins University Press, 1979. 272p.	Third
Third	Freeman, A. Myrick, Robert H. Haveman, and Allen V. Kneese. The Economics of Environmental Policy. Malabar, Fla.; Krieger Pub. Co., 1984. 184p. (Originally published Santa Barbara, Calif.; Wiley, 1973.)	Second
Second	Fritts, Harold C. Tree Rings and Climate. London and New York; Academic Press, 1976. 567p.	

Developed Countries Ranking		Third World Ranking
Second	Frome, Michael. The Forest Service. 2d ed., rev. and updated. Boulder, Colo.; Westview Press, 1984. 364p.	
Second	Fujimori, Takao, and David Whitehead, eds. Crown and Canopy Structure in Relation to Productivity; Proceedings of an International Workshop, Japan, Oct. 1985. Ibaraki, Japan; Forestry and Forest Products Research Institute, 1986. 448p.	
Third	Funk, A. Parasitic Microfungi of Western Trees. Victoria, British Columbia; Canadian Forestry Service, Pacific Forest Research Centre, 1981. 190p.	
Second	Furniss, Robert L., and V. M. Carolin. Western Forest Insects. Washington, D.C.; USDA Forest Service, 1980. 654p. (USDA Miscellaneous Publication no. 1339) (Earlier ed., 1977.)	
	Furtado, Jose I., ed. Tropical Ecology and Development; Proceedings of the 5th International Symposium of Tropical Ecology, Kuala Lumpur, Malaysia, April 1979. Kuala Lumpur; International Society of Tropical Ecology, 1980. 2 vols.	First
Third	Furtado, Jose I., et al., eds. Tropical Resources: Ecology and Development. New York; Harwood Academic, 1990. 306p.	Second
	Gamser, Matthew S. Power from the People: Innovation, User Participation, and Forest Energy Development. London; IT Publications, 1988. 159p.	Third
	Garner, Willa Y., and John Harvey, eds. Chemical and Biological Controls in Forestry; Based on a Symposium sponsored by the Division of Pesticide Chemistry, 185th Meeting of the American Chemical Society, Seattle, Mar. 1983. Washington, D.C.; ACS, 1984. 406p.	Third
Third	Garrett, H. E. Gene, ed. Proceedings of the 2d Conference on Agroforestry in North America, Springfield, Mo., Aug. 1991. Columbia, Mo.; School of Natural Resources, University of Missouri, 1991. 403p.	Third
First	Gates, David M. Biophysical Ecology. New York; Springer-Verlag, 1980. 611p.	Second
Third	George, Edwin F., and Paul D. Sherrington. Plant Propagation by Tissue Culture: Handbook and Directory of Commercial Laboratories. Eversley, Basingstoke, U.K.; Exegetics, 1984. 709p.	
First	Gerhold, H. D., et al., eds. Disease and Insect Resistance of Forest Trees; Proceedings of a NATO/NSF Advanced Study Institute on Genetic Improvement for Disease and Insect Resistance of Forest Trees, University Park, Pa., 1964. Oxford, U.K. and New York; Symposium Pub. Div., Pergamon Press, 1966. 505p.	
Second	Gholz, Henry L., ed. Agroforestry: Realities, Possibilities, and Potentials. Dordrecht and Boston; M. Nijhoff in cooperation	First

Developed Countries Ranking		Third World Ranking
	with ICRAF; Hingham, Mass.; Kluwer Academic, 1987. 227p.	
	Ghosh, Rathis C. Handbook on Afforestation Techniques. Delhi, India; Controller of Publications, 1977. 411p.	Third
	Gibson, I. A. S. Diseases of Forest Trees Widely Planted as Exotics in the Tropics and Southern Hemisphere. Kew, U.K.; Commonwealth Mycological Institute, 1975.	Third
	Gibson, I. A. S., and Rodolfo Salinas Quinard. Notas sobre Enfermedades forestales y su Manejo = Noted on Forestry Diseases and Their Control. Mexico City; Instituto Nacional de Investigaciones Forestales, 1985. 196p. (INIF Boletin tecnico no. 106)	Third
Third	Giles, Robert H. Wildlife Management. San Francisco, Calif.; W. H. Freeman, 1978. 416p.	
Second	Givnish, Thomas J., ed. On the Economy of Plant Form and Function; Proceedings of the 6th Maria Moors Cabot Symposium, Evolutionary Constraints on Primary Productivity, Adaptive Patterns of Energy Capture in Plants, Harvard Forest, Aug. 1983. Cambridge, U.K. and New York; Cambridge University Press, 1986. 717p.	
Third	Glover, Nancy, and Norman Adams, eds. Tree Improvement of Multipurpose Species. Arlington, Va.; Winrock International Institute for Agricultural Development, 1990. 112p. (Multipurpose Tree Species Network Technical Series no. 2)	Second
	Goldsmith, Edward, and Nicholas Hildyard, eds. The Earth Report: The Essential Guide to Global Ecological Issues. Los Angeles; Price, Stern, Sloan, 1988. 240p.	Third
	Golley, Frank B., ed. Tropical Rain Forest Ecosystems. Amsterdam and New York; Elsevier, 1983–1989. 2 vols. (Ecosystems of the World no. 14A-B)	First
	Golley, Frank B., et al. Mineral Cycling in a Tropical Moist Forest Ecosystem. Athens, Ga.; University of Georgia Press, 1975. 148p.	Second
Second	Golley, Frank B., and Ernesto Medina, eds. Tropical Ecological Systems; Trends in Terrestrial and Aquatic Research. New York; Springer-Verlag, 1975. 398p.	Second
First	Gordon, J. C., and C. T. Wheeler, eds. Biological Nitrogen Fixation in Forest Ecosystems: Foundations and Applications. The Hague and Boston; M. Nijhoff/W. Junk, 1983. 342p.	First
First	Gordon, J. C., C. T. Wheeler, and D. A. Perry, eds. Symbiotic Nitrogen Fixation in the Management of Temperate Forests. Corvallis, Oreg.; Forest Research Laboratory, Oregon State University, 1979. 501p.	
Third	Gordon, John C., and William R. Bentley. A Handbook on the Management of Agroforestry Research. New Delhi, India; Winrock International and Oxford University, 1990. 72p.	

Developed Countries Ranking		Third World Ranking
Third	Gordon, William. The Law of Forestry. London, U.K.; H. M. Stationary Office, 1955. 574p.	
	Graaf, N. R. de. A Silvicultural System for Natural Regeneration of Tropical Rain Forest in Suriname. Wageningen; Agricultural University, 1986. 250p. (Ecology and Management of Tropical Rain Forests in Suriname no. 1)	Second
First	Grace, J., E. D. Ford, and P. G. Jarvis, eds. Plants and Their Atmospheric Environment; Proceedings of the 21st Symposium of the British Ecological Society. Oxford, U.K. and Boston; Blackwell Scientific, 1981. 419p.	Second
	Grainger, A. The Threatening Desert: Controlling Desertification. London; Earthscan, 1990. 382p.	Third
	Grant, William E. Systems Analysis and Simulation in Wildlife and Fisheries Sciences. New York; J. Wiley, 1986. 338p.	Third
	Gray, John W. Forest Revenue Systems in Developing Countries: Their Role in Income Generation and Forest Management Strategies. Rome; Food and Agriculture Organization, 1983. 261p. (FAO Forestry Paper no. 43)	Second
Third	Greeley, William B. Forest Policy. New York; McGraw-Hill, 1953. 278p.	
	Greenland, D. J., and R. Lal, eds. Soil Conservation and Management in the Humid Tropics; Proceedings of an International Conference . . . Ibadan, Nigeria, 1975. London and New York; J. Wiley, 1977. 233p.	First
	Gregersen, H. M., et al. Guidelines for Economic Appraisal of Watershed Management Projects. Rome; Food and Agriculture Organizatin, 1987. 144p. (FAO Conservation Guide no. 16)	Third
Third	Gregersen, H. M., and Arnoldo Contreras. U.S. Investment in the Forest-Based Sector in Latin America: Problems and Potentials. Baltimore, Md.; Published for Resources for the Future by Johns Hopkins University Press, 1975. 113p.	
Second	Gregerson, Hans, Sydney Draper, and Dieter Elz, eds. People and Trees: The Role of Social Forestry in Sustainable Development. Washington, D.C.; World Bank, 1989. 273p.	First
Third	Gregory, G. Robinson. Forest Resource Economics. New York; J. Wiley, 1972. 548p.	
First	Gregory, G. Robinson. Resource Economics for Foresters. New York; J. Wiley, 1987. 477p.	Second
Third	Grey, Gene W., and Frederick J. Deneke. Urban Forestry. 2d ed. New York; J. Wiley, 1986. 299p. (1st ed., 1978. 279p.)	Second
First	Guldin, Richard, and James P. Barnett, eds. Proceedings of the Southern Containerized Forest Tree Seedling Conference, Savannah, Ga., Aug. 1981. New Orleans, La.; USDA Forest Service, 1982. 156p. (General Technical Report SO no. 37)	
Third	Gunter, John E., and Harry L. Haney. Essentials of Forestry	Third

Developed Countries Ranking		Third World Ranking
	Investment Analysis. Corvallis, Oreg.; Oregon State University Book Stores, 1984. 337p.	
	Gupta, Avijit. Ecology and Development in the Third World. London and New York; Routledge, 1988. 80p.	Third
	Haig, Irvine T., M. A. Huberman, and U. Aung Din, preparers. Tropical Silviculture. Rome; Food and Agriculture Organization, 1958. 3 vols. (FAO Forestry and Forest Products Studies no. 13)	First
Third	Hakkila, P. Utilization of Residual Forest Biomass. London; Springer-Verlag, 1989. 568p.	
First	Halevy, Abraham H., ed. CRC Handbook of Flowering. Boca Raton, Fla.; CRC Press, 1985. 6 vols.	
	Hall, D. O., N. Myers, and N. S. Margaris, eds. Economics of Ecosystem Management. Dordrecht and Boston; W. Junk, 1985. 244p.	Third
	Hall, D. O., and R. P. Overend. Biomass: Regenerable Energy. Chichester, U.K.; J. Wiley, 1987. 1 vols.	Second
	Halle, Francis, R. A. A. Oldeman, and P. B. Tomlinson. Tropical Trees and Forests: An Architectural Analysis. Berlin and New York; Springer-Verlag, 1978. 441p.	First
Third	Halls, Lowell K., ed. White-Tailed Deer: Ecology and Management. Harrisburg, Pa.; Stackpole Books, 1984. 870p.	
	Hallsworth, E. G., ed. Socio-Economic Effects and Constraints in Tropical Forest Management: The Results of an Enquiry; Papers presented at the Workshop . . . , Dehra Dun, India, 1981. Chichester U.K. and New York; J. Wiley, 1982. 233p.	Second
Third	Hamel, Dennis R. Forest Management Chemicals: A Guide to Use When Considering Pesticides for Forest Management. Washington, D.C.; USDA Forest Service, 1983. 645p. (USDA Agriculture Handbook no. 585) (Earlier ed., 1981. 512p.)	
Third	Hamel, Margaret P., ed. Current Challenges to Traditional Wood Procurement Practices. Madison, Wis.; Forest Products Research Society, 1988. 96p.	
	Hamilton, Lawrence S., ed. Forest and Watershed Development and Conservation in Asia and the Pacific. Boulder, Colo.; Westview Press, 1983. 559p.	Third
	Hamilton, Lawrence S., and Michael Bonell, eds. Country Papers on Status of Watershed Forest Influence Research in Southeast Asia and the Pacific. Honolulu, Hawaii; East-West Center, 1982. 1 vol.	Third
	Hamilton, Lawrence S., and Peter N. King. Tropical Forested Watersheds: Hydrologic and Soils Response to Major Uses or Conversions; Report of the Watershed Forest Influence Workshop, East-West Center, Honolulu, Sept. 1981. Boulder, Colo.; Westview Press, 1983. 168p.	Second

Developed Countries Ranking		Third World Ranking
	Hamilton, Lawrence S., and Samuel C. Snedaker, eds. Handbook for Mangrove Area Management. Honolulu, Hawaii; Environment and Policy Institute, East-West Center, 1984. 123p.	Third
Third	Handbook of Plant Cell Culture.New York; Macmillan, 1983–1990. 6 vols.	Third
Third	Hanxi, Yang, et al., eds. The Temperate Forest Ecosystem; Proceedings on an International Symposium, Changbai Mountain Research Station, Academia Sinica, Antu, Jilin Province, People's Republic of China, July 1986. Grange-over Sands, UK; Institute of Terrestrial Ecology, 1987. 189p.	
	Hardy, Frederick. Edafologia Tropical. Mexico City; Herrero Hermanos, 1979. 416p.	Third
Second	Harlow, William M., et al. Textbook of Dendrology: Covering the Important Forest Trees of the United States and Canada. 7th ed. New York; McGraw-Hill, 1991. 501p. (1st ed., 1937. 557p.)	
Second	Harper, John L. Population Biology of Plants. London and New York; Academic Press, 1977. 892p.	
Third	Harrington, T. C., and F. W. Cobb, eds. Leptographium Root Diseases on Conifers. St. Paul, Minn.; American Phytopathological Society Press, 1988. 149p.	
Third	Harris, Esmond, and Jeanette Harris. Wildlife Conservation in Managed Woodlands and Forests. Oxford, U.K. and Cambridge, U.K.; Blackwell, 1991. 358p.	
First	Harris, Larry D. The Fragmented Forest: Island Biogeography Theory and the Preservation of Biotic Diversity. Chicago; University of Chicago Press, 1984. 211p.	
	Harris, Richard W. Arboriculture: Integrated Management of Landscape Trees, Shrubs, and Vines. 2d ed. Englewood Cliffs, N.J.; Prentice-Hall, 1991. 674p. (1st ed., 1983, as Arboriculture: Care of Trees, shrubs, and Vines in the Landscape. 688p.)	Third
	Harrison, Paul. The Greening of Africa: Breaking Through in the Battle for Land and Food. London; Paladin for International Institute for Environment and Development-Earthscan, 1987. 380p.	Second
	Hart, Robert A. de J. Forest Gardening. Green Books, 1991. 211p.	Third
Second	Hartig, Robert. Important Diseases of Forest Trees: Contributions to Mycology and Phytopathology for Botanists and Foresters . . . trans. by William Merrill, David H. Lambert, and Walter Liese. St. Paul; American Phytopathological Society, 1975. 120p.	
	Hartley, Charles W. S. The Oil Palm (Elaeis Guineensis Jacq.). 3d ed. London; Longman, 1988. 761p. (1st ed., 1967. 706p.)	Third
First	Hartmann, Hudson T., et al. Plant Propagation: Principles and	First

Developed Countries Ranking		Third World Ranking
	Practices. 5th ed. Englewood Cliffs, N.J.; Prentice-Hall, 1990. 647p. (1st ed., 1959. 559p.)	
	Haugen, Christine, Lee Medema, and Celso B. Lantican, eds. Multipurpose Tree Species Research for Small Farms: Strategies and Methods; Proceedings of an International Conference, Jakarta, Indonesia, Nov. 1989. Arlington, Va.; Forestry/Fuelwood Research and Development Project; Ottawa; International Development Research Centre of Canada, 1990. 217p. (Multipurpose Tree Species Network Proceedings Series no. 7)	Second
Second	Hawksworth, Frank G., and Robert F. Scharpf. Biology of Dwarf Mistletoes; Proceedings of a Symposium, Colorado State University, Fort Collins. Fort Collins, Colo.; USDA Forest Service, Rocky Mountain Forest and Range Experiment Station, 1984. 131p. (General Technical Report RM no. 111)	
Second	Hawksworth, Frank G., and Delbert Wiens. Biology and Classification of Dwarf Mistletoes (Arceuthobium). Washington, D.C.; U.S. Forest Service, 1972. 234p. (USDA Agriculture Handbook no. 401)	
Third	Hay, Ronald L., Frank W. Woods, and Hal DeSelm, eds. Central Hardwood Forest Conference, 6th, Knoxville, 1987; Proceedings . . . Knoxville, Tenn.; Dept. of Forestry, Wildlife and Fisheries, 1987. 526p.	
Third	Haygreen, John G., and Jim L. Bowyer. Forest Products and Wood Science: An Introduction. Ames, Iowa; Iowa State University Press, 1982. 495p.	
First	Haynes, Richard W. An Analysis of the Timber Situation in the United States: 1989–2040. Fort Collins, Colo.; USDA Forest Service, 1991. 268p. (General Technical Report RM no. 199) (Earlier ed., 1980. 821p.)	
	Hecht, Susanna B., ed. Amazonia: Agriculture and Land Use Research; Proceedings of an International Conference sponsored by the Rockefeller Foundation. Cali, Colombia; Centro Internacional de Agricultura Tropical, 1982. 428p.	First
	Hecht, Susanna B., and Alexander Cockburn. The Fate of the Forest: Developers, Destroyers, and Defenders of the Amazon. London and New York; Verso, 1989. 266p.	Second
Second	Hedlin, Alan F., et al. Cone and Seed Insects of North American Conifers. Ottawa; Environment Canada, Canadian Forestry Service; Washington, D.C.; USDA Forest Service, 1980. 122p.	
	Hedstrom, Ingemar, ed. La Situacion ambiental en Centroamerica y el Caribe. San Jose, C.R.; Departamento Ecumenico de Investigaciones, 1989. 318p.	Third
	Hegde, N. G., and P. D. Abhyankar, ed. The Greening of Waste-	Third

lands; Proceedings of a National Workshop on Utilisation of Wastelands for Bio-Energy, Pune, April 1985. Pune, India; Bharatiya Agro Industries Foundation, 1986. 204p.

Third   Heilman, Paul E., Harry W. Anderson, and David M. Baumgartner, compilers and eds. Forest Soils of the Douglas-Fir Region. Pullman, Wa.; Washington State University, 1981. 289p.

Hemming, John, ed. Change in the Amazon Basin. Manchester, U.K. and Dover, N.H.; Manchester University Press, 1985. 2 vols.   Third

Second   Hendee, John C., George H. Stankey, and Robert C. Lucas. Wilderness Management. 2d ed., rev. Fort Collins, Colo.; International Wilderness Leadership Foundation, 1990. 546p. (1st ed., Washington, D.C.; USDA Forest Service, 1978. 381p.)   Third

Second   Henke, Randolph R., et al., eds. Tissue Culture in Forestry and Agriculture. New York; Plenum Press, 1985. 390p.   Third

Second   Hennessy, Thomas, et al. Stress Physiology and Forest Productivity. Boston; Nijhoff Publishers, 1986. 239p.

Second   Hepting, George H. Diseases of Forest and Shade Trees of the United States. Washington, D.C.; USDA Forest Service, 1971. 658p. (USDA Agriculture Handbook no. 386)

Third   Herfindahl, Orris C., and Allen V. Kneese. Economic Theory of Natural Resources. Columbus, Ohio; Merrill, 1974. 405p.   Third

Third   Hesketh, J. D., and James W. Jones, eds. Predicting Photosynthesis for Ecosystem Models. Boca Raton, Fla.; CRC Press, 1980. 2 vols.

Heuveldop, J., and J. Lagemann, eds. Agroforestry; Proceedings of a Seminar held at CATIE, Turrialba, Feb.-Mar. 1981. Turrialba, Costa Rica; CATIE, 1984. 112p.   Third

Second   Hewitt, Oliver H., ed. The Wild Turkey and Its Management. Washington, D.C.; Wildlife Society, 1967. 589p.

Third   Hewlett, John D. Principles of Forest Hydrology. Athens, Ga.; University of Georgia Press, 1982. 183p.

Second   Heybroek, H. M., B. R. Stephan, and K. von Weissenberg, eds. Resistance to Diseases and Pests in Forest Trees; Proceedings of the 3d International Workshop on the Genetics of Host-Parasite Interactions in Forestry, Wageningen, Sept. 1980. Wageningen; Centre for Agricultural Pub. & Documentation, 1982. 503p.   Third

Third   Higuchi, Takayoshi, ed. Biosynthesis and Biodegradation of Wood Components. Orlando, Fla.; Academic Press, 1985. 679p.

Third   Hillis, W. E. Heartwood and Tree Exudates. Berlin and New York; Springer-Verlag, 1987. 268p.

Third   Hillis, W. E., ed. Wood Extractives and Their Significance to the

Developed Countries Ranking		Third World Ranking
	Pulp and Paper Industries. New York; Academic Press, 1962. 513p.	
Second	Hillis, W. E., and A. G. Brown, eds. Eucalypts for Wood Production. East Melbourne, Australia; Commonwealth Scientific and Industrial Research Organization; Sydney and Orlando; Academic Press, 1984. 434p.	Second
Third	Hinckley, Thomas M., J. P. Lassoie, and S. W. Running. Temporal and Spatial Variations in the Water Status of Forest Trees. Washington, D.C.; Society of American Foresters, 1978. 72p.	
Third	Hocker, Harold W. Introduction to Forest Biology. New York; J. Wiley, 1979. 467p. (Introduction to Forest Biology: A Manual in Silvics. Durham, N.H.: University of New Hampshire, 1973. 201p.)	
	Holdgate, M. W., and M. J. Woodman, eds. The Breakdown and Restoration of Ecosystems; Proceedings of a Conference on the Rehabilitation of Severely Damaged Land and Freshwater Ecosystems in Temperate Zones, Reykjavik, Iceland, July 1976. New York; Plenum Press, 1978. 496p. (NATO Conference Series. I, Ecology no. 3)	Third
	Holdridge, L. R., et al. Forest Environments in Tropical Life Zones; A Pilot Study. Oxford, U.K. and New York; Pergamon Press, 1971. 747p.	First
	Holdridge, Leslie R., and Luis J. Poveda. Arboles de Costa Rica. San Jose, Costa Rica; Centro Cientifico Tropical, 1975. 2 vols.	Third
Third	Holland, Israel I., G. L. Rolfe, and David A. Anderson. Forests and Forestry. 4th ed. Danville, Ill.; Interstate, 1990. 476p. (1st ed., 1970. 357p.)	
	Holm-Nielsen, L. B., et al. Tropical Forests: Botanical Dynamics, Speciation and Diversity. New York; Academic Press, 1989. 390p.	Third
	Holzer, W., M. J. A. Werger, and I. Ikusima, eds. Man's Impact on Vegetation. The Hague and Boston; W. Junk, 1983. 370p.	Third
Third	Hon, David, and Nobuo Shiraishi, editors. Wood and Cellulosic Chemistry. New York; M. Dekker, Inc., 1991. 1020p.	
Third	Hoover, Robert L., and Dale L. Wills, eds. Managing Forested Lands for Wildlife. Denver, Colo.; Colorado Division of Wildlife, 1984. 459p.	
Second	Horn, Henry S. The Adaptive Geometry of Trees. Princeton, N.J.; Princeton University Press, 1971. 144p. (Monographs in Population Biology no. 3)	
	Horsfall, James G., and Ellis B. Cowling, eds. Plant Disease: An Advanced Treatise. New York; Academic Press, 1977–1988. 3 vols.	Third
	Howes, K. M. W., and R. A. Rummery, eds. Integrating Agri-	Second

Developed		Third
Countries		World
Ranking		Ranking

	culture and Forestry; Proceedings of a Workshop, Bunbury, W.A., Sept. 1977. Dehra Dun, India; Natraj Pub., 1984. 238p.	
	Hueck, Kurt. Los Bosques de Sudamerica: Ecologia, Composicion e Importancia Economica. Eschborn, Germany; Sociedad Alemania de Cooperacion Tecnica, 1978. 476p. (Translation of Die Walder Sudamerikas)	Third
	Hufschmidt, Maynard M., et al. Environment, Natural Systems, and Development: An Economic Valuation Guide. Baltimore and London; Johns Hopkins University Press, 1983. 338p.	Third
Third	Hughes, M. K. et al., eds. Climate from Tree Rings; Based on the 2d International Workshop on Global Dedroclimatology, Universitt of East Anglia, Norwich, U.K., July 1980. Cambridge, U.K. and New York; Cambridge University Press, 1982. 233p.	
Third	Hummel, F. C. Forestry Policies in Europe: An Analysis. Rome; Food and Agriculture Organization, 1989. 120p. (FAO Forestry Paper no. 92)	
Third	Hunt, George M., and George A. Garratt. Wood Preservation. 3d ed. New York; McGraw-Hill, 1967. 433p. (1st ed., 1938. 457p.)	
	Hunter, Malcolm L. Wildlife, Forests, and Forestry: Principles of Managing Forests for Biological Diversity. Englewood Cliffs, N.J.; Prentice-Hall, 1990. 370p.	Second
Second	Husch, Bertram, Charles I. Miller, and Thomas W. Beers. Forest Mensuration. 3d ed. New York; J. Wiley, 1982. 402p. (1st ed., as Forest Mensuration and Statistics. New York; Ronald Press Co., 1963. 474p.)	
Second	Hutchinson, T. C., and K. M. Meema, eds. Effects of Atmospheric Pollutants on Forests, Wetlands, and Agricultural Ecosystems; Proceedings of a NATO Advanced Research Workshop, Toronto, 1985. Berlin and New York; Springer-Verlag, 1987. 652p. (NATO ASI Series G, Ecological Sciences no. 16)	
Third	Hutchison, B. A., and B. B. Hicks, eds. The Forest-Atmosphere Interaction; Proceedings of a Forest Environmental Measurements Conference, Oak Ridge, Tenn., Oct. 1983. Dordrecht and Boston; D. Reidel, 1985. 684p.	
	Huxley, Peter A., ed. Methodology for the Exploration and Assessment of Multipurpose Trees. Nairobi, Kenya; International Council for Research in Agroforestry, 1983. 1 vols.	First
	Huxley, Peter A., ed. Plant Research and Agroforestry; Proceedings of a Consultative Meeting, Nairobi, April 1981. Nairobi, Kenya; International Council for Research in Agroforestry, 1983. 617p.	First
	Huxley, Peter A., and Sidney B. Westley, eds. Multipurpose	Second

Developed Countries Ranking		Third World Ranking
	Trees: Selection and Testing for Agroforestry. Nairobi, Kenya; International Council for Research in Agroforestry, 1989. 120p.	
First	Hyde, William F., and D. H. Newman. Forest Economics and Policy Analysis: An Overview. Philadelphia, Pa.; World Bank, 1991. 92p.	
	Hyde, William F., ed. Economic Evaluation of Investments in Forestry Research. Durham, N.C.; Acorn Press, 1983. 106p.	Third
Second	Hyde, William F. Timber Supply, Land Allocation, and Economic Efficiency. Baltimore; Published for Research for the Future by the Johns Hopkins University Press, 1980. 224p.	Second
	International Board for Soil Research and Management. Tropical Land Clearing for Sustainable Agriculture; Proceedings of an IBSRAM Inaugural Workshop, Jakarta and Bukittinggii, Indonesia, Aug.-Sept. 1985. Bangkok, Thailand; IBSRAM, 1987. 226p. (IBSRAM Proceedings no. 3)	Second
	International Conference on Biomass, 1st, Brighton, East Sussex, Eng., 1980. Energy from Biomass; Proceedings . . . , edited by W. Palz, P. Chartier, and D. O. Hall. London; Applied Science Publishers, 1981. 982p.	Third
Third	International Poplar Commission. Poplars and Willows in Wood Production and Land Use. Rome; Food and Agriculture Organization, 1979. 328p. (FAO Forestry Series no. 10) (Earlier ed., 1958, as Poplars in Forestry and Land Use. FAO Forestry and Forest Products Studies no. 12)	
	International Seminar on Shelterbelts. Proceedings . . . Tunis, Tunisia, Oct.-Nov. 1983. Ottawa; International Development Research Centre, 1985. 251p. (IDRC Manuscript Report no. 117e,f)	Third
	International Symposium on Remote Sensing of Environments, 20th, Nairobi, Kenya, 1986. Proceedings . . . Ann Arbor, Mich.; Environmental Research Institute of Michigan, 1986. 1502p. 3 vols.	Third
	International Symposium on Tropical Forests Utilization and Conservation, Yale University, New Haven, Conn., 1980. Proceedings., edited by Francois Mergen. New Haven, Conn.; Yale University, School of Forestry and Environmental Studies, 1981. 199p.	Third
	Islam, A. S., ed. Improvement of Tropical Crops through Tissue Culture; Proceedings of an International Workshop, Dhaka, Bangladesh, Mar. 1981. Dhaka, Bangladesh; Office of the Bangladesh Journal of Botany, 1981. 137p.	Third
Third	IUFRO Conference, Williamsburg, Va., 1986. Proceedings of a Joint Meeting of Working Parties on Breeding Theory, Progeny Testing, Seed Orchards. Raleigh, N.C.; North Carolina State University, 1986. 673p.	Second

Developed Countries Ranking		Third World Ranking
Third	IUFRO Division 4 Conference, Thessaloniki, Greece, 1984. Policy Analysis for Forestry Development; Proceedings . . . Thessaloniki, Greece; International Union of Forestry Research Organizations, 1984. 2 vols.	Third
Third	IUFRO Joint Meeting of Working Parties on Genetics about Breeding Strategies including Multiclonal Varieties, Sensenstein, Germany, 1982. Proceedings . . . Staufenberg-Escherode, Germany; Lower Saxony Forest Research Institute, Dept. of Forest Tree Breeding, 1982. 238p.	
	IUFRO Symposium on Breeding and Yield of Fast Growing Trees, Aguas de Sao Pedro, Brazil, 1980. Fast Growing Trees (IUFRO em Melhoramento Genetico e Produtividade de Especies Florestais de Rapido Crescimento: Anais). Sao Paulo, Brazil; Sociedade Brasileira de Silvicultura, 1983. 4 vols. in 1 (Text in Portuguese, English, or French)	Third
Third	IUFRO Symposium on Flowering and Seed Bearing in Forest Seed Orchards, Kornik, Poland, 1985. Proceedings . . . Kornik; Institute of Dendrology, Polish Academy of Sciences, 1985. n.p.	
Third	IUFRO Symposium on Forest Management Planning and Managerial Economics, University of Tokyo, 1984. Forest Planning for Improved Management: Research and Practice; Proceedings of a Symposium. Tokyo; Forest Management Laboratory, Dept. of Forestry, University of Tokyo, 1984. 812p.	Second
Second	IUFRO World Congress, 17th, Kyoto, 1981. Proceedings . . . Ibaraki, Japan; Japanese International Union of Forest Research Organizations Congress Committee, 1981. 7 vols.	
Second	IUFRO World Congress, 18th, Ljubljana, Yugoslavia, 1986. Congress Report. Ljubljana, Yugoslavia; IUFRO, 1986. 683p.	
	IUFRO/MAB Conference. Research on Multiple Use of Forest Resources; Proceedings . . . Flagstaff, Ariz., May 1980. Washington, D.C.; USDA Forest Service, 1980. 161p. (USDA General Technical Report, WO no. 25)	Third
	Ives, J., and D. C. Pitt, eds. Deforestation: Social Dynamics in Watersheds and Mountain Ecosystems. London and New York; Routledge, 1988. 247p.	Second
Third	Jackson, David H. The Microeconomics of the Timber Industry. Boulder, Colo.; Westview Press, 1980. 136p.	
	Jackson, J. K., ed. Social, Economic, and Institutional Aspects of Agro-Forestry. Tokyo, Japan; United Nations University, 1984. 97p.	Third
	Jacobs, Marius. The Tropical Rain Forest: A First Encounter . . . edited by Remke Kruk, et al. Berlin and New York; Springer-Verlag, 1988. 295p. (Translation of Tropisch Regenwoud, een eerste Kennismaking.)	First

Developed Countries Ranking		Third World Ranking
	Jacobs, Maxwell R. Eucalypts for Planting. Rev. ed. Rome; Food and Agriculture Organization, 1979. 677p. (FAO Forestry Paper no. 11) (Earlier ed., by Andre Metro.)	Third
Second	Jacobson, Jay S., and A. Clyde Hill, eds. Recognition of Air Pollution Injury to Vegetation: A Pictorial Atlas. Pittsburgh, Pa.; Air Pollution Control Association, 1970. 109p. (APCA Informative Report TR-7 no. 1)	
	Janzen, Daniel H. Ecology of Plants in the Tropics. London; E. Arnold, 1975. 66p. (Institute of Biology's Studies in Biology no. no 58)	Third
Third	Japanese Forest Agency, Ministry of Agriculture, Forestry and Fisheries. Forest Technology in Japan. Tokyo; Forest Agency, 1981. 199p.	
Third	Jarvis, P. G., ed. Agroforestry: Principles and Practices; Proceedings of an International Conference, University of Edinburgh, July 1989. Amsterdam; Elsevier, 1991. 356p.	First
Second	Jarvis, P. G., and T. A. Mansfield, eds. Stomatal Physiology. Cambridge, U.K. and New York; Cambridge University Press, 1981. 295p. (Society for Experimental Biology, Seminar Series no. 8)	
Third	Jennings, D. H., ed. Integration of Activity in the Higher Plant. Cambridge, U.K. and New York; Cambridge University Press, 1977. 539p.	Third
Second	Johansson, Per-Olov, and Karl-Gustaf Lofgren. The Economics of Forestry and Natural Resources. Oxford, U.K. and New York; Basil Blackwell, 1985. 292p.	Third
Second	Johnson, Dale W., and Steven E. Lindberg, eds. Atmospheric Deposition and Forest Nutrient Cycling: A Synthesis of the Integrated Forest Study. New York; Springer-Verlag, 1992. 707p.	
Third	Johnson, Jay A., and W. Ramsay Smith, eds. Forest Products Trade: Market Trends and Technical Developments; Proceedings of the 3d International Symposium on World Trade in Forest Products, University of Washington, 1987. Seattle; University of Washington Press, 1988. 244p.	
Third	Johnson, Rebecca L., and Gary V. Johnson, eds. Economic Valuation of Natural Resources: Issues, Theory, and Applications. Boulder, Colo.; Westview Press, 1990. 220p.	
First	Johnson, Warren T., and Howard H. Lyon. Insects That Feed on Trees and Shrubs. 2d ed., rev. Ithaca, N.Y.; Comstock Pub. Associates, 1991. 560p. (1st ed., 1976. 464p.)	
Third	Johnston, George M., and Peter M. Emerson, eds. Public Lands and the U.S. Economy: Balancing Conservation and Development; Proceedings of a Conference . . . sponsored by the Wilderness Society. Boulder, Colo.; Westview Press, 1984. 309p.	

Developed Countries Ranking		Third World Ranking
Second	Jones, Hamlyn G., T. J. Flowers, and M. B. Jones, eds. Plants under Stress. Cambridge, U.K.; Cambridge University Press, 1989. 1 vols. (Society for Experimental Biology Seminar Series no. 39)	
	Jonkers, W. B. J. Vegetation Structure, Logging Damage and Silviculture in a Tropical Rain Forest in Suriname. Wageningen; Agricultural University, 1987. 172p. (Ecology and Management of Tropical Rain Forests in Suriname no. 3) (Summary in Dutch and English.)	Third
	Jordan, Carl F. Amazonian Rain Forests: Ecosystem Disturbance and Recovery: Case Studies of Ecosystem Dynamics under a Spectrum of Land Use-Intensities. New York; Springer-Verlag, 1987. 133p.	Third
	Jordan, Carl F. Nutrient Cycling in Tropical Forest Ecosystems: Principles and Their Application in Management and Conservation. Chichester U.K. and New York; J. Wiley, 1985. 190p.	Second
Third	Jozsa, L. A. et al. How Climate Effects Tree Growth in the Boreal Forest. Edmonton, Alta.; Northern Forest Research Centre, Canadian Forestry Service, Environment Canada, 1984. 67p.	
Third	Jubenville, Alan, Ben W. Twight, and Robert H. Becker. Outdoor Recreation Management: Theory and Application. Rev. and enlarged ed. State College, Pa.; Venture Pub., 1987. 219p.	
Third	Justice, Oren L., and Louis N. Bass. Principles and Practices of Seed Storage. Washington, D.C.; U.S. Govt. Print. Off., 1978. 289p. (USDA Agriculture Handbook no. 506)	
	Kairiukstis, L., S. Nilsson, and A. Straszak, eds. Forest Decline and Reproduction: Regional and Global Consequences; Proceedings of a Workshop, Krakow, Poland, Mar. 1987. Laxenburg, Austria; International Isntitute for Applied Systems Analysis, 1987. 670p.	Third
Second	Kallio, Markku et al., eds. Systems Analysis in Forestry and Forest Industries. Amsterdam and New York; North-Holland, 1986. 487p. (TIMS Studies in the Manamgement Sciences no. 21)	
First	Kallio, Markku, Dennis P. Dykstra, and Clark S. Binkley, eds. The Global Forest Sector: An Analytical Perspective. Chichester U.K.; J. Wiley, 1987. 717p.	First
Third	Kallman, Harmon et al., eds. Restoring America's Wildlife, 1937–1987: The First 50 Years of the Federal Aid in Wildlife Restoration (Pittman-Robertson) Act. Washington, D.C.; U.S. Dept. of the Interior, Fish and Wildlife Service, 1987. 394p.	
Second	Kang, B. T., and J. van der Heide, eds. Nitrogen Management in Farming Systems in Humid and Subhumid Tropics. Haren,	Second

Developed Countries Ranking		Third World Ranking
	Netherlands; Institute for Soil Fertility; Ibadan, Nigeria; International Institute of Tropical Agriculture, 1985. 362p.	
	Kang, B. T., and L. Reynolds, eds. Alley Farming in the Humid and Subhumid Tropics; Proceedings of an International Workshop, Ibadan, Nigeria, Mar. 1986. Ottawa, Canada; International Development Research Centre, 1989. 251p.	First
Second	Kaufman, Herbert. The Forest Ranger, A Study in Administrative Behavior. Baltimore, Md.; Published for Resources for the Future by John Hopkins University Press, 1967. 259p. (Earlier ed., 1960. 259p.)	
Third	Keating, William G., and Eleanor Bolza. Characteristics, Properties, and Uses of Timbers. Melbourne, Australia; Inkata Press, 1982.	
	Keller, Reiner, ed. Hydrology of Humid Tropical Regions: Aspects of Tropical Cyclones, Hydrological Effects of Agriculture and Forestry Practice; Proceedings of a Symposium held during the 18th General Assembly of the International Union of Geodesy and Geophysics, Hamburg, Aug. 1983. Washington, D.C.; International Association of Hydrological Sciences, 1983. 468p.	Third
Third	Kelty, Matthew J., Bruce C. Larson, and Chadwick D. Oliver, eds. The Ecology and Silviculture of Mixed-Species Forests. Boston; Kluwer Academic Press, 1992. 287p. (Forestry Sciences no. no. 40)	
	Kerkhof, Paul. Agroforestry in Africa: A Survey of Project Experience. London; Panos Pub., 1990. 216p.	First
	Kershaw, Kenneth A., and John H. H. Looney. Quantitative and Dynamic Plant Ecology. 3d ed. London; E. Arnold, 1985. 282p. (Earlier ed., 1973. 308p.)	Second
	Keswani, C. L., and B. J. Ndunguru, eds. Intercropping; Proceedings of the 2d Symposium on Intercropping in Semi-Arid Areas, Norogoro, Tanzania, Aug. 1980. Ottawa, Canada; International Development Research Centre, 1982. 168p. (IDRC no. 186e)	Third
Second	Khan, A. A., ed. The Physiology and Biochemistry of Seed Development, Dormancy, and Germination. Amsterdam and New York; Elsevier Biomedical Press, 1982. 547p. (Earlier ed., 1977, as The Physiology and Biochemistry of Seed Dormancy and Germination.)	Second
Second	Khosla, P. K., ed. Advances in Forest Genetics. New Delhi, India; Ambika, 1981. 375p.	First
	Khosla, P. K., and D. K. Khurana, eds. Agroforestry for Rural Needs; Proceedings of a Workshop. Solan, India; Indian Society of Tree Scientists, 1987.	Second
	Khosla, P. K., and Sunil Puri. Agroforestry Systems, a New	Third

Developed Countries Ranking		Third World Ranking
	Challenge; Proceedings of the Satellite Seminar on Agroforestry . . . organized at the Birla Institute of Technology, Ranchi, Jan. 1984. Solan, India; Indian Society of Tree Scientists, 1986. 286p.	
	Khosla, P. K., and R. N. Sehgal, eds. Trends in Tree Sciences; Proceedings of the National Symposium on Advances in Tree Sciences, Forest Research Institute and Colleges, Dehra Dun, April 1983. Solan, India; Indian Society of Tree Scientists, 1988. 350p.	Third
Second	Kimmins, J. P. Forest Ecology. New York and London; Macmillan & Collier, 1987. 531p.	Third
Third	Kneese, Allan V., and James L. Sweeney, eds. Handbook of Natural Resource and Energy Economics. Amsterdam and New York; North-Holland, 1985. 2 vols.	
First	Knight, Fred B., and Herman J. Heikkenen. Principles of Forest Entomology. 5th ed. New York; McGraw-Hill, 1980. 461p. (1st ed., 1929, by Samuel A. Graham. 339p.)	
	Knight, W. J., and J. D. Holloway, eds. Insects and the Rain Forests of South East Asia (Wallacea). London; Royal Entomological Society, 1990. 343p.	Third
Third	Koch, Peter. Utilization of Hardwoods Growing on Southern Pine Sites. Washington, D.C.; USDA Forest Service, 1985. 3 vols. (USDA Agriculture Handbook no. 605)	
First	Koch, Peter. Utilization of the Southern Pines. Washington, D.C.; USDA Forest Service, Southern Forest Experiment Station, 1972. 2 vols. (USDA Agriculture Handbook no. 420)	
	Kolari, Kimmo K., ed. Growth Disturbances of Forest Trees; Proceedings of an International Workshop, Jyvaskyla and Kivisuo, Finland, Oct. 1982 = Metsapuiden kasvuhairiot. Helsinki, Finland; Finnish Forest Research Institute, 1983. 208p.	Third
Second	Kollmann, Franz F. P. Principles of Wood Science and Technology. Berlin and New York; Springer-Verlag, 1984. 592p. (Reprint of 1968 ed.)	First
Third	Kossuth, Susan V., and Steve D. Ross, eds. Hormonal Control of Tree Growth; Proceedings of the Physiology Working Group Technical Session, Society of American Foresters National Convention, Birmingham, Ala., Oct. 1986. Dordrecht and Boston; Nijhoff; Hingham, Mass.; Kluwer Academic, 1987. 243p. (Reprinted from Plant Growth Regulation, vol. 6, nos. 1–2.)	
	Kotschi, Johannes, ed. Ecofarming Practices for Tropical Smallholdings. Weikersheim, Germany; J. Margraf, 1990. 185p. (Tropical Agroecology no. 5)	Second

Developed Countries Ranking		Third World Ranking
Second	Kozlowski, T. T., ed. Flooding and Plant Growth. Orlando, Fla.; Academic Press, 1984. 356p.	
First	Kozlowski, T. T., ed. Seed Biology. New York; Academic Press, 1972. 3 vols.	First
Third	Kozlowski, T. T. Tree Growth and Environmental Stresses. Seattle, Wa.; University of Washington Press, 1979. 192p.	
First	Kozlowski, T. T., ed. Water Deficits and Plant Growth. New York; Academic Press, 1968. 7 vols.	
First	Kozlowski, T. T., and C. E. Ahlgren, eds. Fire and Ecosystems. New York; Academic Press, 1974. 542p.	
Second	Kozlowski, T. T., Paul J. Kramer, and Stephen G. Pallardy. The Physiological Ecology of Woody Plants. San Diego, Calif.; Academic Press, 1991. 657p.	Second
	Kozub, Jacques J., Norman Meyers, and Emmanuel D'Silva, preparers. Land and Water Resources Management; Report of a Seminar . . . Washington, D.C., Nov. 1986. Washington, D.C.; Economic Development Institute of the World Bank, 1987. 76p. (EDI Policy Seminar Report no. 6)	Third
	Kral, David M., et al., eds. Soil Erosion and Conservation in the Tropics; Proceedings of a Symposium. Madison, Wis.; American Society of Agronomy and Soil Science Society of America, 1982. 149p. (ASA Special Publication no. 43)	Third
First	Kramer, Paul J. Water Relations of Plants. New York; Academic Press, 1983. 489p.	Second
First	Kramer, Paul J., and Theodore T. Kozlowski. Physiology of Woody Plants. New York; Academic Press, 1979. 811p. (Rev., expanded and updated ed. of Kramer's Physiology of Trees and Kozlowski's Growth and Development of Trees.)	First
Third	Kreeb, K. H., H. Richter, and T. M. Hinckley, eds. Structural and Functional Responses to Environmental Stress: Water Shortage; Proceedings of the 15th International Botanical Congress, Berlin, July-Aug. 1987. The Hague; SPB Academic Pub., 1989. 308p.	
Third	Krischer, Otto. Die Wissenschaftlichen Grundlagen der Trocknungstechnik. 3d ed. Berlin and New York; Springer, 1978. 489p. (1st ed., 1956. 400p.)	
	Krutilla, John V., and Anthony C. Fisher. The Economics of Natural Environments: Studies in the Valuation of Commodity and Amenity Resources. Washington, D.C.; Resources for the Future, 1985. 300p. (1st ed., 1976. 292p.)	Third
Second	Kuijt, Job. The Biology of Parasitic Flowering Plants. Berkeley, Calif.; University of California Press, 1969. 246p.	
Third	Kula, Erhun. The Economics of Forestry: Modern Theory and Practice. London; Croom Helm and; Portland, Ore.; Timber Press, 1988. 185p.	
	Kumar, Raj. The Forest Resources of Malaysia: Their Eco-	Third

Developed Countries Ranking		Third World Ranking
	nomics and Development. Singapore and Oxford, U.K.; Oxford University Press, 1986. 268p.	
Third	Kyle, D. J., C. B. Osmond, and C. J. Arntzen, eds. Photoinhibition. Amsterdam and New York; Elsevier, 1987. 315p.	Third
Second	Laarman, Jan G., and Roger A. Sedjo. Global Forests: Issues for Six Billion People. New York; McGraw-Hill, 1992. 337p.	Second
	Laban, P., ed. Workshop on Land Evaluation for Forestry; Proceedings . . . International Workshop of the IUFRO/ISSS, Wageningen, Netherlands, Nov. 1980. Wageningen, Netherlands; International Institute for Land Reclamation and Improvement, 1981. 355p.	Second
Third	Lal, R., ed. Soil Erosion Research Methods. Ankeny, Iowa; Soil and Water Conservation Society, 1988. 244p.	First
Third	Lal, R., and D. J. Greenland, eds. Soil Physical Properties and Crop Production in the Tropics. Chichester, U.K., and New York; J. Wiley, 1979. 551p.	Second
	Lal, R., P. A. Sanchez, and R. W. Cummings, eds. Land Clearing and Development in the Tropics; Proceedings of an International Symposium . . . Rotterdam and Boston; A. A. Balkema, 1986. 450p.	First
	Lamb, D. Exploiting the Tropical Rain Forest: An Account of Pulpwood Logging in Papua New Guinea. Paris; UNESCO; Carnforth, U.K., and Park Ridge, N.J.; Parthenon Pub., 1990. 259p.	Second
Third	Lamb, F. Bruce. Mahogany of Tropical America; Its Ecology and Management. Ann Arbor, Mich.; University of Michigan Press, 1966. 220p.	
	Lamprecht, Hans. Silviculture in the Tropics: Tropical Forest Ecosystems and Their Tree Species: Possibilities and Methods for Their Long-Term Utilization . . . trans. by John Brose, Stephen Conn, and Gregory G. Woods-Schank. English ed. Eschborn, Germany; Deutsch Gesellschaft fur Technische Zusammenarbeit, 1989. 296p. (Translation of Waldbau in den Tropen.)	First
First	Landsberg, J. J. Physiological Ecology of Forest Production. London and Orlando, Fla.; Academic Press, 1986. 198p.	
Second	Landsberg, J. J., and C. V. Cutting, eds. Environmental Effects on Crop Physiology; Proceedings of the 5th Long Ashton Symposium, University of Bristol, April 1975. London and New York; Academic Press, 1977. 388p.	
Second	Landsberg, J. J., and W. Parsons, eds. Research for Forest Management; Proceedings of a Conference organized by the Division of Forest Research CSIRO, Cambridge, Australia, May 1984. East Melbourne, Australia; Commonwealth Scientific and Industrial Research, 1985. 296p.	
First	Lange, O. L., et al., eds. Physiological Plant Ecology I: Re-	

Developed Countries Ranking		Third World Ranking
	sponses to the Physical Environment. Berlin and New York; Springer-Verlag, 1981. 625p. (Encyclopedia of Plant Physiology; New series no. 12A)	
First	Lange, O. L., et al., eds. Physiological Plant Ecology II: Water Relations and Carbon Assimilation. Berlin and New York; Springer-Verlag, 1982. 747p. (Encyclopedia of Plant Physiology; New series no. 12B)	
Second	Lange, O. L., et al., eds. Physiological Plant Ecology III: Responses to the Chemical and Biological Enivornment. Berlin and New York; Springer-Verlag, 1983. 799p. (Encyclopedia of Plant Physiology no. 12c)	
First	Lange, O. L., et al., eds. Physiological Plant Ecology IV: Ecosystem Processes, Mineral Cycling, Productivity and Man's Influence. Berlin and New York; Springer-Verlag, 1983. 644p.	Third
Second	Lange, O. L., L. Kappen, and E.-D. Schulze, eds. Water and Plant Life: Problems and Modern Approaches. Berlin and New York; Springer-Verlag, 1976. 536p.	
	Lanly, Jean-Paul, preparer. Manual of Forest Inventory with Special Reference to Mixed Tropical Forests. Rome; Food and Agriculture Organization, 1973. 200p.	Third
	Lanly, Jean-Paul. Tropical Forest Resources. Rome; Food and Agriculture Organization, 1982. 106p. (FAO Forestry Paper 30	Second
First	Larcher, Walter. Physiological Plant Ecology . . . trans. by M. A. Biederman-Thorson. 2d ed., rev. Berlin and New York; Springer-Verlag, 1980. 303p. (Translation of Okologie der Pflanzen. 1st ed., Stuttgart; Ulmer, 1973. 320p.)	
Second	Lassoie, James P., and Thomas M. Hinckley, eds. Techniques and Approaches in Forest Tree Ecophysiology. Boca Raton, Fla.; CRC Press, 1991. 599p.	
Third	Lauchli, A., and R. L. Bieleski, eds. Inorganic Plant Nutrition. Berlin and New York; Springer-Verlag, 1983. 2 vols. (Encyclopedia of Plant Physiology, New Series no. 15)	
	Lauenroth, William K., Gaylord V. Skogerboe, and Marshall Flug, eds. Analysis of Ecological Systems: State-of-the-Art in Ecological Modelling; Proceedings of a Symposium, Colorado State University, Fort Collins, May 1982. Amsterdam and New York; Elsevier Scientific, 1983. 992p.	Third
Third	Lavender, Denis P., compiler and ed. Woody Plant Growth in a Changing Chemical and Physical Environment; Proceedings of a Workshop of IUFRO Working Party on Shoot Growth Physiology, Vancouver, July 1987. Vancouver, British Columbia; University of British Columbia, Forest Sciences Dept., 1989. 316p.	
	Leach, Gerald, and Robin Mearns. Beyond the Woodfuel Crisis:	Second

Developed Countries Ranking		Third World Ranking
	People, Land and Trees in Africa. London; Earthscan, 1988. 309p.	
Second	Leaf, Albert L. Impact of Intensive Harvesting on Forest Nutrient Cycling. Syracuse; State University of New York, College of Environmental Science and Forestry, 1979. 421p.	First
Third	Leary, Rolfe A. Interaction Theory in Forest Ecology and Management. Dordrecht and Boston; M. Nijhoff, W. Junk; Hingham, Mass.; Kluwer Academic, 1985. 219p.	
	Lee, David W. The Sinking Ark: Enviromental Problems in Malaysia and Southeast Asia. Kuala Lumpur, Malaysia; Heinemann, 1980. 85p.	Third
Third	Lee, Richard. Forest Hydrology. New York; Columbia University Press, 1980. 349p.	
Third	Lee, Richard. Forest Microclimatology. New York; Columbia University Press, 1978. 276p.	
Third	Lee, Robert G., Donald R. Field, and William Burch, eds. Community and Forestry: Continuities in the Sociology of Natural Resources. Boulder, Colo.; Westview Press, 1990. 301p.	Third
	Leigh, Egbert G., Stanley A. Rand, and Donald M. Windsor, eds. The Ecology of a Tropical Forest: Seasonal Rythms and Long-Term Changes. Washington, D.C.; Smithsonian Institution Press, 1982. 468p.	Second
Third	LeMaster, Dennis C., David M. Baumgartner, and Roger C. Chapman, compilers and eds. Forestry Predictive Models: Problems in Application. Pullman, Wa.; Washington State University Cooperative Extension, 1981. 116p.	
Third	Leonard, Kurt J., and William E. Fry, eds. Plant Disease Epidemiology. New York; Macmillan; London; Collier Macmillan, 1986–1989. 2 vols.	
Third	Leopold, Aldo. Game Management. Madison, Wis.; University of Wisconsin Press, 1986, 1933. 481p. (Reprint. Originally pub., New York: Scribner, 1933.)	
Second	Leopold, Aldo. A Sand County Almanac, and Sketches Here and There. New York; Oxford University Press, 1987, 1949. 228p. (Reprint.)	
Third	Leopold, Aldo C., and Paul E. Kriedemann. Plant Growth and Development. 2d ed. New York; McGraw-Hill, 1975. 545p. (1st ed., Carl A. Leopold, 1964. 466p.)	
Third	Leuschner, William A. Introduction to Forest Resource Management. Malabar, Fla.; Krieger Pub. Co., 1991. n.p. (Earlier ed., New York; Wiley, 1984. 298p.)	Second
First	Levitt, J. Responses of Plants to Environmental Stresses. 2d ed. New York; Academic Press, 1980. 2 vols. (1st ed., 1972. 697p.)	First
Third	Lewis, Trevor. Thrips: Their Biology, Ecology and Economic Importance. London and New York; Academic Press, 1973. 349p.	

Developed Countries Ranking		Third World Ranking
Second	Li, P. H., and A. Sakai, eds. Plant Cold Hardiness and Freezing Stress: Mechanisms and Crop Implications. New York; Academic Press, 1978–. 2 vols.	
	Liese, Walter, ed. Biological Transformation of Wood by Microorganisms; Proceedings of the Sessions on Wood Products Pathology, 2d International Congress of Plant Pathology, Minneapolis, Sept. 1973. Berlin and New York; Springer-Verlag, 1975. 203p.	Third
Third	Lieth, Helmut F. H., ed. Patterns of Primary Production in the Biosphere. Stroudsburg, Pa.; Dowden, Hutchinson & Ross, 1978. 342p.	
	Lieth, Helmut, and M. J. A. Werger, eds. Tropical Rain Forest Ecosystems: Biogeographical and Ecological Studies. Amsterdam and New York; Elsevier Science Pub., 1989. 713p.	Third
	Lieth, Helmut, and Robert H. Whittaker, eds. Primary Productivity of the Biosphere. New York; Springer-Verlag, 1975. 339p.	Third
First	Likens, Gene E., et al. Biogeochemistry of a Forested Ecosystem. New York; Springer-Verlag, 1977. 146p.	Second
Third	Lillesand, Thomas M., and Ralph W. Kiefer. Remote Sensing and Image Interpretation. 2d ed. New York; J. Wiley, 1987. 721p. (1st ed., 1979. 612p.)	Third
Third	Lin, Bo-Qun, ed. Forest Soils and Modern Forest Management; Proceedings of the 1st International Symposium on Forest Soils: A Satellite Meeting of the 14th International Congress of Soil Science, Harbin, July 1990. Harbin, China; Publishing House of Northeast Forestry University, 1990. 497p.	Third
Third	Linnartz, Norwin E., and Mark K. Johnson, eds. Agroforestry in the Southern United States; Proceedings of the 33rd Forestry Symposium, Louisiana State University, Baton Rouge, 1984. Baton Rouge; Louisiana Agricultural Experiment Station, 1984. 183p.	Third
First	Little, C. H. A., compiler and ed. Control of Shoot Growth in Trees; Proceedings of a Joint Workshop of IUFRO Working Parties on Xylem Physiology and Shoot Growth. Fredericton, New Brunswick; Maritimes Forest Research Centre, 1980. 357p.	
Third	Little, Elbert L. Checklist of United States trees (Native and Naturalized). Washington, D.C.; Forest Service, USDA, 1979. 375p. (USDA Agriculture Handbook no. 541)	
Third	Little, Elbert L. Common Fuelwood Crops: A Handbook for Their Identification. Morgantown, W.Va.; Communi-Tech Associates, 1983. 354p.	
	Little, Peter D., Michael M. Horowitz, and A. Endre Nyerges, eds. Lands at Risk in the Third World: Local-Level Perspectives. Boulder, Colo.; Westview Press, 1987. 416p.	Third

Developed Countries Ranking		Third World Ranking
Third	Mai, W. F. Pictorial Key to Genera of Plant-Parasitic Nematodes. 4th ed., rev. Ithaca, N.Y.; Comstock Pub. Associates, 1975. 219p. (1st ed., 1960. 1 vol.)	
	Major, Michael, Gerardo Budowski, and Rolain Borel. Manual of Teaching Methods for Use in Agroforestry Intensive Short Courses. Turrialba, Costa Rica; Centro Agronomico Tropical de Investigacion y Ensenanza, CATIE, Dept. de Recursos Naturales Renovables, 1985. 72p.	Third
	Maler, Karl-Goren. Environmental Economics; A Theoretical Inquiry. Baltimore; Published for Resources for the Future by the Johns Hopkins University Press, 1974. 267p.	Third
	Maloney, George T. Chemicals from Pulp and Wood Waste: Production and Applications. Park Ridge, N.J.; Noyes Data, 1978. 289p. (Chemical Technology Review no. 101)	Third
Third	Maloney, Thomas M. Modern Particleboard and Dry-Process Fiberboard Manufacturing. San Francisco; Miller Freeman Pub., 1977. 672p.	
First	Manion, Paul D. Tree Disease Concepts. 2d ed. Englewood Cliffs, N.J.; Prentice-Hall, 1991. 402p. (1st ed., 1981. 399p.)	Second
	Mansfield, Edwin. Technology Transfer, Productivity, and Economic Policy. New York; Norton, 1982. 243p.	Third
First	Marks, G. C., and T. T. Kozlowski, eds. Ectomycorrhizae. New York; Academic Press, 1973. 444p.	
First	Marschner, Horst. Mineral Nutrition of Higher Plants. London and Orlando, Fla.; Academic Press, 1986. 674p.	
	Marsh, Clive W., and Russell A. Mittermeier, eds. Primate Conservation in the Tropical Rain Forest. New York; A. R. Liss, 1987. 365p.	Third
Second	Marshall, B., and F. I. Woodward, eds. Instrumentation for Environmental Physiology. Cambridge, U.K. and New York; Cambridge University Press, 1985. 241p. (Society for Experimental Biology no. 22)	
Second	Marshall, J. K., ed. The Belowground Ecosystem: A Synthesis of Plant-Associated Process. Fort Collins, Colo.; Colorado State University, 1977. 351p. (CSU Range Science Series no. 26)	Third
	Martin, Claude. The Rainforests of West Africa: Ecology, Threats, Conservation . . . trans. by Linda Tsardakas. Basel and Boston; Birkhauser Verlag, 1991. 235p. (Translation of Regenwalder Westafrikas.)	Third
	Martin, Franklin W., and M. Rubert. Edible Leaves of the Tropics. Washington, D.C.; U.S. Department of Agriculture, 1975. 235p.	Third
Third	Martineau, R. Insects Harmful to Forest Trees. Ottawa, Canada; Multiscience Pub., 1984. 261p.	

Developed Countries Ranking		Third World Ranking
Third	Mather, Alexander S. Global Forest Resources. London; Belhaven, 1990. 341p.	Third
Third	Mathy, Pierre, ed. Air Pollution and Ecosystems; Proceedings of the Symposium on Effects of Air Pollution on Terrestrial and Aquatic Ecosystems, Grenoble, France, May 1987. Dordrecht and Boston; D. Reidel Pub. Co., 1987. 981p.	
Third	Matthews, John D. Silvicultural Systems. Oxford, U.K.; Clarendon Press; New York; Oxford University Press, 1989. 284p.	Second
Third	Mattson, William J., Jean Levieux, and C. Bernard-Dagan, eds. Mechanisms of Woody Plant Defenses against Insects: Search for Pattern. New York; Springer-Verlag, 1988. 416p.	
Second	Mayer, A. M., and A. Poljakoff-Mayber. The Germination of Seeds. 4th ed. Oxford, U.K. and New York; Pergamon Press, 1989. 270p. (1st ed., 1963. 236p.)	Second
	McDermott, Melanie J., ed. The Future of the Tropical Rain Forest; Proceedings of an International Conference. Oxford, U.K.; Oxford Forestry Institute, Dept. of Plant Sciences, University of Oxford, 1988. 110p.	Second
	McDouggall, E. Ann, ed. Sustainable Agriculture in Africa; Proceedings of the Agricultural Systems and Research Workshop and Canadian Association of African Studies Meeting, University of Alberta, Edmonton, May 1987. Trenton, N.J.; Africa World Press, 1990. 335p.	Third
	McGaughey, Stephen E., and Hans M. Gregersen, eds. Forest-Based Development in Latin America: An Analysis of Investment Opportunites and Financing Needs. Washington, D.C.; Inter-American Development Bank, 1983. 215p.	Second
Third	McGaughey, Stephen E., and Hans M. Gregersen. Investment Policies and Financing Mechanisms for Sustainable Forestry Development. Washington, D.C.; Inter-American Development Bank, 1988. 126p.	Second
Third	McIntosh, Robert P. The Background of Ecology: Concept and Theory. Cambridge, U.K. and New York; Cambridge University Press, 1985. 383p.	
	McNeely, Jeffery A., et al. Conserving the World's Biological Diversity. Gland, Switzerland; International Union for Conservation of Nature and Natural Resources, 1990. 193p.	Third
	McNeely, Jeffrey A. Economics and Biological Diversity: Developing and Using Incentives to Conserve Biological Resources. Gland, Switzerland; International Union for Conservation of Nature and Natural Resources, 1988. 236p.	Second
	McNeely, Jeffrey A., and Kenton R. Miller, eds. National Parks, Conservation, and Development: The Role of Protected Areas in Sustaining Society; Proceedings of the World Congress on	Second

Developed Countries Ranking		Third World Ranking
	National Parks, Bali, Indonesia, Oct. 1982. Washington, D.C.; Smithsonian Institution Press, 1984. 825p.	
	Medina, E., H. A. Mooney, and C. Vazquez-Yanes, eds. Physiological Ecology of Plants of the Wet Tropics; Proceedings of an International Symposium, Oxatepec and Los Tuxtlas, Mexico, June-July 1983. The Hague and Boston; W. Junk, 1984. 254p.	Second
Third	Meehan, William R., ed. Influences of Forest and Rangeland Management on Salmonid Fishes and Their Habitats. Bethesda, Md.; USDA Forest Service, 1991. 751p.	
	Meggers, Betty J., Edward S. Ayensu, and W. Donald Duckworth, eds. Tropical Forest Ecosystems in Africa and South America; A Comparative Review; Proceedings of a Symposium planned by the Association for Tropical Biology, Ghana, West Africa, Feb. 1971. Washington, D.C.; Smithsonian Institution Press, 1973. 350p.	Second
Third	Megraw, R. A. Wood Quality Factors in Loblolly Pine: The Influence of Tree Age, Position in Tree, and Cultural Practice on Wood Specific Gravity, Fiber Length, and Fibril Angle. Atlanta, Ga.; Tappi Press, 1985. 88p.	
Third	Mengel, Konrad, and Ernest A. Kirkby. Principles of Plant Nutrition. 4th ed. Bern, Switzerland; International Potash Institute, 1987. 687p. (1st ed., 1978. 593p.)	Third
	Mergen, Francois, ed. International Symposium on Tropical Forests Utilization and Conservation: Ecological, Sociopolitical and Economic Problems and PotentialProceedings of an International Symposium, Yale University, School of Forestry and Environmental Studies, New Haven, Conn., April 1980. New Haven, Conn.; Yale University Press, 1981. 199p.	Second
	Mergen, Francois, and Jeffrey R. Vincent, eds. Natural Management of Tropical Moist Forests: Silvicultural and Management Prospects of Sustained Utilization. New Haven, Conn.; Yale University, School of Forestry and Environmental Studies, 1987. 212p.	First
Third	Meyer, Robert W., and Robert M. Kellogg. Structural Use of Wood in Adverse Environments. New York; Van Nostrand Reinhold, 1982. 510p.	
Third	Midgley, S. J., J. W. Turnbull, and R. D. Johnston, eds. Casuarina Ecology Management and Utilization; Proceedings of an International Workshop, Canberra, Australia, Aug. 1981. Melbourne; Commonwealth Scientific and Industrial Research Organization, 1983. 286p.	Third
Second	Miksche, J. P., ed. Modern Methods in Forest Genetics. Berlin and New York; Springer-Verlag, 1976. 288p.	Second
Third	Miller, Kenton, and Laura Tangley. Trees of Life: Saving Tropi-	Third

Developed Countries Ranking		Third World Ranking
	cal Forests and Their Biological Wealth. Boston, Mass.; Beacon Press, 1991. 218p.	
	Miller, Marc L., Richard P. Gale, and Perry J. Brown, eds. Social Science in Natural Resource Management Systems. Boulder, Colo.; Westview Press, 1987. 265p.	Third
	Mills, Edwin S., and Philip E. Graves. The Economics of Environmental Quality. 2d ed. New York; Norton, 1986. 368p. (1st ed., 1978. 304p.)	Second
Second	Mirov, Nicholas T. The Genus Pinus. New York; Ronald Press Co., 1967. 602p.	
Third	Mirov, Nicholas T., and Jean Hasbrouck. The Story of Pines. Bloomington, Ind.; Indiana University Press, 1976. 148p.	
	Mitchell, Colin W. Terrain Evaluation: An Introductory Handbook to the History, Principles, and Methods of Practical Terrain Assessment. 2d ed. Harlow, Eng.; Longman Scientific and Technical and New York: J. Wiley, 1991. 441p. (1st ed., London: Longman, 1973. 221p.)	Third
	Moeller, G. H., and D. T. Seal, eds. Technology Transfer in Forestry; Proceedings of a Meeting of the International Union of Forestry Research Organization, Edinburgh University, July-Aug. 1983. London; H.M.S.O., 1984. 113p. (Forestry Commission Bulletin no. 61)	Third
Third	Moffitt, Francis H., and Edward M. Mikhail. Photogrammetry. 3d ed. New York; Harper & Row, 1980. 648p. (1st ed., Scranton, Pa.; International Textbook Co., 1959. 455p.)	
Third	Moldenhauer, W. C., and N. W. Hudson, eds. Conservation Farming on Steep Lands. Ankeny, Iowa; Soil and Water Conservation Society, 1988. 296p.	Third
	Mongi, H. O., and P. A. Huxley, eds. Soils Research in Agroforestry; Proceedings of an Expert Consultation held at the International Council for Research in Agroforestry, Nairobi, March 1979. Nairobi, Kenya; ICRAF, 1979. 584p.	First
Third	Monsi, Masami, and T. Saeki, eds. Ecophysiology of Photosynthetic Production. Tokyo; University of Tokyo Press, 1978. 272p.	
	Monsoon Asia Agroforestry Joint Research Team. Comparative Studies on the Utilization and Conservation of the Natural Environment by Agroforestry Systems; Report . . . Kyoto, Japan; MAAJRT, 1986. 453p.	Second
First	Monteith, John L., ed. Vegetation and the Atmosphere. London and New York; Academic Press, 1975. 2 vols.	First
First	Monteith, John L., and M. H. Unsworth. Principles of Environmental Physics. 2d ed. London and New York; E. Arnold, 1990. 291p. (1st ed., 1973. 241p.)	Second
	Monyo, J. H., A. D. R. Ker, and Marilyn Campbell, eds. Inter-	Third

Developed Countries Ranking		Third World Ranking
	cropping in Semi-Arid Areas; Report of a Symposium held at the University of Dar es Salaam, Morogoro, Tanzania, May 1976. Ottawa, Canada; International Development Research Centre, 1976. 72p. (IDRC no. 076e)	
Third	Mooney, H. A., et al. Fire Regimes and Ecosystems Properties; Proceedings of a Conference, Honolulu, Dec. 1978. Washington, D.C.; U.S. Forest Service, 1981. 594p. (USDA Forest Service General Technical Report WO no. 26)	
Second	Mooney, Harold A., William E. Winner, and Eva J. Pell, eds. Response of Plants to Multiple Stresses. San Diego, Calif.; Academic Press, 1991. 422p.	
Third	Moore, Thomas C. Biochemistry and Physiology of Plant Hormones. 2d ed. New York; Springer-Verlag, 1989. 330p. (1st ed., 1979. 274p.)	
	Morgan, R. P. C., ed. Soil Conservation: Problems and Prospects. Chichester, U.K. and New York; J. Wiley, 1981. 576p.	Second
Third	Morgan, R. P. C. Soil Erosion and Conservation . . . edited by D. A. Davidson. Essex, U.K.; Longmans Scientific; New York; J. Wiley, 1986. 298p. (Earlier ed. as Soil Erosion, 1979. 113p.)	Third
Third	Moriarty, F. Ecotoxicology: The Study of Pollutants in Ecosystems. 2d ed. London and San Diego; Academic Press, 1988. 289p. (1st ed., 1983. 233p.)	
Third	Morrison, Ian K. Mineral Nutrition of Conifers with Special Reference to Nutrient Status Interpretation: A Review of Literature. Ottawa, Canada; Environment Canada, Forestry Service, 1974. 74p.	
	Moss, Rowland P. Fuelwood and Rural Energy Production and Supply in the Humid Tropics; Report for the United Nations University with Special Reference to Tropical Africa and South-East Asia. Dublin; United Nations University by Tycooly International, 1981. 224p.	Second
First	Mudd, J. Brian, and T. T. Kozlowski, eds. Responses of Plants to Air Pollution. New York; Academic Press, 1975. 383p.	
Third	Mueller-Dombois, Dieter, and Heinz Ellenberg. Aims and Methods of Vegetation Ecology. New York; J. Wiley, 1974. 547p.	
Third	Mullins, E. J., and T. S. McKnight, eds. Canadian Woods: Their Properties and Uses. 3d ed. Toronto; University of Toronto Press, 1981. 389p. (1st ed., Ottawa; J. O. Patenaude, 1935. 345p.)	
	Munslow, Barry, et al. The Fuelwood Trap: A Study of the SADCC Region. London; Earthscan, 1988. 181p.	Third
	Myers, Norman. Conversion of Tropical Moist Forests; A Report for the Committee on Research Priorities in Tropical Biology	First

Developed Countries Ranking		Third World Ranking
Third	National Academy of Sciences (U.S.). Firewood Crops: Shrub and Tree Species for Energy Production; Report of an Ad Hoc Panel of the Advisory Committee on Technology Innovation, Board of Science and Technnology for International Development, Commission on International Relations. Washington, D.C.; National Academy of Sciences, 1980, 1983. 2 vols.	First
Second	National Academy of Sciences (U.S.). Leucaena: Promising Forage and Tree Crops for the Tropics; Report of a Study. Washington, D.C.; National Academy Sciences, 1977. 115p. (Summary in French and Spanish.)	First
Third	National Forest Products Association. National Design Specification for Wood Construction: Structural Lumber, Glued Laminated Timber, Timber Piles, Connections, NDS: Recommended Practice. Washington, D.C.; National Forest Products Association, 1991. 125p.	
	National Research Council (U.S.). Agroforestry in the West African Sahel; Report of the Advisory Committee on the Sahel, Board on Science and Technology for International Development. Washington, D.C.; National Academy Press, 1984. 86p.	Second
Third	National Research Council (U.S.). Biologic Markers of Air-Pollution Stress and Damage in Forests. Washington, D.C.; National Academy Press, 1989. 363p.	
	National Research Council (U.S.). Calliandra: A Versatile Small Tree for the Humid Tropics; Report of an Ad Hoc Panel of the Advisory Committee on Technology Innovation, Board on Science and Technology for International Development. Washington, D.C.; National Academy Press, 1983. 52p.	Second
	National Research Council (U.S.). Casuarinas, Nitrogen-Fixing Trees for Adverse Sites; Report of an Ad Hoc Panel of the Advisory Committee on Technology Innovation, Board on Science and Technology for International Development. Washington, D.C.; National Academy Press, 1984. 118p.	Second
	National Research Council (U.S.). Sustainable Agriculture and the Environment in the Humid Tropics. Washington, D.C.; National Academy Press, 1993. 702p.	Third
	National Research Council (U.S.). Diffusion of Biomass Energy Technologies in Developing Countries. 2d ed. Washington, D.C.; National Academy Press, 1984. 120p. (1st ed., 1982.)	Second
Second	National Research Council (U.S.). Ecological Aspects of Development in the Humid Tropics. Washington, D.C.; National Academy Press, 1982. 297p.	First
Second	National Research Council (U.S.). Firewood Crops: Shrub and Tree Species for Energy Production; Report of an Ad Hoc Panel of the Advisory Committee on Technology Innovation, Board on Science and Technology for International Develop-	

Developed Countries Ranking		Third World Ranking

	ment. Washington, D.C.; National Academy of Sciences, 1980–1983. 2 vols.	
Third	National Research Council (U.S.). Impacts of Emerging Agricultural Trends on Fish and Wildlife Habitat. Washington, D.C.; National Academy Press, 1982. 303p.	
	National Research Council (U.S.). Managing Global Genetic Resources: Forest Trees. Washington, D.C.; National Academy Press, 1991. 228p.	Second
	National Research Council (U.S.). Mangium and Other Fast-Growing Acacias for the Humid Tropics; Report of an Ad Hoc Panel of the Advisory Committee on Technology Innovation, Board on Science and Technology for International Development. Washington, D.C.; National Academy Press, 1983. 62p.	Second
	National Research Council (U.S.). Tropical Legumes: Resources for the Future; Report of an Ad Hoc Panel of the Advisory Committee on Technology Innovation, Board on Science and Technology for International Development. Washington, D.C.; National Academy of Sciences, 1979. 331p.	First
	National Research Council (U.S.). Underexpoloited Tropical Plants with Promising Economic Value; Report . . . Washington, D.C.; National Academy of Sciences, 1975. 188p.	First
Third	Naveh, Zeev, and Arthur S. Lieberman. Landscape Ecology: Theory and Application. New York; Springer-Verlag, 1984. 356p.	
	Nectoux, Francois. Timber from the South Seas: An Analysis of Japan's Tropical Timber Trade and its Environmental Impact. Gland, Switzerland; World Wildlife Fund International, 1990. 134p.	Third
	Nectoux, Francois, and Nigel Dudley. A Hard Wood Story: An Investigation into the European Influence on Tropical Forest Loss. London, UK; Friends of the Earth Trust, Earth Resources Research, 1987. 118p.	Third
Third	Neergaard, Paul. Seed Pathology. London; MacMillan, 1977. 2 vols.	
	Nelson, Michael. The Development of Tropical Lands: Policy Issues in Latin America. Baltimore, Md.; Published for Resources for the Future by Johns Hopkins University Press, 1973. 306p.	Third
Third	Nelson, P. E., T. A. Toussoun, and R. J. Cook, eds. Fusarium, Diseases, Biology, And Taxonomy. University Park, Pa.; Pennsylvania State University Press, 1981. 457p.	
Third	Nemetz, Peter N., ed. Emerging Issues in Forest Policy. Vancouver, British Columbia; University of British Columbia Press, 1992. 573p.	
Third	Newton, Michael, and Fred B. Knight. Handbook of Weed and	

Developed Countries Ranking		Third World Ranking
	Insect Control Chemicals for Forest Resource Managers. Beaverton, Oreg.; Timber Press, 1981. 213p.	
Third	Nicholas, Darrel D., ed. Wood Deterioration and Its Prevention by Preservative Treatments. Syracuse, N.Y.; Syracuse University Press, 1973. 2 vols. (Syracuse Wood Science Series no. 5)	
Second	Nickell, Louis G., ed. Plant Growth Regulating Chemicals. Boca Raton, Fla.; CRC Press, 1983. 2 vols.	First
Third	Niesslein, Erwin, and Gerhard Voss. Was Wir Uber das Waldsterben Wissen. Koln, Germany; Deutscher Instituts-Verlag, 1985. 259p.	
	Nitrogen Fixing Tree Association. Leucaena Research in the Asian-Pacific Region; Proceedings of a Workshop, Singapore, Nov. 1982. Ottawa, Canada; International Development Research Centre, 1983. 192p. (IDRC no. 211e)	Second
First	Nobel, Park S. Biophysical Plant Physiology and Ecology. San Francisco, Calif.; W. H. Freeman, 1983. 488p.	Second
Third	Norse, Elliott A. Conserving Biological Diversity in Our National Forests. Washington, D.C.; Wilderness Society, 1986. 116p.	
Second	North American Forest Biology Workshop, 5th, University of Florida. Proceedings . . . , Mar. 1978, edited by Charles A. Hollis and Anthony E, Squillace; sponsored by School of Forest Resource and Conservation, University of Florida. Gainesville; School of Forest Resources and Conservation, University of Florida, 1978. 430p.	
Third	North American Forest Biology Workshop, 6th, University of Alberta, 1980. Proceedings . . . , edited by B. P. Dancik, K. O. Higginbotham. Edmonton, Alberta; University of Alberta, 1980. 153p.	
Third	North American Forest Biology Workshop, 8th, Logan, Utah, 1984. Proceedings . . . , edited and compiled by Ronald M. Lanner. Logan, Utah; Utah State University, 1984. 196p.	
Third	North American Forest Biology Workshop, 9th, Stillwater, Okla., 1986. Forest Decline: Assessing Impacts of a Changing Environment on Forest Productivity; Proceedings . . . , edited by Thomas C. Hennessey, Phillip M. Dougherty, and Charles G. Tauer. Victoria, Canada; Heron Publishing, 1987. 102p.	
	Norton, Bryan G. Why Preserve Natural Variety? Princeton, N.J.; Princeton University Press, 1987. 281p.	Third
	Novoa, Andres R., and Joshua L. Posner, eds. Seminario Internacional sobre Produccion Agropecuaria y Forestal en Zonas de Ladera de America Tropical, Turrialba, Dec. 1980. Turrialba, Costa Rica; Centro Agronomico Tropical de Investigacion y Ensenanza and Rockefeller Foundation, 1981. 357p. (Serie Tecnica. Informe Tecnico, CATIE no. 11)	Third

Developed Countries Ranking		Third World Ranking
	Ntima, O. O., compiler. The Auracarias. Oxford, U.K.; Commonwealth Forestry Institute, 1968. 139p.	Third
Second	Nutman, P. S., ed. Symbiotic Nitrogen Fixation in Plants. Cambridge, U.K. and New York; Cambridge University Press, 1976. 584p.	Second
	Nwoboshi, Louis C. Tropical Silviculture: Principles and Techniques. Ibadan, Nigeria; Ibadan University Press, 1982. 333p.	Second
	Nye, Peter H., and D. J. Greenland. The Soil Under Shifting Cultivation. Farnham Royal, U.K.; Commonwealth Agricultural Bureaux, 1965. 156p. (CBSS Technical Communication no. 51) (Earlier ed., 1960.)	First
	O'Loughlin, Colin L. The Effects of Forest Land Use on Erosion and Slope Stability; Report of a Seminar . . . Honolulu; East-West Environment and Policy Institute, 1985.	First
First	Odum, Eugene P. Basic Ecology. Philadelphia; Saunders College Pub., 1983. 613p. (Previous eds. as Fundamentals of Ecology. 1st ed., 1953. 384p.)	
	Odum, Howard T., ed. A Tropical Rain Forest; A Study of Irradiation and Ecology at El Verde, Puerto Rico. Oak Ridge, Tenn.; Office of Information Services, U.S. Atomic Energy Commission, 1970. 3 vols.	Second
Second	Old, K. M., G. A. Kile, and C. P. Ohmart, eds. Eucalypt Dieback in Forests and Woodlands; Proceedings of a Conference, Commonwealth Scientific and Industrial Research Organization Division of Forest Research, Canberra, Aug. 1980. Melbourne, Australia; CSIRO, 1981. 285p.	Third
Third	Olien, Charles R., and Myrtle N. Smith, eds. Analysis and Improvement of Plant Cold Hardiness. Boca Raton, Fla.; CRC Press, 1981. 215p.	
Second	Oliver, Chadwick D., and Bruce C. Larson. Forest Stand Dynamics. New York; McGraw-Hill, 1990. 467p.	
Second	Oosting, Henry J. The Study of Plant Communities; An Introduction to Plant Ecology. 2d ed. San Francisco; W. H. Freeman, 1956. 440p. (1st ed., 1948. 389p.)	
	Openshaw, Keith. Cost and Financial Accounting in Forestry: A Practical Manual. Oxford, U.K. and New York; Pergamon Press, 1980. 188p.	Third
	Organizacion para Estudios Tropicales, Centro Agronomico Tropical de Investigacion y Enseanza. Sistemas Agroforestales: Principios y Aplicaciones en los Tropicos. San Jose, Costa Rica; OTS, CATIE, 1986. 818p.	Third
Third	Orians, G. H., et al., eds. The Preservation and Valuation of Biological Resources. Seattle, Wash.; University of Washington Press, 1991. 301p.	Third
First	Overholts, Lee O. The Polyporaceae of the United States, Alaska and Canada . . . prepared by Josia L. Lowe. Ann Arbor,	

Developed Countries Ranking		Third World Ranking
	Mich.; University of Michigan Press, 1953. 466p. (UM Studies, Scientific Series no. 19) (3d printing, 1977.)	
Third	Pacific Northwest Research Station. Wildlife and Management of Unmanaged Douglas-Fir Forests. Portland, Oreg.; USDA Forest Service, 1991. 533p.	
	Paine, David P. Aerial Photography and Image Interpretation for Resource Management. New York; J. Wiley, 1981. 571p.	Third
Third	Paleg, L. G., and D. Aspinall, eds. The Physiology and Biochemistry of Drought Resistance in Plants. Sydney and New York; Academic Press, 1981. 492p.	
	Palo, Matti, and Gerardo Mery, editors. Deforestation or Development in the Third World. Helsinki, Finland; Mesantutkimuslaitos, 1990. 349p. 3 vols. (Scandinavian Forest Economics no. 32)	Third
	Panayotou, Theodore, and Peter S. Ashton. Not by Timber Alone: Economics and Ecology for Sustaining Tropical Forests. Washington, D. C.; Island Press, 1992. 282p.	Third
Second	Panshin, Alexis J., et al. Forest Products: Their Sources, Production and Utilization. 2d ed. New York; McGraw-Hill, 1962. 538p. (1st ed., 1950. 549p.)	
First	Panshin, Alexis J., and Carl de Zeeuw. Textbook of Wood Technology: Structure, Identification, Properties, and Uses of the Commercial Woods of the United States and Canada. 4th ed. New York; McGraw-Hill, 1980. 722p. (1st ed., 1949–1952. 2 vols.)	
	Papendick, R. I., P. A. Sanchez, and G. B. Triplett. Multiple Cropping; Proceedings of a Symposium . . . Madison, Wis.; American Society of Agronomy, 1976. 378p. (ASA Special Publication no. 27)	First
Third	Patton, David R. Wildlife Habitat Relationships in Forested Ecosystems. Portland, Oreg.; Timber Press, 1991. 392p.	
Third	Paul, Eldor A., and F. E. Clark. Soil Microbiology and Biochemistry. San Diego; Academic Press, 1989. 273p.	
Second	Peace, Thomas R. Pathology of Trees and Shrubs, with Special Reference to Britain. Oxford, U.K.; Clarendon Press, 1962. 753p.	
First	Pearcy, R. W., et al., eds. Plant Physiological Ecology: Field Methods and Instrumentaiton. London and New York; Chapman & Hall, 1989. 457p.	
Second	Pearse, Peter H. Introduction to Forestry Economics. Vancouver, British Columbia; University of British Columbia Press, 1990. 226p.	Third
Third	Peel, Alan J. Transport of Nutrients in Plants. London; Butterworths, 1974. 258p.	
Third	Penka, Miroslav, et al. Floodplain Forest Ecosystem. Amster-	Third

Developed Countries Ranking		Third World Ranking
Third	The Pinyon-Juniper Ecosystem. Proceedings of a Symposium, Utah State University, College of Natural Resources, May 1975. Logan, Utah; Utah Agricultural Experiment Station, 1975. 194p.	
Second	Pirone, P. P., et al. Tree Maintenance. 6th ed. New York; Oxford University Press, 1988. 514p. (1st-2d eds. published as Maintenance of Shade and Ornamental Trees.)	
Second	Pirson, A., and M. H. Zimmermann, eds. Encyclopedia of Plant Physiology. New series. Berlin and New York; Springer-Verlag, 1975. 17 vols. (Earlier ed., by W. Ruhland, 1955–1967, as Handbuch der Pflanzenphysiologie.)	
Third	Plochmann, Richard. Wood as Resource in World Economics: Proceedings of the 3d International Congress in conjunction with INTERFORST, Munich, 1978. Munster-Hiltrup, Germany; Landwirtschaftsverlag, 1979. 504p.	
	Plotkin, Mark, and Lisa Famolare, eds. Sustainable Harvest and Marketing of Rain Forest Products. Washington, D.C.; Island Press, 1992. 325p.	Third
	Poels, R. L. H. Soils, Water and Nutrients in a Forest Ecosystem in Suriname. Wageningen; Agricultural University, 1987. 253p. (Ecology and Management of Tropical Rain Forests in Suriname no. 2)	Third
Second	Poore, Duncan, Peter Burgess, John Palmer, Simon Rietberger, and Timothy Synott. No Timber without Trees: Sustainability in the Tropical Forest. London; Earthscan Publications, 1989. 252p.	First
	Poore, M. E. D., and C. Fries. The Ecological Effects of Eucalyptus. Dehra Dun, India; Natraj Pub., 1987. 98p. (Earlier ed., Rome; Food and Agriculture Organization, 1985. 87p. FAO Forestry Paper no. 59)	First
Third	Postgate, J. R. The Fundamentals of Nitrogen Fixation. Cambridge, U.K. and New York; Cambridge University Press, 1982. 252p.	Third
	Poulsen, Gunnar. Malawi: The Function of Trees in Small Farmer Production Systems. Rome; Food and Agriculture Organization, 1981. 65p.	Third
	Poulsen, Gunnar. Man and Tree in Tropical Africa: Three Essays on the Role of Trees in the African Environment. Ottawa, Canada; International Development Research Centre, 1978. 31p.	Third
Third	Pound, B., and L. Martinez Cairo. Leucaena: Its Cultivation and Uses. London; Overseas Development Adminsitration, 1983. 287p.	
	Prance, Ghillean T., ed. Biological Diversification in the Tropics; Proceedings of the 5th International Symposium of	First

Developed
Countries
Ranking

Third
World
Ranking

	the Association for Tropical Biology, Macuto Beach, Caracas, Venezuela, Feb. 1979. New York; Columbia University Press, 1982. 714p.	
	Prance, Ghillean T., ed. Tropical Rain Forests and the World Atmosphere. Boulder, Colo.; Westview Press, 1986. 105p. (AAAS Selected Symposium no. 101)	Second
	Prance, Ghillean T., and Thomas S. Elias, eds. Extinction is Forever; Proceedings of a Symposium held at the New York Botanical Garden, May 1976. New York; Botanical Garden, 1977. 437p.	First
Third	Price, Colin. The Theory and Application of Forest Economics. Oxford, U.K. and New York; Blackwell, 1989. 402p.	
	Prinsley, Roslyn T., ed. Agroforestry for Sustainable Production: Economic Implications; Papers presented at the Commonwealth Science Council Meeting, Swaziland, 1989. London; Commonweatlh Science Council, 1990. 417p.	First
First	Pritchett, William L., and Richard F. Fisher. Properties and Management of Forest Soils. 2d ed. New York; J. Wiley, 1987. 494p. (1st ed., by W. L. Pritchett, 1979. 500p.)	First
	Proctor, J., ed. Mineral Nutrients in Tropical Forest and Savanna Ecosystems. Oxford, U.K. and Boston; Blackwell Scientific; Brookline Village, Mass.; Publisher's Business Services, 1989. 473p. (British Ecological Society, Special Publication no. 9)	Second
Third	Puri, G. S., et al. Forest Ecology. 2d ed. New Delhi, India; Oxford University Press & IBH, 1983. 2 vols. (1st ed., 1960.)	Third
Third	Purseglove, J. W. Tropical Crops. Dicotyledons. New York; J. Wiley, 1968. 2 vols. (Reprinted, 1987. 791p.)	First
Third	Pyne, Stephen J. Introduction to Wildland Fire: Fire Management in the United States. New York; J. Wiley, 1984. 455p.	
Third	Raedeke, Kenneth J., ed. Streamside Management: Riparian Wildlife and Forestry Interactions. Seattle; University of Washington, College of Forest Resources, Institute of Forest Resources, 1988. 277p. (UW Contribution no. 59)	
Third	Raghavendra, A. S., ed. Physiology of Trees. New York; Wiley-Interscience, 1991. 509p.	Third
	Raintree, John B., compiler and ed. D & D User's Manual: An Introduction to Agroforestry Diagnosis and Design. Nairobi, Kenya; International Council for Research in Agroforestry, 1987. 110p.	First
	Raintree, John B., ed. Land, Trees and Tenure; Proceedings of an International Workshop on Tenure Issued in Agroforestry, Nairobi, May 1985. Nairobi, Kenya; ICRAF and; Madison; Land Tenure Center, University of Wisconsin, 1987. 412p.	First
	Rao, Y. S., Napoleon T. Vergara, and George W. Lovelace, eds.	Third

Developed Countries Ranking		Third World Ranking
	Community Forestry: Socio-Economic Aspects. Bangkok, Thailand; Regional Office for Asia and the Pacific, Food and Agriculture Organization, 1985. 420p.	
Third	Rasmussen, Edmund F., ed. Dry Kiln Operator's Manual. Rev. ed. Washington, D.C.; USDA Forest Service, 1983. 274p. (USDA Agriculture Handbook no. 188) (Earlier ed., 1961. 197p.)	
	Redford, Kent H., and Christine Padoch, eds. Conservation of Neotropical Forests: Working from Traditional Resource Use. New York; Columbia University Press, 1992. n.p.	Third
Third	Reichle, David E., ed. Analysis of Temperate Forest Ecosystems. Berlin; Springer-Verlag, 1970. 304p.	
First	Reichle, David E. Dynamic Properties of Forest Ecosystems. New York; Cambridge University Press, 1981. 683p.	First
	Reid, Rowan, and Geoff Wilson. Agroforestry in Australia and New Zealand: The Growing of Productive Trees on Farms. 2d ed., rev. Victoria, British Columbia; Goddard & Dobson, 1986. 255p.	Third
	Reifsnyder, William E., and T. O. Darnhofer, eds. Meteorology and Agroforestry; Proceedings of an International Workshop on the Application of Meteorology to Agroforestry Systems Planning and Management, Nairobi, Kenya, Feb. 1987. Nairobi; ICRAF, 1989. 546p.	First
Second	Repetto, Robert C. The Forest for the Trees? Government Policies and the Misuse of Forest Resources. Amended Feb. 1990. Washington, D.C.; World Resources Institute, 1990. 105p. (Originally published, 1988.)	Second
Third	Repetto, Robert, et al. Wasting Assets: Natural Resources in the National Income Accounts. Washington, D.C.; World Resources Institute, 1989. 68p.	
First	Repetto, Robert, and Malcolm Gillis, eds. Public Policies and the Misuse of Forest Resources. Cambridge, U.K. and New York; Cambridge University Press, 1988. 432p.	First
Third	Rice, Elroy L. Allelopathy. 2d ed. Orlando, Fla.; Academic Press, 1984. 422p. (1st ed., 1974. 353p.)	Second
Third	Richards, E. G., ed. Forestry and the Forest Industries: Past and Future. Major Developments in the Forest and Forest Industry Sector Since 1947 in Europe, the USSR, and North America. Dordrecht and Boston; M. Nijhoff, 1987. 428p.	
Second	Richards, John F., and Richard P. Tucker, eds. World Deforestation in the Twentieth Century. Durham, N.C.; Duke University Press, 1988. 321p.	Second
First	Richards, Paul W. The Tropical Rain Forest; An Ecological Study. Cambridge, U.K.; Cambridge University Press, 1952. 450p. (Reprinted, 1979.)	First

Developed Countries Ranking		Third World Ranking
Third	Richardson, S. D. Forests and Forestry in China: Changing Patterns of Resource Development. Washington, D.C.; Island Press, 1990. 352p.	Third
Third	Roberts, Eric H. Viability of Seeds. London; Chapman & Hall, 1972. 448p.	
Third	Robinson, Glen O. The Forest Service: A Study in Public Land Management. Baltimore; Published for Resources for the Future by Johns Hopkins University Press, 1975. 337p.	
Third	Robinson, J. B. D., ed. Diagnosis of Mineral Disorders in Plants. 1st Amer. ed. New York; Chemical Pub. Co., 1984. 5 vols. (Earlier ed., by T. Wallace, as The Diagnosis of Mineral Deficiencies in Plants. First published in 1943 by H.M.S.O.)	
	Roche, M. M., and R. Hodder. Social and Economic Perspectives on Forestry and Agroforestry; Proceedings of the Forestry Section of the 56th ANZAAS Congress, Massey University, Palmerston North, Jan. 1987. Palmerston North; New Zealand Forest Service, 1987. 1 vols.	Second
	Rocheleau, Dianne, Fred Weber, and Alison Field-Juma. Agroforestry in Dryland Africa. Nairobi, Kenya; International Council for Research in Agroforestry, 1988. 311p. (Science and Practice of Agroforestry no. 3)	First
Third	Rodiek, Jon E., and Eric G. Bolen, eds. Wildlife and Habitats in Managed Landscapes. Washington, D.C.; Island Press, 1991. 219p.	
Third	Rohrig, Ernst, et al. Ecological Basis for Silviculture. 5th ed. Hamburg, Germany; P. Parey, 1980–1982. 2 vols. (4th ed., 1971–1972. Translation of Waldbau auf okologischer Grundlage.)	
Third	Rolston, Holmes. Environmental Ethics: Duties to and Values in the Natural World. Philadelphia, Pa.; Temple University Press, 1988. 391p.	
Third	Rose, Arthur H., and O. H. Lindquist. Insects of Eastern Hardwood Trees. Ottawa; Canadian Forest Service, 1982. 304p. (Forestry Technical Report no. 29)	
Second	Rose, Arthur H., and O. H. Lindquist. Insects of Eastern Larch, Cedar and Juniper. Rev. ed. Sault Sainte-Marie, Ont.; Great Lakes Forest Research Centre, 1991. 100p. (Forest Technical Report no. 28) (1st ed., 1980.)	
Second	Rose, Arthur H., and O. H. Lindquist. Insects of Eastern Pines. Rev. ed. Ottawa; Canadian Forestry Service, 1984. 127p. (CFS Publication no. 1313) (1st ed., 1973.)	
Third	Rose, Arthur H., and O. H. Lindquist. Insects of Eastern Spruces, Fir, and Hemlock. Rev. ed. Ottawa; Forestry Canada, 1989. 159p. (Forestry Technical Report no. 23) (1st ed., 1977.)	

Developed Countries Ranking		Third World Ranking
Third	Rosen, Howard N., ed. The Impact of Energy and Environmental Concerns on Chemical Engineering in the Forest Products Industry. New York; American Institute of Chemical Engineers, 1984. 106p. (AIChE Symposium Series no. 239)	
Third	Ross, I. Ukhan. The Radiation Regime and Architecture of Plant Stands. The Hague and Boston; W. Junk, 1981. 391p. (Translation of Radiatsionny i reshim i arkhitektonika rastitel nogo pokrova.)	
	Rosswall, T., ed. Nitrogen Cycling in West African Ecosystems; Proceedings of a Workshop. Stockholm, Sweden; SCOPE/UNEP International Nitrogen Unit, Royal Swedish Academy of Sciences, 1980. 450p.	Third
	Round, Philip D. Resident Forest Birds in Thailand: Their Status and Conservation. Cambridge, U.K.; International Council for Bird Conservation, 1988. 211p. (ICBP Monograph no. 2)	Third
Third	Rowe, John W., ed. Natural Products of Woody Plants: Chemicals Extraneous to the Lignocellulosic Cell Wall. Berlin and New York; Springer-Verlag, 1989. 2 vols.	
Third	Rowell, Roger, ed. The Chemistry of Solid Wood. Washington, D.C.; American Chemical Society, 1984. 614p.	Third
First	Royer, Jack P., and Christopher D. Risbrudt, eds. Nonindustrial Private Forests: A Review of Economic and Policy Studies; Proceedings of a Symposium, April 1983. Durham, N.C.; Duke University, School of Forestry and Environmental Studies, 1983. 398p.	
Second	Russell, Edward W. Russell's Soil Conditions and Plant Growth . . . edited by Alan Wild. 11th ed. Harlow; Longmans, 1988. 991p. (1st ed., as Soil Conditions and Plant Growth, London and New York; Longmans, Green, 1912. 168p.)	Third
Second	Russell, G., B. Marshall, and P. G. Jarvis, eds. Plant Canopies: Their Growth, Form, and Function. Cambridge, U.K. and New York; Cambridge University Press, 1989. 178p. (Society for Experimental Biology Seminar Series no. 31)	
Third	Russell, Robert S. Plant Root Systems: Their Functions and Interaction with the Soil. London and New York; McGraw-Hill, 1977. 298p.	
	Ruthenberg, Hans. Farming Systems in the Tropics. 3d ed. Oxford, U.K.; Clarendon Press; and New York; Oxford University Press, 1980. 424p. (1st ed., 1971. 313p.)	First
	Ruttan, Vernon W. Agricultural Research Policy and Development. Rome; Food and Agriculture Organization, 1987. 249p. (FAO Research and Technology Paper no. 2)	Third
Third	Sabins, Floyd F. Remote Sensing: Principles and Interpretation. 2d ed. New York; W. H. Freeman, 1986. 449p. (1st ed., 1978. 426p.)	Third

Developed Countries Ranking		Third World Ranking
Second	Sakai, Akira, and Larcher W. Frost Survival of Plants: Responses and Adaptation to Freezing Stress. Berlin and New York; Springer-Verlag, 1987. 321p.	
	Salas, Gonzalo de las. Suelos y Ecosistemas Forestales: Con Enfasis America Tropical. San Jose, Costa Rica; Instituto Interamericano de Cooperacion para la Agricultura, 1987. 447p. (Coleccion de Libros y Materiales Educativos no. 80)	Third
	Salas, Gonzalo de las, ed. Workshop Agro-Forestry Systems in Latin America, Turrialba, Costa Rica, March 1979; Proceedings . . . sponsored by CATIE. Turrialba, Costa Rica; Centro Agronomico Tropical de Investigacion y Ensenanza, Program of Natural Renewable Resources, 1979. 220p.	First
	Sanchez, Pedro A. Properties and Management of Soils in the Tropics. New York; J. Wiley, 1976. 618p. (Available in Spanish as Suelos del Tropico: Caracteristicas y Manejo. San Jose, Costa Rica; Instituto Interamericano de Cooperacion para la Agricultura, 1981.)	First
	Sanchez, Pedro A., C. A. Palm, and T. J. Smyth. Approaches to Mitigate Tropical Deforestation by Sustainable Soil Management Practices. Amsterdam; Elsevier, 1990. 220p.	Second
	Saxena, Ghanshyam. The Forest Crisis. Dehra Dun, India; Natraj Pub., 1990. 182p.	Third
	Scarratt, J. B. Tree Improvement Symposium; Proceedings. Toronto, Sept. 1978. Sault Ste. Marie, Ont.; Canadian Forestry Service, Dept. of the Environment, 1979. 233p.	Third
Second	Schenck, N. C., ed. Methods and Principles of Mycorrhizal Research. St. Paul, Minn.; American Phytopathalogical Society, 1982. 244p.	
	Schippers, B., and W. Gams, eds. Soil-Borne Plant Pathogens; Proceedings of an International Symposium on Factors Determining the Behavior of Plant Pathogens in Soil, 4th, Munich, 1978. London and New York; Academic Press, 1979. 686p.	Third
	Schlabrendorff, Fabian von. The Legal Structure of Transnational Forest-Based Investments in Developing Countries: A Comparitive Study of the Contractual Practice in Selected Host of the Tropical Asia-Pacific Region. Zurich, Switzerland; ETH Zurich, Fachberich Forstokonomie und Forspolitik, 1987. 337p. (Forstwissenschaftliche Beitrage no. 5) (Avail)	Third
Third	Schmidt, John L., and Douglas L. Gilbert, compilers and eds. Big Game of North America: Ecology and Management. Rev ed. Harrisburg, Pa.; Stackpole Books, 1980. 494p.	
	Schmidt-Haas, Paul, ed. Inventorying and Monitoring Endangered Forests = Inventaire et Surveillance des Forets Menacees = Inventur und Uberwachung von Gefahrdeten Waldern; IUFRO Conference, Zurich, Switzerland, Aug. 1985.	Third

Developed Countries Ranking		Third World Ranking
	Birmensdorf, Switzerland; Eidgenossische Anstalt fur das Forstliche Versuchswesen, 1985. 405p. (English, French and German.)	
Third	Schmithusen, Franz. Forest Legislation in Selected African Countries: Based on the Review and Analysis of Forest Legislation in Eleven Member Countries of the African Timber Organization. Rome; Food and Agriculture Organization, 1986. 345p. (FAO Forestry Paper no. 65)	
Third	Schmithusen, Franz. Forest Utilization Contracts on Public Land. Rome; Food and Agriculture Organization, 1977. 197p. (FAO Forestry Paper no. 1) (1st ed., 1971, as Handbook on Forest Utilization Contracts on Public Land.)	Third
Third	Schmithusen, Franz, ed. Forstgesetzgebung; Bericht der IUFRO Arbeitsgruppe S4.08–03 = Forestry Legislation; Report of the IUFRO Working Party S4.08.03 = Legislation forestiere; Rapport du groupe de travail IUFRO S4.08–03. Zurich, Switzerland; ETH Zurich, 1988. 262p. (Forstwissenschafliche Beitrage no. 6)	
	Schmutterer, H., K. R. S. Ascher, and H. Rembold, eds. Natural Pesticides from the Neem Tree (Azadirachta Indica A. Juss); Proceedings of the 1st International Neem Conference, Rottach-Egern, Federal Republic of Germany, June 1980. Eschborn, Federal Republic of Germany; German Agency for Technical Cooperation (GTZ), 1981. 297p.	Third
	Schreuder, Ferard F., ed. Global Issues and Outlook in Pulp and Paper; Proceedings of a Symposium, University of Washington, 1987. Seattle, Wash.; University of Washington Press, 1988. 302p.	Third
Third	Schreuder, Gerard F., ed. World Trade in Forest Products, 2. Seattle, Wash.; University of Washington Press, 1986. 362p.	Third
First	Schulze, E.-D., O. L. Lange, and R. Oren, eds. Forest Decline and Air Pollution: A Study of Spruce (Picea abies) on Acid Soils. Berlin and New York; Springer-Verlag, 1989. 475p.	
Third	Schweingruber, Fritz H. Anatomie europaischer Holzer = Anatomy of European Woods. Bern, Switzerland; Verlag P. Haupt, 1990. 800p.	
Third	Schweingruber, Fritz Hans. Tree Rings: Basics and Applications of Dendrochronology. 1st English ed., rev. and expanded. Dordrecth and Boston; D. Reidel Pub. Co., 1988. 276p. (1st ed. entitled, Der Jahrring: Standort, Methodik, Zeit und Klima in der Dendrochronologie. Bern and Stuttgart; P. Haupt, 1983. 234p.)	
First	Sedjo, Roger A. The Comparative Economics of Plantation Forestry: A Global Assessment. Washington, D.C.; Resources for the Future, 1983. 161p.	First
Second	Sedjo, Roger A., ed. Governmental Interventions, Social Needs,	

Developed Countries Ranking		Third World Ranking
	and the Management of U.S. Forests. Washington, D.C.; Resources for the Future, 1983. 300p.	
First	Sedjo, Roger A., ed. Investments in Forestry: Resources, Land Use, and Public Policy. Boulder, Colo.; Westview Press, 1985. 246p.	
Second	Sedjo, Roger A., and Kenneth S. Lyon. The Long-Term Adequacy of World Timber Supply. Washington, D.C.; Resources for the Future, 1990. 230p.	Second
First	Seppala, Risto Row, Clark, and Anne Morgan, eds. Forest Sector Models; Proceedings of the 1st North American Conference. Berkhamsted, U.K.; Academic, 1983. 354p.	
	Seth, Shiam K. India and Sri Lanka Agroforestry. Rome; Food and Agriculture Organization, 1981. 116p.	Third
Third	Sethuraj, M. R., and A. S. Raghavendra, eds. Tree Crop Physiology. Amsterdam; Elsevier, 1987. 361p.	Third
Third	Shands, William E., and John S. Hoffman, eds. The Greenhouse Effect, Climate Change, and U.S. Forests. Washington, D.C.; Conservation Foundation, 1987. 304p.	
Third	Sharma, Narendra P. Managing the World's Forests: Looking for Balance Between Conservation and Development. Dubuque, Iowa; Kendall/Hunt Pub. Co., 1992. 605p.	
Third	Sharpe, Grant W., Clare W. Hendee, and Wenonah F. Sharpe. Introduction to Forestry. 5th ed. New York; McGraw-Hill, 1986. 629p. (1st ed., by Shirley W. Allen. New York and London; 1938. 402p.)	
Third	Shaw, James H. Introduction to Wildlife Management. New York; McGraw-Hill, 1985. 316p.	
Third	Shepherd, K. R. Plantation Silviculture. Dordrecht and Boston; M. Nijhoff; Norwell, Mass.; Kluwer Academic, 1986. 322p.	
Second	Shigo, Alex L. Tree Decay: An Expanded Concept. Washington, D.C.; USDA Forest Service, 1979. 72p. (USDA Agriculture Information Bulletin no. 419)	
Third	Shigo, Alex L., and Harold G. Marx. Compartmentalization of Decay in Trees. Washington, D.C.; USDA Forest Service, 1977. 73p. (USDA Agriculture Information Bulletin no. 405)	
	Sholto Douglas, James, and Robert A. Hart. Forest Farming: Towards a Solution to Problems of World Hunger and Conservation. London; Intermediate Technology Pub., 1984. 207p. (1st ed., London; Watkins, 1976. 197p.)	Third
Second	Shugart, Herman H. A Theory of Forest Dynamics: The Ecological Implications of Forest Succession Models. New York, etc.; Springer-Verlag, 1984. 278p.	
Third	Shugart, Herman H., Rik Leemans, and Gordon B. Bonan, eds. A Systems Analysis of the Global Boreal Forest. Cambridge, U.K. and New York; Cambridge University Press, 1992. 565p.	

Developed Countries Ranking		Third World Ranking
Third	Siau, John F. Transport Processes in Wood. Berlin and New York; Springer-Verlag, 1984. 245p.	Third
	SIDA/FAO Consultation on Intermediate Technology in Forestry, New Delhi and Dehra Dun, India, 1981. Appropriate Technology in Forestry; Report . . . Rome; Food and Agriculture Organization, 1982. 137p. (FAO Forestry Paper no. 31)	Third
	Siderius, W., ed. Land Evaluation for Land-Use Planning and Conservation in Sloping Areas; Proceedings of an International Workshop, Enschede, Netherlands, Dec. 1984. Wageningen; International Institute for Land Reclamation and Improvement, 1986. 334p. (ILRI Publication no. 40)	Third
Third	Siebert, Horst. Economics of the Environment: Theory and Policy. 3d ed. Berlin and New York; Springer-Verlag, 1992. n.p. (1st ed., Lexington, Mass.; Lexington Books, 1981. 230p.)	Second
	Sim, D., and H. A. Hilmi. Forestry Extension Methods. Rome; Food and Agriculture Organization, 1987. 155p. (FAO Forestry Paper no. 80)	Third
	Simon, Julian L., and Herman Kahn, eds. The Resourceful Earth: A Response to Global 2000. Oxford, U.K. and New York; Blackwell, 1984. 585p.	Third
Third	Sinclair, Steven A. Forest Products Marketing. New York; McGraw-Hill, 1992. 403p.	
First	Sinclair, Wayne A., Howard H. Lyon, and Warren T. Johnson. Diseases of Trees and Shrubs. Ithaca, N.Y.; Comstock Pub. Associates, 1987. 574p.	
	Singh, R. V., ed. Silviculture, Management, and Utilization of Poplars; Proceedings of a Symposium, Forest Dept., Jammu and Kashmir State and Forest Research Institute Conifers Research Centre, Simla, Oct. 1979. Delhi, India; Controller of Publications, 1981. 181p.	Third
	Sioli, Harold, ed. The Amazon: Limnology and Landscape Ecology of a Mighty Tropical River and its Basin. Dordrecht, Netherlands; Junk, 1984. 763p.	Second
	Sisam, John W. B. Forestry and Forestry Education in a Developing Country: A Canadian Dilemma. Toronto, Canada; Faculty of Forestry, University of Toronto, 1982. 167p.	Third
	Sistemas Agroforestales: Principios y Aplicaciones en los Tropicos. San Jose, Costa Rica; Organizacion para Estudios Tropicales, Centro Agronomico Tropical de Investigacion y Enseanza, 1986. 818p.	Third
	Sjogren, Erik, ed. Forests of the World: Diversity and Dynamics (Abstracts). Stockholm, Sweden; Almqvist & Wiksell International, 1989. 295p.	Third
Third	Sjostrom, Eero. Wood Chemistry: Fundamentals and Applications. New York; Acadmic Press, 1981. 223p.	

Developed Countries Ranking		Third World Ranking
Third	Skaar, Christen. Water in Wood. Syracuse, N.Y.; Syracuse University Press, 1972. 218p. (Syracuse Wood Science Series no. 4)	
Third	Slatyer, R. O. Plant-Water Relationships. London and New York; Academic Press, 1967. 366p.	
	Smil, Vaclav, and William E. Knowland, eds. Energy in the Developing World: The Real Energy Crisis. Oxford, U.K. and New York; Oxford University Press, 1980. 386p.	Third
First	Smith, David M. The Practice of Silviculture. 8th ed. New York; J. Wiley, 1986. 527p. (Previous eds. by Ralph C. Hawley. 1st ed., 1921. 352p.)	First
Third	Smith, Keith A., ed. Soil Analysis: Instrumental Techniques and Related Procedures. New York; M. Dekker, 1983. 562p.	
Third	Smith, Keith A., and Chris E. Mullins, eds. Soil Analysis: Physical Methods. New York; M. Dekker, 1991. 620p.	Second
	Smith, Kenneth M. A Textbook of Plant Virus Diseases. 3d ed. London; Longmans, 1972. 684p. (1st ed., London; J. A. Churchill, 1937. 615p.)	Third
Third	Smith, V. Kerry, and John V. Krutilla, eds. Explorations in Natural Resource Economics. Baltimore; Published for Resources for the Future by Johns Hopkins University Press, 1982. 352p.	
Second	Smith, William H. Air Pollution and Forests: Interactions between Air Contaminants and Forest Ecosystems. New York; Springer-Verlag, 1981. 379p.	
	Smith, William H. Tree Pathology: A Short Introduction; The Mechanisms and Control of Pathological Stresses of Forest Trees. New York; Academic Press, 1970. 309p.	Third
Third	Smook, G. A., and M. J. Kocurek, eds. Handbook for Pulp and Paper Technologists. Atlanta, Ga.; TAPPI; and Montreal; Canadian Pulp and Paper Association, 1982. 395p.	
	Smucker, Glenn R., ed. Planting Trees with Small Farmers: A Planning Workshop; Proceedings. Washington, D.C.; Pan American Development Foundation, 1987. 189p.	Third
Second	Sociedade Brasileira de Silvicultura. Fast Growing Trees; Proceedings of a Symposium . . . IUFRO em Melhoramento Genetico e Produtividade de Especies Florestais de Rapido Crescimento, Sao Paulo. Sao Paulo, Brazil; SBS, 1983. 4 vols. (Text in Portuguese, English, or French, with summaries in all three languages.)	First
First	Society of American Foresters. Forestry Handbook . . . edited by Karl F. Wenger. 2d ed. New York; J. Wiley, 1984. 1335p. (1st ed., edited by Reginald D. Forbes, New York; Ronald Press, 1955. 1140p.)	
	Soepadmo, E., A. N. Rao, and D. J. Macintosh, eds. Asian Symposium on Mangrove Environment Research and Manage-	Third

Developed Countries Ranking		Third World Ranking
	ment, Kuala, Lumpur, Malaysia, 1980; Proceedings . . . Kuala Lumpur, Malaysia; University of Malaya and UNESCO, 1984. 828p.	
Third	Solomon, J. D., et al. Oak Pests: A Guide to Major Insects, Diseases, Air Pollution and Chemical Injury. New Orleans, La.; USDA Forest Service, 1987. 68p. (Protection Report R8 no. PR 7)	
Third	Sonnenfeld, Jeffrey A. Corporate Views of the Public Interest: Perceptions of the Forest Products Industry. Boston; Auburn House, 1981. 285p.	
First	Sopper, William E., ed. Forest Hydrology; Proceedings of a National Science Foundation Advanced Science Seminar, Pennsylvania State University, University Park, Aug.-Sept. 1965. 1st ed. Oxford, U.K. and New York; Symposium Publications, Pergamon Press, 1967. 813p.	
First	South, David B., ed. International Symposium on Nursery Management Practices for the Southern Pines; Proceedings . . . Montgomery, Ala., Aug. 1985. Auburn, Ala.; Auburn University, Dept. of Research Information, 1986. 594p.	
	Southgate, Douglas D., and John F. Disinger, eds. Sustainable Resource Development in the Third World. Boulder, Colo.; Westview Press, 1987. 177p.	Second
Third	Sprent, Janet I., and Peter Sprent. Nitrogen Fixing Organisms: Pure and Applied Aspects. London and New York; Chapman & Hall, 1990. 256p.	Third
Second	Spurr, Stephen H. Forest Inventory. New York; Ronald Press, 1952. 476p.	
First	Spurr, Stephen H., and Burton V. Barnes. Forest Ecology. 3d ed. Malabar, Fla.; Krieger Pub., 1991. n.p. (Earlier ed., Ann Arbor, Mich.; Ann Arbor Pub., 1962. 290p.)	
	Srivastava, P. B. L., et al., eds. Tropical Forests: Source of Energy through Optimisation and Diversification; Proceedings of an International Forestry Seminar, Serdang, Selangor, Malaysia, Nov. 1980. Serdang, Malaysia; Universiti Pertanian Malaysia, 1982. 397p.	Second
Third	Stamm, Alfred J. Wood and Cellulose Science. New York; Ronald Press, 1964. 549p.	
	Starfield, A. M., and A. L. Bleloch. Building Models for Conservation and Wildlife Mangement. 2d ed. Edina, Minn.; Burgess international Group, 1991. 253p. (1st ed., 1986.)	Third
	Steiner, Kurt G. Intercropping in Tropical Smallholder Agriculture with Special Reference to West Africa. 2d ed. Eschborn, Germany; Germany Agency for Technical Cooperation, 1984. 304p. (1st ed., 1982. 303p.)	Second
Third	Stenzel, George, Thomas A. Walbridge, and J. Kenneth Pearce.	Third

Logging and Pulpwood Production. 2d ed. New York; J.
Wiley & Sons, 1985. 358p. (1st ed., 1972. 453p.)

Steppler, Howard A., and P. K. R. Nair, eds. Agroforestry: A   First
Decade of Development. Nairobi, Kenya; International Council for Research in Agroforestry, 1987. 335p.

Third   Stern, Klaus, and Laurence Roche. Genetics of Forest Ecosystems. Berlin and New York; Springer-Verlag, 1974. 330p.

Third   Stevens, William C. Wood Bending Handbook. Woburn, Mass.;
Woodcraft Supply Corp., 1978, 1970. 110p. (Reprint. 1st ed.
by W. C. Stevens and N. Turner entitled, Solid and Laminated
Wood Bending. London: H. M. Stationery Off., 1948. 71p.)

Third   Stevenson, F. J. Cycles of Soil: Carbon, Nitrogen, Phosphorus,
Sulfur, Micronutrients. New York; J. Wiley, 1986. 380p.

Second   Stipes, R. Jay, and Richard J. Campana, eds. Compendium of
Elm Diseases. St. Paul, Minn.; American Phytopathological
Society, 1981. 96p.

Third   Stoddard, Charles H., and Glenn M. Stoddard. Essentials of Forestry Practice. 4th ed. New York; J. Wiley, 1987. 407p. (1st
ed., 1959. 258p.)

Second   Stone, Earl L., ed. Forest Soils and Treatment Impacts; Proceedings of the 6th North American Forest Soils Conference, University of Tennessee, Knoxville, June 1983. Knoxville, Tenn.;
Dept. of Forestry, Wildlife and Fisheries, University of Tennessee, 1984. 454p.

Stonehouse, B., ed. Biological Husbandry: A Scientific Ap-   Third
proach to Organic Farming. London and Boston; Butter-
worths, 1981. 352p.

Third   Suchsland, Otto, and George E. Woodson. Fiberboard Manufacturing Practices in the United States. Updated ed. Washington,
D.C.; USDA Forest Service, 1990. 262p. (Earlier ed., 1987.
263p.)

Third   Sundberg, Ulf. Harvesting Man-Made Forests in Developing   Second
Countries: A Manual on Techniques, Roads, Production and
Costs. Rome; Food and Agriculture Organization, 1976. 185p.

Third   Sutherland, Jack R., Thomas Miller, and Rodolfo S. Quinard,
eds. Cone and Seed Diseases of North American Conifers.
Vancouver, British Columbia; Ministry of Supply and Services, Canada, 1987. 77p. (North American Forestry Commission Publication no. 1)

Third   Sutton, R. F. Form and Development of Conifer Root Systems.
Farnham Royal, U.K.; Commonwealth Agricultural Bureaux,
1969. 131p. (Commonwealth Forestry Bureau, Technical
Communication no. 7)

First   Sutton, S. L., T. C. Whitmore, and A. C. Chadwick, eds. Tropi-   First
cal Rain Forest: Ecology and Management. Oxford, U.K. and

Developed Countries Ranking		Third World Ranking
	Boston; Blackwell Scientific; and St. Louis, Mo.; Blackwell Mosby, 1983. 498p.	
Second	Swank, Wayne T., and D. A. Crossley, eds. Forest Hydrology and Ecology at Coweeta. Berlin and New York; Springer-Verlag, 1988. 496p.	
Second	Swift, Michael J., O. W. Heal, and J. M. Anderson. Decomposition in Terrestrial Ecosystems. Oxford, U.K.; Blackwell Scientific, 1979. 372p.	Third
Third	Symposium on the Loblolly Pine Ecosystem. East Region. Proceedings . . . Raleigh, N.C., Dec. 1982. Raleigh, N.C.; School of Forest Resources, North Carolina State University, 1983. 335p.	
First	Tennessee Valley Authority. Forest Fertilization: Theory and Practice; Proceedings of a Symposium . . . University of Florida, 1967. Muscle Shoals, Ala.; TVA, National Fertilizer Development Center, 1968. 306p.	
Third	Thielges, Bart A., compiler and ed. Physiology and Genetics of Intensive Culture; Proceedings of the 7th North American Forest Biology Workshop, Lexington, Ky., July 1982. Lexington, Ky.; Dept. of Forestry, University of Kentucky, 1982. 467p.	
Third	Thirgood, J. V. Man and the Mediterranean Forest: A History of Resource Depletion. London and New York; Academic Press, 1981. 194p.	
Second	Thomas, Jack W., ed. Wildlife Habitats in Managed Forests: The Blue Mountains of Oregon and Washington. Washington, D.C.; U.S. Govt. Print. Off., 1979. 512p. (USDA Agriculture Handbook no. 553)	
Third	Thomas, Jack W., and Dale E. Toweill, compilers and eds. Elk of North America: Ecology and Management. Harrisburg, Pa.; Stackpole Books, 1982. 698p.	
Third	Thorpe, Trevor A., ed. Plant Tissue Culture: Methods and Application in Agriculture. New York; Academic Press, 1981. 379p.	Second
Second	Tigerstedt, Peter M. A., Pasi Puttonen, and Veikko Koski, compilers and eds. Crop Physiology of Forest Trees; Proceedings of an International Conference on Managing Forest Trees as Cultivated Plants, Finland, July 1984. Helsinki, Finland; University of Helsinki, Dept. of Plant Breeding, 1985. 336p.	
Second	Tillman, David A. Forest Products: Advanced Technologies and Economic Analyses. Orlando, Fla.; Academic Press, 1985. 283p.	
Third	Tillman, David A. Wood as an Energy Resource. New York; Academic Press, 1978. 252p.	Third
Third	Tinus, Richard W., and Stephen E. McDonald. How to Grow Tree Seedlings in Containers in Greenhouses. Bottineau,	

Developed Countries Ranking		Third World Ranking
	N.Dak.; USDA Forest Service, Rocky Mountain Forest and Range Experiment Station, 1979. 256p. (USDA General Technical Report RM-60)	
Third	Tiwari, K. M. Social Forestry in India. Dehra Dun, India; Natraj, 1983. 296p.	
Third	Tober, James A. Wildlife and the Public Interest: Nonprofit Organizations and Federal Wildlife Policy. New York; Praeger, 1989. 220p.	
	Toda, Ryookitit, ed. Forest Tree Breeding in the World. Meguro, Japan; Government Forest Experimental Station of Japan, 1974. 205p.	Third
	Tomlinson, P. B., and Martin H. Zimmermann, eds. Tropical Trees as Living Systems; Proceedings of the 4th Cabot Symposium, Harvard Forest, Petersham, Mass., April 1976. Cambridge, U.K. and New York; Cambridge University Press, 1978. 675p.	First
	Tomlinson, Philip B. The Botany of Mangroves. Cambridge, U.K. and New York; Cambridge University Press, 1986. 413p.	Second
Third	Tranquillini, Walter. Physiological Ecology of the Alpine Timberline: Tree Existence in High Altitudes with Special Reference to the European Alps . . . trans. by Udo Benecke. Berlin and New York; Springer-Verlag, 1979. 137p.	
Third	Treshow, Michael. Environment and Plant Response. New York; McGraw-Hill, 1970. 422p.	
	Troup, Robert S. Troup's The Silviculture of Indian Trees. Delhi, India; Controller of Publications, 1975. 6 vols. (1st ed., Oxford; Clarendon Press, 1921. 3 vols.)	Second
	Turnbull, John W., ed. Australian Acacias in Developing Countries; Proceedings of an International Workshop, Forestry Training Centre, Gympie, Queensland, Aug. 1986. Canberra; Australian Centre for International Agricultural Research, 1987. 196p. (ACIAR Proceedings no. 16)	Second
	Turnbull, John W., ed. Tropical Tree Seed Reasearch; Proceedings of an International Workshop, Forestry Training Centre, Gympie, Australia, Aug. 1989. Canberra; Australian Centre for International Agricultural Research, 1990. 156p. (ACIAR Proceedings no. 28)	Second
First	Turner, Neil C., and Paul J. Kramer, eds. Adaptation of Plants to Water and High Temperature Stress. New York; J. Wiley, 1980. 482p.	
Second	Ulrich, B., and J. Pankrath, eds. Effects of Accumulation of Air Pollutants in Forest Ecosystems; Proceedings of a Workshop, Gottingen, Germany, May 1982. Dordrecht and Boston; D. Reidel Pub. Co., 1983. 389p.	

Developed Countries Ranking		Third World Ranking
	UNESCO. Tropical Forest Ecosystems: A State-of-Knowledge Report . . . prepared by UNESCO, UNEP, FAO. Paris; UNESCO, 1978. 683p. (Natural Resources Research no. 14)	First
Third	United Nations Economic Commission for Europe, Food and Agriculture Organization. European Timber Trends and Prospects to the Year 2000 and Beyond. New York; United Nations, 1986. 2 vols.	
Third	U.S. Bureau of Land Management. Manual of Instructions for the Survey of the Public Lands of the United States, 1973. Washington, D.C.; U.S. Dept. of Interior, 1983. 333p. (Technical Bulletin no. 6) (1st ed., 1931. 530p.)	
Third	U.S. Bureau of Outdoor Recreation. Outdoor Recreation; A Legacy for America. Washington, D.C.; The Bureau, 1973. 89p.	
	U.S. Congress, Office of Technology Assessment. Technologies to Sustain Tropical Forest Resources. Washington, D.C.; U.S. Govt. Print. Off., 1984. 344p.	Second
Third	U.S. Forest Service. Atlas of United States Trees. Washington, D.C.; U.S. Dept. of Agriculture, 1971–1981. 6 vols. (USDA Misc. Pub. no. 1146)	
	U.S. Forest Service. Evaluacion de Tierras y Recursos para la Planeacion nacional en las Zonas tropicales = Land and Resource Evaluation for National Planning in the Tropics; Proceedings of an International Conference and Workshop, Chetumal, Mexico, Jan. 1987. Washington, D.C.; USDA Forest Service, 1987. 524p. (General Technical Report WO no. 39)	First
Third	U.S. Forest Service. Fire Regimes and Ecosystems Properties; Proceedings of a Conference, Dec. 1978, Honolulu. Washington, D.C.; USDA Forest Service, 1981. 594p. (General Technical Report WO no. 26)	
Second	U.S. Forest Service. Insects of Eastern Forests. Washington, D.C.; USDA Forest Service, 1985. 608p. (USDA Miscellaneous Publication no. 1426)	
Third	U.S. Forest Service. National Forest Log Scaling Handbook. 3d ed. Washington, D.C.; USDA Forest Service, 1985. 247p. (1st ed., 1964. 193p.) (2d ed., 1977. 184p.)	
Third	U.S. Forest Service. The Outlook for Timber in the United States. Washington, D.C.; U.S. Govt. Print. Off., 1973. 367p. (Forest Resource Report no. 20)	
Second	U.S. Forest Service. The Principal Laws Relating to Forest Service Activities. Rev. ed. Washington, D.C.; USDA Forest Service, 1983. 591p. (USDA Agriculture Handbook no. 453) (Earlier ed., 1974.)	
Third	U.S. Forest Service. RPA Assessment of the Forest and Rangeland Situation in the United States, 1989. Washington, D.C.;	

Developed Countries Ranking		Third World Ranking
	USDA Forest Service, 1989. 72p. (Forest Resource Report no. 26)	
Second	U.S. Forest Service. Seeds of Woody Plants in the United States; prepared by the Forest Service; technical coordinator, C.S. Schopmeyer. Washington, D.C.; Forest Service, U.S. Dept. of Agriculture, 1974. 883p. (USDA Agriculture Handbook no. 450)	
Third	U.S. Forest Service. The South's Third Forest: Alternatives for the Future. Washington, D.C.; USDA Forest Service, 1988. 512p. (Forest Resource Report no. 24)	
Third	U.S. Forest Service. Symposium on Dwarf Mistletoe Control through Forest Management, Berkeley, April 1978; Proceedings . . . Berkeley, Calif.; USDA Forest Service, Pacific Southwest Forest and Range Experiment Station, 1978. 190p. (General Technical Report PSW no. 31)	
Third	U.S. Forest Service, Southern Region. Southern Pine Nursery Handbook. Atlanta, Ga.; U.S. Dept. of Agriculutre, 1984. 1 vols.	
Third	U.S. Forest Products Laboratory. Effect of Growth Acceleration on the Properties of Wood; Proceedings of a Symposium, University of Wisconsin, 1971. Madison, Wis.; USDA Forest Service, Forest Products Laboratory, 1972. 1 vols.	
Third	U.S. Forest Products Laboratory. How the Environment Affects Lumber Design: Assessment and Recomendations; Proceedings of a Workshop sponsored by Society of Wood Science and Technology. Madison, Wis.; USDA Forest Service, 1982. 146p.	
	U.S. Forest Products Laboratory. Proceedings of the Conference on Improved Utilization of Tropical Forests, Madison, Wisc., May 1978, sponsored by Forest Products Laboratory, Forest Service, U.S. Dept. of Agriculture, and Agency for International Development, U.S. Dept. of State. Madison, Wisc.; USDA, Forest Products Laboratory, 1978. 569p.	Third
Third	U.S. Forest Products Laboratory. Wood Engineering Handbook. 2d ed. Englewood Cliffe, N.J.; Prentice-Hall, 1990. 1 vols. (Earlier ed., 1974, as Wood Handbook.)	
Second	U.S. Forest Products Laboratory. Wood Handbook: Wood as an Engineering Material. Rev. ed. Washington, D.C.; USDA Forest Service, 1987. 1 vols. (USDA Agricultural Handbook no. 72) (Earlier ed., 1974. 431p.)	
	U.S. Forest Products Laboratory PASA Study Team. Forestry Activities and Deforestation Problems in Developing Countries. Madison, Wis.; Forest Products Laboratory, 1980. 115p.	Third
	U.S. Forest Products Research and Industries Development Commission. Guidelines for the Improved Utilization and	Third

Developed Countries Ranking		Third World Ranking
	Marketing of Tropical Wood Species. College, Laguna, Philippines; FORPRIDECOM, 1980. 153p.	
	U.S. Interagency Task Force on Tropical Forests. The World's Tropical Forests: A Policy, Strategy, and Program for the United States; Report to the President. Washington, D.C.; U.S. Govt. Print. Off., 1980. 53p. (International Organization and Conference Series no. 145)	Third
Third	U.S. Soil Conservation Service. Soil Taxonomy: A Basic System of Soil Classification for Making and Interpreting Soil Surveys. Washington, D.C.; U.S. Govt. Print. Off., 1975. 754p. (USDA Agriculture Handbook no. 436)	
	Vainio-Mattila, Arja. Bura Fuelwood Project: Domestic Fuel Economy. Helsinki, Finland; Institute of Development Studies, University of Helsinki, 1987. 164p.	Third
Third	Valentine, Fredrick A., ed. Forest and Crop Biotechnology: Progress and Prospects. New York; Springer-Verlag, 1988. 466p.	Third
	Van Beusekom, C. F., C. P. Van Goor, and P. Schmidt, eds. Wise Utilization of Tropical Rain Forest Lands. Ede, Netherlands; Tropenbos Programme, 1987. 154p.	Third
Second	Van der Plank, J. E. Plant Diseases: Epidemics and Control. New York; Academic Press, 1963. 1p. 349 vols.	
Third	Vandermeer, John H. The Ecology of Intercropping. Cambridge, U.K. and New York; Cambridge University Press, 1989. 237p.	Second
	Vergara, Napoleon T., and Nicomedes D. Briones, eds. Agroforestry in the Humid Tropics: Its Protective and Ameliorative Roles to Enhance Productivity and Sustainability. Honolulu, Hawaii; Environment and Policy Institute, East-West Center, 1987. 259p.	First
	Vogel, E. F. de. Seedlings of Dicotyledons: Structure, Development, Types: Descriptions of 150 Woody Malesian Taxa. Wageningen, Netherlands; Centre for Agricultural Pub. & Documentation, 1980. 465p.	Third
	Von Carlowitz, Peter G. Multipurpose Tree and Shrub Seed Directory. Nairobi, Kenya; International Council for Research in Agroforestry, 1986. 265p.	Second
Third	Wagenfuhr, Rudi. Anatomie des Holzes: unter besonderer Beruchsichtigung der Holztechnik = Wood Anatomy. 3d ed. Leipzig, Germany; VEB Gachbuchverlag, 1984. 320p.	
Third	Walstad, John D., and Peter J. Kuch, ed. Forest Vegetation Management for Conifer Production. New York; J. Wiley, 1987. 523p.	
Third	Walter, Heinrich. Die Vegetation der Erde in Oko-Physiologisher Betrachtung. 3d ed. Stuttgart, Germany; Fischer, 1973 +.	

Developed Countries Ranking		Third World Ranking
Second	Wardlaw, I. F., and J. B. Passioura, eds. Transport and Transfer Processes in Plants. New York; Academic Press, 1976. 484p.	
Second	Wareing, P. F., ed. Plant Growth Substances 1982; Proceedings of the 11th International Conference, Aberystwyth, July 1982. London and New York; Academic Press, 1982. 683p.	
Third	Wareing, P. F., and D. J. Phillips. Growth and Differentiation in Plants. 3d ed. Oxford, U.K. and New York; Pergamon Press, 1981. 343p. (Originally titled: The Control of Growth and Differentiation in Plants. 1st ed., 1970. 303p.)	
Second	Waring, Richard H., ed. Forests, Fresh Perspectives from Ecosystem Analysis; Proceedings of the 40th Annual Biology Colloquium, Oregon State University, 1979. Corvallis, Oreg.; Oregon State University Press, 1980. 199p.	First
First	Waring, Richard H., and William H. Schlesinger. Forest Ecosystems: Concepts and Management. Orlando, Fla.; Academic Press, 1985. 340p.	
Third	Watt, Allan D., et al., eds. Population Dynamics of Forest Insects; Proceedings of a Conference, Edinburgh, Sept. 1989. Andover, U.K.; Intercept, 1990. 408p.	
	Wattle Research Institute. Handbook on Eucalypt Growing: Notes on the Management of Eucalypt Plantations Grown for Timber in the Wattle-Growing Regions of South Africa. Pietermaritzburg, South Africa; Wattle Research Institute, 1972. 164p.	Third
	Webb, Derek B., et al. A Guide to Species Selection for Tropical and Sub-Tropical Plantations. 2d ed., rev. Oxford, U.K.; Unit of Tropical Silviculture, Commonwealth Forestry Institute, University of Oxford, 1984. 256p. (Tropical Forestry Papers no. 15) (1st ed., 1980. 342p.)	First
	Weber, Edward, Barry Nestel, and Marilyn Campbell, eds. Intercropping with Cassava; Proceedings of an International Workshop, Trivandrum, India, Nov.-Dec. 1978. Ottawa, Canada; International Development Research Centre, 1979. 142p.	Third
Third	Weber, Fred R. Reforestation in Arid Lands. Arlington, Va.; Volunteers in Technical Assistance, 1986. 335p. (Earlier ed, 1977. 248p.)	Third
	Weber, Fred R., and Marilyn Hoskins. Agroforestry in the Sahel. Blacksburg, Va.; Dept. of Sociology, Virginia Polytechnic Institute and State University, 1983. 102p.	Third
	Weidelt, Hans J., and Valeriano S. Banaag. Aspects of Management and Silviculture of Philippine Dipterocarp Forests. Eschborn, Germany; Deutsche Gesellschaft fur Technische Zusammenarbeit, 1982. 302p.	Second
Second	Wein, Ross W., and David A. MacLean, eds. The Role of Fire in Northern Circumpolar Ecosystems. Chichester, U.K. and New	

Developed Countries Ranking		Third World Ranking
	York; Published for Scientific Committee on Problems of the Environment,International Council of Scientific Unions by J. Wiley, 1983. 322p.	
	Werner, D., and P. Muller, eds. Fast Growing Trees and Nitrogen Fixing Trees; Proceedings of the 1st International Conference . . . Marburg, Oct. 1989. Stuttgart and New York; G. Fischer Verlag, 1990. 396p.	Third
First	West, Darrell C., Herman H. Shugart, and Daniel B. Botkin, eds. Forest Succession: Concepts and Application. New York; Springer-Verlag, 1981. 517p.	
Second	Westcott, Cynthia. Westcott's Plant Disease Handbook . . . rev. by R. Kenneth Horst. 5th ed. New York; Van Nostrand Reinhold, 1990. 953p. (1st ed., 1950. 746p.)	
Second	Westoby, J. C. Introduction to World Forestry: People and Their Trees. Oxford, U.K. and New York; Blackwell, 1989. 228p.	Second
First	Westoby, Jack. The Purpose of Forests: Follies of Development. Oxford, U.K.; Blackwell, 1987. 343p.	Second
	Weyerhaeuser Science Symposium. Forest Plantations: The Shape of the Future; Proceedings of a Symposium, Tacoma, Wash., April-May 1979. Tacoma, Wash.; Weyerhaeuser Co., 1980. 224p. (Weyerhaeuser Science Symposium no. 1)	Third
Third	White, Timothy L., and Gary R. Hodge. Predicting Breeding Values with Applications in Forest Tree Improvement. Dordrecht and Boston; Kluwer Academic, 1989. 367p.	
Second	Whitmore, T. C. An Introduction to Tropical Rain Forests. Oxford, U.K.; Clarendon Press; New York; Oxford University Press, 1990. 226p.	
First	Whitmore, T. C. Tropical Rain Forests of the Far East. 2d ed. Oxford, U.K.; Clarendon Press, 1984. 352p. (1st ed., 1975. 282p.)	First
First	Whittaker, Robert H. Communities and Ecosystems. 2d ed. New York; Macmillan, 1975. 385p. (1st ed., 1970. 162p.)	
	Wiersum, K. F. Forestry Aspects of Stabilizing Shifting Cultivation in Africa. Wageningen, Netherlands; Forestry Dept., Wageningen Agricultural University, 1985. 196p.	Second
Second	Wiersum, K. F., ed. Strategies and Designs for Afforestation, Reforestation and Tree Planting; Proceedings of an International Symposium . . . Wageningen, Sept. 1983. Wageningen, Netherlands; Centre for Agricultural Publishing and Documentation, 1984. 432p.	First
	Wiersum, K. F., ed. Viewpoints on Agroforestry II: Reader on Aspects of Agroforestry based on Review Articles and Lecture Notes of Staff, Agricultural University Wageningen. 2d ed., rev. Wageningen; Dept. of Forestry "Hinkeloord," Agri-	First

Developed Countries Ranking		Third World Ranking
	cultural University Wageningen, 1988. 256p. (Earlier ed., 1981. 185p.)	
	Wilde, Sergius A., et al. Soil and Plant Analysis for Tree Culture. 5th ed., rev. New Delhi, India; Oxford University Press & IBH Pub. Co., 1979. 224p. (3d ed., rev., Calcutta; 1964. 209p.)	Third
Third	The Wilderness Society. Below-Cost Sales; Proceedings of a Conference on the Economics of National Forest Timber Sales, Spokane, Wash., Feb. 1986. Washington, D.C.; Wilderness Society, 1987. 266p.	
Third	The Wilderness Society. National Forest Planning: A Conservationist's Handbook. Washington, D.C.; Wilderness Society, 1983. 121p.	
Third	The Wilderness Society. National Forests: Policies for the Future. Washington, D.C.; Wilderness Society, 1988. 5 vols.	
Third	The Wilderness Society. Protecting Roadless Lands in the National Forest Planning Process: A Citizen Handbook. Washington, D.C.; Wilderness Society, 1985.	
Second	Wilkins, Malcolm B., ed. Advanced Plant Physiology. London and Marshfield, Mass.; Pitman, 1984. 514p.	
	Williams, C. N., and K. T. Joseph. Climate, Soil and Crop Production in the Humid Tropics. Rev. ed. Kuala Lumpur, Malaysia and New York; Oxford University Press, 1973. 177p. (1st ed., 1970. 177p.)	Third
Third	Williams, M. R. W. Decision-Making in Forest Management. Chichester, U.K. and New York; Research Studies Press, 1981. 143p.	
Third	Williams, Michael. Americans and Their Forests: A Historical Geography. Cambridge, U.K. and New York; Cambridge University Press, 1989. 599p.	
Third	Williams, Peter, ed. Agroforestry in North America; Proceedings of the 1st Conference . . . University of Guelph, Aug. 1989. Guelph; Ontario, Ministry of Agriculture and Food, 1991. 262p.	Third
Third	Williston, Ed M. Lumber Manufacturing: The Design and Operation of Sawmills and Planer Mills. San Francisco, Calif.; Miller Freeman Pub., 1976. 512p.	
Third	Williston, Ed M. Manufacturing Lumber from Small Logs; Proceedings . . . Seattle, Wash.; University of Washington, Institute of Forest Resources, 1982. 158p.	
Third	Wilson, Brayton F. The Growing Tree. Rev. ed. Amherst, Mass.; University of Massachusetts Press, 1984. 138p. (1st ed., 1970. 152p.)	
Third	Wilson, E. O., and Frances M. Peter, eds. Biodiversity; National	Third

Developed Countries Ranking		Third World Ranking
	Forum on Biodiversity, Washington, D.C., 1986. Washington, D.C.; National Academy Press, 1988. 521p.	
	Winterbottom, Robert. Taking Stock: The Tropical Foresty Action Plan After Five Years. Washington, D.C.; World Resources Institute, 1990. 59p.	Third
	Withington, Dale, et al., eds. Multipurpose Tree Species for Small-Farm Use; Proceedings of an International Workshop, Pattaya, Thailand, Nov. 1987. Arlington, Va.; Winrock International Institute for Agricultural Development; Ottawa, International Development Research Centre of Canada, 1988. 282p.	Third
	Withington, Dale, Nancy Glover, and James L. Brewbaker, eds. Gliricidia sepium (JACQ.) WALP.: Management and Improvement; Proceedings of a Workshop, Turrialba, Costa Rica, June 1987. Waimanalo, Hawaii; Nitrogen Fixing Tree Association, 1987. 255p. (NFTA Special Publication no. 87–01)	Third
Third	Wolf, Paul R. Elements of Photogrammetry, with Air Photo Interpretation and Remote Sensing. 2d ed. New York; McGraw-Hill, 1983. 628p. (1st ed., 1974. 562p.)	
Second	Wood, Gene W., ed. Prescribed Fire and Wildlife in Southern Forests. Georgetown, S.C.; Belle W. Baruch Forest Science Institute, Clemson University, 1981. 170p.	
	Workshop Agro-Forestry Systems in Latin America, Turrialba, Costa Rica, 1979. Proceedings . . . Turrialba, Costa Rica; Centro Agronomico Tropical de Investigacion y Ensenanza, Program of Natural Renewable Resources, 1979. 220p.	Third
Third	The World Bank. The Forest Sector. Washington, D.C.; World Bank, 1991. 98p. (A World Bank Policy Paper)	Third
Third	World Commission on Environment and Development. Our Common Future. Oxford, U.K. and New York; Oxford University Press, 1987. 400p.	Second
	World Consultation on Forest Genetics and Tree Improvement, Stockholm, 1963. Proceedings . . . Rome; Food and Agriculture Organization, 1963. 2 vols.	Second
	World Consultation on Forest Tree Breeding, 2d, Washington, D.C., 1969 Rome; Food and Agriculture Organization, 1970. 2 vols.	Second
	World Consultation on Forest Tree Breeding, 3d, Canberra, Australia, Mar. 1977. Proceedings . . . = Troisi'em Consultaion Mondiale sur L'Amelioration des Arbes Foreti'eres = Tercera Consuta Mondiale Sobre Mejora de Arboles Forestales. Canberra, Austarlia; Commonwealth Scientific and Industrial Research Organization, 1978. 3 vols.	Second
	World Forestry Congress, 7th, Buenos Aires, 1972. Proceedings . . . Buenos Aires, Argentina; WFC, 1972. 2 vols.	Second

Developed Countries Ranking		Third World Ranking
	World Forestry Congress, 8th, Jakarta, Indonesia, 1978. Forests for People. Hamburg, Germany; Kommissionverlag Buch, 1979. 211p.	First
	World Forestry Congress, 9th, Mexico City, 1985. Proceedings . . . Mexico; WFC, 1985. 1 vol.	Second
Second	World Resources Institute. Tropical Forests, a Call for Action; Report of an International Task Force. Washington, D.C.; WRI, 1985. 3 vols.	First
Third	Worrell, Albert C. Economics of American Forestry. New York; J. Wiley, 1963. 441p. (Originally published, 1959.)	
Second	Worrell, Albert C. Principles of Forest Policy. New York; McGraw-Hill, 1970. 243p.	
Third	Wright, Henry A., and Arthur W. Bailey. Fire Ecology, United States and Southern Canada. New York; J. Wiley, 1982. 501p.	
	Wright, Jonathan W. Genetics of Forest Tree Improvement. Rome; Food and Agriculture Organizaiton, 1962. 399p. (FAO Forestry and Forest Products Studies no. 16)	Third
First	Wright, Jonathan W. Introduction to Forest Genetics. New York; Academic Press, 1976. 463p.	First
Third	Wright, R. Gerald. Wildlife Research and Management in the National Parks. Urbana, Ill. and Chicago; University of Illinois Press, 1992. 223p.	
	Young, Anthony. Agroforestry for Soil Conservation. Wallingford, U.K.; CAB International, 1989. 276p. (Science and Practice of Agroforestry no. 4)	First
	Young, Anthony. The Potential of Agroforestry for Soil Conservation. Nairobi, Kenya; International Council for Research in Agroforestry, 1986–1987. 3 vols. (ICRAF Working Paper no. 42–44)	Second
	Young, Anthony. Tropical Soils and Soil Survey. Cambridge, U.K. and New York; Cambridge University Press, 1976. 468p. (Cambridge Geographical Studies no. 9)	Second
First	Young, James A., and Cheryl G. Young. Seeds of Woody Plants in North America. Rev. and enl. ed. Portland, Oreg.; Dioscorides Press, 1992. n.p. (Earlier ed. as Seeds of Woody Plants in the United States, 1974.)	
Second	Young, Raymond A., and Ronald L. Giese, eds. Introduction to Forest Science. New York; J. Wiley, 1990. 586p. (1st ed., 1982. 554p.)	
	Zhaohua, Zhu, Cai Mantang, Wang Shiji, and Jiang Youxu, eds. Agroforestry Systems in China. Beijing; Chinese Academy of Forestry; Ottawa; International Development Research Centre, 1991. 216p.	Second
	Ziller, Wolf G. The Tree Rusts of Western Canada. Ottawa; Information Canada, 1974. 272p.	Third

Developed Countries Ranking		Third World Ranking
First	Zimmerman, Martin H. Xylem Structure and the Ascent of Sap. Berlin and New York; Springer-Verlag, 1983. 143p.	
	Zimmerman, Robert C. Environmental Impact of Forestry: Guidelines for Its Assessment in Developing Countries. Rome; Food and Agriculture Organization, 1982. 85p. (FAO Conservation Guide no. 7)	Second
First	Zimmermann, M. H., and J. A. Milburn, eds. Transport in Plants, 1: Phloem Transport. Berlin and New York; Springer-Verlag, 1975. 535p.	
First	Zimmermann, Martin H., and Claud L. Brown. Trees: Structure and Function. New York; Springer-Verlag, 1971. 336p.	
Third	Zobel, Bruce, and Johannes P. van Buijtenen. Wood Variation: Its Causes and Control. Berlin and New York; Springer-Verlag, 1989. 363p.	
First	Zobel, Bruce, and John Talbert. Applied Forest Tree Improvement. New York; J. Wiley, 1984. 505p. (Reprinted, 1991.)	
Second	Zobel, Bruce, Gerrit Van Wyk, and Per Stahl. Growing Exotic Forests. New York; J. Wiley, 1987. 508p.	Second
Third	Zottl, H. W., and R. F. Huttl, eds. Management of Nutrition in Forests under Stress; Proceedings of an International Symposium, Freiburg, Germany, Sept. 1989. Dordrecht and Boston; Kluwer Academic, 1991. 668p.	Third
	Zulberti, Ester, ed. Professional Education in Agroforestry; Proceedings of an International Workshop. Nairobi, Kenya; International Council for Research in Agroforestry, 1987. 378p.	Second

H. Core Monograph Holdings in Four Libraries

An effort was made to determine the improvements in library collections that can be expected when the compact disks with full text of the core literature are issued. To provide this determination, the Core Agricultural Literature Project asked selected libraries to check the core monograph lists against their library holdings. Three libraries supplied these data:

N = 1,054 titles (both lists)	Owned	Earlier eds. only
Michigan State University	83%	3.0%
Institut Penyelidikan Perhutanan Malaysia (Forest Research Institute)	33	.5
University of British Columbia MacMillan Library	81	1.0

The MacMillan Library with a strong forestry collection lacked only 8.4% of the developed country titles (n = 669). Most of the missing titles, 141, were Third World titles only.

The list of core Third World monographs only (n = 528) was checked by the library of the International Centre for Research in Agroforestry (ICRAF) in Nairobi, Kenya. The center and library have existed fifteen years which matches the dates of the core monograph list where 75% of the titles were published in the 1980s. This institution has had good funding and receives a quantity of literature by gift particularly from U.N. agencies. The ICRAF library probably has the best literature collection for agroforestry and Third World forestry in a developing country. ICRAF had 55.9% of all titles in the Third World list of core monographs. Clearly the compact disk library would be of major substantive assistance to the ICRAF library as well as for the Malaysian Forest Research Institute.

12. Primary Journals and Serials in Forestry and Agroforestry

PETER MCDONALD

New York State Agricultural Experiment Station, Geneva

This chapter examines the most valuable journals and serials cited in the literature of forestry and agroforestry science. The chapter covers most serially published material, such as annual reviews and conferences as well as popular magazines and scholarly journals published in a numbered series but lacking issues with a unique or changing title.

Annuals such as *Annual Review of Plant Physiology* are included in this chapter since they fit the same criteria. Proceedings volumes require clarification. Annual proceedings generally belong to the category of serials and were so designated in the Core Agricultural Literature Project. They follow much the same pattern as annual reviews, since they have uniform titles and represent the continuing deliberations of an organization or society. On the other hand, international and specialized conference proceedings which are held irregularly and which have distinctive titles or subject foci with each conference were categorized as monographs. Of the numerous conference proceedings identified in the citation analysis, approximately 32% were tallied and evaluated as journals; the rest were considered monographs for the reasons outlined. Report series, although published serially, are considered monographs in this study because they cover one topic per issue.

A. Source Documents and Methodology

The thirty monographs used by the Core Agricultural Literature Project as source documents to ascertain citation patterns to monographs in forestry also served to gather data on journals and serials. The same methods and caveats apply to journals as outlined for monographs in Section A of Chapter 11. All journal and serial titles cited in the source documents were recorded, and were tabulated by title and date of publication.

These lists were then analyzed using a variety of criteria. Seperate tabulations were kept depending on whether the source document pertained to temperate forestry or aspects of tropical forestry. Fourteen source documents dealt primarily with temperate forestry (see Subject Categories in Chapter 3) and sixteen on tropical forestry and agroforestry. It should be noted that the sixteen source documents dealing with the Third World had a double focus. On the one hand, tropical silviculture and agroforestry were analyzed by the Core Agricultural Literature Project in ten of the sixteen Third World source documents, while the study of tropical forests as natural ecosystems was covered by the rest, typified by *An Introduction to Tropical Rainforests* (1990).[1] In combination, it was felt that analysis of these intertwining disciplines would best represent the entire spectrum of forest science currently practiced in developing countries.

By analyzing source documents dealing with temperate regions seperately from tropical regions, this study gathered data to compare and contrast journal citation patterns from developed countries and the Third World. The top cited journals on both lists were then compared to data gathered from the *Science Citation Index: Journal Citation Reports* and other sources.

B. Journal Literature Findings

The source documents yielded 21,486 citations, of which 12,507 (58.2%) were to journal articles. These 12,500 references provided citations to 648 distinct journal titles. Such a wide scattering is not uncommon in a discipline such as forestry which has close ties to economics, international policy, environmental science, natural resource management, and a host of laws, regulations, and standards.

Counts were made each time a journal was cited. Just under 30% of the journals in the entire analysis were cited only once each. The top four journals, each with 400 citations or more, accounted for only 15.9% of all journal citations. Bradford's law of distribution postulates that a small core of journals will garner the majority of citations while the remainder will be scattered nearly evenly among a large number of journals.[2] In previous volumes of the *Literature of the Agricultural Sciences* series, this law of distri-

---

1. T. C. Whitmore, *An Introduction to Tropical Rainforests* (Oxford, U.K.: Clarendon Press, 1990).
2. B. C. Vickery, "Bradford's Law of Scattering," *Journal of Documentation* 4 (1948): 198–203.

bution applied, notably in agricultural engineering,[3] soil science,[4] and in food science and human nutrition where a single society journal garnered almost four times the number of citations as the second-ranked journal.[5]

This was clearly not the case with forestry. The scattering was far more evenly distributed as Figure 12.1 shows. Although four journals are ranked in a cluster at the top, the rest, beginning with the *Canadian Journal of Botany*, are ranked evenly downward. The average count difference is roughly five between succeeding top-ranked journals in an unusually even spread from 287 to fifty citations, which is the lowest statistically valid count for journal ranking in this study.

A common measure of the currency of a journal is its cited half-life which is defined as the number of journal-publication years going back

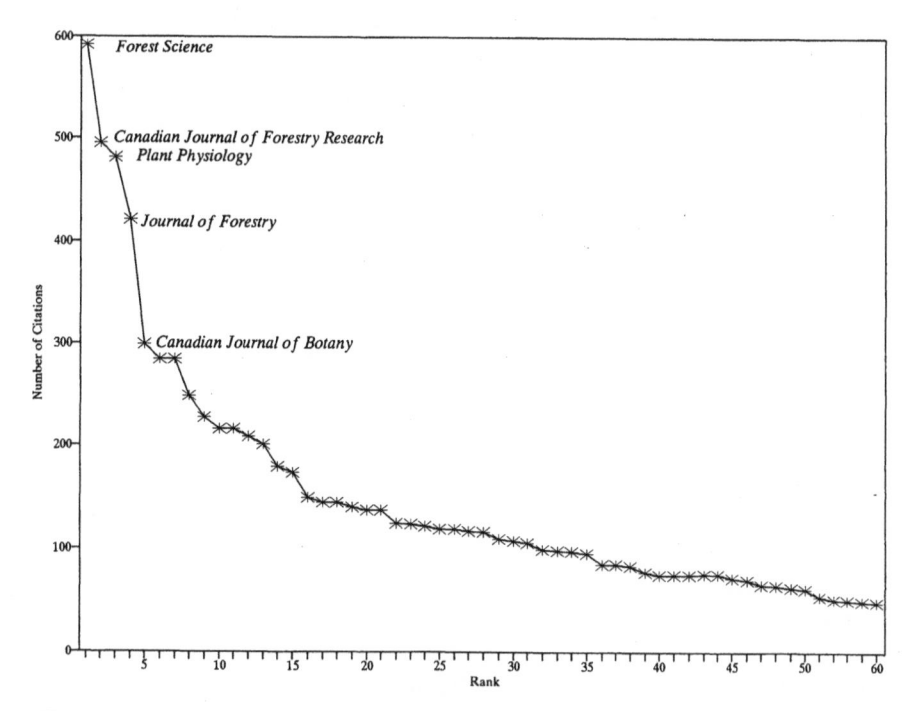

Figure 12.1. Distribution of journals and serials.

3. Wallace C. Olsen, "Primary Journals and Reports," in *The Literature of Agricultural Engineering*, ed. Carl W. Hall and Wallace C. Olsen (Ithaca, N.Y.: Cornell University Press, 1993), pp. 318–324.

4. Peter McDonald, "Primary Journals in Soil Science," in *The Literature of Soil Science*, ed.Peter McDonald (Ithaca, N.Y.: Cornell University Press, 1994).

5. Wallace C. Olsen and Jennie Brogdon "Primary Journals and Serials," in *The Literature of Food Science and Human Nutrition*, ed. Jennie Brogdon and Wallace C. Olsen (Ithaca, N.Y.: Cornell University Press, 1994).

from the current year which account for 50% of the total citations given. The cited half-life for journals was 8.3 years from Third World source documents, and 9.0 years for the developed countries. Surprisingly, the cited half-life of journal articles was .7 years longer than for journals in the Third World source document analysis. Clearly, seminal articles have longer currency than the topical research articles which made up the bulk of journal citations in the tropical forestry and agroforestry source monographs.

One unique source for the comparative analysis of agricultural serials is the database *World List of Agricultural Serials* (*WLAS*) distributed by the National Agricultural Library on CD-ROM. *WLAS* contains all current and ceased serials in the NAL collection, as well as serial titles indexed by CAB International, the Food and Agriculture Organization, and other institutions. It contains bibliographic data on roughly 54,000 serials, including some report series and annual proceedings. Since 1980, the *WLAS* database has added 354 new serial titles in forestry. Table 12.1 gives a count of the number of forestry serial publication types as identified in the *WLAS* database.

Table 12.1. Forestry publication types in *World List of Agricultural Serials*

Journals	2,798	73.4%
Report series	450	11.8%
Bulletins	202	5.3%
Annual reports	133	3.4%
Newsletters	64	1.6%
Annual proceedings & Society deliberations	41	1.0%
Abstracts	17	0.4%
Miscellaneous[a]	106	2.7%

[a]Statistical summaries, accession/trade lists, catalogs, supplements, etc.

Serials in the *WLAS* database have been subject coded using *AGRICOLA* category codes. Using the forestry codes as delimiters, the universe of serials pertaining to forestry stands at 3,811. Of these, 215 or 5.6% are known to have ceased publication, and 530 or 13.9% were superseded by a new serial publication. Ceased publications are not often known and only posted in library databases when complete information is available which raised the prospect of at least 5% more of the 3,811 titles being inactive or dead. Some of the titles, although listed as forestry serials, have only a small portion of forestry articles in them. This leads to duplicate subject coding, which is particularly true in the plant sciences (*AGRICOLA* category codes F000–F851), where it is estimated that 20% of the forestry-coded titles are not predominantly forestry but plant science titles. This additionally reduces the number of predominantly forestry titles. Removing these as well as the ceased and superseded titles leaves the number of currently active titles as represented in the *WLAS* near 2,000 serial titles.

Almost 28% of this total are annual publications including annual reviews, annual reports, and annual deliberations of professional societies. Table 12.2 lists place of publication of the serials for the top ten countries.

Table 12.2. Place of publication of forestry serials in *WLAS*

United States	40.2%
Canada	8.1%
United Kingdom	4.3%
Australia	3.2%
Germany	3.2%
Japan	3.0%
Sweden	2.8%
China	2.1%
India	2.1%
France	2.0%

Third World countries with the highest numbers of forestry serial publications were: India—81; Indonesia—51; Brazil—35; Mexico—34; Philippines—29; Nigeria—26; Malaysia—24.

Primary languages of these serials were English 74.4%, German 6.0%, French 5.5%, Spanish 3.2%, and Swedish at 2.4%. All other languages comprised the remaining 8.5%. In *WLAS*, republics of the former Soviet Union have been listed separately. The low count of thirty-two forestry publications for Russia is probably due to the fact that the majority of Russian forestry serials were not obtained by the National Agricultural Library or indexed by the other two cooperators.

C. Top-Ranked Journals

Of the 648 journal titles identified in the citation analysis, only 10% were ranked heavily enough or statistically valid to be considered as potential core journals. A cutoff level was reached for the developed countries' journals which was slightly higher in actual count than that of journals from the Third World analysis. Special attention had to be given to tropical ecology and botany for the Third World since some aspects of tropical forestry and dendrology were not cited in forestry journals. Therefore, the scattering of subjects for the Third World literature in journals was greater than with the developed countries' literature. Two such journals had great relevance to a wide region or continent that otherwise had little or no representation. The combined list of statistically valid and relevant titles has fifty-six unique titles. These are itemized in Table 12.3. Rankings from the citation analysis

Table 12.3. Top journals ranked by number of citations to them

Journal titles	Beginning date	Core citation analysis rankings		*Science Citation Index* data	
		Developed countries A	Third World B	Total citations rank 1988 C	Impact factor rank 1986–88 D
Forest Science	1955–	1	10	26	33
Canadian Journal of Forestry Research	1971–	2	24–26	27	27
Plant Physiology	1926–	3	24–26	3	6
Journal of Forestry	1917–	4	17–18	33	40
Agroforestry Systems	1982–		1	50	48
Canadian Journal of Botany	1951–	5		7	23
Ecology	1920–	6	12	4	8
Physiologia Plantarum	1948–	7	27–28	8	11
Forest Products Journal	1950–	8		38	38
Southern Journal of Applied Forestry	1977–	9			
Silvae Genetica	1951–	10	20–21	40	41–42
New Phytologist	1902–	11		12–13	16
Oecologia	1968–	12	16	10	13
American Journal of Botany	1914–	13		11	18
Journal of the American Society for Horticultural Science	1875–	14		20	30
Science	1883–	15	8	2	3
Soil Science Society of America Journal	1936–	16		6	20
Water, Air, and Soil Pollution	1971–	17		31	24
Journal of Experimental Botany	1950–	18		14	12
Annual Review of Plant Physiology	1950–	19		12–13	2
Annals of Botany	1937–	20		42	22
Forest Ecology and Management	1976–	21	19	45	32
Botanical Gazette	1875–	22		23	25
Journal of Horticultural Science	1924–	23		36	37
Plant, Cell, and Environment	1978–	24		29	9
New Zealand Journal of Forestry Science	1971–	25		44	47
Tappi Journal	1917–	26		25	43
Australian Journal of Botany	1953–	27		30	31
Plant and Soil	1948–	28–29	15	17	26
Forestry Chronicle	1925–	28–29		43	44
Nature	1869–	30	11	1	1
Unasylva	1950–		2		
Ambio	1972–		3	34	28
Biotropica	1969–		4–5	37	36
Turrialba	1950–		6	46	49
Malaysian Forester	1973–		7		
Forestry	1927–	31		47	34
Journal of Applied Ecology	1964–	32	22	22	19

Table 12.3. Continued

Journal titles	Beginning date	Core citation analysis rankings		Science Citation Index data	
		Developed countries A	Third World B	Total citations rank 1988 C	Impact factor rank 1986–88 D
Commonwealth Forestry Review	1922–		9	49	45
HortScience	1966–	33		21	35
Australian Journal of Plant Physiology	1974–	34		24	14
Ecological Monographs		35		18	4
Journal of Ecology	1913–	36	4–5	16	15
Planta; Archiv fur Wissenschaft-liche	1925–	37		5	7
Indian Forester	1875–		14		
Acta Horticulturae	1963–	38–39		32	
Journal of Wildlife Management	1937–	38–39		15	29
Agronomy Journal	1907–	40		9	17
BioScience	1964–	41		28	10
Environmental Pollution	1987–	42		39	21
Annals of the Missouri Botanical Garden	1914–		17–18	35	41–42
Experimental Agriculture	1925–		20–21	41	39
Bois et Forêts des Tropiques	1947–		23		
Journal of Tropical Ecology	1985–		24–26		
Australian Forestry	1936–		27–28	48	46
Annual Review of Ecology and Systematics	1970–	43	13	19	5

of the Core Agricultural Literature Project are given in columns A and B, ranked separately in both cases; those for the developed countries, and those for the Third World.

Column A ranks the journals by citation counts from the developed country source documents analyzed by the Core Agricultural Literature Project. The subjects covered by these source documents broadly include forest biology, forest protection and pathology, forest management and production, and forest policy. These subject and their divisions are more fully identified in Chapters 2 and 11.

Column B ranks the journals by the sixteen agroforestry and tropical rainforest source documents also in Chapter 11. Of the top twenty-nine journals resulting from the tropical or Third World source document analysis, thirteen do not appear in Column A. There is little overlap between Columns A and B. The highest ranked journal in Column B also found in Column A is the *Journal of Ecology* ranked at four to five for the Third

World but thirty-sixth for the developed. In Column A, 69.7% of the titles appear in that column only, in Column B, 46.4%. The majority of the developed world titles (Column A) not found in Column B, deal with botany, horticulture, and environmental pollution.

Column C is a ranking for the same titles based on counts from the *Science Citation Index; Journal Citation Reports* indicate the number of times that journal was cited by other journals in 1988. The *SCI* counts each time a journal is cited in the references of all citing journals for a given year. For a publication such as *Agroforestry Systems* this citation count may be as low as thirty-eight as in 1988. *SCI* does not have citing information for the five top-ranked tropical forestry journals: *Agroforestry Systems, Unasylva, Malaysian Forester, Bois et Forêts des Tropiques,* and *Indian Forester.* One criticism of the *SCI* has been its short coverage of literature related to non-industrialized countries.[6] Only a fraction of all journals indexed in the *SCI* are from Third World countries and these are not commonly cited. The *SCI* lack of Third World coverage in forestry is partly explained by this practice.

Impact factors computed by the *Science Citation Index* for 1986–1988 are given in Column D. The number is derived by dividing the number of times a journal is cited in any given year by the number of citations published in that journal over the same period. Generally, the higher the impact factor the greater the short-term impact of a journal. Impact factors are useful since they reduce the heavy influence of publications with great quantities of articles or pages.

Comparisons of the Core Agricultural Literature Project citation analysis with impact factors for top ranked journals in the *Science Citation Index* (see Chapter 5, Table 5.1) revealed that many titles ranked highly by ISI did not garner sufficient citations in the core analysis to warrant inclusion in Table 12.3. In fact, nine of the seventeen leading journals ranked by ISI impact factor presented by Tombaugh (Chapter 5) do not even appear in Table 12.3. Analysis was conducted on three years of *Forest Science*, the highest cited journal in the discipline, for 1987, 1990, and 1993. Of the 5,312 citations in *Forest Science* analysis, 55.7% were to journals. *Tree Physiology*, which is ranked topmost in Table 5.1, garnered fewer than 2% of the total. This is undoubtedly an important journal, but it began in 1986, and its preeminence is not yet proven over a long period. Other titles in Table 5.1, such as *European Journal of Forest Pathology* and *Wood and Fiber Science*, did not even appear in the *Forest Science* analyses. These

6. Yvon Chatelin and Rigas Arvanitis, *Stratégies Scientifiques et Dévelopment-Sols et Agriculture des Régions Chaudes* (Paris: ORSTOM, 1988).

differences tend to demonstrate identification of journals of longer value, and to delay decisions on those of which are more recent and as yet not fully proven.

The Core Agricultural Literature Project made a conscientious effort to select monographic titles for analysis which would reflect broad trends in the literature, with a firm historical perspective. Furthermore, every effort was made to select monographs which would assist in the analyses of Third World forestry and agroforestry literature, and give Third World scientists and librarians a firm basis on broad trends. A selection of monographs to represent long-lasting, broadly defined trends, are not going to reflect the most recent trends in a discipline, which are more apt to be shown by ISI impact factors. Also, ISI impact factors look at citation patterns across disciplines. Titles such as *Wood and Fiber Science* are more commonly cited in chemistry and mechanical sciences literature than in forestry. This particular title, while not listed in Table 12.3, is part of the core journal list at the end of this chapter. Similarly, *Tree Physiology* seems to be more heavily cited in the biology and environmental science literature than in *Forest Science*.

Eighteen titles (32.1%) on the combined list of fifty-six temperate and tropical journals in Table 12.3 deal specifically with forestry or wood science, which demonstrates that scientists working in forestry deal with a host of allied disciplines. Of the remaining thirty-eight titles in Table 12.3, the division of non-forestry journals are as follows:

fourteen deal with aspects of botany,
seven deal with aspects of ecology, including human ecology
five titles, including *Science* and *Nature,* are in general science,
five cover general agriculture,
four are horticultural,
two cover pollution, and
one journal is on wildlife management.

Nine of these fifty-six titles deal with tropical subjects, of which six are devoted entirely to forestry. Table 12.3 provides a variety of indicators which help explain the value of individual journals.

Professional societies and nonprofit organizations account for 71.4% of the fifty-six titles. Commercial publishing houses account for 23.3% and governments publish the remaining 5.3%. These percentages follow a similar pattern found in other agricultural disciplines analyzed by the Core Agricultural Literature Project, where societies generally play the key role in publishing the most journals, of most influence. Universities notably publish few top-ranked agricultural journals, and in the case of forestry, none.

Country of publication is another point of interest for these fifty-six journals. The United States dominates the publishing field with twenty-four titles, followed by the United Kingdom with ten, the Netherlands and Australia with five each, Canada and Germany with three apiece, and one each from New Zealand, Malaysia, India, France, Denmark, Italy, and Costa Rica.

D. Third World Forestry and Agroforestry Journals

Only three journals on the core list are published by developing countries, *Malaysian Forester*, *Turrialba* in Costa Rica, and *Indian Forester* in Dehra Dun, India. Six other titles deal specifically with tropical subjects but are published in developed countries, notably the top-ranked Third World journal *Agroforestry Systems* which is published by a commercial publisher in the Netherlands in cooperation with International Centre for Research in Agroforestry (ICRAF) in Nairobi. The remaining journals in Column B are from developed countries with little or no tropical area. As with general forest science, tropical forestry and agroforestry literature appears in a wide range of journals.

The analysis of tropical forestry and agroforestry source documents revealed that the majority of citations were to monographs, not to journals. Journals accounted for only 46% of source document citations. Conferences, seminal monographs, and important report series were more heavily cited and explain the high monographic count. This is exemplified by the Core Agricultural Literature Project's analysis of a typical agroforestry source document, *Plant Research and Agroforestry*, which cited the agroforestry report series of the Centro Agriconomico Tropical de Investigacion y Ensenanza (CATIE) seventeen times, ICRAF reports twenty-five times, and the International Institute of Tropical Agriculture (IITA) nine times, for a total of fifty-two references, but cited the journal *Agroforestry Systems* only twice.[7] The a reliance on reports in agroforestry is great. There is presently, only one refereed journal, *Agroforestry Systems,* which is specific to the subject. A newsletter, *Agroforestry Today,* while widely read, as are most newletters, is rarely cited.[8]

Only two other journals in Table 12.3 pertain to tropical forestry specifically. This paucity of journal titles pertaining to the development of forestry

7. Peter Huxley, ed., *Plant Research and Agroforestry* (Nairobi, Kenya: International Council for Research in Agroforestry, 1983).

8. Anthony Young, "Change and Constancy: An Analysis of Publications in Agroforestry Systems, Volumes 1–10," in *Agroforestry Systems* 13 (1991): 195–202, 1991.

in the Third World is not surprising given that the management of dwindling tropical forest resources and agroforestry are relatively recent scientific developments (see Chapter 4).

It may be some time before more quality journals are forthcoming in this field. Journals covering broad subjects such as botany and ecology regularly publish material on tropical forest ecosytems and seem to fill this publishing niche. Agroforestry as an emerging discipline has only one journal as a primary vehicle for articles, hence the influence of *Agroforestry Systems* is widespread. Table 12.4 shows the wide geographic spread of authors of the journal's articles. Its prominence when compared to other journals garnered from the Third World source documents analysis more nearly conforms to Bradford's law of distribution, having three and a half times as many citations as the second ranked journal, *Unasylva*.

Table 12.4. Agroforestry Systems; Residences of all authors (1988–1991)

Country of residence	No. of authors	Country of residence	No. of authors
United States	67	China	5
India	45	Indonesia	5
Nigeria	26	Rwanda	5
Kenya	23	Philippines	4
Costa Rica	21	Zimbabwe	4
Netherlands	18	Bolivia	3
Australia	17	France	3
United Kingdom	16	Japan	3
Germany	14	Malaysia	3
Mexico	8	Senegal	3
Chile	5	Thailand	3

Five countries, Ethiopia, Portugal, Sierre Leone, Togo, and Zambia, had two authors apiece, and eight countries, Brazil, Honduras, Italy, Ivory Coast, Madagascar, Niger, Spain, and Sweden had only one. These counts do not necessarily represent where reasearch is being done, since many United States, United Kingdom, and German authors wrote about experiences in developing countries, but it gives a rough estimate of which countries are currently supporting agroforestry research. Thirty-six countries are listed as residences of authors. Those in Africa, Asia, and Central America predominate and one study showed that these continents accounted for 34%, 28% and 19.6% of the article counts in *Agroforestry Systems* respectively.[9] North America was the subject of 8% of the articles, mostly on temperate agroforestry; surprisingly, only 6.3% of the articles in this four

9. Ibid.

year period dealt with South America. Low counts for South American countries may be in part, because forestry management in the heavily forested Amazon and the Orinoco Basin are based more on traditional extractive tropical silviculture than agroforestry systems.[10] Other forestry journals published in South America also account for a quantity of the literature. Traditional silviculture is not covered in much depth by *Agroforestry Systems* whose focus is on sustainable systems of tree-crop integration. Nor does *Agroforestry Systems* publish in Spanish or Portuguese, but almost exclusively in English with a few articles in French.

Unasylva and *Ambio*, the second and third most cited publications among Third World journals, have similar broad-focused formats usually covering a single topic per issue. The social and environmental costs of human impact on ecosystems are the main topics pointing out that environmental costs of development in Third World countries are in the forefront of the scientific literature pertaining to forest management.

By contrast, the two highest cited journals in Column A of Table 12.3, *Forest Science* and *Canadian Journal of Forest Research*, published few articles on aspects of forestry or the lumber trade in developing countries. In Column B (Third World) they are tenth and 24–26 respectively. A three year count of articles in *Forest Science* (1989–1991) revealed that only 5% of the articles dealt with Third World topics; in the *Canadian Journal of Forest Research* for the same period the percentage was lower at 3.8%. Yet it appears from their rank in Column B, that the few articles they publish pertinent to tropical forestry topics are widely cited. A similar situation exists with multi-disciplinary journals such as *Science* and *Nature,* which consistently rank high in Third World counts. The prestige and high regard in which these journals are held, along with their wide circulation, provide a heavy impact given their low coverage of topics dealing with scientific disciplines in developing countries. Tropical forestry and rainforest ecology were no exceptions.

E. Core Journals and Serials List

Sixty titles comprise the core journals in forestry and agroforestry. Thirteen are unique to developing countries, thirty-three to developed, and fourteen are common to both.

10. Humberto Jimenez Saa, "El Aporte de las Ciencas Forestales a la Literatura Agricola Latinamericana." in *Simposio Internacional sobre las Ciencias Forestales y su Contribucion al Desarollo de la America Tropical*, ed. Manuel Chavarria (San Jose, Costa Rica, 1981).

The core list is not a precise match of the overall rankings in Table 12.3. Some titles did not receive sufficient citation counts, from either the developed world source documents or from tropical forestry source documents, to be ranked in either Column A or B. But when the counts for each column were combined, their overall standing indicated that a journal deserved to be on the core list. Additional titles have been added to reflect influences in the changing subject coverage of the last ten years, correlating data from other sources, and the reaction of scholars in the field.

As noted, *Wood and Fiber Science*, while not heavily cited in the overall literature, has been included because analysis by the Core Agricultural Literature Project of wood science source documents was slight compared to forest biology where most forest science research is heavy today. In the forest products literature *Wood and Fiber Science* was widely cited, this skew was accommodated. Lumber products are an important component of forest science and are represented on the core list by only one other top-ranked journal, *Forest Products Journal,* which is ninth in Column A of Table 12.3. Although ranked lower overall than some botany titles which were not included, *Wood and Fiber Science* broadens the disciplinary coverage of the core journals.

A similar case is made for *Holzforschung* which is here considered core. This journal did not rank high in the Core Agricultural Literature Project citation counts. However, examination of the *Science Citation Index* for 1986 and 1988 reveals it was listed as the top journal in forestry with an average impact factor of 0.966. This discrepancy of ranking is easily explained. *Holzforschung* is not a forestry journal *per se,* but rather falls into the category of material science as the *SCI* clearly states, listing it both with forestry and also under wood and pulp chemistry. Published bimonthly, most articles in *Holzforschung* are on applied aspects of the discipline with little emphasis on university level instruction which was of great interest to the Core Agricultural Literature Project. Citation patterns in the *SCI* reveal that other chemistry journals primarily cite *Holzforschung* and not forest science journals. As noted elsewhere, research in wood chemistry forms only a small fraction of the forestry citations in AGRICOLA, and therefore has but slight impact on forest science as a whole. The few source documents on wood products chosen by the Core Agricultural Literature Project proportionately reflect this impact. *Holzforschung*, with its strong focus on aspects of wood science, chemistry, and pulp technology, is added as core on its usual strengths and topranking in the *SCI's Journal Citation Report.*

The aim of the list is to provide libraries with the titles of currently published journals which comprise a core academic collection. Titles that have ceased publication are not included. As noted earlier, few titles are published in developing countries and few deal specifically with tropical forestry topics. In Third World countries much of the literature pertinent to tropical forestry is included in general scientific and agricultural journals. A good example is *Turrialba*, published in Costa Rica, which covers all of agriculture and a good deal in forestry.

Of the eighteen forestry-specific journals, eight were published prior to 1950, the oldest being *Indian Forester* inaugurated in 1875. Five journals started publication since 1970, the most recent being *Agroforestry Systems* in 1982. Almost 60% of the titles began publication on or before 1950, which represents little fluctuation or change in forestry journals. Journals in the core list have been put within each community in a top rank (1) or a second rank (2).

Core Forestry and Agroforestry Journals for Developed and Third World Countries (N = 60)

	Developed Countries	Third World
Acta Horticulturae. Vol. 1 (1963) +. The Hague; International Society for Horticultural Science.	2	–
Agroforestry Systems. Vol. 1 (1982) +. Dordrecht, Neth.; Kluwer Academic. Monthly 1991 + , 9 issues in 1990, Bimonthly 1988–1989, Quarterly 1982–1987.	–	1
Agronomy Journal. Vol. 1 (1907/09) +. Madison, Wis.; American Society of Agronomy. Bimonthly.	2	–
Ambio. Vol. 1 (1972) +. Stockholm; Royal Swedish Academy of Sciences. Bimonthly.	–	1
American Journal of Botany. Vol. 1 (1914) +. Lawrence, Kan.; Botanical Society of America. Monthly.	1	–
Annals of Botany. New Series. v. 1 (1937) +. London and New York; Academic Press. Bimonthly. Continues: *Annals of Botany* (v. 1–50, 1887–1936).	1	–
Annals of the Missouri Botanical Garden. Vol. 1 (1914) +. St. Louis, Mo.; Missouri Botanical Garden Press. Quarterly.	–	2
Annual Review of Ecology and Systematics. Vol. 1 (1970) +. Palo Alto, Calif.; Annual Reviews Inc. Annual.	2	–
Annual Review of Plant Physiology and Plant Molecular Biology. Vol. 39 (1988) +. Palo Alto, Calif.; Annual Reviews Inc. Annual. Continues: *Annual Review of Plant Physiology*.	1	–
Australian Forestry. Vol. 1 (1936) +. Perth; Institute of Foresters of Australia. Quarterly.	–	2
Australian Journal of Botany. Vol. 1 (1953) +. Melbourne;	2	–

Commonwealth Scientific and Industrial Research Organization. Bimonthly.

Australian Journal Plant Physiology. Vol. 1 (1974)+. Melbourne; Commonwealth Scientific and Industrial Research Organization. Bimonthly. — 2 —

Biological Reviews of the Cambridge Philosophical Society. Vol. 1 (1923)+. Cambridge; Cambridge University Press. Quarterly. — 2 —

BioScience. Vol. 1 (1951)+. Arlington, Va.; American Institute of Biological Sciences. Monthly. — 2 —

Biotropica. Vol. 1 (1969)+. Washington, D.C.; Association for Tropical Biology. Quarterly. — — 1

Bois et Forets des Tropiques. n. 1 (1947)+. Paris; Societe pour le Developpement de l'Utilisation des Bois Tropicaux. Quarterly. — — 2

Canadian Journal of Botany = Journal Canadien de Botanique. Vol. 29 (1951)+. Ottawa; National Research Council of Canada. Monthly 1981– , Semimonthly 1975–80, Monthly 1951–75. Continues: *Canadian Journal of Research*. Bimonthly 1929–50. — 1 —

Canadian Journal of Forest Research = Journal Canadien de la Recherche Forestiere. Vol. 1 (1971)+. Ottawa; National Research Council of Canada. Quarterly. — 1 2

The Commonwealth Forestry Review. Vol. 1 (1922)+. Oxford; Commonwealth Forestry Association. Quarterly. — — 1

Ecological Monographs. Vol. 1 (1931)+. Durham, N.C.; Duke University Press. — 2 —

Ecology. Vol. 1 (1920)+. Durham, N.C.; Ecological Society of America and the Duke University Press. 6 no. a year. Supersedes: *Plant World*. — 1 1

Environmental Pollution. Vol. 1 (1970)+. Barkingham, U.K.; Elsevier Applied Science Pub. Semimonthly. — 2 —

Experimental Agriculture. Vol. 1 (1965)+. London and New York; Cambridge University Press. Quarterly. Supercedes: *Empire Journal of Experimental Agriculture*. (1933–1964) — — 2

Forest Ecology and Management. Vol. 1 (1976/77)+. Amsterdam; Elsevier Scientific Pub. Co. Quarterly. — 1 2

Forest Products Journal. Vol. 1 (1951)+. Madison, Wis.; Forest Products Research Society. Monthly. — 1 —

Forest Science. Vol. 1 (1955)+. Bethesda, Md.; Society of American Foresters. Quarterly. — 1 1

Forestry. Vol. 1 (1927)+. London; Oxford University Press. Semiannual. — 2 —

The Forestry Chronicle. Vol 1. (1925)+. Sainte-Anne-de-Bellevue, Quebec; Canadian Institute of Forestry. Bimonthly — 2 —

Holzforschung. Vol. 1 (1947) + . Berlin; Walter de Gruyter. Bimonthly.	2	–
HortScience. Vol. 1 (1966) + . Mt. Vernon, Va.; American Society for Horticultural Science. Bimonthly.	2	–
The Indian Forester. Vol. 1 (1875) + . Dehra Dun; Indian Forester. Monthly.	–	1
International Journal of Plant Sciences. Vol. 1 (1875) + . Chicago; University of Chicago Press. Quarterly. Continues: *Botanical Gazette.*	1	–
Journal of Applied Ecology. Vol. 1 (1964) + . Oxford; Blackwell Scientific Pub. Three times a year.	2	2
The Journal of Ecology. Vol. 1 (1913) + . Oxford; Blackwell Scientific Pub. Three times per year.	2	1
Journal of Experimental Botany. Vol. 1 (1950) + . Oxford; Clarendon Press. Three times a year.	1	–
Journal of Forestry. Vol. 1 (1905) + . Bethesda, Md.; Society of American Foresters. Monthly. Continues: *Proceedings of the Society of American Foresters.*	1	2
The Journal of Horticultural Science. Vol. 1 (1919) + . Ashford, U.K.; Headley Brothers. Quarterly.	1	–
Journal of the American Society for Horticultural Science. Vol. 94 (1969) + . Alexandria, Va.; American Society for Horticultural Science. Bimonthly. Continues: Proceedings of the American Society for Horticultural Science.	1	–
Journal of Tropical Ecology. Vol. 1 (1985) + . Cambridge and New York; Cambridge University Press.	–	2
The Journal of Wildlife Management. Vol. 1 (1937) + . Washington, D.C.; Wildlife Society. Quarterly.	2	–
The Malaysian Forester. Vol. 1 (1931) + . Kepong, Selangor, Malaysia; Officers of the Forest Report. Quarterly.	–	1
Nature. Vol. 1 (1869) + . London; Macmillan. Weekly.	2	1
New Phytologist. Vol. 1 (1902) + . London and New York; Academic Press. Irregular.	1	–
New Zealand Journal of Forestry Science. Vol. 1 (1971) + . Rotorua, N.Z.; Forest Research Institute. Three times a year.	1	–
Oecologia. Vol. 1 (1968) + . Berlin; Springer-Verlag. Monthly.	1	2
Physiologia Plantarum. Vol. 1 (1948) + . Copenhagen; Munksgaard. Monthly.	1	2
Plant and Soil. Vol. 1 (1948) + . The Hague; Kluwer Academic. Irregular, 1948–, Bimonthly, 1976– .	2	1
Plant, Cell and Environment. Vol. 1 (1978) + . Oxford; Blackwell Scientific Pub. Quarterly.	1	–
Plant Physiology. Vol. 1 (1926) + . Rockville, Md.; American Society of Plant Physiologists. Monthly; 3 vols. a year.	1	2

Planta; Archiv fur Wissenschaftliche Botanik. Vol. 1 (1925)+. Berlin; J. Springer-Verlag. Irregular.	2	–
Science. New Series, Vol. 1, no. 1 (1895)+. Washington, D.C.; American Association for Advancement in Science. Weekly.	1	1
Silvae Genetica. Vol. 1 (1957)+. Frankfurt; J. D. Sauerlander's Verlag. Bimonthly.	1	2
Soil Science Society of America Journal. Vol. 1 (1936)+. Madison, Wis.; Soil Science Society of America. Bimonthly. Continues: *Soil Science Society of America. Proceedings.*	1	–
Southern Journal of Applied Forestry. Vol. 1 (1977)+. Bethesda, Md.; Society of American Foresters. Quarterly.	1	–
Tappi Journal. Vol. 1 (1918)+. Atlanta, Ga.; Technical Association of the Pulp and Paper Industry. Monthly.	2	–
Turrialba. Vol. 1 (1950)+. Turrialba, Costa Rica; Inter-American Institute of Agricultural Sciences. Quarterly.	–	1
Unasylva. Vol. 1 (1947)+. Rome; Food and Agriculture Organization. Quarterly.	–	1
Vegetatio. Vol. 1 (1948)+. The Hague; W. Junk. Bimonthly.	2	–
Water, Air, and Soil Pollution. Vol. 1 (1971)+. Dordrecht; D. Reidel Pub. Co. Quarterly.	1	–
Wood and Fiber Science: Journal of the Society of Wood Science and Technology. Vol. 1 (1968)+. Lawrence, Kan.; Society of Wood Science and Technology. Continues: *Wood Science.*	2	–

F. Selected Journals Published since 1980

As with any discipline, several important journals have appeared in the last fifteen years that should be given mention. Many are obviously too recent to be regularly cited in the monographic or journal literature and therefore do not appear on the core list. However, the titles given below can be considered up-and-coming journals of prominence. Changes of titles without new volume numbering are not included, nor are slight changes of name.

Recently Begun Journals in Forestry and Agroforestry

Environmental Studies & Practices. no. 1 (1991)+ New Haven, Conn.: Yale School of Forestry and Environmental Studies. Bimonthly.

Forest Pest Management. no. 1 (1993)+ Missoula, Montana: U.S. Forest Service. Irregular

International Journal of Geographical Information Systems. Vol. 1 (1987)+ London and New York: Taylor & Francis. Quarterly.

International Journal of Remote Sensing. Vol. 1 (1980)+ London: Taylor & Francis. Monthly. An official journal of the Remote Sensing Society.

The International Tree Crops Journal. Vol. 1 (1980)+ Berkhamsted, UK: A B Academic Publishers. Quarterly.

Journal of Forest Engineering. Vol. 1 (1989)+ Fredericton, New Brunswick: Dept. of Forest Engineering, University of New Brunswick. Semiannual.

Journal of Sustainable Forestry. Vol. 1 (1993)+ Binghamton, NY: Food Products Press. Quarterly.

Journal of Tree Sciences. Vol. 1 (1982)+ Solan, India: Indian Society of Tree Scientists. Semiannual.

Journal of Tropical Forestry. Vol. 1 (1985)+ Jabalpur, India: Society of Tropical Forestry Scientists. Quarterly.

The Journal of World Forest Resource Management. Vol. 1 (1984)+ Berkhamsted, UK: A B Academic Publishers. Semiannual. An international journal on "biology, biotechnology, and management of afforestation and reforestation."

New Forests. Vol. 1 (1986)+ Dordrecht and Boston: M. Nijhoff. Quarterly.

Northern Journal of Applied Forestry. Vol. 1 (1984)+ Bethesda, Md.: Society of American Foresters. Quarterly.

Scandinavian Journal of Forest Research. Vol. 1 (1986)+ Stockholm, Sweden: Almqvist & Wiksell Periodical Co. Quarterly.

Sylvatrop. Vol. 1 (1991)+ College, Laguna, Phillipines: Phillipine Dept. of Environment and Natural Resources. Quarterly.

Tree Physiology. Vol. 1 (1986)+ Victoria, B.C.: Heron Pub. Quarterly.

Trees; Structure and Function. Vol. 1 (1986)+ Berlin: Springer International. Quarterly.

Urban Forests. Vol. 10 (1990)+ Washington, D.C.: American Forestry Association. Bimonthly. (Supersedes *Urban Forest Forum*)

Western Journal of Applied Forestry. Vol. 1 (1986)+ Bethesda, Md.: Society of American Foresters. Quarterly.

Wild Forest Review. Vol. 1 (1993)+ Portland, Ore.: Save The West. Eleven times a year.

13. Reference Collection Update

PETER MCDONALD

New York State Agricultural Experiment Station, Geneva

Reference tools in forest science are extensive, yet there are few comprehensive guides to sources of forestry information. Bibliographic databases such as *AGRICOLA* and *CAB Abstracts* have done an excellent job of indexing the literature, and CAB's *Forestry Abstracts* in particular, has covered reference tools in its "General Publications" section. But keeping abreast of the continuous stream of new reference works remains a daunting task.

This chapter updates the excellent compilation of 366 agricultural reference tools cited in the *Guide to Sources for Agricultural and Biological Research,* by Blanchard and Farrell, published in 1981.[1] This work serves as the basic guide to the literature of the agricultural sciences. Forestry is included in the section on Plant Sciences, entries B1163–B1529. It should be noted that none of the references cited in the *Guide* are to literature guides.

This update is a supplement and complement to the entries in R. Blanchard and L. Farrell, with an emphasis on emerging topics of critical interest to researchers and practitioners. Dates covered here are 1980 to the present, with no duplication of material listed in the *Guide*. The only exceptions are reference tools which have been reissued or updated in new editions. When a title is a direct successor, the coded entry from the *Guide* (e.g,. B1314) is given with the new citation here; the reader may wish to consult the original volume for a complete annotation. Reference materials cited in this chapter are directed toward forestry scholars, academicians, advanced students, and science librarians.

A brief mention should be made of a rapidly emerging type of resource which this update does not cover. Interactive computer programs and systems are now in regular use in libraries and compliment more traditional

---

1. Richard Blanchard and Lois Farrell, *Guide to Sources for Agricultural and Biological Research* (Berkeley: University of California Press, 1981).

types of paper-based reference material. GIS technology, which is covered in detail in Chapter 6, has extensive applications in forestry and is an information system which is challenging the definition of a reference tool. With its ability to supply maps, data set overlays, numeric files, and some bibliographic data all at one computer work-station, GIS technologies are emerging as a mutlifaceted tool of twenty-first century technology.

Similarly, timber data sets, selected genetic mapping of tree species, online full-text periodical pages, bibliographies and forestry resource updates such as job postings and conference calendars, are now regular features of the Internet. Many universities currently offer access to this information via experimental platform-servers like Gopher, which offer non-hierarchical multi-access interfaces to the Internet. Unfortunately, there are as yet no quality controls and no standards to information loading and retrieval. This makes descriptions difficult since much of the raw data available on the Internet has no accompanying textual reference. Because it is mostly experimental, judgments may be premature.

Currently, the Internet is a limited resource and difficult to use, nevertheless several examples deserve mention. This discussion of Internet resources is selective and makes no claims at being comprehensive. In the vanguard of this electonic information revolution is *Dendrome: Forest Tree Genome Mapping Database*, available on Gopher, compiled by the U. S. Forest Service's Institute of Forest Genetics (IFG) at Albany, California. Internet accessible, this genome mapping project, like so much else on the Internet, is still experimental, but has the potential to serve a wide variety of researchers around the world at their own work-stations. The key information commodity of *Dendrome* is tree species genetic sequencing codes. Species *Pinus*, and specifically the loblolly, are emphasized in the database's current experimental phase. Periodic updates suggest the entire IFG database may soon be available, free of charge, over Gopher.

From the same Gopher menu, Internet prospectors can also access the Forestry Library Gopher at the University of Minnesota, which has compiled numerous forestry-related bibliographies that are keyword searchable over the Internet. Four are currently available: *Social Sciences in Forestry*, an online version of their quarterly indexing journal, the *Trail Planning, Construction and Maintenance Tropical Bibliography, Forest Conservation and Development,* and *Urban Forestry*.

Also available from the same menu, are the Yale Forestry Gopher's EDEX, its collection of ecological data sets of selected forest sites in the Northeast, and LTERnet, the Long-Term Ecological Research Network out of the University of Washington, which contains textual and numeric data on hundreds of long-term projects dealing with forest-land ecology. Since all Gopher menus are cross-referenced, the possibilities of exploration on this distribution network are almost endless.

There are also numerous Listservs on the Internet, several with a forestry focus, which serve as scholarly bulletin boards where newsletters, news services, job postings, calendars of events and research queries are regularly posted to hundreds of members worldwide. In some cases, where direct access to a library is not available, listservs can work as surrogate reference services. Systems such as the BIOSCI Network's Agroforestry Listserv, allow researchers in the field to post queries to literally hundreds of scientists on topics ranging from soil fertility in low latitude countries to questions of seed stocks in equatorial countries. But since these services exist without regulation or quality control there is no way to judge their efficacy.

Government regulators and information scientists are just now beginning to grapple with online systems such as the Internet, specifically in the areas of quality control, uniform standards, archiving, fees and copyright. This brief introduction to Internet resources, is intended only as a preliminary overview to what is essentially an experimental milieu of online platforms and networks compiled haphazardly with few discernable standards. Presently, it is too unclear where this medium will lead the information management profession. For this reason, this current update does not include Internet resources in the listing.

The resources are listed under the following headings:

A. Dictionaries, Encyclopedias, Glossaries and Thesauri
B. Directories and Guides
C. Directories of Research Personnel
D. Handbooks and Manuals
E. Abstracts, Bibliographies and Databases
F. Statistics
G. Atlases
H. Audio-Visual

Update of Reference Sources in Forestry

A. Dictionaries, Encyclopedias, Glossaries and Thesauri

American Plywood Association. *Plywood Encyclopedia*. Tacoma, Wash.: American Plywood Association, 1980. 62p.
A brief introductory guide to the terminology and history of plywood. Includes alphabetical references to roughly 280 terms and concepts.

Bashkirov, I.A. and R. Rubinshtein. *Frantsuzko-Ruskii Lesotekhnicheskii Slovar = French-Russian Dictionary of Wood Technology.* 2nd ed. Moscow: Ruskii Yazyk, 1986. 432p.

A comprehensive dictionary with terminology controlled by the International Standards Organization and its Russian counterpart.

Boutelje, Julius. *Encyclopedia of World Timbers: Names and Technical Literature.* Stockholm: Swedish Forest Products Research Laboratory, 1980. 338p.

This compendium of world timber is divided into three parts: 1) "Scientific Names in Alphabetical Sequence." 2) "Commercial and Local Names in Alphabetical Sequence." 3) "Key to the Technical Literature."

Brown, Curtis L. *Thesaurus of Pulp and Paper Terminology.* Atlanta, Ga.: Institute of Paper Science and Technology, 1991. 561p.

This work of related terms is organized to exercise "vocabulary control", with the intention of assisting the retrieval of information from archived documents without obtaining unwanted material. The definitive grouping of related terms under each main entry serves two purposes: 1) it offers an array of approved expressions for related ideas; 2) provides allied concepts to consider during document retrieval. Standard library abbreviations such as NT—narrower term, BT—broader term, RT—related term, among others are used with each main entry. Contains more than 20,000 entries, along with over 5,000 scope and explanatory notes. More than 750 homonyms have been resolved. Conforms to usage in the PAPERCHEM database.

Burns, Russell and Barbara Honkala, eds. *Silvics of North America.* Washington, D.C.: USDA Forest Service, 2 vols., 1990. Vol. 1 Conifers, p.675 ; Vol. 2 Hardwoods, p.877. [*USDA Agriculture Handbook* No. 654]   [B1383]

A massive compendium updates the 1965, *Silvics of Forest Trees of the United States*, (Agriculture Handbook No. 271). The two volumes describe the silvical characteristics of some 200 conifers and hardwood trees in the conterminous U.S., Alaska, Hawaii and Puerto Rico. After a brief introduction with general notes and selected references, both volumes are arranged alphabetically using genus/species entries. The individual entries were researched and written by knowledgeable Forest Service staff and university scholars in appropriate fields. Each volume, contains many graphs, tables and photographs. Both volumes have Appendices, including: Glossary; Summary of Tree Characteristics; Checklist of Insects and Mites; Checklist of Organisms Causing Tree Diseases; Checklist of Birds; Index of Authors and Tree Species and an Index of Trees by Common Name. Of value as a teaching tool for researchers, educators and practicing foresters.

Chudnoff, Martin. *Tropical Timbers of the World*. Washington, D.C. USDA Forest Service, 1984. 464p [*USDA Agriculture Handbook* no. 607]

This compendium is divided into four parts with a lengthy introduction. Each part is followed by a bibliography. Part 1: "Tropical American Species." Part 2: "African Species." Part 3: "Southeast Asian and Oceanian Species." Part 4 has "Comparative Tables of Properties and End Uses." There are five appendices: A) Selected Forest Products Reference Material; B) Generic Synonyms; C) Generic Groupings; D) Derivation of Comparative Toughness Values; E) Kiln Schedules. The Index covers trade and common names.

Davis, Richard C. ed. *Encyclopedia of American Forest and Conservation History*. New York: Macmillan; London: Collier, 1983. 2 vols.

This encyclopedia of American forestry has excellent alphabetical entries by over 100 forest scientists. Photographs accompany the text. Five appendices are included: 1) National Forests; 2) National Parks; 3) Chronology of Federal Legislation in Forestry; 4) Chronology of Presidential Administrations and their forestry appointments; 5) Atlas. The encyclopedia is indexed. The main entries are somewhat cumbersome to use since they reflect the interests of the authors rather than a coherent editorial plan.

Evans, David S., ed. *Random Lengths: Terms of the Trade*. 3rd ed. Eugene, Ore.: Random Lengths Publications, 1993. 243p. (1st ed., 1978)

[B1314]

This book contains words and phrases used in forestry, logging, manufacturing and marketing, with the intention of offering adequate access to specialized terms in varied fields of work. With over 1000 entries, the work is divided into four major sections: 1) Terms of the trade; 2) Abbreviations; 3) Patterns and Sizes; 4) Organization Guide. Contains illustrations and historical references. A work of particular utility to beginners.

Ferlin, Guy. *Elsevier's Dictionary of the World's Game and Wildlife*. Amsterdam: Elsevier, 1989. 426p.

This volume is dived by animal order: mammalia, aves amd reptilia. Within these divisions the wildlife is described by continent, as well as by coastal sea species. A section on commonly hunted game follows, divided into mammals and game birds, with a compendium of hunting terms. There are numerous essays on hunting methods, including falconry and trapping. Ethics of wildlife management are discussed and there is a section on game diseases. There is no bibliography. Glossary in English, Latin, French, German, Dutch and Spanish.

Ford-Robertson, F.C. ed. *Terminology of Forest Science Technology Prac-*

tice and Products. English-Language Version. 2nd printing. Bethesda, Md.: Society of American Foresters, 1983. 370p. (Authorized by FAO/ IUFRO.)   [B1290]

This 300 page vocabulary is the most extensive available with over 2000 entries. Appendices include: 1) Key to Published Sources; 2) Figures Illustrating Vocabulary; 3) Families of Connected Terms; 3) List of Deprecated Key Terms with preferred synonyms; 5) Consolidated Key to Abbreviations; 6) Summary Flow-chart of the Project's Consultative Procedure. A lengthy Addendum is included for this 2nd printing.

Gorse, Jean, E. *Terminology of Forestry and Related Subjects, English-French = Terminologie Forestière et Vocabulaire Français-Anglais*. 2nd rev. ed. Washington, D.C.: World Bank, 1987. 53p.

This is a straight-forward alphabetical dictionary of forestry terms including terms in on forest soils, forest protection and some engineering. The first half is English into French, the second, French into English. Of particular value for researchers in sub-Saharan Africa where French is widely spoken.

Hillier, Harold. *The Hillier Colour Dictionary of Trees and Shrubs*. New York: Van Nostrand Reinhold Co., 1982. 323p.

Considerable front matter is given to the Introduction, a listing of symbols and abbreviations, the dictionary description and a detailed explanation of plant names and classification. There is also a section on choosing trees and shrubs for a variety of habitats. The dictionary proper covers "Trees & Shrubs", "Climbers" and "Conifers". A glossary of botanical names is included. Of particular value to landscape planners.

James, N.D.G. *An Historical Dictionary of Forestry and Woodland Terms*. Oxford: Blackwell, 1991. 235p.

A compilation of terms, many antiquated and in disuse. Quotations from old sources give these quaint terms their context and meaning. The references at the end are arranged in numerical order under respective letters of the alphabet.

Kireev, D.M. *Ekologo-Geograficheskie Terminy v Lesovedenii = Ecological and Geological Terms in Forest Science: A Reference Dictionary*. Novosibirsk, USSR: Nauka, Sibirskoe Otdelenie, 1984. 182p.

A Russian language dictionary using landscape unit concepts to group the detailed definitions of over 2,000 terms. Aimed primarily at aerial-photo interpretation of forest types, soil classification and important tree species. The index is cross-referenced. Introduction only in English.

Lavigne, John. *Pulp and Paper Dictionary*. 2d ed. San Francisco: Miller Freeman Publ., 1993. 488p.

A brief historical overview serves as the "Introduction" to this list of pulp and paper definitions. There is a four page bibliography of conversion tables.

Medina, A.L. *An English-Spanish Glossary of Terminology Used in Forestry, Range, Wildlife, Fishery, Soils and Botany.* Tempe, Ariz: Rocky Mountain Forest and Range Experiment Station, USDA Forest Service, 1988. 54p. (*USDA General Technical Report* no. RM-152)

This brief dictionary is grouped by subject. Translations are given in English to Spanish and Spanish to English. Of value for field researchers in the Southwest.

Mozhaev, D.V., et al., eds. *Anglo-Russkii Lesotekhnicheskii Slovar = English-Russian Dictionary of Forestry and Forest Industries.* Moscow: Russkii Yazyk, 1983. 669p.

This is a comprehensive glossary of some 60,000 terms and their Russian equivalents. Scope notes and origin descriptions make this a uniquely useful reference tool. Many highly specialized terms are including in the wood-working field, in paper-making and in tree entomology and pest management. Includes a lengthy list of English abbreviations, as well as trade and research organizations.

Negi, S.S. *A Dictionary of Forestry.* Delhi, India: Periodical Expert Book Agency, 1988. 184p.

A comprehensive vocabulary providing succinct explanations for over three thousand forestry terms commonly used in Southeast Asian countries.

Pascovici, N. and R. Vlad-Liteanu. *Dictionar de Silvicultura si Industrai Lemnului German-Roman.* (= German-Rumanian Dictionary of Forestry and the Wood Industry). Bucharest, Romania: Editura Tehnica, 1982. 798p.

This is a comprehensive glossary of over 30,000 German terms with their Rumanian equivalents. A short bibliography is included.

Petrokofsky, G. *Forestry and Forest Products Dictionaries, Terminologies, Glossaries and Vocabularies—II.* Wallingford, U.K.: CAB International, 1985. 76p. (*CAB Annotated Bibliography* no. F39).

An extraction of 358 titles from the *CAB Abstracts* database entered between 1973–1984, arranged by subject and indexed by author and language.

Priszter, S., ed. *European Trees and Shrubs: An Octolingual Dictionary = Arbores Fruticesque Europea - Vocabularium Octo Linguis Redactum.* Budapest, Hungary: Akademiai Kiado, 1983. 300p.

This volume is based on the 1980 compendium *Flora Europea.* The Latin names of all European trees are given with their names and descriptions in English, French, German, Hungarian, Italian, Spanish and Russian. Indexes for each language included.

Prusinkiewicz, Z. *Wielojezyczny Slownik Terminow z Zakresu Prochnic Lesnych = Multilingual Dictionary of Forest Humus Terms.* 2nd ed. Sienkiewicza, Poland: Dept. of Soil Science, North Copernicus University, 1988. 195p.

Definitions to 126 forest humus terms are given in English, German, French, Russian, Spanish, Polish and Swedish. Some entries are also given equivalences (without description) in Czech, Estonian, Hungarian, Latvian, Lithuanian, Rumanian and Serbo-Croation.

Rabchevsky, G.A., et al., eds *Multilingual Dictionary of Remote Sensing and Photogrammetry. English Glossary and Dictionary with equivalent terms in French-German-Italian-Portuguese-Spanish-Russian.* Falls Curch, Va: American Society of Photogrammetry, 1984. 343p.

The front matter includes a list of abbreviations and acronyms. Almost 2,000 terms are defined in English which are numbered with translations given in the other six languages. The translated terms are then listed alphabetically in separate language lists and cross-referenced.

Reyes, T., et al., eds. *Diccionario Ingles-Espanol de Terminos Forestales y Silvoindustriales = English-Spanish Dictionary of Forestry and Forest Products Industry Terms.* Conception, Chile: Instituto Professional de Chillan y Universidad de Bio-Bio., 1984. 203p.

The terms are given in English with their Spanish equivalents. Includes: 1) Forest and the timber industry; 2) Pulp and fiberboard industry; 3) Paper industry. The appendix contains a thematic list of terms relating to papermaking machines.

Ruokonen, Mirja. *Forestry and Forest Products Vocabulary.* Slough, U.K.: Commonwealth Agricultural Bureaux, 1984. 459p.

This vocabulary was prepared in conjunction with the WESTFORNET office of the U.S. Forest Service. This is an alphabetical English-language vocabulary. Of particular interest because it included British, Australian, New Zealand and American slang covering all aspects of the discipline. Many acronyms are included.

Schmid-Haas, Paul, ed. *Vocabulary of Forest Management.* Vienna, Austria: International Union of Forestry Research Organizations, 1990. 316p. (IUFRO World Series Vol. 1).

This basic six language vocabulary is cross-indexed for English, German, French, Spanish, Italian and Russian.

Schniewind, Arno, ed. *Concise Encyclopedia of Wood and Wood-Based Materials.* Oxford: Pergamon Press; Cambridge, Mass: MIT Press, 1989. 345p.

A comprehensive reference work, this volume covers all aspects of wood science. The listing is arranged alphabetically and extensively cross-referenced. A subject index is included.

Shigo, A.L. *A New Tree Biology Dictionary: Terms, Topics, and Treatments for Trees and Their Problems and Proper Care.* Durham, N.H.: Shigo & Trees, Associates, 1986. 142p.

This dictionary contains more than 200 definitions used to describe the biology, treatment, decay and pathogenic injury of trees.

Staszak, W. *Podreczny Slownik Drzewny Angielsko-Polki = English-Polish Dictionary of Terms Relating to Timber.* Poznan, Poland: Instytut Technologii Drewna, 1985. 499p.

This glossary contains over 10,000 terms, arranged alphabetically in English with codes corresponding to their Polish equivalencies.

Turk, Z., et al., eds. *Recnik iz Oblasti Iskoriscavanja Suma i Sumskih Komunikacija = Dictionary of Forest Utilization and Forest Communication.* Ljubljana, Yugoslavia: Institut za Gozdno in Lesno Gospodarstvo Slovenije, 1980. 287p.

This multilingual glossary also covers forest engineering, harvesting, machinery and transport. The main text is in Serbo-Croatian with equivalents in English, Slovenian and Macedonian. One reviewer said of this dictionary: "This work fills a considerable gap in the international literature in forest products terminology." (CAB International, *Forestry Abstracts,* 1981. #F913105.)

Walker, Aidan, ed. *The Encyclopedia of Wood: A Tree-by-Tree Guide to the World's Most Versatile Resource.* Oxford & New York: Facts on File, 1989. 192p.

This compendium covers: 1) Craftsman. Deals historically with the uses of wood; 2) The Living Tree. Covers ecological aspects of tree propagation; 3) From Tree to Wood. Deals with logging and extraction; 4) The Directory of Wood. Describes how the directory is to be used, the major wood types and their nomenclature; 5) The World of Wood. This is a species by species description. Includes a glossary and an index.

Zimmerman de Olazabal, L. *Vocabulario Tecnico: Industria y Commercializacion de Pasta Celulosica, Papel y Afines, Ingles-Espanol, Espanol-Ingles = Technical Dictionary: The Pulp and Paper Industry and Trade and Associated Industries, English-Spanish, Spanish-English.* Resistencia, Argentina: Centro de Informacion Bioagropecuaria Y Forestal, Universidad Nacional de Nordeste, 1983. 156p.

A comprehensive terminology of paper industry terms in English to Spanish, and Spanish to English.

B. Directories and Guides

California Dept. of Forestry and Fire Protection. *Forestry and Tree Education Catalogue*. Sacramento, Calif: State of California, Resource Agency, Dept. of Forestry and Fire Protection, 1990. 981p.
A listing of 4,500 "best" forestry and tree education resource materials available for educational purposes. The Description of Educational Resources, comprising 841 pages, alphabetically lists books, journals, serials, films, videos, audio-tapes, slide-shows, workshops, booklets, posters and teachers' aides. The catalogue also includes: a Glossary of Relevant Terms; Conceptual Categories; Organizational Resources; State Tree and Arbor Days and an index of key descriptors.

Castner, J.L. *Rainforests: A Guide to Research and Tourist Facilities at Selected Tropical Forest Sites in Central and South America*. Gainesville, Fla.: Entomology Dept., University of Florida, 1991. 380p.
This handguide describes and evaluates tropical forest sites in Central and South America. Specific information is given for Peru, Ecuador, Costa Rica, French Guiana, Panama Trinidad and Venezuela, including contacts, addresses, logistics, seasonality, environment and forest type. An annotated rainforest bibliography is included, as is a chapter on funding sources for tropical forest research.

Cordell, Charles, et al., coordinators. *Forest Nursery Pests*. Washington, D.C.: USDA Forest Service, 1989. 184p. [*USDA Agriculture Handbook no. 680*; Supersedes *Handbook no. 470* (1975).]
Preliminary chapters of this handbook cover "Diagnosis of Pest Problems", "Integrated Nursery Pest Management", Evaluation of Nursery Losses Due to Pests", "Soil-Pest Relationships" and "Pesticide Regulations". The main text is divided into conifer and hardwood sections, which are subdivided by fungal and viral pests as well as injurious insects accompanied by color photographs. Miscellaneous pest problems are covered at the end. Contains an index of nursery pests and their host trees, a list of registered pesticides and a glossary.

Deshprabhu, C.N., et al. *Directory of the Indian Plywood and Other Wood Based Panel Industry*. Bangalore, India: Indian Plywood Industries Research Institute, 1986. 105p.
This compendium includes Indian standards relating to the industry, a list of species used for plywood and special forest resources.

Elias, Thomas. *Field Guide to North American Trees*. 2nd ed. Danbury, Conn.: Grolier Book Clubs, 1989. 948p.
A classic guide with easy identification of over 750 North American Trees. Arranged by natural relationships, particularly by conifers and

flowering trees. The hardwoods, which contain the largest number of species, are arranged with less specialized, primitive groups first. Both common and scientific names are given. The illustrations are monochrome. Each species has a distribution map. Fully indexed.

Evans, Cynthia. *A Guide to Exporting Solid Wood Products*. Rev. ed. Washington, D.C.: USDA Forest Products Division, 1990. 116p. [*USDA Agriculture Handbook* No. 662]

An excellent introductory overview of the timber export market. Contains information on supply consideration, financing, shipping, business organization, export market information and assistance, USDA export programs and major markets. Ten appendices cover: Selected Publications; Shipper's Export Declaration Samples; Foreign Importers Listing; Addresses of State Foresters; Desk Offices of the International Trade Administration; U.S. and Foreign Commercial Service District Offices; Small Business Administration Offices; Glossary of Forest Products and Exporting Terms; Forest Products Volume Conversion Units; Forest Products Shipping Weights and Volumes.

Evans, Peter A. and Mark Arizmendi, compilers. *International Directory of Forestry and Forest Products Libraries*. 3rd ed. Berkeley, Calif.: Forestry Library, University of California, Berkeley, 1991. 108p.   [B1334] The typeset and usability of this compendium is greatly improved over the original 1982 edition. The directory is divided by regions, with 400 + libraries listed alphabetically. Each entry contains information on the name, address and telecommunications of the library, as well as staff, areas of subject strength, special services and name of supporting institution. Includes Index of Countries and an Index of Librarians.

Fahl, Kathryn, compiler. *Forest History Museums of the World*. Santa Cruz, Calif.: Forest History Society, 1983. 75p.

This directory includes museums and forestry-related collections of artifacts from around the world. Entries are arranged alphabetically by country and by city. The U.S. and Canada are also arranged by state and province. Entries contain location information, contact person, hours open, a brief description of the collection and additional information on restaurants, shops and publications. Indexed alphabetically.

Food and Agriculture Organization. *Databook on Endangered Tree and Shrub Species and Provenances*. Rome, Italy: FAO, 1986. 524p. [*FAO Forestry Paper* no. 77]

This red databook includes detailed analysis and status of 82 species of woody plants, listed alphabetically by genus/species. Of particular value to conservationists and forest planners.

Food and Agriculture Organization. *Directory of Forestry Research Organ-*

izations. Rome, Italy: FAO, 1993. 172p. [*FAO Forestry Paper* no. 109].
This directory is based on a questionaire sent to all organizations dealing
with forestry research on the mailing lists of FAO and IUFRO. The work
has been made "deliberately simple," as the Preface states, in order to
make the work manageable in length, and to prevent error in data entry.
Each of the 1318 entries contains the address, phone number, FAX, staff
size, general subject coverage and primary contact for each organization.
The list is alphabetical by country.

Food and Agriculture Organization. *World List of Forestry Schools.* 3rd rev
ed. Rome, Italy: FAO, 1986. 524p. [*FAO Forestry Paper* no. 3 rev./2]
[B1327]
This compendium is an update of the 1977 edition. Contains: 1) "World
List of University Level Schools.." 2) "World List of Non-University
Level Schools . . . " Each section is divided by continent, with countries
and their institutions listed alphabetically.

Friends of the Earth, (UK). *The Good Wood Guide.* London: Friends of the
Earth, 1988. 125p.
An environmental guide to retailers, manufacturers and institutional users
of tropical hardwood timber, perceived by Friends of the Earth as actively helping to save rainforests by obtaining timber from ecologically
benign sources. The work is color-coded by tree species. Of particular
value to builders, architects and conservationists.

Genino, Angela, ed. *Amazonia: Voices From the Rainforest. A Resource
and Action Guide.* San Francisco: Rainforest Action Network, 1990. 93p.
Despite its regional focus, this guide lists facts, figures, support groups,
and action agendas from 27 countries interested in Amazonia. The Directory is divided by continent and by country. Contains an extensive bibliography of books and videos. It is well indexed. Of particular interest to
individuals and institutions involved with rainforest protection and conservation.

Gilluly, David. ed. *Directory of Forestry and Natural Resources Computer
Software.* 5th ed. Florence, Ala.: Forest Resources Systems Institute
FORS, 1989. 171p.
Compiled by the Forest Research Systems staff between 1982 and 1988,
this guide is divided into 13 subjects: Economics, Engineering & Mapping, Forest Protection, Harvesting, Marketing, Mensuration, Networks
and Databases, Operations Research, Recreation, Soils & Hydrology,
Urban Forestry, Utilities, and Wildlife & Fisheries. The directory is indexed by title, keyword, computer type, and corporate source.

Hamel, Dennis. *Forest Management Chemicals: A Guide to Use When
Considering Pesticides for Forest Management.* Rev. ed. Washington,

D.C.: U.S. Dept. of Agriculture, Forest Service, 1983. 645p. [*USDA Agricultural Handbook* no. 585]

An indepth compendium covering all aspects of chemical pest management in U.S. forestry. The main text covers the major pesticide types. All sections have detailed strategies for pesticide application and a list of references. Sections A) Insecticides and Acaricides; B) Fungicides and Nematicides; C) Herbicides; D) Animal-Damge Management Chemicals. Appendices include: Glossary, Abbreviations, Formulae, Conversion Tables, References, and Technical Assistance. Index included.

Henley, Russell and Paul Ellefson. *State Forest Practice Regulation in the U.S.: Administration, Cost, and Accomplishment*. St. Paul, Minn.: Agricultural Experiment Station, University of Minnesota, 1986. [*Station Bulletin* AD-SB-3011]

Using literature reviews, questionnaires and interviews, the authors of this guide gathered information on trends in the use of public programs to influence private forest management, and on the administration, costs, and effectiveness of state forest practice regulations. Contains an introductory chapter on the historical and legal background of early state forest laws. Seven states, including Massachusetts, Nevada, Alaska, Idaho, Oregon, Washington and California are extensively covered. Contains a ten page bibliography. Of great value to state forest planners.

Hilmi, H.A. *World Compendium of Forestry and Forest Products Research Institutions*. Rome, Italy: Food and Agriculture Organization, 1986. 613p. [*FAO Forestry Paper* no. 71]

This excellent guide is divided by continents, providing information on the names and addresses of institutions, their status, the number of staff members, the number of field research stations, principal fields of activity, special facilities, publications and services, and official working language. The lists are in English, French and Spanish. Particular emphasis on the Third World makes this a unique resource.

Hoover, W.L. *Forest Owners' Guide to Timber Investments, the Federal Income Tax, and Tax Recordkeeping*. Washington, D.C.: USDA, Forest Service, 1989. 96p. [*USDA Agriculture Handbook* no. 681.] (Supersedes 1982 *USDA Agriculture Handbook* no. 596)

Earlier edition was *A Guide to Federal Income Tax for Timber Owners*. Includes chapters on timber investment considerations, tax planning, general tax considerations, cost considerations, income, other taxes, searching a tax question, sources of tax assistance, forest records, and a glossary. Appendices include summaries of select revenue rulings, form T (timber), and forms for recording transactions. Indexed. Harold E. Burghart an IRS timber tax law specialist reviewed and made recommendations for the manuscript.

International Pulp and Paper Directory, 1994. San Francisco: Miller Freeman, 1994. 1072p. Biennial.   [B1347]

A comprehensive world-wide guide to the pulp and paper industry. Sections include a General Index, a worldwide index to executive offices, mills and paper merchants; Pulp and Paper Mills, a complete listing arranged by continent and then by country of all producers and mills; Mill Grades, divided into pulp and paper; Market Pulp, divided by grades and producers by country; Paper Merchants, divided by European-based and North American importers and exporters; Industry Information Sources; Trade Shows and Conferences; Buyer's Guide; and two indexes to buyers and advertisers. The directory covers 101 countries.

International Union for Conservation of Nature and Natural Resources. *1990 United Nations List of National Parks and Protected Areas; Liste des Nations Unies des Parcs Nationaux et des Aires Protégées 1990*. Gland, Switzerland, and Cambridge, U.K.: IUCN, 1990.

Kelly, R.T. *Profiles of USA Forestry Schools and Consortia*. Washington, D.C.: Forestry Support Program, USDA Forest Service, 1984. 314p.

Funded by US Agency for International Development, this directory describes educational, research and extension resources available to students. Each entry covers curriculum options, areas of academic concentration, supporting research institutions and involvement in international development. Included is detailed information on 42 US universities with forestry programs.

Krutilla, K. *Guide to Investment and Trade in the Forest Products Sectors of Southeast Asia: Indonesia, Malaysia and the Philippines*. Athens, Ga.: Southeast Center for Forest Economics Research, 1988. 189p. (FPEI-Working Paper no. 32)

The forest management policies, timber resources and status of wood-processing industries are reviewed in detail for the three targeted countries. Other chapters analyze exporting/importing regulations, product quality and standards, shipping, financing, and trade barriers. An annotated bibliography is included.

Lockwood-Post's Directory of the Pulp, Paper and Allied Trades. 121st Year, 1994. Edward, Dyer, et al., eds. San Francisco: Miller Freeman, 1994. 754p.   [B1349]

The definitive source to the United States and Canadian pulp and paper manufacturing and converting industries, completely updated each year. All the important mills in the US and Canada are listed with maps. Covers each mill, its exectutives, their distributers and paper merchants. Sections include paper stock and rag users, watermarks, trade associations, wood pulp agents, industry statistics and a buyer's guide. Advertisers are indexed. A comprehensive reference tool for anyone interested with the current paper industry in North America.

Malaysian Timber Industry Board. *Directory of Timber Trade 91/92*. Kuala Lumpur: Malaysian Timber Industry Board, 1991. 433p.

Full addresses and background information is given on the MTIB, other local and overseas timber associations, and useful contacts within Malaysia and abroad. There is a section of general and technical information listing commercial timbers, export statistics and existing status of resources. Includes a comprehensive index.

Melamed-Gonzales, R and R. Giasson, compilers. *A Directory: NGOs in the Forest Sector*. 2nd African ed. New York: International Tree Project Clearinghouse, 1987. 256p.

This English language listing of Africa contains: 1) Profiles of indigenous and externally based NGOS, arranged alphabetically by country and by organization; 2) Practical, technical and policy information sources relating to agroforestry and environmental activities in Africa; 3) A keyword index.

Mohlenbrock, R.H. *Field Guide to U.S. National Forests*. New York: Congdon and Weed, 1984. 336p.

This is a comprehensive guide to the national forests of the United States, arranged by region. Each entry gives a full description, including address, superintendent, facilities and services, habitat and current status and use.

National Wildlife Federation. *Conservation Directory, 1994*. 39th ed. Washington, D.C.: National Wildlife Federation, 1994. Annual.

This annual lists organizations, agencies and officials concerned with natural resource use and management. Sections include: 1) Committees of United Staztes Congress; 2) Congressional Members; 3) US Government Executive Branch; 4) US Government Independent Agencies; 5) International, National and Regional Commissions; 6) State and Territorial Agencies and Citizen Groups, by states; 7) Canadian Federal and Provincial Agencies, Citizen Groups. The Appendices include: Colleges and Universities in the US and Canada; National Wildlife Refuges; National Forests; Bureau of Land Management Districts; Periodicals of Interest; Directories of Interest; Conservation Offices of Foreign Governments; Audio-Visual Sources, and several indexes by name, publication and subject.

O'Hara, Tim, et al., eds. *Directory of Wood Products Computer Software.* Florence, Ala.: Forest Research Systems Institute, 1991. 65p.

Eighty software applications for the wood products industry are covered. Each is extensively described in alphabetical sequence. Indexes include: Title, Keyword, Computer and Contact.

Petrides, George A. *A Field Guide to Western Trees.* Illustrated by Olivia Petrides. New York: Houghton Mifflin Co., 1992. 306p. (The Peterson Field Guide Series).

Because of its small size, Petrides has created a useful field tool for the identification of western species. All native and naturalized trees of western North America are included. The work is divided into six basic sections according to the leaf structure of the plant, from needle bearing conifers to parallel-veined plants such as palms. These leaf types are illustrated on the endpapers. Each species is accompanied by a computer generated map. Lavishly illustrated, the color plates are found grouped in the middle of the book. Contains several appendices including a key to winter trees, plant relationships, a glossary, and extensive references. The work is well indexed.

Shigo, A.L. *Tree Defects: A Photo Guide.* Durham, N.H.: USDA, Northeast Forest Service Experiment Station, 1983. 178p. [General Technical Report no. NE-82]

Common lumber defects are shown in a series of 110 photographs, illustrating discoloration, decay, knots, insect damage and other problems related to blemishes of finished wood products.

Thompson, Claudia. *Recycled Papers: The Essential Guide.* Cambridge, Mass.: MIT Press, 1992. 162p.

This directory gives a comprehensive overview of the recycled paper industry with emphasis on fine quality paper. Appendices include a glossary, a bibliography, lists of recycled paper distributors and a synopsis of the pulping and papermaking processes.

United Kingdom Overseas Development Administration. *Natural Resources Development Directory: A Guide to Sources of British Expertise and Training in Renewable Natural Resources Available to Developing Countries.* Rev. ed. London: CAB International, 1987. 109p.

This ODA publication updates previous versions and is designed to give those involved in development of renewable natural resources overseas a guide to experts and technicians in the United Kingdom. Covers agriculture, forestry, fisheries and other natural resource sectors. Entries are in alphabetical order.

U.S. Forest Service. *Directory of Forest Tree Nurseries in the United*

States. Washington, D.C.: American Association of Nurserymen in cooperation with the USDA Forest Service, 1987. 78p.   [B1377]
This 1987 update of the 1981 edition contains 356 entries. Divided by state. Each entry is further divided by type of nursery. The sections on bareroot operations and container operations list available acreage for planting and seedlings shipped. A reference of notable importance to persons involved with reforestation.

U.S. Forest Service. *Insects of Eastern Forests*. Washington, D.C.: USDA Forest Service, 1985. 608p. (*U.S.D.A. Miscellaneous Publ*. no. 1426)
This guide is divided into: 1) General introduction; 2) Forest insect control; 3) Insect and tree diseases; 4) Insects and related organisms; 5) Keys to orders and families of forest insects; 6) Selected list of important orders of forest insects. Contains an extensive bibliography. Includes several indexes. Common and scientific names of host plants; Index to insects by host plant; Insect index.

Vetorino, Frank, ed. *Crow's Buyer's and Seller's Directory of the Forest Products Industry, 1992–1993*. Portland, Ore.: Crow Publications, 1992. 731p.
Useful biennial publication which lists: US Wholesalers, Canadian Wholesalers, Lumber Products Index, Plywood/Panel Products Index, Treaters, Importers/Exporters, Associations, Remanufactures, Key Personnel and a Company Index.

Walker, L.C. *Forests: A Naturalist's Guide to Trees and Forest Ecology*. New York: J. Wiley and Sons, 1990. 288p.
After a brief introductory chapter on the changing sensibilities of forest management and ecology, this popular guide is arranged by forest type and the dominant tree species of the landscape.

Willan, R.L., compiler. *A Guide to Forest Seed Handling with Special Reference to the Tropics*. Rome, Italy: Food and Agriculture Organization, 1985. 379p. [FAO Forestry Paper no. 20/2]
A comprehensive handbook on seeds and all aspects of seed handling. Includes a 162 word glossary and index.

C. Directories of Research Personnel

Arnott, J.T., et al., compilers. *Directory of Tropical Silviculturalists*. Washington, D.C.: USDA Forest Service, 1986. 71p.
This directory covers North America and Mexico only. Several hundred individuals are listed by country alphabetically, with field of expertise, geographical familiarity and tree species specialties.

Association of Consulting Foresters of America, *1993–94 Membership Spe-*

cialization Directory. Bethesda, Md.: Assoc. of Consulting Foresters, 1994. 296p. Annual.

This directory has shoddy editorial oversight. The Table of Contents does not list many important parts of the book and the pagination given is flawed throughout. A work likely to be of use to researchers interested in forestry contacts with particular skills or geographic distribution. With almost 1000 entries, this directory is divided into sections covering states or provinces of residence, international consultants, people with languages skills, subject skills and a list of retired members.

Batzar, H.O. and D. Skilling, compilers. *World Directory of Forest Pathologists and Entomologists*. St. Paul, Minn.: North Central Forest Experiment Station, 1985. 269p.

This directory of 1700+ names was produced under the direction of IUFRO Working Party S2.06.08. Countries and individual names are arranged alphabetically. A coding system links these entries to tree genera, pathogen genera and subject categories. Indexes provide access to 131 tree species, 225 pathogens, and 467 species of insects. Personal names are indexed.

Forest Ecosystem Research Network. *Directory of European Research Groups Active in Forest Ecosystem Research Network (FERN)*. Strassbourg, France: European Science Foundation, 1987. 228p.

An alphabetical listing of groups and individuals involved with FERN and ecosystem research in Europe. Includes selected world-wide activity.

Horsley, Stephen B. and Corinne A. Weldon, compilers *International Directory of Woody Plant Physiologists*. Warren, Penn: U.S. Dept. of Agriculture, Forest Service, Northeast Forest Experiment Station, 1995. 217p.   [B1320]

This volume contains a straight alphabetical listing of plant physiologists from around the world, giving their addresses and research interests. The latter are numerically coded along with 1,278 plant species. A second part divides the physiologists by country; a third by 3 by research interest, and the last by tree species expertise.

Tikkala, Willard, compiler. *Directory of Urban and Community Forestry Professionals in the U.S. and Canada*. Washington, D.C.: American Forestry Association, 1983. 144p.

This publication is divided into three sections: 1) Forestry Practitioners; 2) Researchers in Urban and Community Forestry: 3) Foreign Forestry Professionals. Appendix A is a complete index of individuals alphabetized by last name. Appendix B lists "Urban and Community Forestry Organizations and Associations." Appendix C lists appropriate periodicals and journals.

D. Handbooks and Manuals

Burns, Russell M. *Silvicultural Systems for the Major Forest Types of the United States*. Rev. ed. Washington D.C.: Forest Service, USDA, 1983. [*USDA Agriculture Handbook* no. 445]. 194p.

This handbook constitutes a revision of the material contained in the 1973 edition of the same title. The handbook is a compilation of 48 individual manuscripts written by USDA foresters. The work is divided into two major sections, western forest cover types, and eastern types. It contains a glossary, lists of common names, with a forest type and author index. This work will be of particular interest to reserachers interested in a concise overview of particular forest types in North America.

Falconer, Julia. *Forestry and Nutrition: A Reference Manual*. Bangkok: FAO Regional Office for Asia and the Pacific, 1989. 114p.

This English language manual is the first of a planned series by FAO on the relations between forestry, food security and nutrition. Contains an extensive annotated bibliography. The sections include: 1) Farm Tree Foods and Nutrition; 2) Forest Food Diets in Rural Areas; 3) Forest Resources and Nutrition; 4) Forest Foods; 5) Closing Remarks; 6) A Literature Review. Geographical and subject indexes included.

Hayes, Jack, et al., eds. *Using Our Natural Resources: 1983 Yearbook of Agriculture*. Washington, D.C.: USDA, 1983. 572p.

Although all of the natural resources of the United Sates are covered, forestry and the national forest system are the main focus. It is divided into 4 parts: 1) Introduction; 2) Physical, Biological and Social Components; 3) Managing Natural Resource Systems; 4) Human Dimensions in Resources Management. A glossary and index are included.

Krussman, Gerd. *Manual of Cultivated Broad-Leaved Trees and Shrubs*. Translated by Michael Epp. Beaverton, Ore.: Timber Press, 1984. 3 vols.

This guide was produced in cooperation with the American Horticultural Society. A "General Guide to Terminology", an "Alphabetical Guide to Botanical Terms in 5 Languages (English, Latin, German, French and Dutch)", as well as explanatory symbols, hardiness zone maps and abbreviations serve as the front matter to each volume. The main text is an "Alphabetical Listing of Genera" spanning the three volumes. The Appendices in Volume 3 include, an Addenda to Vol. 1, Errata in Vol. 1, a "Taxonomic Outline of Families", an "Index to the Botanical Authors" and a "Registration Authority File for Cultivar Names."

Mies, Wiilard, et al., eds, *Pulp and Paper; 1993 North American Fact Book*. San Francisco: Miller Freeman, 1993. 447p.

This work contains both a general overview of the industry, as well as detailed accounts of its various products, financial status, and world outlook. It is divided into four sections with extensive graphs and maps. Section one is a general overview of the status of the industry in North America, from "Industry Structure" to "Capacity for Expansion". Paper Grades, Paperboard, and Fiber Sources make up the other three sections. Contains a "Glossary of Terms" and extensive company profiles at the end.

Mitchell, Alan. *The Trees of North America*. Illustrated by David More. New York: Facts on File, 1987. 214p.

A large book divided by species, with facing color illustrations to the two column text on each page. Broadly divided by "broadleaf" species, and "conifers". Eighty-one genera are included, with a practical reference section on planting, pruning and disease prevention tips at the end of the book. There are eight pages of distribution maps before the extensive index. A work likely to appeal to the general reader rather than the scholar.

Preston, Richard J. *North American Trees Exclusive of Mexico and Tropical Florida*. 4th ed. Ames, Iowa: Iowa State University Press, 1989. 399p.   [B1409]

This handbook was designed to meet the needs of an interested general public. Concise descriptions of silvic characteristics are included for species that are of particular economic importance or of general interest, while less known species are are only briefly described. Contains 161 full pages of distribution maps and plates. The book is pocket size for easy field use.

Sinclair, Wayne A., Howard Lyon and Warren Johnson. *Diseases of Trees and Shrubs*. Ithaca, N.Y.: Comstock Publishing Associates, Cornell University Press, 1987. 557p.

A comprehensive text, with many full color plates, the title of this compendium is nevertheless somewhat misleading since it pertains almost entirely to species found primarily in temperate climates, notably in North America. Classification of phyto-diseases, as the authors note, is necessarily unsatisfactory since many pathogens cause diverse symptoms. The organization of this work is therefore a blend of schemes, divided into three major groupings: 1) foliage and flowers; 2) twigs, trunks and bark; 3) roots and butt. There is also a section on "entire plant" decline, due to environmental factors. The work is further divided by fungi, bacteria, viruses, nematodes, plant pests, abiotic factors and unknown agents. The references are extensive and come after every section. This work will be of pirimary use to forest scientists and phytopathologists.

U. S. Forest Products Laboratory. *Wood Engineering Handbook*. 2d ed. Englewood Cliffs, N.J.; Prentice-Hall, 1990. Variously paged, 1 vol. no. 72] 230p.   [B1434]
First issued in 1935, and revised in 1939, 1955 1974, and 1987 as a *USDA Agriculture Handbook*. Changed to a commercial publisher with most recent edition. This work provides engineers, architects, and others with a source of information on the physical and mechanical properties of wood. Individual chapters not only describe the wood itself, but also wood-based products, together with the principles of how the wood is dried, fastened, finished and preserved. English and metric systems of measurement are used throughout.

Wenger, Karl. F., ed. *Forestry Handbook*. 2nd. ed. New York: J. Wiley & Sons, 1984. 1335p.
This edition covers all aspects of modern forestry. Divided into 25 chapters, it covers topics as diverse as Forest Ecology, Logging, Recreation Management and Communication and Public Involvement. This handbook provides reference data and methods on all phases of forestry and allied fields, primarily as a tool for foresters in the field. An extensive index is included.

E. Abstracts, Bibliographies and Databases

Agroforestry Abstracts. Wallingford, U.K.: CAB International, 1988 +
A quarterly abstracting journal devoted to agroforestry. 1,500 citations are added yearly from about 600 journals and other litreature. Most records are culled from *Forestry Abstracts* and other *CAB* abstracting journals. Contains news items and occasional review articles. Annual author and subject indexes. Of great value to specialists and researchers in tropical forestry and agroforestry. Searchable online.

Albrecht, Jean, ed. *Social Sciences in Forestry: A Current Selected Bibliography and Index*. St. Paul, Minn.: Forestry Library, University of Minnesota. Revised, 1985 +. Quarterly.   [B1183]
Published since 1963 this bibliography was substantially revised in 1985. The current bibliography is divided into several major sections including: 1) Social Sciences Applied to Forestry at Large; 2) Social Sciences Applied to Forestry Management; 3) Production of Forest Goods and Services; 4) Social Sciences Applied to Forest Industries. Contains an author index with each issue. Since 1985, with no. 62, this bibliography has been searchable via the Internet using Gopher client software and is updated quarterly. In one month, 38,000 connections were made to the file from users throughout the world.

Albrecht, Jean and Patrick Weicherding. *Urban Forestry: A Bibliography*. St. Paul, Minn.: University of Minnesota, 1993. 73p.

This is a an extensive revision of the 1977 and 1980 editions, as well as the 1982 and 1988 supplements. Covering the literature from the late 1970s to the present, this bibliography is divided by broad subject categories followed by an author index. All publications listed are held by the University of Minnesota libraries. A resource widely used by the American and international urban forestry community.

Canadian Forest Industries Database. Produced by Southam, Don Mills, Ontario, M3B 2X7, Canada.

Currently 10,000 bibliographic references since 1989 with updates monthly. Covers operating and administrative management in logging, sawmilling and related wood products. Alerts foresters and industry personnel to new production and operating techniques, new equipment and improved management methods. Emphasis on the Canadian forest industry, with extensive U.S. coverage. Available via Infomart Online for user fee. Call 1-800-668-9215 for online connect details.

Forest Land Use Reports System (FLURS). USDA Forest Service. P.O. Box 96090, Rm. 1010, Washington D.C., 20090-6090. (Tel# 202-235-8212)

FLURS identifies permit holders on federal lands. The file contains the name and address of the permit holder, type of permit, issuing agency, billing information, type of use and location. Each of the eight Forest Service regions maintains their own file and operates their own system. Free searches available through Forest Service regional offices using Freedom of Information Act procedures.

Fortmann, Louise, et al. *Trees and Tenure: An Annotated Bibliography for Agroforesters*. Madison, Wis: Land Tenure Center, Univ. of Wisconsin; Nairobi, Kenya: International Council for Research in Agroforestry, 1985. 135p.

This literature compendium is a comprehensive effort to bring together material from a variety of sources on the subject of land tenure and trees. Divided into five sections, it spans the tropical land masses of the globe, including: Africa, Asia, Latin America and Oceania, with a section on general agroforestry. The Annex lists major monographic works on the subject. The bibliography is indexed by general entry, by country and by tree species.

FS-INFO. USDA Forest Service. Information Systems Staff, 10301 Baltimore Blvd., Beltsville, Md. 20705.

Provides references to all Forest Service publications, including articles written in non-government publications by FS staff. Coverage since

1985, with monthly updates. Size of database is about 75,000 records. Covers all aspects of forest science. Search services available. (301-344-2507).

Indonesian Forestry Abstracts: Dutch Literature until about 1960. Wageningen, the Netherlands: Centre for Agricultural Publishing and Documentation, 1982. 658p.

A magnificent bibliography of historical importance covering the Dutch experience in Indonesia from mid-1870s to 1960. Contains a list of periodicals indexed, a listing of vernacular names of major species of Indonesian trees and a glossary. Several maps are also included. Over 6000 citations are grouped alphabetically under nine subject headings. The text is interspersed with photographs. Contains: Author, Species, Subject Indexes.

National Forest Fire Occurrence Database. USDA Forest Service, Fire and Aviation Management, P.O.Box 96090, Washington, D.C., 20090-6090. Available as a tape.

This database consists of fire report data for the United States. Organized in yearly files with about 14,000 new records per year. Over thirty-five years input. Database searchable by: region, acres burned, dollar damage, size class, number of fires, and statistical cause. Provided to individuals free of charge. A monthly printed forest fire report is also available.

PAPERCHEM. Available online via DIALOG Information Services 1967+. Institute of Paper Science and Technology, 575 14th St. N.W., Atlanta, Ga. 30318.   [I040]

This database is the electronic version of *Abstract Bulletin of the Institute of Paper Science and Technology,* providing citations and abstracts to international patents and journal literature dealing with pulp and paper technology. Mostly in English. Begun in 1967, it now contains over 250,000 records. Updated monthly with 1200+ records. Searchable by author, title, journal, publication date and patent number. Thesaurus available.

Public Domain Forest Inventory System (PD). U.S. Dept. of Interior, Bureau of Land Management, Denver Federal Center Bldg. 50, PO Box 25047, Denver, Colo. 80225.

PD provides information for management of timber resources through analysis of forest inventory data. System contains a broad range of inventory data including county topography, BLM district, forest type, soil and harvesting. Five report series indexed, containing 200+ data elements. The data describe specific tree species, location, watershed and site measurements. Individual searches and printouts available for a fee.

Putz, Francis E. and Michelle Pinard, *Natural Forest Management in the*

American Tropics: An Annotated Bibliography. Washington, D.C.: Office of International Cooperation and Development and USAID, 1991. 340p.

This compendium is divided into two parts. Part 1 contains information on neotropical natural forest management literature. Part 2 contains references to recent and current research projects on natural forest management. Both parts have separate indices for geographical area, general descriptors, silvicultural descriptors and species/ecosystem. Literature entries contain information on author, title, language, geographical location, source, descriptors and abstract. Project entries contain information on authoring agency, project title, geographical location, funding agency, contact person, descriptors, and abstract.

Shedenhelm, Laura. "Latin American Forestry Research: Resources and Strategies." In *SALAM and the Area Studies Community Seminar on the Acquisition of Latin American Library Materials. Papers of the Thirty Seventh Annual Meeting.* Pp. 188–219. Albequerque, N.M.: University of New Mexico, General Library. 1994.

This is an indepth overview of forestry in particular, and agriculture more generally, throughout Latin America. The introductory section provides a broad outline of the important reference resources dealing with agriculture and forestry for the region. An extensive bibliography is divided: "General Sources", "Country-Specific Sources" (covering 27 countries), "International Agencies", and "Databases and Computer Services" listing 21 resources. A quick and handy guide of utility to librarians.

Timber Management System (TM). USDA Forest Service, 12th and Independence Ave., SW, North Bldg., Washington, D.C., 20250.

This database provides national and regional forest information. Contains records identified by land use classification and a file of potential volume yield for all national forests. Provides time harvest volume records for the three most current years. Also available is a file of timber activity programs. Customized computer printouts available; generally no fee. (303-224-1803).

TREECD. Compiled by CAB International. Distributed by Silverplatter Information. Issued in 1993.

This CD-ROM bibliographic database (covering 1937 to 1990), contains 375,000 records and represents an exhaustive collection of forestry literature covering: silviculture, forest management, forest land use and wood products. Most of the references are from CAB's, *Forestry Abstracts, Forest Products Abstracts*, and *Agroforestry Abstracts*. Of great value for its inclusion of over fifty years of literature in one searchable, electronic file.

F. Statistics

American Forest and Paper Association. *Statistics of Paper, Paperboard and Woodpulp, 1993*. Washington, D.C.: American Forest and Paper Association, 1993. 84p.
This is the thirty-first edition of this statistical analysis of the pulp and paper industry. The work is almost entirely composed of charts and figures, with very little text, and is grouped into four functional sections: 1) production and related data; 2) regional data; input trends; 4) financial information. Supplemental tables cover import/exports and analyses of demand. The report provides a comprehensive up-to-date statistical base of annual data.

Clawson, Marion and Carlton Van Doren, eds. *Statistics on Outdoor Recreation*. Washington, D.C.: Resources for the Future, 1984. 368p.
This record of statistics is divided into those through 1956 and since 1956. It covers all aspects of outdoor recreational use in all types of terrain, including State and National Parks, Forests and Wildlife Refuges, as well as lakes, river systems and seashores. Detailed statistics on hundreds of subjects are given. The volume is cross-indexed. Tables and figures are included.

Food and Agriculture Organization. Forestry Policy and Planning Division. *Forest Product Prices, 1971–1990*. Rome, Italy: FAO, 1992. 261p. [*FAO Forestry Paper* No. 104]
A trilingual (French, Spanish and English) price guide to the world of forest products. Contains two major parts: Part 1 covers the various types of wood and forest products worldwide, and Part 2 contains the country tables. Contains an extensive bibliography as well as a brief introduction.

Pease, David A., Ted Blackman, and Jennifer Sowle. *Forest Industries: 1992–1993 North American Factbook. Wood Technology*. San Francisco: Miller Freeman, 1993. 327p.
This biennial review of the lumber production industry covers Canada and the United States. Divided into 13 chapters, it contains many tables and graphs. The sections on lumber production and wood-based panels deal with the product output of specific mills and lumber companies. The chapters include: 1) Year in Review; 2) Lumber Production; 3) Wood-Based Panels; 4) Product Prices; 5) Capital Spending; 6) Niche Marketing; 7) Industry Mergers; 8) Corporate Profiles; 9) Export/Imports; 10) Conversion Guide; 11) Industry Organizations; 12) Government Agencies; 13) Glossary of Industry Terms. An important reference tool for the lumber and wood products industry.

U. S. Forest Service. *U.S. Timber Production, Trade, Consumption and*

Price Statistics, 1960–1988. Washington, D.C.: USDA Forest Service, 1989. 84p. [*USDA Miscellaneous Publication* No. 1486]

An excellent resource containing extensive graphs and tables. Divided into 5 major sections: 1) Timber Production, Trade, Consumption and Prices; 2) Lumber Production, Trade, Consumption and Prices; 3) Plywood and Veneer Production, Trade, Consumption and Prices; 4) Paper, Board, and Woodpulp Production; 5) Particleboard, Hardboard and Insulating Board Production. Contains a brief annotated bibliography.

G. Atlases

Csapody, V. and I. Toth. *A Color Atlas of Flowering Trees and Shrubs.* Budapest, Hungary: Akademiai Kiado, 1982. 146p.

Coverage is almost exclusively Central Europe. Divided into three major sections, Gymnosperms, Dicotyledons, and Monocotyledons. Each family of flowering shrubs is given several paragraphs of botanical and distribution description. Contains no index.

Collins, Mark, ed. *The Last Rainforests: A World Conservation Atlas.* New York: Oxford University Press, 1990. 200p.

This lavishly photographed atlas is an excellent record of our vanishing rain forests. The first five chapters deal with subjects as diverse as: "What is a rain forest?" to "People of the rain forest." The atlas proper is divided into 14 bio-regions. Many maps and photographs reveal the extent of rain forest destruction. The final section is titled "The Challenge of Conservation," and links environmentally sound development to forest preservation. The atlas contains a glossary and is well indexed. Of value to those wanting a brief overview of a complex topic.

Collins, N. Mark, Jeffery Sayer and Timothy Whitmore, eds. *The Conservation Atlas of Tropical Forests: Asia and the Pacific.* New York: Simon and Schuster, 1991. 261p.

Like its companion volume on Africa, this work is lavishly photographed with extensive color maps and graphs. Divided into two broad categories: the issues and country studies. Contains sections on acronyms, a glossary, an index of species and a general index.

Sayer, Jeffery A., Caroline Harcourt and Mark Collins, eds. *The Conservation Atlas of Tropical Forests: Africa.* New York: Simon and Schuster, 1992. 301p.

A glossy, well-photographed reference resource dealing with the loss of the world's tropical forests. It is divided into two major parts. Part 1 has ten chapters dealing with conservation issues of topical importance to Africa. Part 2 is a country by country study of the conservation efforts carried out by the listed countries. Contains a section on acronyms, a glossary, an index of species and a general index.

H. Audio-Visual

U.S. National Agricultural Library. *Forest Service Photo Laser Videodisc.* Beltsville, Md.: National Agricultural Library.

The National Agriculture Library, the Forest Service and the University of Maryland cooperated in the development of a 12-inch optical laser disc to provide word-searchable access to images and image information belonging to the Forest Service, Photo Browsing Collection. The disc contains 34,00 black and white images arranged by 69 subjects, 500 color slides, 55 botanical illustrations and 175 maps. Available for demonstration at NAL. Also in CD-ROM format. (301-504-3876).

Jean Albrecht, forestry librarian at the University of Minnesota, provided extensive assistance and leads for material in this chapter. Her counsel is appreciated.

14. Primary Historical Literature of U.S. Forestry

PETER MCDONALD

New York State Agricultural Experiment Station, Geneva

It has been argued that American forestry began in response to deforestation caused by the expanding industrial base of the nation in the later half of the nineteenth century.[1] The fragility of the eastern timber supply in the years following the American Civil War led to widespread debate among scientists. In a famous address to the American Association for the Advancement of Science (1873), Franklin B. Hough, a renowned physician and director of the New York census (1870), was the first scientist of stature to advocate the establishment of forestry schools and the need for the U.S. government to regulate, manage, and protect the American forest landscape.[2]

Perhaps the single most important text on forestry to come out of this period was George Perkins Marsh's *Man and Nature* (1864). Marsh's classic influenced a generation of scientists to think of the forest environment as a living system and gave considerable attention to forest protection.[3] Marsh argued that the severe depletion of forest cover throughout the Eastern states had a direct effect on flooding, and both he and the Rev. Frederick Starr, a colleague of Marsh's, predicted a "timber famine" within thirty years in the *Report to the Commissioner of Agriculture* of 1865.[4] Several states, notably Wisconsin, Maine, and New York, made half-hearted efforts to address this problem in the 1860s but the commissions appointed on

1. William G. Robbins, *American Forestry: A History of National, State and Private Cooperation* (Lincoln, Neb., and London: University of Nebraska Press, 1985).

2. Franklin B. Hough, "On the Duty of Government in the Preservation of Forests," in *Conservation in the United States: A Documentary History*, ed. Frank E. Smith (New York: Chelsea House, 1971).

3. George Marsh, *Man and Nature or, Physical Geography as Modified by Human Action* (New York: Charles Scribner, 1864).

4. Frederick Starr, "American Forests: Their Destruction and Preservation," in *Report of the Commissioner of Agriculture* (Washington, D.C.: U.S. Dept. of Agriculture, 1865).

timber decline had little impact and no action was taken.[5] It was not until 1876 that an appropriations rider in Congress established a federal forestry agent, Franklin B. Hough, whose "On the Duty of Government in the Preservation of forests" is considered a classic.[6]

In 1875, the American Forestry Association (AFA) was formed in Cincinnati later merging with the American Forestry Congress (AFC) under the leadership of Bernhard E. Fernow. The new organization was active in drafting proposals dealing with forest management and protection for both state and federal legislation. One writer has called the merger of the AFA and the AFC in 1882 as "the real beginning of the conservation movement [in the United States]."[7] It was certainly the first stirrings of a truly American forest science. The *Proceedings of the American Forestry Congress* were inaugurated in 1883 and were published intermittently until 1897 when they ceased publication. A year later, Fernow, who had been a leader in the formation of the American Forestry Congress, left his post at the U.S. Division of Forestry to organize the first four year forestry school in the United States at Cornell University. Commenting on Fernow's departure from government service, W.N. Sparhawk noted that the need for educated foresters was great because "there were probably fewer than a dozen foresters in the country" at that time.[8] Gifford Pinchot replaced Fernow as Chief of the Division of Forestry. It was under Pinchot that American forestry came of age.

A handful of American forestry periodicals appeared in the nineteenth century, but few of them were of scholarly importance. The fanfare which accompanied the organization of the American Forestry Congress (1882) in Cincinnati brought forth the first journal devoted to scientific forestry in the United States, the *American Journal of Forestry*, edited by Franklin Hough. Despite an auspicious beginning, "it survived just one year, vanishing for lack of readers."[9] State forestry associations were the chief serial publishers at this time. Foremost among them was the Pennsylvania Forestry Association which inaugurated its bi-monthly *Forest Leaves* in 1886, a serial which is still in publication. In 1895, the *New Jersey Forester* was launched,

5. W. N. Sparhawk, "The History of Forestry in America," in *The Yearbook of Agriculture 1949* (Washington, D.C.: U.S. Dept. of Agriculture, 1949), pp.702–714.

6. Hough, "On the Duty of Government in the Preservation of Forests".

7. Kenneth B. Pomeroy, "Citizen and Trade Associations in Forestry," in *American Forestry: Six Decades of Growth*, ed. Henry Clepper and Arthur B. Meyer (Washington, D.C.: Society of American Foresters, 1960), p. 274.

8. Sparhawk, "The History of Forestry in America," p. 710.

9. Bernhard E. Fernow, *A Brief History of Forestry in Europe, the United States, and Other Countries* (Toronto: University of Toronto Press; Washington, D.C.: American Forestry Association, 1913), p. 499.

changed its name to *The Forester* a year later, was superseded by *Forestry and Irrigation* in 1900, and after several iterations, became *American Forests* in 1931 under the auspices of the American Forestry Association and is still in publication today.

By 1900, the federal government faced the management crisis in American forests under the able stewardship of Pinchot at the Bureau of Forests. In order to meet this challenge, the first association of professional foresters was formed in Pinchot's Washington office in 1900 under the name of the Society of American Foresters (SAF).[10] By December, 1901, after a year of recruitment, the SAF claimed only 15 members; but from these obscure beginnings, American forestry found its professional voice (see Chapter 5). After publishing sporadic reports and technical discussions, the society inaugurated its *Proceedings of the Society of American Foresters* in 1905, a misnomer, since only a fraction of the articles published were the transactions of the SAF. The major portion of the *Proceedings* consisted of technical articles for publication which were never presented to the society. The other important forestry serial of this period and the first recognized scholarly journal to impact American forestry started as a student publication at Cornell University in 1902 with Fernow as editor. *Forestry Quarterly* in a very real sense "laid the foundation of [American] periodical technical forestry literature," and is considered the forerunner of the important *Journal of Forestry*.[11]

An announcement in the December, 1916, issue of *Forestry Quarterly* declared that the journal had concluded its existence and that with the next issue, it would "begin a new life in amalgamation with the *Proceedings* of the Society of American Foresters under the title *Journal of Forestry*."[12] Thus in January 1917, the *Journal of Forestry* made its debut with Bernhard Fernow as editor-in-chief, and Raphael Zon as managing editor.

Most of the scholarly monographs on forestry which appeared at this time came from Europe. The preponderance were either published in Germany or were by German foresters. Bernhard Fernow, Carl A. Schenck, who founded the Biltmore School of Forestry, in North Carolina, and Arthur B. Recknagel, whose influential *Forest Management* was published in 1919, were all German and their impact on American forestry was considerable. Other German foresters who influenced early American writers, and whom Fernow champions as the founders of German forestry, were Theodor

10. Henry Clepper, "The Society of American Foresters," in *American Forestry: Six Decades of Growth.*

11. Henry Clepper, "The *Journal of Forestry*: An Historical Summary of the First Fifty Years," *Journal of Forestry* 50 (12) (1952): 900.

12. Editorial, *Forestry Quarterly* 14 (4) (1916): 3.

Hartig (1764–1837), Heinrich von Cotta (1763–1844), Friedrich W. Pfiel (1783–1859), Johann C. Hundeshagan (1783–1834), Carl Heyer (1797–1856), and Gottlob Koning (1776–1849).[13] N.D.G. James discusses the works of these German scholars extensively in Chapter 2.

The multi-volume work, *Manual of Forestry*, by Sir William Schlich, a German who taught at Oxford University, was one of the few influential English-language publications of the period, and it reached a fourth edition by 1907.[14] German influence notwithstanding, one writer claims that not until 1896, when Pinchot and Henry S. Graves published *The White Pine*, does a distinctly American scientific report on silviculture appear.[15] In 1882, Franklin Hough wrote *The Elements of Forestry* and three years later the great botanist Charles S. Sargent published his first monograph, *The Woods of the United States*.[16] From these modest beginnings, a truly astounding American forestry literature has evolved. When the U.S. Forest Service published *A Selected Bibliography of North American Forestry* in 1940, it listed almost 22,000 entries on the "permanent literature of forestry."[17] The bibliography gives thirty-eight current U.S. imprint serials as of 1939, with sixty-eight serial titles which had been discontinued.

The most important publisher of forestry technical monographs in the early years of the twentieth century was the U.S. Department of Agriculture. Its *Forest Service Bulletin* series, and the many USDA special publications pertaining to forestry, are an impressive testament to an emerging, uniquely American forestry literature. The influence of the U.S. Forest Service in galvanizing a distinctly North American forest science is often overlooked in studies of the early literature. For many years, the only regularly published account of forestry in the U.S. could be found in the "Report of the Chief of the Division of Forestry" in the annual *Report of the Commissioner of Agriculture*. The 1886 edition is memorable, for this is the year Bernhard Fernow became bureau chief, and his report contains one of the first published literature reviews of American forestry.[18] Other prime examples of U.S. government publications include Pinchot's *A Primer of For-*

13. Fernow, *A Brief History of Forestry.*

14. Sir William Schlich, *A Manual of Forestry* (London: Bradbury, Agnew, 1889–1896).

15. Gifford Pinchot and Henry S. Graves, *The White Pine: A Study, with Tables of Volumes and Yield* (New York: Century, 1896).

16. (a) Franklin B. Hough, *The Elements of Forestry* (Cincinnati, Ohio: R. Clarke, 1882). (b) Charles S. Sargent, *The Woods of the United States* (New York: Appleton, 1885).

17. E. N. Munns, *A Selected Bibliography of North American Forestry* (Washington, D.C.: U.S. Dept. of Agriculture, 1940 [USDA Miscellaneous Publication no. 364]).

18. Bernhard E. Fernow, "Report of the Chief of the Division of Forestry," in *Report of the Commissioner of Agriculture 1886* (Washington, D.C.: GPO, 1887), p. 226.

estry, and Henry S. Graves's *The Woodman's Handbook* (see Chapter 7).[19]

But the government was not the only important publisher. Harper and Appleton, both of New York, were two of the earliest American publishers to distribute books on the subject of forestry. By 1900, John Wiley had numerous forestry titles in its catalog, a tradition they have maintained to the present. Macmillan, Scribner's, Harcourt and Brace, and Ronald Press were also strong. However, the most influential accumulation of forestry texts by one publisher is probably that of the McGraw-Hill Book Company. Early titles included Arthur Koehler's *The Properties and Uses of Wood* (1924), and John Weaver's *Plant Ecology* (1929). In 1931, McGraw-Hill inaugurated its American Forestry Series which published over fifteen titles by 1950; many are today considered classics.

From 1860 through 1910, dwindling forest resources in the East steadily moved the hub of forest management activity westward to the large federal land holdings in Montana, Idaho, Washington, Oregon and California. Despite this trend, the two earliest forestry schools in the U.S. were East Coast institutions, at Cornell and Yale universities. These forestry schools were established in 1898 and 1900 respectively (see Chapter 9). As noted, *Forestry Quarterly* (*FQ*) was first published at Cornell, although the journal's ties to the university were severed when the forestry school at Cornell moved to Syracuse University in 1903. Thereafter *FQ* became an independent publication. Yale is better known for two important publications, its forestry *Bulletin* series, begun in 1916, and *Tropical Woods*, inaugurated in 1925. By 1915, twenty-one universities had forestry departments or schools, and ten more were added between World Wars I and II. Also influential in creating an American forestry literature were the many agricultural experiment station publications from land-grant institutions. These pamphlets were almost exclusively concerned with applied forestry and generally dealt with aspects of silviculture of interest to owners of private forest land. Several university presses, notably Yale, Harvard, and the University of Connecticut were strong in publishing forestry-related titles in the first half of this century.

The growth and influence of American forestry education is represented in Table 14.1 showing the increase in the number of forestry degrees awarded at U.S. universities. A general trend has been noted in the literature concerning U.S. university curricula in forestry, stressing scientific and technical know-how through the 1940s, evolving to give greater emphasis

19. (a) Gifford Pinchot, *A Primer of Forestry* (Washington, D.C.: Div. of Forestry, 1905 [USDA Forestry Bulletin no. 24]). (b) H. S. Graves, *The Woodman's Handbook* (Washington, D.C.: Forest Service, 1910 [USDA Forest Service Bulletin no. 36]).

Table 14.1. Number of U.S. forestry degrees granted

Year	Bachelors	Masters	Doctorate
1909	47	44	n/d
1919	53	6	n/d
1929	291	54	n/d
1939	1102	112	8
1949	1443	205	12
1959	1470	241	43

to management in the 1950s and 1960s, to the current demands for ecosystems analysis of natural resources.[20] The oldest doctoral thesis in forestry from an American university appears to have been done at Yale. Awarded in 1900, it bears the title, *A Treatise on Hickories, with Special Reference to Their Uses in the Arts.*[21]

Several distinct trends in the literature can be traced as the concept of forestry matured in the United Sates between 1850 and 1950. The early literature dealt heavily with tree identification and taxonomy, and the trade periodical literature discussed issues important to the pulp and lumber industries. Daniel J. Browne's *The Trees of America: Native and Foreign,* published by Harper in 1846, is an example of typical identification work for the general public. The first Arbor Day in Nebraska in 1872 instigated a trend in both the periodical and monographic literature of framing the problem of forest depletion in terms of vigorous tree planting programs and state management of resources. The Minnesota State Forestry Association published Leonard B. Hodge's *The Forest Tree Planter's Manual* in 1879, and in New York, Orange Judd published *Forest Planting: A Treatise on the Care of Timber Lands and the Restoration of Denuded Wood-Lands on Plains and Mountains,* by Henry N. Jarchow (1893). Similar titles were published by local forestry associations from several plains states where trees were scarce.

By 1920, forestry literature in the United States included forest botany, silviculture, economics, wood technology, fire protection, timber utilization, forest range management, mensuration, and the importance of the recreational use of forest lands. Influential journals of this period included *Timberman* published in Portland, Oregon; *Southern Lumberman* from Tennessee; *Paper Trade Journal* from Chicago; *Forestry Chronicle* from Canada; and in forest ecology and botany, the *Botanical Gazette,* also from Chi-

---

20. D. P. Duncan and F. H. Kaufert, "Education for the Profession," in *American Forestry: Six Decades of Growth.*

21. H. R. A. Bristol, *A Treatise on Hickories, with Special Reference to Their Uses in the Arts* (Dissertation accepted by Yale University, School of Forestry, 1900).

cago. *Ecology*, begun in 1920 as the official publication of the Ecological Society of America, has many articles on forest ecosystems. The Torrey Botanical Club of New York also published extensively in forest botany; its *Bulletin*, inaugurated in 1870, is still in publication. The USDA's *Journal of Agricultural Research* also covered forestry extensively and was well-represented by authors in the Forest Service.

The Great Depression of the 1930s assisted forestry in the United States on a number of fronts. As the economic crisis deepened, the New Deal administration of Franklin D. Roosevelt was committed to putting people to work. It envisioned a variety of conservation-oriented projects, notably forestry, to accomplish this end. Fundamental to this proposal was the establishment of the Civilian Conservation Corps (CCC) which helped reduce unemployment by constructing roads and trails in forested wilderness areas, by building fire lookouts, by replanting denuded forest land, and by clearing fire lines and fire breaks. This massive effort focused foresters' attention on management problems, resulting in many classics on forestry management from this period. Frederick S. Baker wrote *Theory and Practice of Silviculture* in 1934. *Forest Mensuration* by Donald Bruce and Francis X. Schumacher was published by McGraw-Hill (1935), which also published Hugh H. Bennett's *Soil Conservation* in 1939 with a strong critique of soil erosion from clearcutting. And Aldo Leopold, an outstanding leader in land-use ethics, and a Yale forestry school graduate, came by much of his knowledge of conservation both as a U.S. Forest Service employee and as a state wildlife professional in Wisconsin working closely with the CCC.

One clear indication of the growth of United States forestry during the first half of this century can be seen by tracking the circulation of the *Journal of Forestry*, from its inception in 1917 to 1950. The *Journal*, as foresters call it, is the most important forestry journal of this period and can be seen as a bell-weather of the profession. In 1917, the circulation of the *Journal* stood at 1001, including Society of America Forester members and subscribers. By 1929 it had risen to 2353, at the height of the depression (1933) to 2681; in 1937 it stood at 4783; and in 1950 at 9512, an almost tenfold increase in just thirty-three years.[22] By any measure, the United States was a key leader in forest science by 1950.

A. The Historical Preservation Project

The objective of this historical component is to identify the U.S. forestry literature worthy of long-term preservation. Changes in the processes of

•

22. Henry Clepper, "The *Journal of Forestry*."

papermaking in the 1860s, when rag paper was replaced by pulp, have created a great need for libraries to address the problem of brittle paper in their collections. Because wood fibers and bleaching leave an acidic residue in pulp paper, it self-destructs over time. Conservative estimates suggest that upwards of 40% of nineteenth-century publications made from pulp paper are in ruinous condition in U.S. collections.

The historical preservation component of the Core Agricultural Literature Project concentrated on the forestry literature published in English between 1850 and 1950. Due to their age and rarity, works published prior to 1850 were deemed intrinsically important, to be preserved as a matter of course. Emphasis was on the literature of importance in the United States and, to a lesser extent, Canada. Three distinct types of publications were identified and analyzed: (a) monographs, including bulletin series; (b) scholarly and professional journals; and (c) popular and trade periodicals. While data was kept on land-grant publications in series and federal documents, these publications have been excluded from the final historical lists because they have already been preserved or plans are underway. Many of the important scholarly serials have been recorded on microfilm, and plans are currently being made for preservation of the remaining titles.[23] The document identified in this chapter are the first component of the U.S. preservation plan for forestry literature.

B. Source Documents

The historical forestry literature was identified by analysis of the citations in landmark or overview works for the period 1850–1950. These important source documents were recommended by historians and foresters with many years of experience, and in consultation with the Core Agricultural Literature Project's Forestry and Agroforestry Steering Committee. The major headings in *Selected Bibliography of North American Forestry* (1940) and *A Classification for Forestry Literature*, by the faculty of the Yale Forestry School, served as guides to determine the scope of subjects to be covered in the historical analysis and to assist in identifying source documents.[24] Forest

---

23. Nancy E. Gwinn, *A National Preservation Program for Agricultural Literature* (Washington, D.C., 1993). 20p. Prepared for and accepted by the U.S. Agricultural Information Network.
24. (a) Munns, *A Selected Bibliography of North American Forestry*, pp.1–2. (b) *A Classification for Forestry Literature* (New Haven, Conn.: Yale University Press, 1912. [*Yale Forestry School Bulletin* no. 1]).

engineering was omitted since it was extensively covered in a previous volume in this series.[25]

Major subject areas for the historical analysis of forestry included:

General Forestry	Forest Pathology and Protection
Forest Botany and Ecology	Logging and Lumbering
Silviculture	Wood Technology
Forest Influences	Forest Economics
Management and Mensuration	Forest Policy
Dendrology	

Citation analysis (explained in Chapters 11 and 12) involves a count of each journal, report series, dissertation and monograph, as well as date of each, cited in the documents analyzed. These citations also provided a basic list of monographs which were then evaluated by forest scientists or correlated with other studies. The following list of source documents were analyzed to obtain the basic lists and quantitative counts. Care was taken to select titles which reflected historical trends, notably in the areas of tree taxonomy (dendrology), which was germane to nineteenth century science but which is today mostly of historical interest, and mensuration, which has largely fallen out of current university curricula.

Source Documents for Historical Forestry

All titles had monographs extracted for evaluation; items with [a] were also analyzed.

Baker, Fredrick S., *Theory and Practice of Silviculture*. New York: McGraw-Hill, 1934. 502p.

[a]Boyce, John S., *Forest Pathology*. New York: McGraw-Hill, 1938. 600p.

Brockman, C. F., *Recreational Use of Wild Lands*. New York: McGraw-Hill Co., 1959. 346p.

[a]Brown, H. P., *Textbook of Wood Technology*. New York: McGraw-Hill, 1952.

Brown, Nelson C., *The American Lumber Industry*. New York: J. Wiley & Sons, 1923. 279p.

Brown, Nelson C., *Forest Products*. New York: J. Wiley & Sons, 1919. 471p.

[a]Bryant, Ralph C., *Logging*. New York: J. Wiley & Sons, 1923. 273p.

Bryant, Ralph C., *Lumber; Its Manufacture and Distribution*. New York: J. Wiley & Sons, 1938. 535p.

---

25. J. L. Fridley et al., "Development Forest Engineering and Its Literature," in *The Literature of Agricultural Engineering* ed. Carl W. Hall and Wallace C. Olsen (Ithaca: Cornell University Press, 1992), pp. 126–143.

[a]Chapman, Herman H., *Forest Management*. Bristol, Conn.: Hildrith Press, 1950. 582p.

Chapman, Herman H., and Walter H. Meyer, *Forest Valuation*. New York: McGraw-Hill, 1947. 521p.

[a]Cheyney, Edward G., *American Silvics and Silviculture*. Minneapolis, Minn.: University Press, 1942. 472p.

[a]Dana, Samuel T., *Forest and Range Policy*. New York: McGraw-Hill, 1956. 455p.

Forestry Almanac. Washington, D.C.: The American Tree Association, 1943.

Graves, Henry S., *Forest Mensuration*. New York: J. Wiley & Sons, 1906. 458p.

Guise, Cedric H., *The Management of Farm Woodlands*. New York: McGraw-Hill, 1950. 356p.

[a]Hawley, Ralph C., and David M. Smith, *The Practice of Silviculture*, 6th ed., New York: J. Wiley & Sons, 1954. 525p.

Hawley, Ralph C., and Paul W. Stickel, *Forest Protection*, 2d ed., New York: J. Wiley & Sons, 1948. 355p.

Ise, John, *The United States Forest Policy*. New Haven: Yale University Press, 1920. 395p.

Leopold, Aldo., *Game Management*. New York: C. Scribner's Sons, 1933. 481p.

Lutz, Harold J. and Robert F. Chandler, *Forest Soils*. New York: J. Wiley & Sons, 1946. 514p.

Marquis, Ralph W., *Economics of Private Forestry*. New York: McGraw-Hill, 1939. 219p.

Matthews, Donald M., *Management of American Forests*. New York: McGraw-Hill, 1935. 495p.

Meyer, Arthur H., Arthur B. Recknagel, and Donald D. Stevenson, *Forest Management*. New York: Ronald Press, 1952. 290p.

Record, Samuel J., *Identification of the Timbers of Temperate North America*. New York: J. Wiley & Sons, 1934. 196p.

Richards, P. W., *Tropical Rain Forest; An Ecological Study*. Cambridge: Cambridge University Press, 1952. 450p.

Sherwood, Malcolm H., *From Forest to Furniture*. New York: W. W. Norton & Co., 1936. 284p.

[a]Toumey, James W., *Foundations of Silviculture Upon an Ecological Basis*. 2d ed., edited by Clarence F. Korstian, New York: J. Wiley & Sons, 1947. 468p.

Toumey, James W., *Seeding and Planting in the Practice of Forestry*. New York: J. Wiley & Sons, 1931. 507p.

Trees: The Yearbook of Agriculture. Washington, D.C.: U.S. Gov't Printing Office, 1949. 944p.

[a]Westveld, R. H., *Applied Silviculture in the United States*. New York: J. Wiley & Sons, 1949. 590p.

Zon, Raphael, *Forests and Water in the Light of Scientific Investigation*. Washington, D.C.: U.S. Govt. Printing Office, 1927. 106p.

C. Citation Analysis and Compilation

Much of the early literature in forestry is problematic since it is devoid of citations. First editions of several seminal source documents had no references. Consequently, it was necessary to use second or later editions in order to find citations. In some cases, notably in the area of dendrology, many nineteenth century texts which were deemed worthy of analysis, lacked both references and later editions. Efforts were then made to overcome this lacunae by using secondary works or historical overviews from later periods.

The source documents yielded 5,806 citations, of which 3,994 (68.7%) were to journal articles and 1,878 (31.3%) to monographs, including reports. The journal citations were scattered among 139 different titles, twenty of which were non-English. The citations to monographs provided 567 distinct titles of historical importance in English, 132 titles in languages other than English, and over 140 reports and bulletins from the U.S. and Canadian forest services, including agricultural and forestry experiment station publications; over 800 titles in all.

Comparisons with current trends in forestry literature (journals 58.2%, monographs 41.8%; see Chapters 11 and 12) reveal that journals were more commonly cited prior to 1950 than after. Several factors may account for this. From 1850 until about 1910, journals were the predominant source of scholarly forestry information. Monographs were not common and were almost all from Germany or England. Second, although report series on forestry appeared soon after World War I, notably from the U.S. Forest Service, analysis of this literature suggests that not until the 1930s and 1940s were these publications regularly cited. Their importance in current forestry literature is noted in Chapters 3, 4, and 11, accounting for almost 20% of the total post-1950 citation count. Finally, since 1930, the number of forest monographs published in the United States has risen dramatically. Of the 451 titles on the core historical list for evaluation, 43% were published after 1930.

Neverless, monographs were an important source of scholarly literature prior to 1950. Citations to material published before 1900 are almost all to German publications, which taper off by 1910. A few monograph titles cited before 1900 came from the United Kingdom, and fewer than 5% were U.S. imprints, notably the works of F.B. Hough, C.S. Sargent, and E.W. Hilgard, along with several important U.S. government publications. From 1900 to 1920, the monograph count increases dramatically with titles in U.S. government report series superseding commercially published material by the latter date. After 1920, the citation pattern to monographs falls into

Table 14.2. Types of publishers

Publisher	% of monograph citations
Government	47.5%
Commercial	25.6
University	18.1
Organization	3.6
Miscellaneous[a]	5.2

[a]This includes patents, laws, maps, measurement tables, and some archival material.

an even pattern: just under 50% are to government monographs, roughly 25% to commercial publishers, and almost 20% to universities. On the other hand, dissertations, standards, and conferences were rarely cited and account for less than 4% of the total monographic count for the 1850–1950 period. Surprisingly, there were only twenty-four citations to seven separate conferences. Two were notable, the *Proceedings of the American Forestry Congress* (1905), and the *3rd International Congress of Soil Science* held in Oxford, U.K., in 1935. Table 14.2 gives the types of publisher and their percentages for all the monographs cited.

With the decline of the German influence by 1910, the United States clearly dominated with 88.3% of the historical count with almost all the citations coming after 1920. Germany was second, the United Kingdom third. Canada was a distant fourth with less than 1%, followed by thirteen additional countries. Table 14.3 gives the percentages for the place of publication.

In almost every source document, the U.S. government was the most important cited publisher of early forestry literature. Table 14.4 gives the citation counts, by category, for the nine source documents submitted to citation analysis.

The top-cited commercial publishers were McGraw-Hill (89), John Wiley

Table 14.3. Place of publication

Country	% of monograph publications
United States	88.3%
Germany	5.1
United Kingdom	5.0
Canada	.8
Others[a]	.8

[a]In descending order: India, France, Japan, Netherlands, Austria, Sweden, Finland, Denmark.

Table 14.4. Citations to journals and monographs by publisher type[a]

Source document	Journals	Monographs		Government	Commercial	University	Organization
R.C. Bryant (1923)	261	150	=	82	47	11	8
H. H. Chapman (1950)	905	141	=	82	25	24	7
H. P. Brown et al. (1952)	203	179	=	73	65	16	22
R. C. Hawley et al. (1954)	279	272	=	157	43	57	3
R. H. Westveld (1956)	535	199	=	132	7	55	5
S. T. Dana (1956)	132	139	=	38	49	35	13
J. S. Boyce (1938)	959	266	=	106	83	58	—
E. G. Cheney (1942)	276	228	=	167	17	42	1
J. W. Toumey (1947)	444	251	=	44	139	37	7

[a]The discrepancies in the totals are accounted for by references to patents, laws, standards and other miscellaneous publications, such as maps, which do not fall clearly into the two categories.

(68), Macmillan (28) and Springer-Verlag (24); a total of twenty-two publishers were represented. The three top-cited organization publishers were the Society of American Foresters (SAF), the American Forestry Association (AFA), and the National Lumber Manufacturers Association (NLMA), an umbrella organization representing over twenty smaller associations.[26] In agricultural disciplines, notably soils, crops, and engineering, societies played a role in monographic publication; this was not the case with forestry. The SAF still had few scholarly monographs, fifty years after incorporation. Other than the occasional terminology or other handbook, the SAF was not noted for its monographs in the historical period.[27]

University publications were cited 347 times, of which 171 were to agricultural experiment station (AES) bulletins or other titles. The highly cited *Bulletin* of the Forestry School at Yale University does not fall into the latter category, although the Yale Forestry School shares state land-grant status with the University of Connecticut. Yale's *Bulletin*, an irregular publication inaugurated in 1912, is really a report series with a single topic per issue. With eighty-nine citations, it received the most references of any university publication. A similar series, *Harvard Forest Bulletin*, inaugurated in 1920 was ranked second with thirty-one citations. It ceased publication in 1963. As with the Yale series, each issue of *Harvard Forest Bulletin* covered a single topic and was generally of pamphlet length.

26. For a full listing of U.S. lumber associations, see: *Forestry Almanac: 1933 Edition* (Washington, D.C.: American Tree Association, 1933), pp. 448–55.

27. Notable exceptions are: (a) Ralph C. Hawley et al., eds. *Forestry Terminology* (Washington, D.C.: SAF, 1944). (b)Henry S. Graves and Cedric H. Guise, *Forest Education* (Washington, D.C.: SAF, 1934). (c) Robert K. Winters et al., eds., *Fifty Years of Forestry in the USA* (Washington, D.C.: SAF, 1950).

The most highly cited agricultural experiment station bulletins came from the University of Minnesota and Cornell University, with fifteen citations apiece. The AES bulletins at the universities of Connecticut, Michigan, Wisconsin, and Pennsylvania State followed closely. In total, thirty-one state AES's were cited. Topics of AES bulletins varied, but broadly fell into two general categories: aspects of silviculture pertaining to specific species of trees, notably white pine or northern hardwoods, and treatises on disease and insect infestation. Many were applied offering practical tips on seeding, growing, and protecting timber stands.

The U.S. government was the single most important publisher of monographic literature during the historical period with a total of 813 citations. The USDA (exclusive of the Forest Service series) had 456 citations and the Forest Service 311. Other U.S. federal agencies and the U.S. Congress accounted for the remaining citations. Of the 813 U.S. government citations, report series accounted for 75.1% of the total. Table 14.5 lists the top-cited USDA and Forest Service report series.

Table 14.5. Most cited U.S. government agency publications in citation analysis

Name of publication	Date begun	Number of times cited
USDA Bulletin	1913	167
USDA Technical Bulletin	1927	142
USDA Circular	1927	84
USDA Miscellaneous Publication	1927	55
Farmers Bulletin (USDA)	1889	28
Forest Service Circular	1886	25
Forest Service Bulletin	1887	21
USDA Weather Bureau Bulletin	1892	14
USDA Agriculture Handbook	1949	6
USDA Yearbook	1894	6

By the 1920s, the Forest Service had established twelve forest experiment stations (FES), including the Forest Products Laboratory in Madison, Wisconsin. It was left to the discretion of each station to publish the results of research, annual reports, statistical summaries, and station notes. By 1935, all twelve FES were publishing reports in series, and by 1940 sponsoring an occasional conference. Series publications by the FES included: *Research Notes* (cited 57 times), *Station Papers* (50), *Technical Notes* (24), *Mimeo Reports* (3) and, miscellaneous publications not in series (5). Table 14.6 gives the top-cited Forest Service experiment stations and the number of times their publications were cited. For a detailed analysis of Forest Service publications see Chapter 7.

The only other important government source of publications came from the Canadian Forest Service which had twelve citations.

Table 14.6. U.S. Forest Service experiment stations citations in citation analysis

Forestry experiment station	No. of times cited
Forest Products Laboratory	28
Lake States Forestry Experiment Station (FES)	26
Southern FES	17
Pacific Northwest FES	16
Northeastern FES	14
Northern Rocky Mountain	10
Southeastern FES	10
Central States FES	6
Rocky Mountain FES	5
California FES	5
Intermountain FES	2

D. Identifying and Ranking the Monographs of Historical Importance

The citations from source documents provided the basis for the monographic list which was evaluated. These monographs were identified when cited in a source document, added to a cummulative list, and counted each time they were cited. After the key historical monographs were identified, the second step involved sending the monograph list to reviewers for evaluation using these criteria:

1) The work is the first of its kind published, or records a major advance in the field.
2) The work has had an important influence in the subject field.
3) The work embodies an historical record of changes in the field.
4) The title is valuable because it is a superior work of a leader in forestry.
5) The volume has intrinsic value because of its binding, illustrations, advertisements, or other important artifactual qualities.

The scholars graded these monographs using a three-category scale: Category 1 = an important historical title worthy of preservation; Category 2 = worth preserving, but of secondary importance; Category 3 = a title of marginal scholarly value. Instructions to reviewers stressed that low marks in the review process in no way implied that these titles would be discarded or ignored, but that a priority was being established so the most valuable would be preserved first.

Subject specialists with extensive research and teaching experience reviewed the list; most were faculty members of land-grant colleges. They were chosen because of their many years of academic forestry experience, an interest in forestry history, and knowledge of early literature. Reviewers also made suggestions of titles for inclusion; although these were few.

Historical Forestry Monograph Reviewers

James L. Bowyer
    College of Natural Resources
    University of Minnesota, St. Paul

Raymond P. Guries
    Department of Forestry
    University of Wisconsin-Madison

Logan Norris
    College of Forestry
    Oregon State University

Ralph Nyland
    College of Environmental Science and Forestry
    State University of New York at Syracuse

David Smith
    College of Forestry and Wildlife Resources
    Virginia Polytechnic Institute and State
        University

Harold Steen
    Forest History Society
    Durham, North Carolina

Each title was tallied combining the number of times it was cited in the literature with the reviewers' rankings, each of which was weighted. Category 1 rankings were weighted by a factor of 3; Category 2 by a ranking factor of 2; Category 3 had a weight of one. Titles universally ranked 3 by the reviewers were dropped from the list if no artifactual value could justify retaining it. The final monograph ranking was calculated using the formula:

Final ranking = (1 × # citation hits) + (3 × 1st ranking) + (2 × 2nd ranking) + (1 × 3rd ranking)

The 351 remaining in the list were divided into First, Second, and Third ranks depending on their scores. Titles in the first rank have the highest priority for preservation.

Although the list of core monographs for preservation is predominantly of scholarly imprints, several titles of a popular nature are included. Many national and state associations published popular monographs as did conservation and nature organizations, and universities. These often constituted handy pocket manuals on common tree or wildlife identification, or were titles on tree planting and care. Some, like Ernest H. Wilson's *The Romance of Our Trees*, have a poetic charm and are important because of their early photographs of North American trees. These titles, mostly about trees, were carefully selected for reasons which included, though were not limited to, their photographs or color plates, their bindings, their uniqueness, or because they were the first books of their genre covering a particular region, species, or landscaping application.

Trade and wildlife associations were best known for their popular serials, but they did publish monographs on occasion. A good example was the publication in 1927 of a series of popular titles by the American Tree Association on the trees of New York, Massachusetts, New Jersey, Ohio, Indiana and Michigan, and the widely read *Tree Planting Book* (1927). Concerning wildlife, the American Nature Association published many excellent booklets with such titles as, *A Plea for Our Hawks, Our Disappearing Fur-bearers*, and *Birds of the States* one of the first bird identification pocket guides. Works of this genre, with their quaint illustrations and historical provenance certainly deserve preservation attention.

Primary Core Historical Monographs Worthy of Preservation (N = 351)

A

Third	Allen, Edward T. Practical Forestry in the Pacific Northwest. Portland, Ore.; Western Forestry & Conservation Association, 1911. 130p.
First	Allen, Shirley Walter. An Introduction to American Forestry. 1st ed. New York and London; McGraw-Hill, 1938. 402p. (3d ed., 1960. 466p.)
Third	American Society of Photogrammetry. Manual of Photogrammetry. New York and Chicago; Pitman Corp., 1945. 841p.
Third	Anderson, Mark L. The Selection of Tree Species: an Ecological Basis of Site Classification for Conditions Found in Great Britain and Ireland. 1st ed. Edinburg; Oliver and Boyd, 1950. 151p.
Second	Arthur, Joseph C. Manual of the Rusts in United States and Canada . . . illustrated by George B. Cummins. Lafayette, Ind.; Purdue Research Foundation, 1934. 438p. (Reprinted, New York; Hafner Pub. Co., 1962.)

B

Second	Bailey, Liberty H. The Cultivated Conifers in North America. New York; Macmillan, 1933. 404p.
Second	Bailey, Liberty H. The Holy Earth. Ithaca, N.Y.; L. H. Bailey, 1915. 171p.
First	Baker, Frederick S. Principles of Silviculture. 1st ed. New York; McGraw-Hill, 1950. 414p.

First   Baker, Frederick S. Theory and Practice of Silviculture. New York; McGraw-Hill, 1934. 502p.

First   Baldwin, Henry I. Forest Tree Seed of the North Temperate Regions. Waltham, Mass.; Chronica Botanica Company, 1942. 240p. (*Chronica Botanica* 8)

Third   Barr, Percy M. A General Outline of Forestry with Special Reference to the United States. Berkeley; University of California, 1935. 152 leaves.

Third   Baterden, James R. Timber. London; Constable, 1908. 351p.

Second   Belyea, Harold C. Forest Measurement. New York; Wiley, 1931. 319p.

Second   Bennett, Hugh H. Soil Conservation. 1st ed. New York and London; McGraw-Hill, 1939. 993p.

First   Bently, J. and A. B. Recknagel. Forest Management. New York; John Wiley & Sons, 1919. 269p. (2d ed., 1926. 328p.)

Third   Berry, Edward W. Tree Ancestors; A Glimpse into the Past. Baltimore, Md; Williams & Wilkins Co., 1923. 270p.

Third   Berry, James B. Farm Woodlands. Yonkers-on-Hudson, New York; World Book Co., 1923. 425p.

Second   Betts, Harold S. Timber; Its Strength, Seasoning, and Grading. 1st ed. New York; McGraw-Hill, 1919. 234p.

Second   Blake, Ernest G. The Seasoning and Preservation of Timber. London; Chapman & Hall, 1924. 132p.

Third   Blenkarn, John. British Timber Trees: Their Rearing and Subsequent Management in Woods, Groves, & Plantations. London and New York; Routledge, Warne, & Routledge, 1859. 275p.

Third   Boerker, Richard H. Our National Forests. New York; Macmillan Co., 1918. 238p.

Second   Boulger, George S. Wood; A Manual of the Natural History and Industrial Application of the Timbers of Commerce. 1st ed. London; E. Arnold, 1902. 369p. (2d ed. 1908. 348p.)

First   Bowman, Isaiah. Forest Physiography; Physiography of the Unites States and Principles of Soils in Relation to Forestry. 1st ed. New York; J. Wiley & Sons, 1911. 759p.

First   Boyce, John S. Forest Pathology. 1st ed. New York; McGraw-Hill, 1938. 600p. (3d ed., 1961. 572p.)

First   Brandis, Dietrich. Indian Trees: An Account of Trees, Shrubs, Woody Climbers, Bamboos, and Palms Indigenous or Commonly Cultivated in the British Indian Empire. London; A. Constable & Co., 1906. 767p. (Reprinted at Dehra Dun, India; Bishen Singh Mahendra Pal Singh, 1971.)

First   Braun, E. Lucy. Deciduous Forests of Eastern North America. New York; Hafner Press, 1950. 596p. (Reprinted, New York; Free Press, 1985.)

First   Braun-Blanquet, Josias. Plant Sociology: The Study of Plant Communities . . . trans., rev., and edited by George D. Fuller and Henry S. Conard. New York; Stechert-Hafner Service Agency, 1932. 439p. (Translation of Pflanzensoziologie. Reprinted, 1972.)

Third   Brereton, Bernard. The Practical Lumberman. Seattle; B. Brereton, 1908. 119p. (6th ed. 1940. 44p.)

First   Britton, Nathaniel L. North American Trees; Being Descriptions and Illustrations of the Trees Growing Independently of Cultivation in North

America, North of Mexico and the West Indies. New York; H. Holt and Co., 1908. 894p.

Third   Broun, Alfred F. Sylviculture in the Tropics. London; Macmillan, 1912. 309p.

First   Brown, Harry P., H. P. Panshein, and C. C. Forsaith. Textbook of Wood Technology. 1st ed. New York; McGraw-Hill, 1949. 266p.

Second   Brown, James, and James Nisbet. The Forester: A Practical Treatise on the Planting and Tending of Forest Trees and the General Management of Woodland Estates. 6th ed., enl. Edinburgh; W. Blackwood, 1894. 2 vols.

Third   Brown, John P. Practical Arboriculture; How Forests Influence Climate. Chicago; The Henneberry Press, 1906. 454p.

First   Brown, Nelson C. The American Lumber Industry. New York; J. Wiley & Sons, 1923. 279p.

First   Brown, Nelson C. Forest Products, Their Manufacture and Use. 1st ed. New York; J. Wiley & Sons, 1919. 471p. (3d ed., 1927. 447p.)

Second   Brown, Nelson C. A General Introduction to Forestry in the United States. New York; J. Wiley & Sons, 1935. 293p.

Second   Brown, Nelson C. Lumber; Manufacture, Conditioning, Grading, Distribution, and Use. New York; J. Wiley & Sons, 1947. 344p.

Second   Brown, Nelson C. Timber Products and Industries. New York; J. Wiley & Sons, 1937. 316p.

Second   Browne, Daniel J. The Sylva Americana. Boston; W. Hyde & Co., 1832. 408p.

Third   Browne, Daniel J. The Trees of America: Native and Foreign. New York; Harper, 1846. 520p.

First   Bruce, Donald, and Francis X. Schumacher. Forest Mensuration. 1st ed. New York and London; McGraw-Hill, 1935. 360p. (3d ed. 1950. 483p.)

Second   Bruncken, Ernest. North American Forests and Forestry. 1st ed. New York and London; G. P. Putnam's Sons, 1900. 265p. (2d ed. 1902. 265p.)

First   Bryant, Ralph Clement. Logging: The Principles and General Methods of Operation in the United States. 1st ed. New York; J. Wiley & Sons, 1913. 590p. (2d ed. rev. 1923. 556p.)

First   Bryant, Ralph Clement. Lumber; Its Manufacture and Distribution. New York; J. Wiley & Sons, 1922. 539p.

Third   Bunbury, H. M. The Destructive Distillation of Wood. London; Benn, 1923. 320p.

Second   Burtt-Davy, Joseph. The Classification of Tropical Woody Vegetation-Types. Oxford; Imperial Forestry Institute, 1938. 85p.

First   Busgen, Moritz. The Structure and Life of Forest Trees . . . trans. by Thomas Thomson. 3d ed. rev. and enlarged. London; Chapman & Hall, 1929. 496p. (Translation of Bau und Leben unserer Waldbaume.)

Second   Butler, Ovid M. ed. American Conservation in Picture and in Story. Washington, D.C.; The American Forestry Association, 1935. 144p.

Second   Butler, Ovid M. ed. Youth Rebuilds; Stories from the C.C.C. Washington, D.C.; American Forestry Association, 1934. 189p.

First   Buttrick, P. L. Forest Economics and Finance. New York; John Wiley & Sons, Inc., 1942. 484p.

C

First Cameron, Jenks. The Development of Governmental Forest Control in the United States. Baltimore, Md.; Johns Hopkins Press, 1928. 471p.

First Cameron, Jenks. The National Park Service; Its History, Activities, and Organization. New York; D. Appleton and Co., 1922. 172p.

Third Carpenter, John R. An Ecological Glossary. 1st ed. Norman; University of Oklahoma Press, 1938. 306p.

First Cary, Austin. A Manual for Northern Woodsmen. Cambridge; Harvard University Press, 1909. 250p. (5th ed. as Woodman's Manual. 1935. 366p.)

First Champion, Harry G. and A. L. Griffith. Manual of General Silviculture for India. Calcutta and New York; Oxford University Press, 1948. 330p. (First published in 1938 as part of Manual of Indian Silviculture, by G. Champion and Sir Gerald Trevor)

First Chapman, Herman H. Forest Valuation; With Special Emphasis on Basic Economic Principles. 1st ed. New York; John Wiley & Sons, Inc., 1915. 310p. (McGraw-Hill, 1947. 521p.)

First Chapman, Herman H. Forest Finance. 1st ed. New Haven, Conn.; Tuttle, Morehouse, & Taylor Co., 1926. 352p. (Continued as Forest Valuation; NY; Mc-Graw Hill, 1947. 521p.)

First Chapman, Herman H. Forest Managment. Albany, N.Y.; J. B. Lyon, 1931. 544p. (Revised ed. Bristol, Conn.; Hildreth Press, 1950. 582p.)

Third Chapman, Herman H. Forestry, An Elementary Treatise. Chicago; American Lumberman, 1912. 79p.

First Chapman, Herman H. and Dwight B. Demeritt. Elements of Forest Mensuration. Albany, N.Y.; J. B. Lyon Co., 1932. 452p.

First Chapman, Herman H. and Walter H. Meyer. Forest Mensuration. 1st ed. New York; McGraw-Hill, 1949. 522p.

Second Charpentier, Paul. Timber; A Comprehensive Study of Wood in All Its Aspects. London; Scott, Greenwood & Co., 1902. 437p. (Translated from Le Bois)

First Cheyney, Edward G. American Silvics and Silviculture. Minneapolis; University of Minnesota Press, 1942. 472p.

Second Cheyney, Edward G. The Farm Woodlot; A Handbook of Forestry. New York; Macmillan Company, 1914. 343p.

Second Chittenden, Hiram M. The Yellowstone National Park, Historical and Descriptive. Cincinatti; Hiram Matin, 1895. 199p. (New and Enlarged ed. in 1915. 350p.)

First Clapp, Earle H., preparer. A National Program of Forest Research. Washington, D.C.; American Tree Association, 1926. 232p.

Third Cleveland, Horace W. S. The Culture and Management of Our Native Forests for Development as Timber or Ornamental Wood. Springfield, H. W. Bokker, 1882. 16p.

Third Coker, William C. Trees of the Southeastern States. 1st ed. Chapel Hill, N.C.; University of North Carolina Press, 1934. 399p. (3d ed, 1945. 419p.)

Third Collingwood, G. H. Knowing Your Trees. 1st ed. Washington, D.C.; The American Forestry Association, 1937. 109p. (3rd ed. 1974. 374p.)

Second Compton, Wilson M. The Organization of the Lumber Industry. Chicago; American Lumberman, 1916. 153p.

Third　　Connery, Robert H. Governmental Problems in Wild Life Conservation. New York; Columbia University Press, 1935. 250p.

Third　　Coutlas, Harland. What May Be Learned From A Tree. 1st ed. Philadelphia; C. Sherman & Son, 1859. 192p. (2d ed. 1860. 190p.)

Third　　Crocker, William. Growth of Plants: Twenty Years' Research at Boyce Thomson Institute. New York; Reinhold, 1948. 459p.

D

Second　　D'Arcy, W. E. Forest Working Plans, India. Calcutta, India; Office of Supt. of Government Printing, 1898. 160p.

Third　　Darwin, Charles. The Power of Movement in Plants. London; J. Murray, 1880. 592p.

Second　　Defebaugh, James E. History of the Lumber Industry of America. Chicago; American Lumberman, 1906. 2 vols.

Third　　Dixon, Henry H. Transpiration and the Ascent of Sap. Washington, D.C., 1911. 425p. (Later ed. London; Macmillan, 1914. 216p.)

Second　　Doane, Rennie W., E. C. Van Dyke, W. J. Chamberlin, and H. E. Burke. Forest Insects. New York; McGraw-Hill, 1936. 463p.

Second　　DuPuy, William A. Green Kingdom; The Way of Life of a Forest Ranger. Evanston, Ill.; Row, Peterson and Co., 1940. 64p.

E

Third　　Eliot, Willard A. Forest Trees of the Pacific Coast. New York; Putnam, 1948. 565p.

Third　　Elwes, Henry J. and Augustine Henry. The Trees of Great Britain & Ireland. Edinburgh; Private Printing, 1906–1913. 7 vols.

Second　　Ely, Richard T., Mary L. Shine, and George S. Wehrewin. Outlines of Land Economics. 1st ed. Ann Arbor, Mich.; Edwards Brothers, 1922. 3 vols. (2d ed. New York; MacMillan Co., 1940. 512p.)

First　　Emerson, George B. A Report on the Trees and Shrubs Growing Naturally in the Forests of Massachusetts. Boston; Dutton and Wentworth, 1846. 547p. (By 1894 2 vols. were included under this title.)

Third　　Erdtman, Gunnar. An Introduction to Pollen Analysis. New York; Ronald Press, 1943. 239p. (2d revised printing, Waltham, Mass.; Chronica Botanica, 1954.)

F

First　　Ferguson, John A. Farm Forestry. 1st ed. New York; J. Wiley & Sons, 1916. 241p.

First　　Fernow, Bernhard E. A Brief History of Forestry in Europe, the United States, and Other Countries. New Haven, Conn; Price, Lee & Adkins. Co., 1907. 374p.

First　　Fernow, Bernhard E. Economics of Forestry: A Reference Book for Students of Political Economy and Professional and Lay Students of Forestry. New York; Thomas Y. Crowell, 1902. 520p.

Third　　Folweiler, Alfred D. Fire in the Forests of the United States. Baton Rouge; Louisiana State University, 1939. 164p. (Last ed. 1953 by A. A. Brown. 223p.)

Third　　Forest Farmers Association Cooperative. Forest Farmer Manual; A Guide-

	book for Southern Timber Owners. Atlanta; Forest Farmers Association, 1953. 207p. (5th ed. 1957. 168p.)
Third	Forbes, A. C. The Development of British Forestry. London; E. Arnold, 1910. 274p.
Third	Fuller, Andrew S. Practical Forestry. New York; Orange Judd Co., 1884. 299p. (Revised, 1900. 299p.)
First	Fry, Walter, and John R. White. Big Trees. Palo Alto; Stanford University Press, 1930. 114p. (Rev. ed. 1938. 126p.)

G

Second	Gabrielson, Ira N. Wildlife Conservation. 1st ed. New York; Macmillan Co., 1941. 250p. (2d ed. 1959. 244p.)
Second	Garratt, George A. The Mechanical Properties of Wood. New York; J. Wiley & Sons, 1931. 276p.
Second	Gaut, Alfred. Seaside Planting of Trees and Shrubs. New York; C. Scribner's Sons, 1907. 101p.
Second	Geiger, Rudolf. The Climate near the Ground. A translation by Milroy N. Stewart, of the 2d German ed. of Das Klima der Bodennahen Luftschicht. Cambridge, Mass.; Harvard University Press, 1950. 482p.
Third	Gibson, Henry H. American Forest Trees. Chicago; Hardwood Record, 1913. 708p.
Second	Gifford, John C. Practical Forestry for Beginners in Forestry. New York; D. Appleton and Co., 1902. 284p.
Second	Gill, Tom. Tropical Forests of the Caribbean. Baltimore, Md.; Tropical Plant Research Foundation, 1931. 317p.
Third	Gillette, Halbert P. Earthwork and its Cost. 1st ed. New York; McGraw-Hill Co., 1912. 238p. (3d ed., 1920. 1346p.)
Second	Graham, Edward H. The Land and Wildlife. New York; Oxford University Press, 1947. 232p.
First	Graham, Samuel A. Principles of Forest Entomology. 1st ed. New York and London; McGraw-Hill Book Co., 1929. 339p. (4th ed., 1965. 417 p.)
Second	Grainger, Allerdale M. Woodsmen of the West. London; E. Arnold, 1908. 206p.
First	Graves, Henry S. Forest Mensuration. 1st ed. New York; J. Wiley & Sons, 1906. 458p.
First	Graves, Henry S. Forest Education. New Haven; Yale University Press, 1932. 421p.
First	Graves, Henry S. The Principles of Handling Woodlands. 1st ed. New York; J. Wiley & Sons, 1911. 325p.
First	Graves, Henry S. The Woodsman's Handbook. Washington, D.C.; Government Printing Office, 1903. 148p. (United States Bureau of Forestry. Bulletin no. 36.)
Second	Green, Charlotte. Trees of the South. Chapel Hill, N.C.; University of North Carolina Press, 1939. 551p.
Second	Green, Samuel Bowdlear. Principles of American Forestry. 1st ed. New York; J. Wiley & Sons, 1903. 334p.
Third	Grimes, John A. and Horace W. Craigue. Principles of Valuation. New York; Prentice-Hall, 1928. 274p.

Third	Grondal, Bror L. and W. W. Woodbridge. Certigrade Handbook of Red Cedar Shingles. 3d ed. Seattle; Red Cedar Shingle Bureau, 1938. 96p.
Second	Guise, Cedric H. The Management of Farm Woodlands. New York; McGraw-Hill, 1939. 352p. (2d ed. 1950. 356p.)

H

Third	Hanna, John. Hanna's Ready Reckoner. Philadelphia; Mackellar, Smith & Jordan, 1867. 75p. (3rd ed. Complete Ready Reckoner and Log, Table, and Form Book. Philadelphia, Lippincott, 1901. 202p.)
Third	Hansen, H. J. Timber Engineer's Handbook. New York; J. Wiley, 1948. 882p.
Second	Hanson, Clarence O. Forestry for Woodmen. 1st ed. Oxford; Clarendon Press, 1911. 222p. (2d ed. 1921. 238p.)
First	Harlow, William M. Textbook of Dendrology; Covering the Important Forest Trees of The United States and Canada. 1st ed. New York; McGraw-Hill, 1937. 557p. (4th ed. 1958. 561p.)
First	Harlow, William M. Trees of the Eastern United States and Canada. New York and London; McGraw-Hill, 1942. 288p. (Whittlesey House Fiels Guide Series
Third	Harrar, Ellwood S. Forest Dendrology. Seattle; Geo. E. Minor Press, 1935–1937. 2vols.
Third	Harrar, Ellwood S. and George Harrar. Guide to Southern Trees. 1st ed. New York and London; McGraw-Hill, 1946. 712p.
Second	Harshberger, John W. Phtogeographic Survey of North America. Leipzig, Germany; W. Engelmann, 1911. 790p.
Third	Hartig, R. Text Book of the Diseases of Trees. London; Geo. Newnes, Ltd., 1894. 331p.
Third	Harvey, A. G. Douglas of the Fir. Cambridge, Mass.; 1947.
Second	Hawley, Lee F. and Louis E. Wise. The Chemistry of Wood. New York; The Chemical Catalog Copany Inc., 1926. 334p.
First	Hawley, Ralph C. Forest Protection. New York and London; J. Wiley & Sons, Inc, 1937. 262p. (2d ed. 1948. 355p.)
Second	Hawley, Ralph C. and Austin Foster Hawes. Forestry in New England. New York; J. Wiley & Sons, 1912. 479p. (2d ed. 1918–1920. 2vols.)
First	Hawley, Ralph C. The Practice of Silviculture. 1st ed. New York; J. Wiley & Sons, 1921. 352p. (4th ed., 1946. 354p.)
Third	Hazard, Joseph T. Our Living Forests: The Story of Their Preservation and Multiple Use. Seattle, Washington; Superior Publishing Co., 1948. 302p.
Third	Heald, Frederick D. Manual of Plant Disease. New York; McGraw-Hill Co., 1926. 891p.
Third	Henderson, Frank. Timber: Its Properties, Pests, Preservation. London; C. Lockwood & Sons, 1939. 185p.
Third	Henry, Augustine. Forests, Woods and Trees in Relation to Hygiene. London; Constable and Co., 1919. 314p. (Chadwick Library
First	Hibbard, Benjamin H. A History of the Public Land Policies. 1st ed. New York; Macmillan Company, 1924. 591p. (3d ed., Madison; University of Wisconsin Press, 1965. 579p.)
First	Hiley, Wilfrid E. Economics of Forestry. Oxford; Clarendon Press, 1930. 254p.

Third	Hiley, Wilfrid E. The Fungal Diseases of the Common Larch. Oxford; Clarendon Press, 1919. 204p.
Third	Hill, Albert F. Economic Botany; A Textbook of Useful Plants and Plant Products. New York and London; McGraw-Hill, 1937. 592p. (2d ed., 1952. 560p.)
Third	Hodges, Leonard B. The Forest Tree Planter's Manual. St. Paul, Minn.; Minnesota State Forestry Association, 1879. 172p.
First	Holbrook, Stewart H. Burning an Empire; The Story of American Forest Fires. New York; Macmillan Co., 1943. 229p.
First	Holbrook, Stewart H. Holy Old Mackinaw: A Natural History of the American Lumberjack. New York; Macmillan, 1938. 278p.
Second	Hornaday, William T. Our Vanishing Wild Life: Its Extermination and Preservation. New York; New York Zoological Society, 1913. 411p.
Third	Hornaday, William T. Wild Life Conservation in Theory and Practice: Lectures Delivered Before the Forest School of Yale University. New Haven, Conn.; Yale University Press, 1914. 240p.
Third	Hosmer, Ralph S. Impressions of European Forestry. Chicago, Ill.; The Lumber World Review, 1922. 80p. (Text was originally published as a series of eight articles in the Lumber World Review.)
Second	Hotchkiss, George W. History of the Lumber and Forest Industry of the Northwest. Chicago; G. W. Hotchkiss & Co., 1898. 754p. (Available from Xerox University Microfilms)
Second	Hough, Franklin B. The Elements of Forestry. Cincinnati, Ohio; R. Clarke, 1882. 381p.
Second	Hough, Romeyn B. The American Woods. Lowville, N. Y.; Romeyn Hough, 1888–1928. 14 vols.
Second	Hough, Romeyn B. Handbook of the Trees of the Northern States and Canada East of the Rocky Mountains. 1st ed. Lowville, N.Y.; Romeyn Hough, 1907. 470p.
Third	Houston, Edwin J. Outlines of Forestry, or The Elementary Principles Underlying the Science of Forestry. Philadelphia and London; J. B. Lippincott, 1912. 254p.
Second	Howard, Alexander L. A Manual of the Timbers of the World, Their Characteristics and Uses. 1st ed. London; Macmillan, 1920. 446p. (3d ed., 1948. 751p. Revised ed., 1934.)
Second	Hoyle, Raymond J. Harvesting and Marketing Timber in New York'. Syracuse; New York State College of Forestry, 1936. 186p.
Second	Hubert, E. E. An Outline of Forest Pathology. New York; J. Wiley & Sons, Inc., 1931. 543p.
First	Hunt, George M. and Garratt George A. Wood Preservation. New York; McGraw-Hill Book Company Inc., 1938. 457p.
Second	Huntington, Annie O. Studies of Trees in Winter; A Description of the Deciduous Trees of Northeastern America. 1st ed. Boston; Knight and Millet, 1902. 198p. (Reprinted in 1910.)

I

Second	Illick, Joseph S. An Outline of General Forestry. New York; Barnes & Noble, 1935. 259p. (3d Rev & enl. 1939).

Second   Illick, Joseph S. Tree Habits; How to Know the Hardwoods. Washington, D.C.; American Nature Association, 1924. 337p.

First   Ise, John. The United States Forest Policy. New Haven, Conn; Yale University Press, 1920. 395p.

Third   Issac, Leo A. Better Douglas Fir Forests from Better Seed. Seattle, Washington; University of Washington Press, 1949. 64p.

Second   Issac, Leo A. Reproductive Habits of Douglas-Fir. Washington, D.C.; U.S. Forest Service, 1943. 107p.

J

Second   Jack, Graham V. and R. O. Whyte. The Rape of the Earth. London; Faber & Faber, 1939. 312p. (Reissued as Vanishing Lands; A World Survey of Soil Erosion, New York; Arno Press, 1972. 332p.)

Third   Jeffrey, Edward C. The Anatomy of Woody Plants. Chicago, Ill.; The University of Chicago Press, 1917. 478p.

Third   Jenny, Hans. Factors of Soil Formation; A System of Quantitative Pedology. 1st ed. New York and London; McGraw-Hill, 1941. 281p.

Second   Jepson, Willis L. The Trees of California. San Francisco, California; Cunningham, Curtis & Welch, 1909. 228p.

Third   Jerram, Martyn R. A Text-Book on Forest Management. London; Chapman & Hall, 1935. 156p.

Third   Jerram, Martyn R. Elementary Forest Mensuration. London; T. Murby & Co., 1939. 124p.

Third   Johnson, J. B. The Theory and Practice of Surveying. 1st ed. New York; J. Wiley & Sons, 1886. 683p. (17th ed., rewritten by L. S. Smith, 1914. 921p.)

K

Second   Keeler, Harriet L. Our Native Trees and How to Identify Them. 1st ed. New York; C. Scribner's Sons, 1900. 533p. (9th ed., 1917. 533p.)

Third   Kellogg, Charles E. The Soil That Supports Us. New York; Macmillan, 1941. 370p.

First   Kellogg, Royal S. Lumber and Its Uses. 1st ed. Chicago, Ill.; The Radford Architectural Co., 1914. 352p.

First   Kellogg, Royal S. Pulpwood and Wood Pulp in North America. 1st ed. New York; McGraw-Hill, 1923. 273p.

Third   Kinney, Abbot. Forest and Water. Los Angeles; Post Publishing Co., 1900. 250p.

First   Kinney, Jay. P. The Essentials of American Timber Law. 1st ed. New York; Wiley, 1917. 279p. (Rev. issue, 1942.)

First   Kinney, Jay P. The Development of Forest Law in America. 1st ed. New York; J. Wiley & Sons, 1917. 254p. (Reprinted New York; Arno Press, 1972. 405p.)

First   Kinney, Jay P. Forest Legislation in America Prior to March 4, 1789. Ithaca, New York; Cornell University Press, 1916. 405p.

First   Kinney, Jay P. Indian Forest and Range; A History of the Administration and Conservation of the Redman's Heritage. Washington, D.C.; Forestry Enterprise, 1950. 357p.

Third   Kirkwood, Joseph E. Northern Rocky Mountain Trees and Schrubs. Stanford, Calif.; Stanford University Press, 1930. 340p.

First   Kittredge, Joseph. Forest Influences; The Effects of Woody Vegetation on Climate, Water, and Soil, with Applications to the Conservation of Water and the Control of Floods and Erosion. New York; McGraw-Hill, 1948. 394p. (Reprinted, New York; Dover, 1973.)

Second   Koehler, Arthur. The Properties and Uses of Wood. 1st ed. New York; McGraw-Hill, 1924. 354p. (Industrial Education Series)

L

Third   Laslett, Thomas. Timber and Timber Trees, Native and Foreign. London; Macmillan, 1875. 352p. (2d ed., 1894. 442p.)

First   Leopold, Aldo. Game Management . . . drawings by Allan Brooks. New York and London; C. Scribner's Sons, 1933. 481p. (Reprinted, Madison; University of Wisconsin Press, 1986.)

First   Leopold, Aldo. A Sand County Almanac, and Sketches Here and There. New York; Oxford University Press, 1949. 226p. (Reissued, 1987. 228p.)

Third   Levison, Jacob J. The Home Book of Trees and Shrubs. New York; Simon and Schuster, 1940. 424p.

Third   Levison, Jacob J. Studies of Trees. 1st ed. New York; J. Wiley & Sons, 1914. 253p.

Second   Licher, Richard. America's Natural Wealth; A Story of the Use and Abuse of Our Resources. New York and London; Harper & Brothers, 1942.

First   Lillard, Richard G. The Great Forest. New York; A. A. Knopf, 1947. 399p.

Second   Lindquist, Bertil. Genetics in Swedish Forestry Practice. Waltham, Mass.; The Chronica Botanica Co., 1948. 173p.

Third   Longyear, Burton O. Trees and Shrubs of the Rocky Mountain Region. New York and London; Putnam's Sons, 1927. 244p.

Second   Lounsberry, Alice. A Guide to the Trees. New York; F. A. Stokes, 1900. 313p.

First   Lutz, Harold J. and Robert F. Chandler. Forest Soils. New York; J. Wiley, 1946. 514p.

First   Luxford, Ronald F. Wood Handbook: Basic Information on Wood as a Material of Construction with Data for its Use in Design and Specification. Washington, D.C.; U.S. Department of Agriculture, 1940. 326p.

Third   Lyon, Thomas L. and Harry O. Buckman. The Nature and Properties of Soils: A College Text of Edaphology. 1st ed. New York; Macmillan, 1922. 588p. (5th ed., 1952. 591p.)

M

Third   Macbride, Thomas H. The Myxomycetes; A Descriptive List of the Known Species With Special Reference to Those Occurring in North America. New York; Macmillan Co., 1934. 339p.

First   Marquis, Ralph W. Economics of Private Forestry. 1st ed. New York; McGraw-Hill, 1939. 219p.

First   Marsh, George P. Man and Nature. 1st ed. New York; C. Scribner's Sons, 1864. 629p. (Reprinted as The Earth as Modified by Human Action. New York; Arno, 1970.)

Third   Marshall, Reginald C. Notes on the Silviculture of the More Important Timber Trees of Trinidad and Tobago, with Information on the Formation of Woods. Trinidad; Govt. Print. Off., 1930. 47p. (Later ed., as Silviculture of the Trees of Trinidad and Tobago, British West Indies. London; Oxford University Press, 1939. 247p.)

First   Marshall, Robert. The People's Forest. New York; H. Smith and R. Haas, 1933. 233p.

First   Mason, Alpheus T. Bureaucracy Convicts Itself; The Ballinger-Pinchot Controversy of 1910. New York; The Viking Press, 1910. 224p.

First   Mason, David T. and Donald Bruce. Sustained Yield Forest Management as a Solution to American Forest Conservation Problems. Portland, Ore.; Mason & Stevens, 1931. 47p.

Third   Mason, Earl G. Forest Mensuration. Corvallis, Ore.; OSC Cooperative Association, 1931. 252p. (5th ed. 1940. 283p.)

Third   Mathews, F. Schuyler. Field Book of American Trees and Shrubs. New York and London; G. P. Putnam's Sons, 1915. 465p.

Second   Matthews, Donald M. Cost Control in the Logging Industry. 1st ed. New York and London; McGraw-Hill, 1942. 374p.

First   Matthews, Donald M. Management of American Forests. 1st ed. New York; McGraw-Hill, 1935. 495p.

First   Maw, Percival T. Complete Yield Tables for British Woodlands and the Finance of British Forestry. London; C. Lockwood and Son, 1912.

Third   Maw, Percival T. The Practice of Forestry; Concerning also the Financial Aspect of Afforestation. Brockenhurst; Hants, Walter and Walter, 1909. 503p.

Third   McFarland, J. H. Getting Acquainted with the Trees. New York; The Outlook Co., 1904. 241p.

Second   McMinn, Howard, and Evelyn Maino. An Illustrated Manual of Pacific Coast Trees. 1st ed. Berkeley, Calif.; University of California Press, 1935. 409p. (Reprinted in 1937.)

Third   Meyer, Bernard S. and Donald B. Anderson. Plant Physiology; A Textbook for Colleges and Universities. New York; Van Nostrand, 1939. 696p. (2d ed., 1955. 784p.)

Third   Millar, Charles E. and L. M. Turk. Fundamentals of Soil Science. New York; J. Wiley & Sons, 1943. 462p. (5th ed., 1972. 454 p.)

First   Moon, Frederick F. The Book of Forestry. New York; D. Appleton and Co., 1920. 315p.

First   Moon, Frederick F. and Nelson C. Brown. Elements of Forestry. 1st ed. New York; J. Wiley & Sons, 1914. 392p. (3d ed., rev. 1937. 397p.)

First   Muir, John. Our National Parks. Boston and New York; Houghton, Mifflin Co., 1901. 365p.

N

Second   National Lumber Manufacturers Association. National Design Specification for Stress-Grade Lumber and its Fastenings. 1st ed. Washington, D.C.; National Lumber Manufacturers Association, 1944. 64p. (1973 ed., 66p.)

Third   Newhall, Charles S. The Trees of Northeastern America. New York and London; Knickerbocker Press, 1890. 245p.

Second   Nisbet, John. British Forest Trees and Their Sylvicultural Characteristics and Treatment. London and New York; Macmillan and Co., 1893. 352p.

Third   Nisbet, John. Studies in Forestry. Oxford; Claredon Press, 1894. 335p.

Third   Norman, Arthur G. The Biochemistry of Cellulose, The Polyronides, Lignin, etc. Oxford; Clarendon Press, 1937. 232p.

Second   Nuttall, Thomas. The North American Sylva; A Description of the Forest Trees of the United States. Philadelphia; D. Rice and A. N. Hart, 1857. 3 vols. (Three volumes in two . . . being the fourth (and fifth) volume of Francios Michaux and Nuttall's North American Sylva)

O

Second   Ott, Emil. Cellulose and Cellulose Derivatives. New York; Interscience Pub., Inc., 1943. 1176p.

Third   Ownes, Charles E. Principles of Plant Pathology. New York; J. Wiley & Sons, 1928. 629p.

P

Second   Pack, Arthur N. Forestry; An Economic Challenge. New York; Macmillan Co., 1933. 161p.

Second   Pack, Arthur N. Our Vanishing Forests. New York; Macmillan, 1923. 189p.

Second   Pack, Charles L. The Forestry Primer. Washington, D.C.; The American Tree Association, 1926. 32p.

Second   Pack, Charles L. Forests and Mankind. New York; Macmillan Co., 1929. 250p.

Second   Pack, Charles L. The School Book of Forestry. Washington, D.C.; American Tree Association, 1922. 159p.

Second   Pack, Charles L. Trees as Good Citizens. Washington, D.C.; The American Tree Association, 1922. 257p.

Third   Parkhurst, Howard E. Trees, Shrubs and Vines of the Northeastern Unites States. New York; C. Scribner's Sons, 1903. 451p. 7 vols.

Third   Parkins, A. E. and J. R. Whitaker, editors. Our Natural Resources and Their Conservation. 1st ed. New York; J. Wiley & Sons, 1936. 634p. (2d ed., 1939. 647p.)

Third   Pearson, Gustaf A. Management of Ponderosa Pine in the Southeast, as Developed by Research and Experimental Practice. Washington, D.C.; U.S. Government Printing Office, 1950. 218p. (United States Department of Agriculture. Agriculture Monograph no. 6.)

Third   Penhallow, D. P. A Manual of the North American Gymnosperms. Boston; Ginn, 1907. 374p.

Second   Perry, Thomas D. Modern Plywood. 1st ed. New York and London; Pitman Pub., 1942. 366p. (2d ed., 1948. 458p.)

Third   Phillips, John C. and Fredrick C. Lincoln. American Waterfowl; Their Present Situation and the Outlook for their Future. Boston and New York; Houghton Mifflin Co., 1930. 312p.

First   Pinchot, Gifford. Breaking New Ground. 1st ed. New York; Harcourt, Brace, 1947. 522p. (Reprinted, Washington, D.C.; Island Press, 1987. 522p.)

First   Pinchot, Gifford. The Fight for Conservation. New York; Doubleday Page & Co., 1910. 152p.

First   Pinchot, Gifford. The Training of a Forester. 1st ed. Philadelphia and London; J. B. Lippincott Co., 1913. 149p. (4th ed. 1937. 129p.)

Third   Pinchot, Gifford, and Henry S. Graves. The White Pine; A Study with Tables of Volume and Yield. New York; The Century Co., 1896. 102p.

Third   Preston, John F. Farm Wood Crops. 1st ed. New York; McGraw-Hill, 1949. 302p. (American Forestry Series)

Second   Preston, Richard J. Rocky Mountain Trees. Ames, Ia; Iowa State College Press, 1940. 285p.

Second   Putter, Stephan A. Looters of the Public Domain. Portland, Ore; Portland Printing House, 1908. 494p.

R

First   Rankin, W. H. Manual of Tree Diseases. New York; Macmillan Company, 1918. 398p.

Second   Raunkiar, Christen. The Life Forms of Plants and Statistical Plant Geography. Oxford; Clarendon Press, 1934. 632p. (Reprinted, New York; Arno Press, 1977.)

Third   Rayner, M. C. Mycorrhiza. London; Wheldon & Wesley Ltd., 1927. 243p.

Third   Read, Arthur D. The Profession of Forestry. New York; Macmillan, 1934. 68p.

First   Recknagel, Arthur B. Forestry: A Study of Its Origin. New York; A.A. Knopf, 1929. 255p.

First   Recknagel, Arthur B. The Theory and Practice of Working Plans. New York; Wiley & Sons, 1913. 235p.

Second   Recknagel, Arthur B. and John Bentley. Forest Management. 1st ed. New York; J. Wiley & Sons, 1919. 269p. (Rev. ed., 1952. 290p.)

First   Record, Samuel J. Identification of the Economic Woods of the United States. 1st ed. New York; J. Wiley & Sons, 1912. 117p. (2d ed., rev. 1919. 157p.)

First   Record, Samuel J. Identification of the Timbers of Temperate North America. New Yorl; J. Wiley & Sons, 1934. 196p.

First   Record, Samual J. The Mechnical Properties of Wood. New York; J. Wiley & Sons, Inc., 1914. 165p.

First   Record, Samuel J. and Clayton D. Mell. Timbers of Tropical America. New Haven, Conn.; Yale University Press, 1924. 610p.

First   Rehder, Alfred. Manual of Cultivated Trees and Shrubs Hardy in North America, Exclusive of the Subtropical and Warmer Temperate Regions. 1st ed. New York; Macmillan, 1927. 930p. (3d ed., rev. and enl., Portland, Oreg.; Dioscorides Press, 1986. 996p.)

Third   Renne, Ronald R. Land Economics: Principals, Problems, and Policies in Utilizing Land Resources. 1st ed. New York; Harper, 1947. (Rev., ed., 1958. 599p.)

Third   Renner, George T. Conservation of National Resources. New York; J. Wiley & Sons, 1942. 228p.

Second   Richards, P. W. Tropical Rain Forest; An Ecological Study. Cambridge; University Press, 1964. 450p.

First   Robbins, Roy M. Our Landed Heritage; The Public Domain, 1776–1936. New York; P. Smith, 1942. 450p.

Third   Rogers, Julia E. The Tree Book; A Popular Guide to a Knowledge of the Trees of North America and to Their Uses and Cultivation. Garden City, N.Y.; Doubleday, Page, 1923. 565p. (The New Nature Library)

Third   Roth, Filibert. First Book of Forestry. Boston and London; Ginn, 1902. 291p.

First   Roth, Filibert. Forest Regulation: The Preparation and Development of Forest Working Plans. Ann Arbor, Mich.; Filibert Roth, 1914. 218p.

Third   Roth, Filibert. Forest Valuation. Ann Arbor, Mich.; Filibert Roth, 1916. 171p. (Also called Michigan Manual of Forestry)

Third   Rowe, Samuel M. Hand Book of Timber Preservation. 1st ed. Chicago; Pettibone, Sawtell, 1904. 203p. (Souvenier ed., rev. 1909. 402p.)

Third   Russell, Carl P. One Hundred Years in Yosemite. 1st ed. Stanford; Stanford University Press, 1932. 258p. (3d ed, 1992. 269p.)

S

Third   Sampson, Arthur W. Range and Pature Management. New York; J. Wiley & Sons, 1923. 421p.

Third   Sargent, Charles S. Forest Flora of Japan. Boston; Houghton, Mifflin, 1894. 93p. (Reprinted from Garden and Forest)

First   Sargent, Charles S. Manual of the Trees of North America (exclusive of Mexico).. with 644 illustrations from drawing by Charles E. Faxon. Boston and New York; Houghton Mifflin, 1905. 826p. (2d corrected. ed., with 783 illustrations by C. E. Faxon and Mary W. Gill, New York; Dover Pub., 1965. 2 vols.)

First   Sargent, Charles S. Report on the Forests of North America (Exclusive of Mexico) Portfolio of 16 Maps. Washington, D.C.; U.S. Govt. Print. Off., 1884.

First   Sargent, Charles S. The Silva of North America; A Description of the Trees which Grow Naturally in North America exclusive of Mexico. Boston and New York; Houghton, Mifflin and Co., 1891–1902. 14 vols.

First   Sargent, Charles S. The Woods of the United States. New York; D. Appleton and Co., 1885. 203p.

First   Schenck, Carl A. The Art of the Second Growth, or American Sylviculture. Albany, New York; Brandow Printing Co., 1912. 206p.

First   Schenck, Carl A. Forest Mensuration. Sewanee, Tenn.; The University Press, 1905. 71p.

First   Schenck, Carl A. Forest Management (Forest Working Plans); Guide to Lectures delivered at the Biltmore Forest School. Asheville, N.C.; Hackney & Mole Co., 1907. 33p.

First   Schenck, Carl A. Forest Finance; Guide to Lectures Delivered at the Biltmore Forest School. Asheville, N.C.; The Inland Press, 1909. 44p.

First   Schenck, Carl A. Forest Protection: Guide to Lectures Delivered at the Biltmore Forest School. Asheville, N.C.; The Inland Press, 1909. 159p.

First   Schenck, Carl A. Forest Policy. 2d ed., rev. and enl. Darmstadt, Germany; Winter, 1911. 168p. (1st ed. was in German).

First   Schenck, Carl A. Logging and Lumbering, or Forest Utilization. Darmstadt, Germany; L. C. Wittich, 1912. 189p.

Third   Schiff, Ashley L. Fire and Water; Scientific Heresy in the Forest Service. Cambridge; Harvard University Press, 1962. 225p.

First   Schlich, Sir William. A Manual of Forestry. 1st ed. London; Bradbury, Agnew & Co., 1889–1896. 5 vols. (5th ed., 1925. 5 vols.)

Third   Schorger, Arlie W. The Chemistry of Cellulose and Wood. New York; McGraw-Hill Book Company, Inc., 1926. 596p.

Second   Schwarz, George F. The Longleaf Pine in Virgin Forest: A Silvical Study. New York; J. Wiley & Sons, 1907. 135p.

Second   Shaw, George R. The Genus Pinus. Cambridge; Riverside Press, 1914. 96p. (Publications of the Arnold Arboretum no. 5)

Third   Shaw, John William. How to Cruise Timber, Adapted for Experienced Cruisers, Loggers, Foresters, Claimants, of for Any One Desiring to Learn to Estimate Timber. Portland, Oregon; 1910. 64p.

Third   Shirley, James C. The Redwoods of Coast and Sierra. 1st ed. Berkeley; University of California Press, 1936. 74p. (4th ed. 1947. 84p.)

Third   Simpson, John. The New Forestry; or The Continental System Adapted to British Woodlands and Game Preservation. Sheffield, England; Pawson & Brailsford, 1900. 202p.

Second   Smith, Darrell H. The Forest Service. Washington, D.C.; Brookings Institution, 1930. 268p. (Service Monographs of the United States Government no. 58)

Third   Smith, Erwin F. An Introduction to Bacterial Diseases of Plants. Philadelphia and London; W. B. Saunders Co., 1920. 688p.

Third   Smith, J. Russell. Tree Crops; A Permanent Agriculture. New York; Harcourt, Brace, 1929. 333p. (Reprinted, Washington, D.C.; Island Press, 1987.)

Second   Snow, Charles H. The Principal Species of Wood; Their Characteristic Properties. 1st ed. New York; J. Wiley & Sons, 1903. 203p. (2d ed., rev. 1908. 212p.)

Third   Snow, Charles H. Wood and Other Organic Structural Materials. 1st ed. New York; McGraw-Hill, 1917. 478p.

First   Society of American Foresters. Forest Terminology. Washington D. C.; Society of American Foresters, 1916. 2 vols.

First   Society of American Foresters. Forestry Terminology: A Glossary of Technical Terms Used in Forestry. Washington, D.C.; SAF, 1944. 84p. (3d ed., rev., 1958. 97p.)

Third   Society of American Foresters. Important Tree Pests of the Northeast; A Revision of Tree Pest Leaflets Nos. 1–50. Boston; Massachusetts Forest and Park Assoc., 1940. 187p. (2d ed., as Important Tree Pests of the Northeast; A Revision of Tree Pest Leaflets Nos. 1–55. Concord, N.H.; Evans Printing Co., 1952. 191p.)

First   Society of American Foresters. A National Program of Forest Research. Washington, D.C.; American Tree Association, 1926. 232p.

Third   Springer, John S. Forest Life and Forest Trees: Comprising Winter Camplife Among the Loggers, and Wild-Wood Adventure. New York; Harper & Brothers, 1851. 259p.

Second   Spruce, Richard. Notes of a Botanist on the Amazon & Andes. London; Macmillan, 1908. 2 vols.

First   Spurr, Stephen H. Aerial Photographs in Forestry. New York; Ronald Press Co., 1948. 340p.

Third   St. Clair Thompson, G. W. The Protection of Woodlands by Natural as Opposed to Artificial Methods. London; H. F. & G. Witherby, 1928. 223p.

Second   Stahl, Rose M. The Ballinger-Pinchot Controversy. Northampton, Mass.; Dept. of History of Smith College, 1926. 138p.

Third   Stebbins, G. Ledyard. Variation and Evolution in Plants. New York; Columbia University Press, 1950. 643p.

Third   Stevens, F. L. The Fungi Which Causes Plant Disease. New York; Macmillan COmpany, 1913. 754p.

Third   Stevenson, J. A. Foreign Plant Diseases. Washington D.C.; U.S. Department of Agriculture, 1926. 198p.

Second   Stevenson, Louis T. The Background and Economics of American Papermaking. 1st ed. New York; Harper & Bros., 1940. 249p.

Third   Stevenson, William. Wood: Its Use as a Constructive Material. London; B. T. Batsford, 1894. 240p.

Third   Stoddard, Herbert L. The Bobwhite Quail; Its Habits, Preservation and Increase. New York; C. Scribner's Sons, 1931. 559p.

First   Stoddart, Laurence A. Range Management. 1st ed. New York and London; McGraw-Hill, 1943. 547p.

Second   Sudworth, George B. Forest Trees of the Pacific Slope. Washington, D.C.; Government Printing Office, 1908. 441p.

Third   Sudworth, George B. Nomenclature of the Arborescent Flora of the United States. Washington, D.C.; Govt. Print. Off., 1897. 419p.

T

Third   Tamm, Oloff F. S. Northern Coniferous Forest Soils; A Popular Survey of the Phenomena Which Determine the Productinve Character of the Forest Soils of Northern Sweden . . . trans. from Swedish by Mark L. Anderson. Oxford; Scrivener Press, 1950. 253p.

Third   Tansley, A. G. The British Islands and Their Vegetation. 1st ed. Cambridge; University Press, 1939. 930p. (Rep. ed., 1953. 2 vols.)

Third   Taylor, Jay L. Handbook for Rangers & Woodsmen. 1st ed. New York; J. Wiley & Sons, 1917. 420p.

Second   Tiemann, Harry D. The Kiln Drying of Lumber. Philadelphia and London; J. B. Lippincott, 1917. 316p.

First   Tiemann, Harry D. Wood Technology; Constitution, Properties and Uses. New York and Chicago; Pitman Pub., 1942. 316p. (3d ed., 1951. 396p.)

First   Toumey, James W. Foundations of Silviculture upon an Ecological Basis. Ann Arbor, Mich.; Edwards Bros., 1925. 2 vols. (2d ed., rev. by Clarence F. Korstian, New York; J. Wiley, 1947. 468p.)

First   Toumey, James W. Seeding and Planting in the Practice of Forestry; A Manual for the Guidance of Forestry Students, Foresters, Nurserymen, Forest Owners, and Farmers. 1st ed. New York; J. Wiley; London; Chapman & Hall, 1916. 507p. (3d. ed. 1942. 520p.)

Third	Townsend, Gilbert. Carpentry. Chicago; American Technical Society, 1935. 436p.
Second	Tracy, John C. Plane Surveying; A Textbook and Pocket Manual. 1st ed. New York; J. Wiley & Sons, 1907. 794p. (Rep. and Enl., 1914. 794p.)
Third	Trayer, George W. Wood in Aircraft Construction; Supply, Suitability, Handling, Fabrication, and Design. Washington, D.C.; National Lumber Manufactures Association, 1930. 276p.
Second	Treat, Payson J. The National Land System, 1785–1820. New York; Payson Jackson Treat, 1910. 426p.
First	Trippensee, Reuben E. Wildlife Management; Upland Game and General Principles. 1st ed. New York; McGraw-Hill, 1948–1953. 2 vols.
Third	Troup, Robert S. Colonial Forest Administration. London and New York; Oxford University Press, 1940. 476p.
First	Troup, Robert S. The Silviculture of Indian Trees. Oxford; Clarendon Press, 1921. 3 vols.
First	Troup, Robert S. Silvicultural Systems. Oxford; Clarendon Press, 1928. 199p. (2d ed., edited by E. W. Jones, 1952.)

U

Third	United States. Bureau of Foreign and Domestic Commerce. The Export Lumber Trade of the United States. Washington, D.C.; U.S. Govt. Print. Off., 1918. 117p.
First	United States Department of Agriculture. Trees: The Yearbook of Agriculture. Washington, D.C.; U.S. Government Printing Office, 1949. 944p.
Third	United States. Division of Forest Economics and Marketing Research. A Selected Bibliography on the Economics of Forestry in the United States. Washington, D.C.; 1941. 172p.
First	United States. Forest Service. Report on Forestry. Washington, D.C.; U.S. Govt. Print. Off., 1878–1884. 4 vols.
First	United States Forest Products Laboratory, Madison, Wis. Wood Handbook: Basic Information on Wood as a Material of Construction with the Data for Its Use in Design and Specification. Washington, D.C.; U.S. Govt. Print. Off., 1935. 326p. (Later rev. ed., 1987, as Wood Handbook: Wood as an Engineering Material. 466p. USDA Agriculture Handbook no. 72)
Third	Unwin, Arthur H. West African Forests and Forestry. London; T. F. Unwin, 1920. 527p.

V

Second	Van Hise, Charles R. The Conservation of Natural Resources in the United States. New York; Macmillan Co., 1910. 413p.
Second	Van Name, Willard G. Vanishing Forest Reserves; Problems of the National Forests nad National Parks. Boston; R. G. Badger, 1929. 190p.

W

Second	Wackerman, Albert E. Harvesting Timber Crops. 1st ed. New York; McGraw-Hill, 1949. 437p. (American Forestry Series)
First	Wahlenberg, William G. Longleaf Pine: Its Use, Ecology, Regeneration,

	Protection, Growth and Management. Washington, D.C.; U.S. Dept. of Agriculture, 1946. 429p.
Second	Wakeley, Philip C. Artificial Reforestation in the Southern Pine Region. Washington, D.C.; U.S. Dept. of Agriculture, 1935. 115p. (USDA Technical Bulletin 492)
Third	Wangaard, Fredrick F. The Mechanical Properties of Wood. New York; Wiley, 1950. 377p.
Third	Ward, H. M. Timber and Some of its Diseases. London; Macmillan & Company, Ltd., 1909. 295p.
Third	Ward, H. M. Trees: A Handbook of Forest-Botany for the Woodlands and the Laboratory. Cambridge; The University Press, 1904–1909. 5 vols.
Third	Warming, Eugenius. Oecology of Plants; An Introduction to the Study of Plant-Communities. Oxford; Oxford University Press, 1909. 422p. (Reprinted, Oxford; Clarendon Press, 1977.)
Second	Weaver, John E. and Frederic E. Clements. Plant Ecology. 1st ed. New York; McGraw-Hill, 1929. 520p. (2d ed., 1938. 601p.)
Third	Webster, Angus D. Tree Wounds and Diseases: Their Prevention and Treatment. Philadelphia; Lippincott, 1916. 215p.
Third	Weed, Clarence M. Our Trees; How to Know Them. Philadelphia; Lippincott, 1908. 295p. (5th ed. Garden City, New York; Garden City Publishing Co., Inc., 1936. 295p; reprinted in 1959).
Third	Western Forestry and Conservation Association. Forty Years of Western Forestry; A History of the Movement to Conserve Forest Resources by Cooperative Effect, 1909–1949. Portland, Ore.; Western Forestry and Conservation Association, 1949. 64p.
Second	Western Forestry and Conservation Association. The Western Fire Fighter's Manual. Portland, Ore.; Western Forestry and Conservation Association, 1924. 1 vol.
First	Westveld, Ruthford H. Applied Siviculture in the United States. 1st ed. Ann Arbor, Mich.; Edwards Brothers, Inc., 1935. 415p. (2d ed., New York: Wiley, 1949.)
Second	Westveld, Ruthford H. and Ralph Peck. Forestry in Farm Management. New York; J. Wiley & Sons, 1941. 339p.
Third	Whetzel, Herbert H. An Outline of the History of Phytopathology . . . with 22 portraits. Philadelphia and London; W. B. Saunders, 1918. 130p. (Reprinted, New York; Arno Press, 1977.)
Second	Wilde, Sergius A. Forest Soils. Madison, Wis., 1941. 384p. (Later ed., as Forest Soils and Forest Growth, Waltham, Mass.; Chronica Botanica, 1946. 241p.)
Second	Wilde, Sergius A. Forest Soils and Forest Growth. Waltham, Mass.; Chronica Botanica Co., 1946. 241p.
Third	Winters, Robert K. ed. Fifty Years of Forestry in the USA. Washington, D.C.; Society of American Foresters, 1950. 385p.
First	Wise, Louis E. ed. Wood Chemistry. New York; Reinhold Pub., 1944. 900p. (2d ed., by Louis E. Wise and Edwin C. Jahn, 1952. 2 vols.)
Second	Wolfe, Linnie M. Son of the Wilderness: The Life of John Muir. 1st ed. Knopf; New York, 1945. 350p. (Reprint. Madison; University of Wisconsin Press, 1978. 364p.)

First	Woolsey, T. S. American Forest Regualtion. New Haven, Conn.; Tuttle, Morehouse, & Taylor Co., 1922. 210p.
Second	Woolsey, T. S. French Forests and Forestry in Tunisia, Algeria, and Corsica. New York; John Wiley & Sons, 1917. 238p.
Second	Woolsey, T. S. Studies in French Forestry. New York; John Wiley & Sons, 1920. 520p.
Third	Wrenn, Charles G. Time on Their Hands: A Report on Leisure, Recreation, and Young People. 1st ed. Washington, D. C.; American Council on Education, 1941. 266p. (reprint, 1974, 266p.)

Z

First	Zon, Raphael and William Sparhawk. Forest Resources of the World. New York; McGraw-Hill Book Co., 1923. 997p. 2 vols.
First	Zon, Raphael. Forests and Water in the Light of Scientific Investigation. Washington, D.C.; U.S. Department of Agriculture, Government Printing Office, 1927. 106p.

E. Scholarly and Professional Journals

The source documents used for the analysis cited 107 journals. Sixteen of these were German, two French, and one Swedish, comprising 17.7% of the basic list. The rest of the titles were in English. Of the total 107 journals, 46.3% were cited only once, and 10.8% were cited no more than two times in a single source document.

Of the original 107, thirty-two titles were core journals of historical importance in forestry. Several titles presented classic skews, having been highly cited in only one source document. Lacking broad correlating data from other sources, these journals were weighted downward.

The most highly cited journal was the SAF's *Journal of Forestry* which superseded the *Proceedings of the Society of American Foresters* in 1916. The *Journal* received seven times the number of citations as the next highest ranking title, *Timberman: An International Lumber Journal* published in Portland, Oregon, from 1900 to 1963. The journal titles receiving the highest counts were rated first rank (A), the rest were rated second (B). This resulted in a statistically valid core list of forty-one scholarly and professional journals worthy of primary preservation attention (see Table 14.7, p. 424).

F. Trade and Popular Periodicals

Citation analysis of seminal works does not assist in identifying and determining the historical value of popular literature because these materials

Table 14.7. Most frequently cited scholarly journals from citations analysis

Journal	Rank	Microfilmed
American Forests	A	UMI
American Journal of Botany	B	UMI
American Lumberman	B	DATAMICS
Botanical Gazette	A	UMI
Bulletin of Torrey Botanical Club	A	UMI
Ecological Monographs	B	Not filmed
Ecology	A	UMI
Empire Forestry Review	A	Not filmed
Forest Industries	B	UMI
Forestry	A	Oxford
Forestry Chronicle	A	Not filmed
Forestry and Irrigation	B	UMI
Forestry Quarterly (Journal of Forestry)	A	UMI
Indian Forester	A	IDC
Industrial & Engineering Chemistry	A	ACS
Journal of Agricultural Research	A	UMI
Journal of Economic Entomology	A	UMI
Journal of Forestry v.15 (1917) +	A	Filmed from 1949 +
Journal of the New York Botanical Garden	B	Chadwyck-Healey Ltd.
Missouri Botanical Garden, Annual Report	B	UMI
Mycologia	A	UMI
Nature	B	UMI
New Phytologist	B	Brookhaven
Paper Industry & Paper World	B	UMI
Paper Trade Journal	B	UMI
Phytopathology	A	UMI
Plant Disease Reporter	B	Princeton
Plant Physiology	B	UMI
The Plant World	B	Not filmed
Proceedings of the Royal Society of London	B	Pergamon and Princeton
Proceedings of the Society of American Foresters	B	Not filmed
Quarterly Journal of Forestry	B	CUBG
Roosevelt Wild Life Annals	B	Spaulding
Science	B	UMI
Soil Science	B	UMI
Southern Lumberman	A	Not filmed
Timberman	A	Not filmed
Transactions of the British Mycological Society	B	UMI
Transactions of the Royal Scottish Arboricultural Society	B	Not filmed
Western Lumberman	B	British Library
Wood	B	UMI

UMI = University Microfilms, Inc.; IDC = Inter Documentation Company; ACS = American Chemical Society; CUBG = University of California at Berkeley; Title not microfilmed where indicated.

are rarely cited. Yet they remain of great interest to historians. Identification of popular periodicals in forestry is a time-consuming problem and a variety of methods had to be used to compile the list which follows.

Journal lists, library catalogs, early bibliographies, and historical writings concerned with agricultural and forestry periodicals and history were used. Notable sources used were:

American Tree Association. *Forestry Almanac*. (Washington, D.C.: ATA, 1933.)

Brown, Nelson C. *The American Lumber Industry*. (New York: J. Wiley, 1923. pp. 232–63.)

Dictionary Catalog of the National Agricultural Library, 1862–1965. (New York: Rowman and Littlefield, 1967.)

Munns, E. N. *A Selected Bibliography of North American Forestry*. 2 vols. (Washington, D.C.: U.S. Dept. of Agriculture, 1940 [*USDA Miscellaneous Publication* No. 364].)

Stuntz, Stephen C. *List of the Agricultural Periodicals of the United States and Canada Published During the Century, July 1810 to July 1910*. (Washington, D.C.: U.S. Dept. of Agriculture, 1941 [*USDA Miscellaneous Publication* No. 398].)

U.S. Dept. of Agriculture. *Catalogue of Publications Relating to Forestry in the Library of the United States Department of Agriculture*. (Washington, D.C.: Gov. Print. Off., 1912 [*USDA Library Bulletin* No. 76].)

No ranking of the popular periodicals was made. However, criteria for preservation priorities should be based on the number of years the publication ran, whether it began publication prior to 1870, if it was national or international in scope, if it was well illustrated and, importantly, if it had notable editors or authors.

U.S. and Canadian Trade and Popular Periodicals in Forestry, 1850–1950

Closing date in the Stuntz list.[28]
// *at end of citation = ceased publication on this date.*

28. Stuntz, Stephen C. *List of the Agricultural Periodicals of the United States and Canada Published During the Century, July 1810 to July 1910* (Washington, D.C.: U.S. Dept. of Agriculture, 1941 [*USDA Miscellaneous Publication* No. 398].)

American forestry. Washington, D.C. m. v.16, no.1–7. Jan.-July 1910.##
   Began as
   (a) The New Jersey forester. v.1, no.1–3. Jan.-May, 1895.
   Continued as
   (b) The forester. v.1, no.4– v.7. July 1895–Dec. 1901.
   Continued as
   (c)Forestry and irrigation. v.8–14, no.8. Jan.1902–Aug. 1908.
   Continued as
   (d) Conservation; forests, waters, soils, and minerals. v.14, no.9–12, v.15,
   no.1–12. Sept. 1908–Dec.1909.
American lumberman. Chicago. w. no.1233–1832. Jan. 7, 1899–July 2, 1910.##
   Began as
   (a) Northwestern lumberman. v.1–[26] Jan. 1873–Dec. 31, 1898.
American lumberman. New Orleans, La. w. v.1–3? 1883–1886?
American paper trade and wood pulp news. New York v.1–19. March 31, 1890–
   1900//.
   Continued as Geyer's stationery.
Barrel and box and packages. Chicago. v.1. 1895 + .
   Began as Barrel and box. v.1–34, no.7. 1885–1929.
Black rock forest papers. Cornwall-on-the-Hudson, N.Y. v.1, no.1. April 1935 + .
Canadian forestry journal. Ottawa, Canadian Forestry Association. quarterly, v.1–
   6, no.2. Jan. 1905–June 1910.##
Colorado forester. Fort Collins, Colorado State Agricultural College, Forestry
   Club. v.1. 1925–1968//.
Connecticut woodlands. New Haven, Connecticut Forest and Park Association. v.1,
   no.1. Feb. 1936 + .
Cornell forester. Ithaca, N.Y. v.1. 1920–1926//.
Crow's lumber digest. Portland, Ore. v.1. 1922 +
   Began as Crow's pacific coast lumber digest and index. 1922–1927.
Dixie wood-worker. Atlanta, Ga. v.1–22. Aug. 1885–1906//.
   Began as Dixie. v.1–21, no.10. Aug. 1885–Aug. 1905//.
Empire forester. Syracuse, N.Y. v.1. 1915–1940//.
Evergreen. Sturgeon Bay, Wis. m. v.4. no.11–v.5. Dec. 1874–July 1876//.
   Began as
   (a) Evergreen and forester tree grower. v.1–4, no.1. Aug. 1871–Jan. 1874.
   Continued as
   (b) Evergreen and fruit-tree grower. v.4, no.2–10. Feb.-Nov. 1874.
Forest and stream. New York. v.1–100, no.7. Aug. 1873–July 1930//.
   Merged into
   (a) Field and stream. v.100, 63rd year.
Forest club annual. Georgia State forest school. Athens, Ga. v.1–2. 1916–1917.
Forest club annual. Lincoln, University of Nebraska. v.1–6. 1909–1915.
Forest club quarterly. Sylvia. v.1–6. 1930–1935.
Forest echoes. Seattle, Wash. v.1. 1898–//.
Forest fish and game. Athens, Ga. v.1–3. 1907–Apr. 1911//?
Forest forage and farm. Ilion, N.Y. v.1. 1880 + ?

Forest leaves. Philadelphia, Pennsylvania Forestry Association. v.1–12, no.4. July 1886– 1910.## irreg. (July 1886–1888), m. (1889–1892), bi-m. (1893–1910).

Forest patrolman. Portland, Ore. v.1–8, no.5. Aug. 1919–1927//.

Forest warden. Raleigh, N.C. v.1, no.1–25. 1924–1931//.

Forest warden news. Harrisburg, Pa. v.1. 1926 +.

Forestry news digest. Washington, D.C. Aug, 1925–1939//.

Gulf coast lumberman. Houston, Texas. v.1. 1913 +.

Hardwood; a journal of the hardwood lumber trade. Chicago. v.1–12. Jan. 1892–July 1897//.

　Merged into

　(a) Chicago hardwood record. v.1–13. April 1902.

　Continued as

　(b) Hardwood record. Chicago. v.1. 1895.

Hawaiian forester and agriculturist. Honolulu. v.1–30, no.1. 1904–Mar. 1933//. m., v.1–19.; quarterly, v.20–30, no.1.

Idaho forester. Moscow. v.1. 1917 +.

Interstate lumberman. Beaumont, Tex. w. v.1. 1899–1900.

Logging. Duluth, Minn. v.1. 1913–1918//.

　Continued as

　(a) Clyde Log. June 1918?-Feb. 1919.

Lumber and veneer consumer. Chicago. v.1–5. 1920–1925//.

Lumber. New York. w. v.1– 1880–1888.

Lumber. New York. m. v.1– 1883–1900. v. 29, no.4 is Jan. 15, 1898; v.30, no.1 is Apr. 15, 1898.

　Began as

　(a) The Saw-mill gazette. New York. m. v.1– 1883?–1894.

Lumber reporter. Indianapolis. April? 1904.

Lumber reporter. St. Louis. m. no.63–71. Dec. 1905–Aug. 1906//.

　Began as

　(a) St. Louis reporter. St Louis, Mo. s-m. (1902–Mar. 15, 1905), m. (Apr.–Nov. 1905). no.1–62. Jan 5, 1903–Nov. 22, 1905.

Lumber trade journal. New Orleans. v.1–143, no.1. 1881–June 1931.

　Merged into Southern lumberman.

Lumber worker. Nashville, Tenn. v.1–8, no.89. Feb. 1925–July 1932//.

Lumber world. Chicago. v.1–13. June 1905–Dec. 1911//. (1905–1906 as Strodes lumber world).

Lumberman's gazette. Bay City, Mich. v.1–29. 1871–1886//?

Lumberman's review. New York. v.1–38, no.453. 1892–May 1930//.

　Absorbed by National lumberman.

Maine hardwood news. Augusta, Maine. v.1. 1928–1929//.

Manufacturers' gazette. Boston. v.1–25. July 1881–1898//?

　Also known as Northeastern lumberman and manufacturers' gazette. March 1895–June 1897.

Minnesota forester. St. Paul. v.1–4, no.1. Jan. 1908–March 1911.

　Continued as North woods.

Mississippi Valley lumberman. Minneapolis. w. v.35–41, no.26. 1904–July 1910.##

Began as Mississippi Valley lumberman and manufacturer. Minneapolis. w. v.1–34. 1876–1903.

Monterey tree grower. Monterey, Calif. v.1, no.1–5. 1909–Feb. 1910//.

National coopers' journal. Philadelphia. v.1. 1885–1951//.

National lumber bulletin. Washington. v.1–12, no.5. 1920–Jan. 1932//.

National lumber news. Washington. no.1–28. 1927–1929//.

National lumberman. St. Louis, Mo. v.1–87, no.1. 1888–1932//.

Began as

(a) St. Louis lumberman. v.1–61, no.2. 1888–1918

Continued as

(b) Lumber. v.61, no.4– v.73. 1918–1924

(c) Lumber manufacturer and dealer. v.73, no.31– v.82, no.3, 1924–1929. New York forestry. N.Y. v.1–10, no.2. 1914–1930//.

New York lumber trade journal. New York. s-m. v.1–47, no.577. 1884–July 1910.##

North woods. St. Paul. v.1–10. 1912–1923//.

Northwestern dealer. Minneapolis. v.1–2, no.5. 1921–1923.

v.1, no.3. as Northwestern lumberman.

Ohio lumber journal. Cleveland, Ohio. w. v.1–2? 1884–1886?

The Pacific lumber trade journal. Seattle, Wash. m. v.1–16, no.3. 1895–July 1910.##

Pacific lumberman, contractor, and electrician. San Francisco. v.1–14. 1885–1894

Began as Pacific lumberman and contractor.

Paper; a journal devoted to the paper industry in all its branches. v.1. 1899–?

Paper industry. Chicago. v.1. 1919–1938//.

Paper maker. Kalamazoo. v.1. 1932.

Paper makers monthly journal. New York. v.1. 1863+.

Paper mill and wood pulp news. New York. v.1. 1876+.

Paper trade journal. New York. v.1. 1872–1986//.

Pennsylvania lumberman. Scranton, Pa. m. v.1– Feb. 11, 1903–July 1910.##

Pine cone. Minneapolis, Manufacturers of Northern Pine Lumber. v.1. no.1–4, no.2. 1912–1915//.

Continued by White pine; Its place in American architecture.

Pioneer western lumberman. San Francisco. s-m. v.1–73, no.7. 1884–1920//.

Began as

(a) Wood and iron. v.1–10, no.4. 1884–Oct. 1889.

Continued as

(b) Pacific coast wood and iron. v.10, no.5– v.52. Nov. 1889–1909.

Merged into

(c)Lumber, later Lumber manufacturer and dealer.

Puget Sound lumberman and northwest trade review. Tacoma, and Seattle, Wash. (1892–1896). s-m. (1890–1891), m.

(1892–1896). v.1–7, no.12. May. 1890–Feb. 1896//; v.4 ended Feb. 1894; Mar. 1894 is v.5, no.1; Apr. 1894–Feb. 1895 are v.6, no.2–12; Mar. 1895 begins v.7.

Combined with West coast lumberman to form West coast and Puget Sound lumberman.

Pulpwood. New York, American Paper and Pulpwood Association. v.1–6. April 1928–1933//.

The Retail lumberman. Minneapolis. m. v.1– June? 1903–

Rivet. Albany, New York Forestry Association. v.1, no.1–14. Dec. 1916–April 1921//.

Superseded by Seed tree.

St. Louis lumber gazette. St. Louis. m. v.1, no.1–6. Nov.20, 1906–April 20, 1907//.

Southern industries and lumber review. Austin, Tex. (1893–1901?), Houston, Tex. (1902?–1910). m. v.1–17. 1893–July–1910##: Has cover titles "Southwest"; Southern industrial and lumber review.

Southern lumber journal. Wilmington, N.C., and Norfolk, Va. s-m. v.1–26 no.1. 1898–July 1, 1910.##

Began as Southern milling and lumber journal. v. 1–15? 1898–1904?

Southern lumberman. Nashville. v.1. 1881–.

Southwestern lumber world. New Orleans. m. v.1–2? 1888–1889.

Southwestern lumberman. Kansas City, Mo. m. v.1 June 1903–1904

Tree ring bulletin. Flagstaff, Arizona. v.1. 1934+.

Tree talk. Stanford, Conn. v.1–8. 1913–March 1928.

Twin city lumberman. Marinette, Wis. w. v.1–4? 1895–1898.

Veneers and plywood; a journal for veneer manufacturers and users. Indianapolis. v.1. 1907–1960//.

Supersedes
(a) Veneers.
Continued as
(a) Wood and Wood Products

Water and Forest. San Francisco, California Water and Forest Association. v.1–7, no.3. 1900–1907//.

West coast lumberman. Seattle. v.1. 1889–1949//.

Also known as West coast and puget sound lumberman, 1904.
Continued as
(a) Lumberman

Western lumberman. Vancouver, British Columbia. m. v.1909–1910.##

Began as
(a) Lumberman and contractor. v.1– 1904–1907.
Continued as
(b) Western Canada lumberman. v. 1908.

Wholesale lumberman. Pittsburgh, Pa. s-m. v.1– 1908–1909.

Wisconsin lumberman. Milwaukee. m. v.1–3? Oct. 1873–Mar. 1875//?

Wood preserving. Baltimore, American Wood Preservers' Association. v.1–5. Jan. 1914–Dec. 1918//.

Also the Association's bulletin. Jan.-Dec. 1914.

Wood preserving news. Chicago, American Wood Preservers' Association. v.1. 1923+.

Wood products. Milwaukee. v.1 Jan. 1925–1951//.
  Supersedes
  (a) Wood turning. v.1–18, no.3. 1908–Dec. 1924//.
Wood Worker; A Journal for Machine Wood-Workers. Indianapolis. v.1. 1882–1960//.
Woodcraft. Cleveland, Ohio. v.1–23, no.2. Mar. 1904–1915
  Also known as Patternmaker. v.1–3. 1904–1905.
Wooden barrel. St. Louis, Mo., Association Cooperage Industries of America. v.1. 1932+.

Index

Literary titles in Chapter 2 (pp. 15–44) are *not* indexed. Authors and titles in the core list of monographs (pp. 262–342), the core list of journals (pp. 357–361), the reference titles (pp. 364–388), and the lists of primary historical literature (pp. 405–430) are *not* included in this index.

431

CPSIA information can be obtained
at www.ICGtesting.com
Printed in the USA
LVHW091940010219
606104LV00008B/12/P